Cognitive Systems Monographs

Volume 3

Editors: Rüdiger Dillmann · Yoshihiko Nakamura · Stefan Schaal · David Vernon

Lorenzo Magnani

Abductive Cognition

The Epistemological and Eco-Cognitive Dimensions of Hypothetical Reasoning

Rüdiger Dillmann, University of Karlsruhe, Faculty of Informatics, Institute of Anthropomatics, Robotics Lab., Kaiserstr. 12, 76128 Karlsruhe, Germany

Yoshihiko Nakamura, Tokyo University Fac. Engineering, Dept. Mechano-Informatics, 7-3-1 Hongo, Bukyo-ku Tokyo, 113-8656, Japan

Stefan Schaal, University of Southern California, Department Computer Science, Computational Learning & Motor Control Lab., Los Angeles, CA 90089-2905, USA

David Vernon, Khalifa University Department of Computer Engineering, PO Box 573, Sharjah, United Arab Emirates

Authors

Lorenzo Magnani

Dipartimento di Filosofia
Universitá di Pavia
Piazza Botta 6
27100 Pavia Italy
E-mail: lmagnani@unipv.it

ISBN 978-3-642-26082-7 e-ISBN 978-3-642-03631-6

DOI 10.1007/978-3-642-03631-6

Cognitive Systems Monographs ISSN 1867-4925

Softcover reprint of the hardcover 1st edition 2010

Typeset & Cover Design: Scientific Publishing Services Pvt. Ltd., Chennai, India.

Printed in acid-free paper

5 4 3 2 1 0

springer.com

To my daughter Giovanna

How was it that man was ever led to entertain that true theory? You cannot say that it happened by chance, because the possible theories, if not strictly innumerable, at any rate exceed a trillion – or the third power of a million; and therefore the chances are too overwhelmingly against the single true theory in the twenty or thirty thousand years during which man has been a thinking animal, ever having come into any man's head. Besides, you cannot seriously think that every little chicken, that is hatched, has to rummage through all possible theories until it lights upon the good idea of picking up something and eating it. On the contrary, you think the chicken has an innate idea of doing this; that is to say, that it can think of this, but has no faculty of thinking anything else. The chicken you say pecks by instinct. But if you are going to think every poor chicken endowed with an innate tendency toward a positive truth, why should you think that to man alone this gift is denied?

Charles Sanders Peirce

Preface

This volume explores abductive cognition, an important but, at least until the third quarter of the last century, neglected topic in cognition. It integrates and further develops ideas already introduced in a previous book, which I published in 2001 (*Abduction, Reason, and Science. Processes of Discovery and Explanation*, Kluwer Academic/Plenum Publishers, New York).

The status of abduction is very controversial. When dealing with abductive reasoning misinterpretations and equivocations are common. What are the differences between abduction and induction? What are the differences between abduction and the well-known hypothetico-deductive method? What did Peirce mean when he considered abduction both a kind of inference and a kind of instinct or when he considered perception a kind of abduction? Does abduction involve only the generation of hypotheses or their evaluation too? Are the criteria for the best explanation in abductive reasoning epistemic, or pragmatic, or both? Does abduction preserve ignorance or extend truth or both? How many kinds of abduction are there? Is abduction merely a kind of "explanatory" inference or does it involve other non-explanatory ways of guessing hypotheses?

The book aims at increasing knowledge about creative and expert inferences. The study of these high-level methods of abductive reasoning is situated at the crossroads of philosophy, logic, epistemology, artificial intelligence, neuroscience, cognitive psychology, animal cognition and evolutionary theories; that is, at the heart of cognitive science. Philosophers of science in the twentieth century have traditionally distinguished between the inferential processes active in the logic of discovery and the ones active in the logic of justification. Most have concluded that no logic of creative processes exists and, moreover, that a rational model of discovery is impossible. In short, scientific creative inferences are irrational and there is no "reasoning" to hypotheses. On the other hand, some research in the area of artificial intelligence has shown that methods for discovery could be found that are computationally adequate for rediscovering – or discovering for the first time – empirical or theoretical laws and theorems.

Moreover, the study of diagnostic, visual, spatial, analogical, and temporal reasoning has demonstrated that there are many ways of performing intelligent and creative reasoning that cannot be described with only the help of classical logic. Abduction is also useful in describing the different roles played by the various kinds of medical reasoning, from the point of view both of human agents and of computational programs that perform medical tasks such as diagnosis. However, non-standard logic has shown how we can provide rigorous formal models of many kinds of abductive reasoning such as the ones involved in defeasible and uncertain inferences. Contradictions and inconsistencies are fundamental in abductive reasoning, and abductive reasoning is appropriate for "governing" inconsistencies. In chapter two many ways of governing inconsistencies will be considered, ranging from the methods activated in diagnostic settings and consistency-based models to the typical ones embedded in some form of creative reasoning, the interpretations in terms of conflicts and competitions to the actions performed on empirical and conceptual anomalies and from the question of generating inconsistencies by radical innovation to the connectionist treatment of coherence.

In 1998 Jaakko Hintikka had already contended that abduction is the "fundamental problem of contemporary epistemology". My aim is to combine philosophical, logical, cognitive, eco-cognitive, neurological, and computational issues, while also discussing some cases of reasoning in everyday settings, in expert inferences, and in science. The main thesis is that abduction is a *basic* kind of human cognition, not only helpful in delineating the first principles of a new theory of science, but also extremely useful in the unification of interdisciplinary perspectives, which would otherwise remain fragmented and dispersed, and thus devoid of the necessary philosophical analysis. In sum, the present book aims at having a strong interdisciplinary nature, encompassing mathematical and logical cases, biological and neurological aspects and analysis of the epistemological impact of the problems caused by the "mathematical physics" of abduction.

The interdisciplinary character of abduction is central and its fertility in various areas of research is evident. The book also addresses the central epistemological question of hypothesis withdrawal in science by discussing historical cases (chapter two), where abductive inferences exhibit their most appealing cognitive virtues. Finally, an interesting and neglected point of contention about human reasoning is whether or not concrete manipulations of external objects influence the generation of hypotheses, for example in science. The book provides an indepth study of what I have called manipulative abduction, showing how we can find methods of constructivity in scientific and everyday reasoning based on external models and cognitive and epistemic mediators.

The book also illustrates the problem of "multimodal abduction", recently pointed out by Paul Thagard, which refers to the various aspects of abductive reasoning, neurological, verbal-propositional, sentential, emotional and manipulative. Multimodal abduction is also appropriate when taking into account the dynamics of the hybrid interplay of the aspects above and the semiotic role played by what I call "semiotic anchors". These anchors constitute ways of favoring hybrid reasoning in various cognitive and epistemic tasks and they play an important role in that event of "externalization of the mind" that researchers such as Andy Clark, Edwin Hutchins, Steven Mithen and others have labelled in various ways, ultimately resorting to the idea of the importance of the external cognitive tools and mediators in cognition. The book provides some case studies derived from the history of discoveries in science, logic, and mathematics, also taking advantage of the agent based perspective" proposed by Dov Gabbay and John Woods.

A central target has been to further study the concept of non-explanatory and instrumental abduction, introduced by Gabbay and Woods in their GW-model of abduction (a model they contrast with the AKM-model, previously proposed by myself, Atocha Aliseda, Theo Kuipers, Ant Kakas and Peter Flach, and Joke Meheus). Non-explanatory and instrumental aspects of abduction (together with the distinction between propositional and strategic plausibility in hypothetical reasoning) have to be discussed and further clarified, especially because they play a crucial role in scientific, mathematical, and logical abduction.

The first chapter, *Theoretical and Manipulative Abduction. Conjectures and Manipulations: the Extra-Theoretical Dimension of Scientific Discovery*, provides an illustration of the main distinctions concerning abductive reasoning concerning creative and selective, theoretical and manipulative abduction, and its primal explanatory character. The significance of the original syllogistic framework proposed by Peirce is also explained together with the status of some recent logical models of abduction. Moreover, some sections introduce the extra-theoretical dimension of scientific reasoning, with the help of the concept of model-based and manipulative abduction. Creativity and discovery are no longer seen as mysterious irrational processes, but, thanks to constructive accounts, they appear as a complex relationship among different inferential steps that can be clearly analyzed and identified. The last part of the chapter is devoted to illustrating the problem of the extra-theoretical explanatory dimension of reasoning and discovery from the perspective of some mathematical cases derived from calculus, where internal and external aspects (optical diagrams) of cognition are at play.

The second chapter, *Non-Explanatory and Instrumental Abduction. Plausibility, Implausibility, Ignorance Preservation*, analyzes and criticizes the difference between GW-model and AKM-model by providing a strict examination of the contrast between explanatory, non-explanatory, and instrumental abduction. Case studies derived from the field of the epistemology of physics, from logic and from mathematics are studied because they are particularly useful to further illustrate the non-explanatory and instrumental aspects of abductive cognition. The issue of instrumental abduction is especially important when intertwined with the exquisite epistemological problem of the role of unfalsifiable hypotheses in scientific reasoning. The role of contradictions, inconsistencies and preinventive forms, and of the computational "automatic abductive scientists" in abductive cognition is also addressed.

The last part of the chapter is devoted to illustrating the problem of the extra-theoretical dimension of cognition from the perspective of the famous discovery of non-Euclidean geometries. This case study is particularly appropriate to the present chapter because it shows relevant aspects of diagrammatic abduction, which involve intertwined processes of both explanatory and non-explanatory abduction acting at the model-based level in what I call mirror and unveiling diagrams. Finally, the last section also deals with the epistemologically some very interesting computational AI applications expressly devoted to the simulation of geometrical reasoning.

The main concern of the third chapter, *Semiotic Brains and Artificial Minds. How Brains Make Up Material Cognitive Systems*, is to furnish an integrated analysis of the abductive processes from an updated epistemological and cognitive/semiotic point of view. Creative abductive reasoning is a risky sort of inference that constitutes a central process in conceptual change in science, mathematics, and logic. Its embodied and distributed aspects and its role in what I call epistemic mediators constitute a central issue of this chapter. Part of the chapter is devoted to the analysis, at a cognitive, neurological, semiotic and epistemological level, of the "externalization of the mind" also considering some classical insights furnished by Turing in the article "Intelligent Machinery"(1948) and some conclusions derived from the paleoanthropological research on what Steven Mithen has called "disembodiment of the mind". The related concepts of mimetic and creative representations and of "mimetic mind" are introduced and explained; a further examination of the problem of on-line and off-line intelligence, in the framework of the relationship between language and inner rehearsal, is provided. The chapter also illustrates abduction from a dynamic perspective and the abductive process of external diagrammatization and iconic brain coevolution, with the help of some mathematical examples. A final scrutiny of the epistemological status of the psychoanalytic concepts of projection and introjection and of psychic externalized "symbols" is accomplished, aided by the concept of manipulative abduction.

In the fourth chapter, *Neuro-Multimodal Abduction. Pre-Wired Brains, Embodiment, Neurospaces*, starting from the results illustrated in the previous chapters regarding the fact that abductive cognition is occurring in a "distributed" framework and in a hybrid way, that is in the interplay between internal and external signs, I contend that we can reconceptualize abduction neurologically. From this perspective abduction is a process in which one neural structure representing the explanatory target generates another neural structure that constitutes a hypothesis. A whole neuro-multimodal framework is depicted, aiming at increasing knowledge about the fact that the classical perspective on abduction, based on logic only, captures limited properties of this cognitive process and considerably disregards model-based aspects. The neuro-multimodal perspective also aims at: i) clarifying the distinction between the hardwired and pre-wired/plastic aspects of abduction; ii) a new understanding of some features of the problem of action and decision in formal reasoning, where a new integrated perspective on action can be worked out, taking advantage of the distinction between thought and motor action, which are both seen as the fruit of brain activity; ii) analyzing the role of abduction in the fundamental mammalian model-based cognitive activities, which relate to representation of object locations within the spatial/pseudo-geometrical framework. A final section is devoted to some philosophical issues arising from the traditions of phenomenology and psychology that are of special interest in elucidating some features of visual and spatial abduction.

Chapter five, *Animal Abduction. From Mindless Organisms to Artifactual Mediators*, is mainly dedicated to clarifying the Peircean originary conflict between the view of abduction as inferential as opposed to instinctual. The first two sections address this puzzling Peircean problem trying to show how his research was anticipatory of central problems and topics of present cognitive science research. Some speculations concerning abduction in terms of the dichotomies between perception and inference, iconicity and logicality, instinct and strategies, should just be admired and closely studied. These basic insights naturally led me to analyze the problem of animal abduction, which represents the other main theme of the chapter. Many animals – traditionally considered "mindless" organisms – make up a series of signs and are engaged in making, manifesting or reacting to a series of signs. Through this semiotic activity – which is fundamentally model-based – they are engaged in "being cognitive agents" and therefore in thinking "intelligently". An important effect of this semiotic activity is a continuous process of "hypothesis generation" that can be seen at the level of both instinctual behavior, as a kind of "hard-wired" cognition, and representation-oriented behavior, where nonlinguistic pseudothoughts drive a plastic model-based cognitive role. This activity is at the root of a variety of abductive performances, which are also analyzed in the light of the concept of affordance, further explored in chapter six. Another important character of the model-based cognitive activity above is the externalization of artifacts that play the role of mediators in animal, languageless, reflexive thinking. The interplay between internal and external representations exhibits a new cognitive perspective on the mechanisms underlying the semiotic emergence of abductive processes in important areas of model-based thinking of mindless organisms. To illustrate this process I also take advantage of the case of affect attunement, which exhibits an impressive case of model-based communication, of the problems of pseudological and reflexive thinking and of the role of pseudoexplanatory guesses in animal plastic cognition.

The title of chapter six is *Abduction, Affordances, and Cognitive Niches. Sharing Representations and Creating Chances through Cognitive Niche Construction.* As a matter of fact, humans continuously delegate and distribute cognitive functions to the environment to lessen their limits. They build models, representations, and other various mediating structures, that are considered to aid thought. In doing these, humans are engaged in a process of cognitive niche construction. In this sense, I argue that a cognitive niche emerges from a network of continuous interplays of hypothetical cognition between individuals and the environment, in which people alter and modify the environment by mimetically externalizing fleeting thoughts, private ideas, etc., into external supports. Hence, cognitive niche construction may also contribute to making a great portion of knowledge available that would otherwise remain simply unexpressed or unreachable. Abductive cognition is a central driver of those designing activities that are closely related to the process of so-called "niche construction". The exploitation of this basically biological concept seems useful to study all those situations that require the transmission and sharing of knowledge, information and, more generally, cognitive resources. Further, some issues concerning the process of transmission and selection of the extragenetic information that is embedded in cognitive niche transformations are considered and their supposed loosely Darwinian character is stressed.

In dealing with the exploitation of cognitive resources embedded in the environment, the notion of affordance, originally proposed by James J. Gibson to illustrate the hybrid character of visual perception, together with the proximal/distal distinction described by Egon Brunswik, are relevant. In order to solve various controversies on the concept of affordance and on the status of the proximal/distal dichotomy, I will take advantage of some useful insights that come from the study on abduction. Abduction may also fruitfully describe all those human and animal hypothetical inferences that are operated through actions made up of smart manipulations to both detect new affordances and to create manufactured external objects that offer new affordances/cues.

Chapter seven, *Abduction in Human and Logical Agents. Hasty Generalizers, Hybrid Abducers, Fallacies*, addresses the problem of logical models of abduction, already introduced in the first chapter. This chapter presents the problem in an agent-based perspective. It is acknowledged that intellectual artifacts like "logical agents" are "ideal" tools for thoughts as is language. These are tools for exploring, expanding, and manipulating our own minds so that creative abductive "new ways of inferring", performed by the "biological" human agents, arise in an unexpected and distributed interplay between brains (and their internal representations) and external representations. The analysis of this issue demonstrates further results regarding the following problems: i) deductive reasoning involves the employment of logical rules in a heuristic manner, even maintaining the truth preserving character: the application of the rules is organized in a way that is able to recommend a particular course of actions instead of another one. Moreover, very often the heuristic procedures of deductive reasoning are performed by means of an "in-formal" (often model-based) abduction; ii) in an agent-based framework fallacies can be redefined and considered as good ways of reasoning; we can hypothesize that what I call manipulative abduction can be re-interpreted as a form of practical reasoning a better understanding of which can furnish a description of human beings as hybrid thinkers in so far they are users of ideal (logical/mathematical) and computational agents; iii) abduction can be seen in an extended eco-logical perspective in so far as it is involved in dialectic processes, where, as a fallacy – from the classical logical perspective, it is exploited in a "distributed" cognitive framework, where epistemic (but also moral) conflicts and negotiations are normally at play.

The last chapter, *Morphodynamical Abduction. Causation of Hypotheses by Attractors Dynamics*, presents some central epistemological, semiotic, and cognitive aspects of what can be called morphodynamical abduction in the perspective of dynamical systems in physics and catastrophe theory in mathematics. Indeed, an integration of the traditional computational view with some ideas developed inside the so-called dynamical approach can suggest some important insights. What is the role of abduction in the dynamical system approach? What is the role of the dichotomy salient/pregnant mathematically depicted by the catastrophe theory with respect to abduction? What is embodied cognition from the point of view – so to say – of its "mathematical physics"? To grasp the role of abduction in these scientific traditions I provide an analysis of the concepts of anticipation, adumbration, attractor, and of the dichotomy salient/pregnant: the result is the description of the abductive generation of new hypotheses in terms of a catastrophic rearrangement of the parameters responsible for the behavior of the system. The main concern of the part of the chapter devoted to the catastrophe theory is to demonstrate that pregnances and saliences provide a further help in increasing knowledge about abductive "hypothesis generation" at the level of both instinctual behavior and representation-oriented behavior, where non-linguistic features drive a "plastic" model-based cognitive role. Furthermore, in terms of dynamic systems and of Thom's mathematical modeling we reach a first sketch of a "physics of abduction", where its cognitive essence is seen in a whole unified naturalistic framework where all phenomena, and so cognition, gain a fundamental eco-physical significance, which also nicely includes some aspects related to a kind of "social epistemology".

A related problem is treated in section 8.6, which illustrates the so-called coalition enforcement hypothesis, which sees humans as self-domesticated animals engaged in a continuous hypothetical activity of building morality, incorporating punishing policies at the same time. Abduction is still at stake, the direct consequence of coalition enforcement being development and the central role of cultural heritage (morality and sense of guilt included). The long-lived and abstract human sense of guilt represents a psychological adaptation which abductively anticipates the appraisal of a moral situation in order to avoid becoming a target of coalitional enforcement.

I started to think upon the research to be exposed in this second book on abduction in 2001 while I was a visiting professor at Georgia Institute of Technology in Atlanta. In addition to my work here in Italy, I further reshaped the manuscript in 2003 as a Weissman Distinguished Visiting Professor at The City University of New York, which provided an excellent work environment, and during visits to the Department of Philosophy of Sun Yat-sen University, in Guangzhou (Canton), P.R. China, where I was visiting professor from 2005 to 2008. These visits added an excellent source of further research and forged strong academic relationships with Asian colleagues, adding to those in the EU and USA. I am grateful to all my colleagues there and in other Universities worldwide for their helpful suggestions and much more. For valuable comments and discussions on a previous draft and about abduction I am particularly grateful to the two anonymous referees and to John Woods, Paul Thagard, Michael Leyton, Dov Gabbay, Claudio Pizzi, Emanuele Bardone, David Gooding, Atocha Aliseda, John Josephson, Walter Carnielli, B. Chandrasekaran, Jon Williamson, Eliano Pessa, Gianluca Introzzi, Douglas Walton, Cameron Shelley, Sami Paavola, Woosuk Park, Giuseppe Longo, Thomas Addis, Diderik Batens, Joke Meheus, Simon Colton, Gerhard Schurz, Ilkka Niniluoto, Theo A. F. Kuipers, Ryan D. Tweney, Peter Flach, Antony Kakas, Oliver Ray, Akinori Abe, Luis A. Pineda, A. Shimojima, P. Langley, Demetris P. Portides, Tommaso Bertolotti. Some sections of chapters one, six, seven, and eight have been written in collaboration with my former Ph.D. students: section 1.7 with Riccardo Dossena, sections 6.1.1, 6.1.2 and 6.2-6.6 with Emanuele Bardone, sections 7.4.2 with Elia Belli, and section 8.1 with Matteo Piazza. The research related to this volume was supported by grants from the Italian Ministry of University, University of Pavia, and the CARIPLO Foundation (Cassa di Risparmio delle Provincie Lombarde). The preparation of the volume would not have been possible without the contribution of resources and facilities of the Computational Philosophy Laboratory (Department of Philosophy, University of Pavia, Italy). This project was conceived as a whole, but as it developed various parts have become articles, which have now been excerpted, revised, and integrated into the current text. I am grateful to Springer for permission to include portions of previously published articles.

Pavia, Italy
June 2009

Lorenzo Magnani

Contents

Chapter 1
Theoretical and Manipulative Abduction
Conjectures and Manipulations: The Extra-Theoretical Dimension of Scientific Discovery

More than a hundred years ago, the American philosopher Charles Sanders Peirce, when working on logical and philosophical problems, suggested the concept of *pragmatism* ("pragmaticism", in his own words) as a logical criterion to analyze what words and concepts express through their practical meaning. Many authors have illustrated creative processes and reasoning, especially in the case of scientific practices. In fact, many philosophers have usually offered a number of ways of construing hypotheses generation, but they aim at demonstrating that the activity of generating hypotheses is paradoxical, obscure, and thus not analyzable.

Those descriptions are often so far from Peircean pragmatic prescription and so abstract to result completely unknowable and obscure. To dismiss this tendency and gain interesting insight about cognitive creativity and the so-called "logic of scientific discovery" we need to build constructive procedures, which could play a role in moving the problem solving process forward by implementing them in some actual models. The "computational turn" gave us a new way to understand creative processes in a strictly pragmatic sense. In fact, by exploiting artificial intelligence, logical, and cognitive science tools, philosophy allows us to test concepts and ideas previously conceived only in abstract terms. It is in the perspective of these *actual computational models* that I have founded the central role of *abduction* in the explanation of creative reasoning in science.

This chapter aims at introducing the distinction between two kinds of *abduction*, *theoretical* and *manipulative*, in order to provide an integrated framework to explain some of the main aspects of both creative and *model-based reasoning* effects engendered by the practice of science and everyday reasoning. The distinction appears to be extremely convenient, after having illustrated the *sentential models* together with their limitations (section 1.4), creativity will be viewed as the result of the highest cases of theoretical abduction demonstrating the role of so-called *model-based abduction* (section 1.5). Moreover, I will delineate what I call *manipulative abduction* (section 1.6) by showing how we can find methods of manipulative constructivity.

From this perspective, creativity and discovery are no longer seen as mysterious irrational processes, but, thanks to constructive accounts, they are viewed as

L. Magnani: Abductive Cognition, COSMOS 3, pp. 1–61.

complex relationships among different inferential steps that can be clearly analyzed and identified. I maintain that the analysis of *sentential*, *model-based* and *manipulative* abduction and of *external* and *epistemic mediators* is important not only to delineate the actual practice of abduction, but also to further enhance the development of programs computationally adequate in rediscovering, or discovering for the first time, for example, scientific hypotheses or mathematical theorems. In this chapter attention will be focused on those particular kinds of abductive cognition that resort to the existence of *extra-theoretical* ways of thinking – *thinking through doing*. Indeed many cognitive processes are centered on *external representations*, as a means to create communicable accounts of new experiences ready to be integrated into previously existing systems of experimental and theoretical practices. The last part of the chapter is devoted to illustrating the problem of the extra-theoretical explanatory dimension of reasoning and discovery from the perspective of some mathematical cases derived from calculus, where internal and external aspects (optical diagrams) of cognition are at play.

1.1 Computational Modeling as a Pragmatic Rule for Clarity

What I call "computational philosophy",[1] aims at investigating many important concepts and problems of the philosophical and epistemological tradition in a new way by taking advantage epistemological, cognitive, and artificial intelligence (AI) computational methodologies. I maintain that the results of computational philosophy meet the classical requirements of Peircean "pragmatic" ambitions and nicely tie together both issues related to the dynamics of information and its systematic embodiment in segments of "knowledge". In the second half of the nineteenth century the great American philosopher Charles Sanders Peirce suggested the idea of *pragmatism* as a logical criterion to analyze what words and concepts express through their practical meaning. In "The fixation of belief" [1877] Peirce enumerates four main methods by means of which it is possible to fix belief: the method of tenacity, the method of authority, the *a priori* method and, finally, the method of science, by means of which, thanks to rigorous research, "[...] we can ascertain by reasoning how things really and truly are; and any man, if he has sufficient experience and he reasons enough about it, will be led to the one True conclusion" [Peirce, 1987, p. 255]. Only the scientific method leads to identify what is "real", that is "true".

Peirce will more clearly explain the public notion of truth here exposed, and the interpretation of reality as the final purpose of the human inquiry, in his subsequent paper "How to make our ideas clear" [1878]. Here Peirce addresses attention on the notions of "clear idea" and "belief". "Whoever has looked into a modern treatise on logic of the common sort, will doubtlessly remember the two distinctions between *clear* and *obscure* conceptions, and between *distinct* and *confused* conceptions" he writes [Peirce, 1987, p. 257]. A clear idea is defined as one which is apprehended so that it will be recognized wherever it is met, and so that no other will be mistaken for it. If it fails to be clear, it is said to be obscure. On the other hand, a distinct idea

[1] Topics and aims of computational philosophy are illustrated in [Magnani, 1997].

is defined as one which contains nothing which is not clear. In this paper Peirce is clearly opposing traditional philosophical positions, such as those by Descartes and Leibniz, who consider clarity and distinction of ideas only from a merely psychological and analytical perspective:

> It is easy to show that the doctrine that familiar use and abstract distinctness make the perfection of apprehension has its only true place in philosophies which have long been extinct; and it is now time to formulate the method of attaining to a more perfect clearness of thought, such as we see and admire in the thinkers of our own time [Peirce, 1987, p. 258].

Where do we have, then, to look for a criterion of clarity, if philosophy has become too obscure, irrational and confusing, if "[...] for an individual, however, there can be no question that a few clear ideas are worth more than many confused ones?" [Peirce, 1987, p. 260]. "The action of thought is excited by the irritation of doubt, and ceases when belief is attained; so that the production of belief is the sole function of thought" [Peirce, 1987, p. 261]. And belief "[...] is something that we are aware of [...] it appeases the irritation of doubt; and, third, it involves the establishment in our nature of a rule of action, or, say for short, a habit" [Peirce, 1987, p. 263]. Hence, the whole function of thought is to produce habits of action. This leads directly to the *methodological* pragmatic theory of meaning, a procedure to determine the meaning of a proposition:

> To develop its meaning, we have, therefore, simply to determine what habits it produces, for what a thing means is simply what habits it involves. Now, the identity of a habit depends on how it might lead us to act, not merely under such circumstances as are likely to arise, but under such as might possibly occur, no matter how improbable they may be. Thus, we come down to what is tangible and conceivably practical, as the root of every real distinction of thought, no matter how subtile it may be; and there is no distinction of meaning so fine as to consist in anything but a possible difference of practice [Peirce, 1987, pp. 265-266].

In this way Peirce creates the equivalence among idea, belief and habit, and can define the rule by which we can reach the highest grade of intellectual clearness, pointing out that is impossible to have an idea in our minds which relates to anything but conceived sensible effects of things. Our idea of something is our idea of its sensible effects: "Consider what effects, that might conceivably have practical bearings, we conceive the object of our conception to have. Then, our conception of these effects is the whole of our conception of the object" [Peirce, 1987, p. 266]. This rule founds the pragmatic procedure thanks to which it is possible to fix our ideas.

1.2 Computational Modeling and the Problem of Scientific Discovery

Peirce's conception of clarity contains the idea that to define the meaning of words and concepts we have to "test" them: the whole conception of some quality lies in its conceivable effects. As he reminds us by the example of the concept of *hardness* "[...] there is absolutely no difference between a hard thing and a soft thing so long as they are not brought to the test" [Peirce, 1987, p. 266]. Hence, we can define the "hardness" by looking at those predictable events that occur every time we think of testing some thing.

This methodological criterion can be useful to solve the problem of *creative* reasoning, and to describe, in rational terms, some aspects of the delicate question of a "logic of discovery": what do we mean by "creative", and how can a "creative process" be described? Much has been said on the problem of *creativity* and hypotheses generation. In the history of philosophy there are at least three important ways for designing the role of hypothesis generation, considered in the perspective of problem solving performances. But all aim at demonstrating that the activity of generating hypotheses is paradoxical, illusory, obscure, implicit, and not analyzable. Plato's doctrine of *reminiscence* can be looked at from the point of view of an epistemological argument about the paradoxical concept of "problem-solving": in order to solve a problem one must in some sense already know the answer, there is no real generation of hypotheses, there is only recollection. The activity of Kantian *schematism* is also implicit, resulting from imagination and completely unknowable, empty, and devoid of any possibility of being rationally analyzed. It is an activity of tacit knowledge, "an art concealed in the depths of the human soul, whose real modes of activity nature is hardly likely ever to allow us to discover, and to have open to our gaze" [Kant, 1929, A141-B181, p. 183]. In turn Polanyi thinks that if all knowledge is explicit and capable of being clearly stated, then we cannot know of the existence of a problem or look for its solution; if problems nevertheless exist, and discoveries can be made by solving them, we can know things that we cannot express: consequently, the role of so-called *tacit knowledge* "the intimation of something hidden, which we may yet discover" is central [Polanyi, 1966].

In all these descriptions, the problem is that the definition of concepts like "creativity" and "discovery" is *a priori*. Following Peirce, the definitions of concepts of this sort are not usually based upon any observed facts, at least not in any great degree; even if sometimes these beliefs are in harmony with natural causes. They have been chiefly adopted because their fundamental propositions seemed "agreeable to reason". That is, we find ourselves inclined to believe them. Usually this frame leads to a proliferating verbosity, in which theories are often incomprehensible and lead to some foresight just by intuition. But a theory which needs intuition to determine what it predicts has poor explanatory power. It just "[...] makes of inquiry something similar to the development of taste" [Peirce, 1987, p. 254].

1.2.1 Abduction and Retroduction

Many philosophical efforts in the last century have been spent in studying the conceptual change in science. In the mid-1960s many critics challenged the comforting picture of conceptual change in terms of continuous and cumulative steps. Contrary to this picture Kuhn claimed that conceptual change in science is analyzable as a kind of irrational and obscure *Gestalt-switch*, that accounts for the inventive processes and the achievement of new scientific theories and paradigms [Kuhn, 1962]. Kuhn argued that major changes in science are best characterized as *revolutions*, involving the over-throw and the replacement of the reigning conceptual systems and world views by means of new ones *incommensurable* with them. Kuhn brought philosophers of science to distinguish between the logic of discovery and the logic of justification (i.e. the distinction between the psychological side of creation and the logic argument of proving new discovered ideas by facts).[2] Most have concluded that no logic of discovery exists and, moreover, that a "rational" model of discovery is impossible. In short, scientific creative reasoning should be non-rational or irrational and there is no reasoning towards hypotheses.

In the last decades philosophers of science have abandoned this attitude. The researchers who work on scientific change tend now to focus attention on the problem of rational choice between competing theories and hypotheses and on the discovery processes. This also leads to the problem of understanding how scientists combine their individual human cognitive abilities with the conceptual resources available to them as members of a scientific community and of a wider natural and social context. It is by means of this synthesis that the creation, elaboration, and communication of a new emerging representation of a scientific domain is made possible.

This chapter aims at introducing and further deepening the distinction, I have already illustrated in my previous book on abduction [Magnani, 2001b], between two kinds of *abduction*, *theoretical* and *manipulative*, in order to provide an integrated framework to explain some of the main aspects of both creative and *model-based reasoning* effects engendered by the practice of science and in everyday reasoning. The distinction appears to be extremely convenient: after having illustrated the *sentential models* together with their limitations (section 1.4), creativity will be viewed as the result of the highest cases of theoretical abduction showing the role of the so-called *model-based abduction* (section 1.5). Moreover, I will delineate what I call *manipulative abduction* (section 1.6) by showing how we can find methods of manipulative constructivity, where the XX century epistemological tradition has settled the most negative effects of theory-ladenness.

Abduction is a popular term in many fields of AI, such as diagnosis, planning, natural language processing, motivation analysis, logic programming, and probability theory. Moreover, abduction is important in the interplay between AI and philosophy, cognitive science, historical, temporal, and narrative reasoning, decision-making, legal reasoning, and emotional cognition.[3] Six volumes (monographs and collections) are currently available [Josephson and Josephson, 1994; Flach and Kakas, 2000b; Kuipers, 2000; Magnani, 2001b; Gabbay and Woods, 2005; Aliseda, 2006; Walton, 2004] and three special issues of international journals (*Philosophica*, 1998 61(1); *Foundations of Science*, 2004, 9; 2008, 13(1); *Logic Journal of the IGPL*, 2006 14(1)). Of course many articles from various disciplinary

[2] A perspective originally established by [Reichenbach, 1938] and [Popper, 1959].

[3] A list of the classical bibliography on abduction is given in [Magnani, 2001b].

fields of research are continually published on this topic.[4] Let us consider the following interesting passage, from an article by [Simon, 1965], dealing with the logic of normative theories:

> The problem-solving process is not a process of "deducing" one set of imperatives (the performance programme) from another set (the goals). Instead, it is a process of selective trial and error, using heuristic rules derived from previous experience, that is sometimes successful in *discovering* means that are more or less efficacious in attaining some end. If we want a name for it, we can appropriately use the name coined by Peirce and revived recently by Norwood Hanson [1958]: it is a *retroductive* process. The nature of this process – which has been sketched roughly here – is the main subject of the theory of problem-solving in both its positive and normative versions [Simon, 1977, p. 151].

Simon states that discovering means that are more or less efficacious in attaining some end are performed by a *retroductive* process. He goes on to show that it is easy to obtain one set of imperatives from another set by processes of discovery or retroduction, and that the relation between the initial set and the derived set is not a relation of logical implication. I completely agree with Simon: retroduction (that is abduction, cf. below) is the main subject of the theory of problem-solving and developments in the fields of cognitive science and artificial intelligence have strengthened this conviction.

[Hanson, 1958, p. 54] is perfectly aware of the fact that an enormous range of explanations (and causes) exists for any event:

> There are as many causes of x as there are explanations of x. Consider how the cause of death might have been set out by a physician as "multiple hemorrhage", by the barrister as "negligence on the part of the driver", by a carriage-builder as "a defect in the brakeblock construction", by a civic planner as "the presence of tall shrubbery at that turning".

The word "retroduction" used by Simon is the Hansonian neopositivistic one replacing the Peircean classical word abduction. Following Hanson's point of view Peirce "[...] regards an abductive inference (such as 'The observed position of Mars falls between a circle and an oval, so the orbit must be an ellipse') and a perceptual judgment (such as 'It is laevorotatory') as being opposite sides of the same coin". It is also well-known that Hanson relates abduction to the role of patterns in reasoning and to the Wittgensteinian "Seeing that" [Hanson, 1958, p. 86].

As Fetzer has stressed, from a philosophical point of view the main modes of argumentation for reasoning from premises to conclusions are expressed by these three general kinds of reasoning: *deductive* (demonstrative, non ampliative, additive), *inductive* (non-demonstrative, ampliative, non additive), *fallacious* (neither, irrelevant, ambiguous). Abduction, which expresses likelihood in reasoning, is a typical form of fallacious inference: "[...] it is a matter of utilizing the principle

[4] General classical considerations on abduction in science and AI can also be found in [Gooding, 1996; Josephson and Josephson, 1994; Kuipers, 1999; Thagard, 1988; Shrager and Langley, 1990].

of maximum likelihood in order to formalize a pattern of reasoning known as 'inference to the best explanation"' [Fetzer, 1990, p. 103].[5] These different kinds of reasoning will be illustrated in the following section.

Many researchers in the area of cognitive science consider scientific thinking (and thinking activity in general), as related to a kind of "representational" system that we can implement in a computational model: thinking is a form of computation. Following the idea that a full understanding of mental processes is possible only from a computational perspective [Johnson-Laird, 1983], these models have been implemented in AI programs where data structures and procedures correspond to assumed mental structures and processes [Thagard, 1992].[6] In the last decades, also to better understand the complex problem of conceptual change in science, besides the ideas elaborated in the AI areas of knowledge representation, problem solving, and machine learning, we needed the important concept of abduction.[7] Scientific theories contain many theoretical hypotheses that cannot be built by simple generalization of observations. Indeed, Peirce presented abduction as a mechanism by which it is possible to account for the generation of new explanatory hypotheses in science.

As I have illustrated in the first section of this chapter, a suggestion that can help to solve the enigma of discovery and creativity comes from the "computational turn" developed in the last years. Recent computational philosophy research in the field of cognitive science make use of tools able to overcome those puzzling speculative problems, or, at least, to redefine them in a strict pragmatical sense. In fact, taking advantage of modern tools of logic, artificial intelligence, and of other cognitive science disciplines, computational philosophy is able to construct actual models of studied processes. It is an interesting constructive rational alternative that, disregarding the most abstract level of philosophical analysis can offer clear and testable architectures of creative processes.

1.3 What Is Abduction?

The development of human society has now reached a technological level in which issues concerning the creation and dynamics of information – especially in science – are absolutely crucial. As I have already said, inside the computational philosophy framework, a new paradigm, aimed at unifying the different perspectives and providing some new design insights, arose by emphasizing the significance of the

[5] On the inference to the best explanation see also [Harman, 1965; Harman, 1968; Thagard, 1987; Lipton, 2004].

[6] It is the so-called "computational-representational understanding of the mind" (CRUM) [Thagard, 1996].

[7] In my previous book on abduction [Magnani, 2001b] I have already illustrated some basic philosophical, logical, cognitive, and computational aspects of the concept of abduction. This book aims at further increasing knowledge taking advantage of other intellectual achievements, not only related to philosophy, logic, and artificial intelligence, but also concerning biology, neurology, anthropology, ecological psychology, dynamical system theory.

concept of *abduction*, in order to illustrate the problem-solving process and to propose a unified and rational epistemological model of scientific discovery, diagnostic reasoning, and other kinds of creative and hypothetical reasoning.

A hundred years ago, Charles Sanders Peirce [1931-1958] coined the concept of abduction in order to illustrate that the process of scientific discovery is not irrational and that a methodology of discovery is possible. Peirce interpreted abduction essentially as an "inferential" *creative process* of generating a new "explanatory" hypothesis. Abduction has a logical form (fallacious, if we model abduction by using classical syllogistic logic)[8] distinct from deduction and induction. Many reasoning conclusions that are not derived in a deductive manner are abductive. For instance, if we see a broken horizontal glass on the floor we might *explain* this fact by postulating the effect of wind blowing shortly before: this is not certainly a deductive consequence of the glass being broken (a cat may well have been responsible for it).[9] Abduction is the process of *inferring*[10] certain facts and/or laws and hypotheses that render some sentences plausible,[11] that *explain* (and also sometimes *discover*) some (eventually new) phenomenon or observation; it is the process of reasoning in which explanatory hypotheses are formed and evaluated.

It is important to note that I adopt in this chapter the view of abduction that it is immediately a generation of plausible hypotheses. Although this claim may be true in various contexts, it is not so generally. I am thinking particularly of Descartes's implausible arguments in the *Meditations*, in which he suggests, for example, that our experiences are caused not by the world stimulating our sensory organs but by the machinations of an evil demon. Such a claim is highly implausible, but that does not deter Descartes given his project of finding any kind of doubt against a belief. Of course, Peirce famously castigated Descartes by remarking that we cannot begin with complete doubt and argued that Descartes's project was a fool's errand. Is *implausible* abduction really abduction? I will treat this problem in the following chapter when dealing with the so-called "instrumental abduction".

Moreover, we have to remember that although explanatory hypotheses can be elementary, there are also cases of composite, multipart hypotheses. Anyway, some hypotheses are empty from the *explanatory* point of view: for example the generalization "every object in the population is female or male" does not explain that Maria is female, since it requires the additional knowledge that Maria is not male.

[8] The abductive inference rule corresponds to the well-known fallacy called affirming the consequent.

[9] This event constitutes in its turn an *anomaly* that needs to be solved/explained but I have to anticipate that surprise or anomalies do not constitute an intrinsic requirement for abduction.

[10] "It must be remembered that abduction, although it is little hampered by logical rules, nevertheless is logical inference, asserting its conclusion only problematically or conjecturally, it is true, but nevertheless having a perfect logical form" [Peirce, 1931-1958, 5.188].

[11] Peirce thinks that humans's capacity to make abductive plausible hypotheses is ultimately based on the *instinct*. His idea is in itself abductive "It is a primary hypothesis underlying all abduction that the human mind is akin to the truth in the sense that in a finite number of guesses it will light upon the correct hypothesis" [Peirce, 1931-1958, 7.220]. I will illustrate the role of instinct in abduction the first two sections 5.1 and 5.2 of chapter five.

The process of finding such generalizations has been called confirmatory (or descriptive) induction :

> A typical form of explanatory induction is concept learning, where we want to learn a definition of a given concept *C* in terms of other concepts. This means that our inductive hypotheses are required to explain (logically entail) why particular individuals are *C*s, in terms of the properties they have. However, in the more general case of confirmatory induction we are not given a fixed concept to be learned. The aim is to learn relationships between any of the concepts, with no particular concept singled out. The formalization of confirmatory hypothesis formation cannot be based on logical entailment, as in Peirce's abduction. Rather, it is a quantitative form of degree of confirmation, which explains its name [Flach and Kakas, 2000a].

Theoretical and manipulative abduction are treated in this chapter in the perspective of the orthodox Peircean *explanatory* view: non-explanatory and instrumental abduction, not clearly considered by Peirce, will be illustrated in detail in the following chapter.

1.3.1 The Syllogistic Framework and the ST-Model

First, it is necessary to show the connections between abduction, induction, and deduction and to stress the significance of abduction to illustrate the problem-solving process. I think the example of diagnostic reasoning is an excellent way to introduce abduction. Some years ago I have developed with others [Lanzola *et al.*, 1990; Ramoni *et al.*, 1992] an epistemological model of medical reasoning, called the *Select and Test Model* (ST-model) [Magnani, 1992; Stefanelli and Ramoni, 1992] which can be described in terms of the classical notions of abduction, deduction and induction; it describes the different roles played by such basic inference types in developing various kinds of medical reasoning (diagnosis, therapy planning, monitoring) but can be extended and regarded also as an illustration of scientific theory change. The model is consistent with the Peircean view about the various stages of scientific inquiry in terms of "hypothesis" generation, deduction (prediction), and induction.

The type of inference called abduction was also studied by Aristotelian syllogistics, as a form of ἀπαγωγή, and later on by medieval reworkers of syllogism. As I have already noted, Peirce,[12] interpreted abduction essentially as an "inferential" *creative process* of generating a new hypothesis: it is extremely important to note the special meaning attributed to the adjective "inferential" by Peirce in its broad philosophical and semiotic perspective, that I will better illustrate below in section 1.5.

Abduction and induction, viewed together as processes of production and generation of new hypotheses, are sometimes called reduction, that is ἀπαγωγή.[13] As

[12] Cf. [Frankfurt, 1958; Reilly, 1970; Fann, 1970; Davis, 1972; Ayim, 1974; Anderson, 1986; Anderson, 1987; Kapitan, 1990; Hookway, 1992; Debrok, 1997; Roesler, 1997; Wirth, 1997].

[13] Sometimes ἀπαγωγή is translated with retroduction, so it is simply referred to abduction (see above in this section).

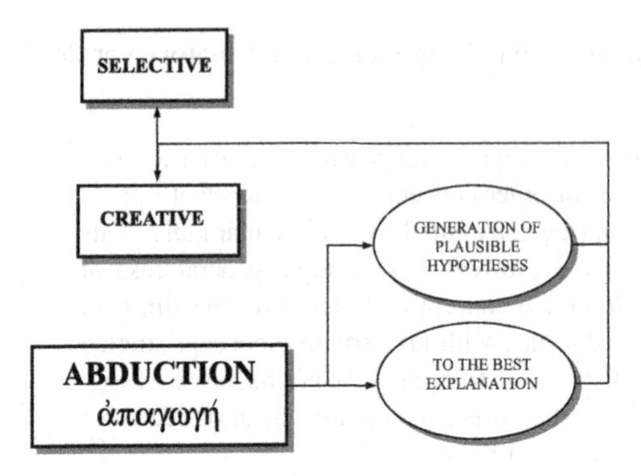

Fig. 1.1 Creative and selective abduction

[Lukasiewicz, 1970] makes clear, "Reasoning which starts from reasons and looks for consequences is called *deduction*; that which starts from consequences and looks for reasons is called *reduction*". The celebrated example given by Peirce is Kepler's conclusion that the orbit of Mars must be an ellipse.[14]

There are two main epistemological meanings of the word abduction [Magnani, 1988; Magnani, 1991; Magnani, 2001b]: 1) abduction that only generates "plausible"[15] hypotheses ("selective" or "creative") and 2) abduction considered as inference "to the best explanation", which also evaluates hypotheses (cf. Figure 1.1) [Harman, 1973; Thagard, 1988; Lipton, 2004].[16] An illustration from the field of medical knowledge is represented by the discovery of a new disease and the manifestations it causes which can be considered as the result of a creative abductive inference. Therefore, "creative" abduction deals with the whole field of the growth of scientific knowledge. This is irrelevant in medical diagnosis where instead the task is to "select" from an encyclopedia of pre-stored diagnostic entities. We can call both inferences ampliative, selective and creative, because in both cases the reasoning involved amplifies, or goes beyond, the information incorporated in the premises [Magnani, 1992].

All we can expect of our "selective" abduction, is that it tends to produce hypotheses for further examination that have some chance of turning out to be the best explanation. Selective abduction will always produce hypotheses that give at least a *partial* explanation and therefore have a small amount of initial plausibility. In the syllogistic view (see below) concerning abduction as inference to the best explanation advocated by Peirce one might require that the final chosen explanation be the *most plausible*.

When we speak of abduction as inference to the best explanation, we have to note that the adjective "best" has to be taken in a Pickwickian sense: actually abduction

[14] A clear reconstruction of Kepler's discovery is given in [Gorman, 1998].

[15] A further analysis of this important concept is illustrated in subsection 2.3.1, chapter two of this book.

[16] Further explanations of this bipolar distinction (and about the use herein of the concept of plausibility) are given below in this chapter.

never reaches the status of best hypothesis, we have to intend the word "best" in a contextual and provisional way. As I will explain in detail in the following chapter (section 2.1), the agent's abduction must be considered as preserving the ignorance that already gave rise to her ignorance-problem. In this perspective, recently suggested and illustrated by [Gabbay and Woods, 2005], abduction does not have to be considered a "solution" of an ignorance problem, but rather a response to it, in which the agent reaches presumptive attainment rather than actual attainment. It is clear that in this framework the inference to the best explanation – if considered as a truth conferring achievement – cannot be a case of abduction, because abductive inference is essentially ignorance preserving. It is also important to note that the emphasis given along this chapter to the "explanatory" character of abductive reasoning is strictly related to the Peircean point of view: Gabbay and Woods have particularly stressed that abduction *is not intrinsically explanatory*, not only, abduction can be merely *radically instrumental*. I will extensively explain these aspects of abductive cognition in the following chapter (sections 2.2 and 2.3).

What I call *theoretical abduction* certainly illustrates much of what is important in creative abductive reasoning, in humans and in computational programs, especially the objective of selecting and creating a set of hypotheses (diagnoses, causes, hypotheses) that are able to dispense good (preferred) explanations of data (observations), but fails to account for many cases of explanations occurring in science and in everyday reasoning when the exploitation of environment is crucial. It fails to account for those cases in which there is a kind of "discovering through doing", cases in which new and still unexpressed information is codified by means of manipulations of some external objects (*epistemic mediators*). I maintain that there are two kinds of theoretical abduction, "sentential", related to logic and to verbal/symbolic inferences, and "model-based", related to the exploitation of internalized models of diagrams, pictures, etc., cf. below in this chapter (cf. Figure 1.2).

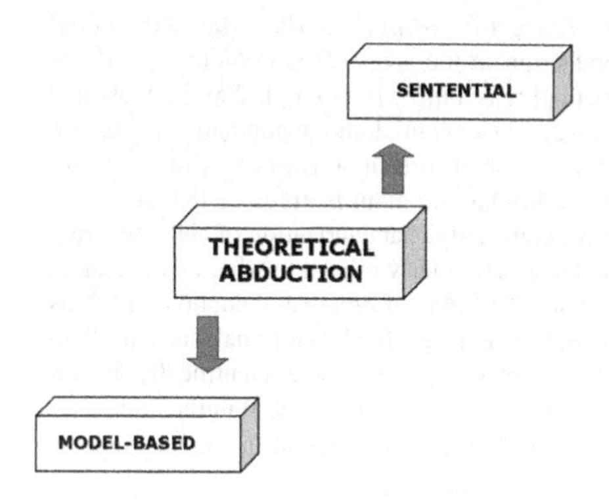

Fig. 1.2 Theoretical abduction

The concept of *manipulative abduction*[17] captures a large part of scientific thinking where the role of action is central, and where the features of this action are implicit and hard to be elicited: action can provide otherwise unavailable information that enables the agent to solve problems by starting and by performing a suitable abductive process of generation or selection of hypotheses.[18]

The epistemological distinction – which I will illustrate and elaborate upon in the following pages – between theoretical and manipulative abduction is certainly based on the possibility of separating the two aspects in real cognitive processes, resorting to the differentiation between off-line (theoretical, when only inner aspects are at stake) and on-line (manipulative, where the interplay between internal and external aspects is fundamental.[19] Some authors have raised doubts about the on-line/off-line distinction on the grounds that no thinking agent is ever wholly on-line or wholly off-line. I think this distinction is at least useful from an epistemological perspective as a way of theoretically illustrating different cognitive levels, which in the following chapters will be further analyzed and seen at work, in human and animal cognition.

We know that throughout his career Peirce defended the thesis that, besides deduction and induction,[20] there is a third mode of inference that constitutes the only method for really improving scientific knowledge, which he called *abduction*. Science improves and grows continuously, but this continuous enrichment cannot be due to deduction, nor to induction: deduction does not produce any new idea, whereas induction produces very simple ideas. New ideas in science are due to *abduction*, a particular kind of non-deductive[21] inference that involves the generation and evaluation of explanatory hypotheses.

I and others [Ramoni *et al.*, 1992] have developed an epistemological model of medical reasoning, called the Select and Test Model (ST-model) which can be described in terms of the classical notions of abduction, deduction and induction. It

---

[17] The concepts of theoretical and manipulative abduction and of epistemic mediators are introduced in [Magnani, 2001b].

[18] I have collected in a recent special issue of the *Logic Journal of the IGPL* [Magnani, 2006a] various contributions regarding research on abduction in the areas of epistemology, artificial intelligence, and of the logic of "so–called practical reasoning". [Patokorpi, 2007] adopts and enriches my distinction between selective, creative, non-sentential and manipulative abduction and applies abduction to the pedagogical problem of analyzing how learners learn in an information society technology (IST): abduction highlights the main features of IST enhanced learning. In a recent paper [Schurz, 2008] further extends my characterization of abductive reasoning proposing a classification of different patterns particularly related to what I call creative abduction. The article illustrates the features of several kinds of creative abductions, such as theoretical model abduction, common cause abduction and statistical factor analysis, and illustrates them by various real case examples. It is also suggested to demarcate scientifically fruitful abductions from purely speculative abductions by using the criterion of causal unification.

[19] The distinction between off-line and on-line thinking is analyzed in detail in chapter three of this book, subsection 3.6.5.

[20] Peirce clearly contrasted abduction with induction and deduction, by using the famous syllogistic model I will describe below in this section. More details on the differences between abductive and inductive/deductive inferences can be found in [Flach and Kakas, 2000b] and [Magnani, 2001b]; cf. also below, subsection 1.4.1.

[21] Non-deductive if we use the attribute "deductive" as designated by classical logic.

describes the different roles played by such basic inference types in developing various kinds of medical reasoning (diagnosis, therapy planning, monitoring) but can be extended and regarded also as an illustration of scientific theory change.[22] The model is consistent with the Peircean view regarding the various stages of scientific inquiry in terms of "hypothesis" generation, deduction (prediction), and induction.

As previously illustrated, I have introduced a distinction between "creative" and "selective" abduction. Selective abduction will always produce hypotheses that give at least a partial explanation and therefore have a small amount of initial plausibility. In the syllogistic view advocated by Peirce (see below) concerning abduction as inference to the best explanation one might require that the final chosen explanation be the most *plausible*. Since the time of John Stuart Mill, the name given to all kinds of non deductive reasoning has been induction, considered as an aggregate of many methods for discovering causal relationships. Consequently induction in its widest sense is an ampliative process of the generalization of knowledge. Peirce [1955a] distinguished various types of induction: a common feature of all kinds of induction is the ability to compare individual statements: by using induction it is possible 1) to synthesize individual statements into general laws – inductive generalizations – in a defeasible way, but 2) it is also possible to confirm or discount hypotheses.

Following Peirce, I am clearly referring here to the latter type of induction, that in the ST-model is used as the process of reducing the uncertainty of established hypotheses by comparing their consequences with observed facts.[23] Some authors stress that abduction and induction derive from a common source, the hypothetical or non-deductive reasoning, others emphasize the various aspects that distinguish them, that is how specifically abduction and induction extend our knowledge. In other cases it is affirmed that all non-deductive reasoning is of the same type, which is called induction [Flach and Kakas, 2000a].

Further classifications of inductive arguments have been proposed, such as arguments based on samples, (that is inductive generalizations), arguments from analogy, and statistical syllogisms [Salmon, 1990]. Finally, we have to remember that in the case of the so-called *inductive logic* [Carnap, 1950] the aim is to solve the problem of knowing the degree of belief we should attribute to the hypothetical conclusion H, given evidence E collected in the premises of an inductive argument, that is identified with the conditional probability $P(H|E)$. This formalization of the inductive support is also called *confirmation theory*: it does not deal with the problem of individuating the ways of "generating" inductive hypotheses but refers to a logic of hypothesis "evaluation". Abduction creates or selects hypotheses; from these hypotheses consequences are derived by deduction that are compared with the available data by induction. This perspective on hypothesis testing in terms of induction is also known in philosophy of science as the "hypothetico-deductive method" [Hempel, 1966] and is related to the idea of confirmation of scientific hypotheses,

[22] I have illustrated the problem of diagnosis, therapy, and monitoring and the related AI computational programs in [Magnani, 2001b, chapter four].

[23] It is possible to treat every good inductive generalization as an instance of abduction [Josephson, 2000].

predominant in neopositivistic philosophy but also present in the anti-inductivist tradition of falsificationism [Popper, 1959].

In summary, it is important to note that if in diagnostic settings and in the classical syllogistic framework I am illustrating in this chapter we basically refer to induction simply as a way of confirming or discounting hypotheses, in various cases of mathematical reasoning, where model-based and manipulative abduction is at play, induction "also" plays the usual generalizing role. In these cases the reasoners have to produce new *hypothetical* knowledge, H, which extends their own preexistent theories such that the observations on which they work can be first of all deduced by the new abductively enriched theories. They abductively provide new individual and *situated* "samples", which offer chances for further knowledge. Each of these abductive situated results can in turn generate further *universal* inductive generalizations possibly to be withdrawn because of disconfirmation; in this last case a further cyclic abductive-inductive process can restart. The "specificity" of the generated abductive hypotheses is related to their ignorance-preserving character (cf the following chapter, section 2.1); the "generality" of inductive hypotheses is related to their truth-conferring/probability-enhancing character, at the same time occasionally endowed with an evaluative function. Abductively building new situated results is in these cases central to make possible an induction able to generate new general knowledge, in these cases not reachable through abduction (further details on the relationship between abduction and induction are illustrated below in subsection 1.4.1).[24] This kind of interplay between abduction and induction is also occurring in the mathematical case I will exploit in chapter three, subsection 3.6.2, to illustrate important aspects of "manipulative" abduction.

Deduction is an inference that refers to a logical implication. Deduction may be distinguished from abduction and induction on the grounds that the truth of the conclusion of the inference is guaranteed by the truth of the premises on which it is based only in deduction. Deduction refers to the so-called non-defeasible arguments. It should be clear that, on the contrary, when we say that the premises of an argument provide partial support for the conclusion, we mean that if the premises were true, they would give us good reasons – but not conclusive reasons – to accept the conclusion. That is to say, although the premises, if true, provide some evidence to support the conclusion, the conclusion may still be false (arguments of this type are called inductive, or abductive, arguments).

[24] I have described this specific kind of abductive/inductive process in [Magnani, 2009b], also illustrating the research provided by [Rivera and Rossi Becker, 2007]. Taking advantage of my concept of manipulative abduction the authors study a pedagogical framework concerning the need of increasing knowledge on the ways in which learners (abducers) in the area of school algebra develop their abilities. They illustrate the case of different subjects [elementary majors] who are given sequences of figural and numerical cues which taken together comprise classes of abstract objects such as even and odd numbers and related diagrams. These sequences are the basis of subjects' subsequent – multimodal (cf. this book, chapter four, section 4.1) – abductions and inductive generalizations and/or evaluations.

All these distinctions need to be exemplified. To describe how the three inferences operate, it is useful to start with a very simple example dealing with diagnostic reasoning and illustrated (as Peirce initially did),[25] in syllogistic terms:

1. If a patient is affected by a pneumonia, his/her level of white blood cells is increased.
2. John is affected by a pneumonia.
3. John's level of white blood cells is increased.

(This syllogism is known as Barbara).

By deduction we can infer (3) from (1) and (2). Two other syllogisms can be obtained from Barbara if we exchange the conclusion (or Result, in Peircean terms) with either the major premise (the Rule) or the minor premise (the Case): by induction we can go from a finite set of facts, like (2) and (3), to a universally quantified generalization – also called categorical inductive generalization, like the piece of hematologic knowledge represented by (1) (in this case we meet induction as the ability to generate simple laws, contrasted with induction as a way to confirm or discard hypotheses, cf. above).[26] Starting from knowing – selecting – (1) and "observing" (3) we can infer (2) by performing a selective abduction.[27] The abductive inference rule corresponds to the well-known fallacy called affirming the consequent (simplified to the propositional case)

$$\frac{\varphi \rightarrow \psi \qquad \psi}{\varphi}$$

It is useful to give another example, describing an inference very similar to the previous one:

1. If a patient is affected by a beta-thalassemia, his/her level of hemoglobin A2 is increased.
2. John is affected by a beta-thalassemia.
3. John's level of hemoglobin A2 is increased.

Such an inference is valid, that is not affected by uncertainty, since the manifestation (3) is pathognomonic for beta-thalassemia (as expressed by the biconditional in $\varphi \leftrightarrow \psi$). This is a special case, where there is no abduction because there is no "selection", in general clinicians very often have to deal with manifestations

[25] Some authors [Flach and Kakas, 2000a; Aliseda, 2000] distinguish between Peircean early syllogistic theory and his later "inferential" one, in which abduction refers to the whole hypothesis formation component of explanatory reasoning (cf. below section 1.5).

[26] We can consider this inference a sort of generalization from a sample of patients [or of beans] to the whole population of them [or of beans in the bag].

[27] We have to remark that at the level of the syllogistic treatment of the subject Peirce calls this kind of argumentation "hypothesis"; he will introduce the term abduction only in his later theory.

which can be explained by different diagnostic hypotheses: in this case the inference rule corresponds to

$$\frac{\begin{array}{c}\varphi \leftrightarrow \psi \\ \psi\end{array}}{\varphi}$$

Thus, *selective abduction* is the making of a preliminary guess that introduces a set of plausible diagnostic hypotheses, followed by deduction to explore their consequences, and by induction to test them with available patient data, (1) to increase the likelihood of a hypothesis by noting evidence explained by that one, rather than by competing hypotheses, or (2) to refute all but one (cf. Figure 1.3.)

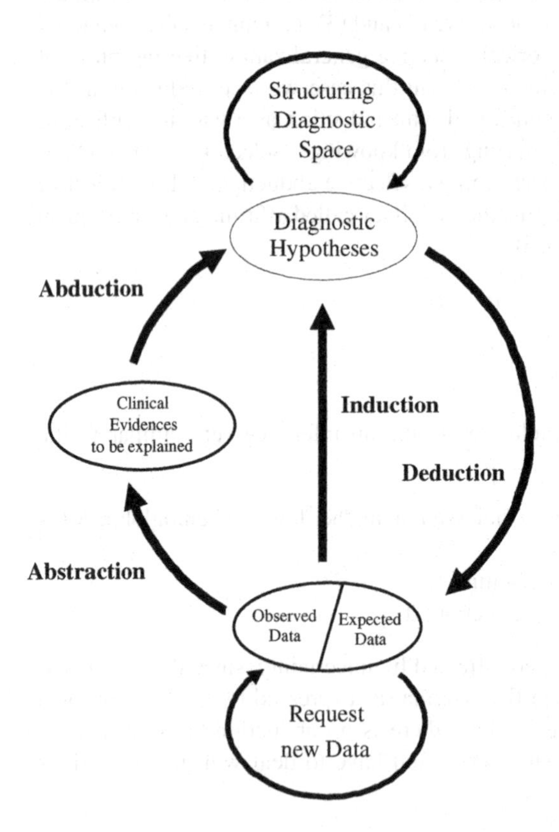

Fig. 1.3 ST-Model. The epistemological model of diagnostic reasoning.

If during this first cycle new information emerges, hypotheses not previously considered can be suggested and a new cycle takes place. In this case the *nonmonotonic* character of abductive reasoning is clear and arises from the logical unsoundness of the inference rule: it draws defeasible conclusions from incomplete information.[28] All recent logical accounts ("deductive") concerning abduction have pointed out that it is a form of nonmonotonic reasoning. It is important to allow the guessing of explanations for a situation, in order to discount and abandon old hypotheses, so as to enable the tentative adoption of new ones, when new information about the situation makes them no longer the best.[29]

As [Stephanou and Sage, 1987] pointed out, uncertainty and imperfect information are fundamental characteristics of the knowledge relative to hypothetical reasoning. The nonmonotonic character of the ST-model arises not only from the above mentioned nonmonotonic character of deductive inference type involved in it, but also from the logical unsoundness of the ascending part of the cycle guessing hypotheses to be tested. [Doyle, 1988] pointed out that, since their unsoundness, these guesses do not exhibit the truth-preservative behavior of ideal rationality characterizing the incremental deduction of classical logic, but the nonmonotonic behavior of limited rationality of commonsense reasoning [Simon, 1969], that allows to discharge and abandon old hypotheses to make possible the tentative adoption of new ones. Notice that this adoption is not merely tentative but rationally tentative, in the sense that, just as abduction, it is based on a reasoned selection of knowledge [Truesdell, 1984] and on some preference criteria which avoid the combinatorial explosion of hypotheses generation.

One of the principal means of limiting rationality is indeed to limit efforts by directing attention to some areas and ignoring others. This character matches exactly with the ability of an expert in generating a small set of hypotheses to be carefully tested. But in such a case, the expert has to be ready to withdraw paths of reasoning when they diverge from the correct path, that is from the path that would have taken the expert had considering the ignored knowledge portions. In such a way, the nonmonotonic character turns out as a foundational epistemological feature of the ST-model of medical reasoning, since this nonmonotonic character is the result not of a mere lack of information but of a reasoned limiting of information imposed by its own logical unsoundness. Finally, we have to remember that in the ST-model the first meaning (see above) of the word abduction is adopted: abduction that only generates "plausible" hypotheses (of course in this case *selective*).

Modern logic allows us to account for this dynamic behavior of abduction also by the concept of *belief revision*. Belief revision [Alchourrón *et al.*, 1985] is a

[28] A logical system is monotonic if the function *Theo* that relates every set of wffs to the set of their theorems holds the following property: for every set of premises S and for every set of premises S', $S \subseteq S'$ implies $Theo(S) \subseteq Theo(S')$. Traditional deductive logics are always monotonic: intuitively, adding new premises (axioms) will never invalidate old conclusions. In a nonmonotonic system, when axioms, or premises, increase, their theorems do not [Ginsberg, 1987; Lukaszewicz, 1970; Magnani and Gennari, 1997]. Following this deductive nonmonotonic view of abduction, we can stress the fact that in actual abductive medical reasoning, when we increase symptoms and patients' data [premises], we are compelled to abandon previously derived plausible diagnostic hypotheses [theorems], as already – epistemologically – illustrated by the ST-model.

[29] The relationship in practical reasoning between nonmonotonicity and scant resources of effort and time is treated in the recent [Gabbay and Woods, 2008].

dynamic notion dealing with the current stage of reasoning. At each stage of reasoning, if it is correct, a belief is held on the basis that that reasoning is justified, even if subsequent stages dictate its retraction. A logic of belief for abduction has been proposed by [Levesque, 1989], and the role of belief revision functions in abduction has already been studied by [Jackson, 1989] (cf. below section 1.4). Clearly abduction in medical diagnostic reasoning can be seen as an example of nonmonotonic deduction.[30]

1.3.2 Abduction as Hypothesis Generation, Abduction as Hypothesis Generation and Evaluation

As stated above, there are two main epistemologico/cognitive meanings of the word abduction: (1) abduction that only generates plausible hypotheses (*selective* or *creative*) – this is the meaning of abduction accepted in my epistemological model – and (2) abduction considered as *inference to the best explanation*, that also evaluates hypotheses by induction. In the latter sense the classical meaning of selective abduction as inference to the best explanation (for instance in medicine, to the best diagnosis) is described by the complete abduction–deduction–induction cycle. This distinction needs further clarification.

It is clear that the two meanings are related to the distinction between hypothesis generation and hypothesis evaluation, so abduction is the process of generating explanatory hypotheses, and induction matches the hypothetico-deductive method of hypothesis testing (1st meaning). However, we have to remember (as we have already stressed) that sometimes in the literature (and also in Peirce's texts) the word abduction is also referred to the whole cycle, that is as an inference to the best explanation (2nd meaning).

As Thagard has pointed out [1988, p. 53] the question was controversial in Peirce's writings too. Before the 1890s, Peirce discussed the hypothesis as follows: "Hypothesis is where we find some very curious circumstance which would be explained by the supposition that it was the case of a certain general rule, and thereupon adopt that supposition" [Peirce, 1931-1958, 2.624]. When Peirce replaced hypothesis with abduction he said that it "furnishes the reasoner with the problematic theory which induction verifies" [Peirce, 1931-1958, 2.776]. Thagard ascribes to the editors of Peirce's work the responsibility for having clouded this change in his thinking by including discussions of hypothesis under the heading of "Abduction", "[...] obscuring his shift from the belief that inference to an explanatory hypothesis can be a kind of justification to the weaker view that it is only a form of discovery". The need for a methodological criterion of justification is caused by the fact that – at least in the Peircean framework – an abduced hypothesis that explains a certain puzzling fact should not be accepted

[30] A more detailed description of abductive reasoning in diagnosis (and in "medical" diagnosis) is provided in my book [Magnani, 2001b, chapter four] and in [Gabbay and Woods, 2005, chapter six]. An interesting recent exploitation of genetic algorithms and computational paradigms inspired by the natural evolution to model abduction in medical diagnosis is illustrated in [Romdhane and Ayeb, 2009].

because of the possibility of other explanations. Having a hypothesis that explains a certain number of facts is far from a guarantee of being true.

In the previous section I have noted that when we speak of abduction as inference to the best explanation, we have to add that the adjective "best" has to be taken in a Pickwickian sense: the idea that the concept of abduction would always also strictly involve its empirical evaluation by induction contrasts with its primitive character of ignorance preserving cognition. In this perspective abduction does not have to be considered a "solution" of a problem, because it only calls for a response to it, with the aim of mitigating ignorance. I think that it is on this basis that Gabbay and Woods contend, in their book on abduction that "A decision to send a proposition (etc.) to experimental trial is neither necessary nor sufficient for its abduction" [Gabbay and Woods, 2005, p. 86]. All the more reason to unlink the idea of abduction as inference to best explanation from the processes of experimental evaluation. Abduced hypotheses are adopted as a positive basis for action in various ways and for various reasons (only subclasses of abductive hypotheses are adopted only after a Peircean process of inductive empirical evaluation).

It is important to note that already at the generation phase of genuine abduction many processes – so to say – of a kind of confrontation (if not exactly of evaluation) with something external to the individual brain, even if not due to experimental tests, can be present, so that we can say (as I noted above in the previous section) that abduction considered as a way of generating hypotheses is often of course immediately a generation of "plausible" hypotheses; that is hypotheses which can be "adopted" in so far as they are considered sufficiently plausible. The presence of these continuous stages of confrontation/coordination with external constrained and fruitful cognitive offerings will be better grasped thanks to the concept of manipulative abduction, which takes into account the external dimension of abductive reasoning (see below in this chapter, subsection 1.5.2 and section 1.6), and later on in chapters two and three:

1. in chapter two (subsection 2.3.1) I will illustrate the various ways *plausibility* is achieved already at the inner level (off-line) of thinking of the abductive human agent; I will also consider the so-called strategical plausibility, where abductions are adopted without any need of the empirical inductive tests;
2. in chapter three (section 3.6) I will describe how the *on-line* interplay between internal and external representations, both *mimetic* and *creative*, guarantees to abductive cognition a continuous *multimodal* confrontation with respect to something external to the individual brain activity, thanks to the dynamical interaction between the meaningful semiotic internal resources and devices and the externalized semiotic materiality already stocked in the environment. It is in this interplay that both the *abductive result* and its *plausibility* grow;
3. in this light the experimental test properly involved in the Peircean evaluation phase, which for many scholars reflects in the most acceptable way the idea of abduction as inference to the best explanation, enters a subclass of the processes of adoption of abductive hypotheses. Hence, this experimental test can be acceptably considered external to the nature of abductive cognition, and inductive in its essence.

Let us come back to the problem of evaluation. In a subclass of cognitive tasks, especially the ones that aim at honoring the rational or epistemic value of empirical evidence and of scientific mentality or methods (for example in science and medical diagnosis), to achieve the best explanation involves having or establishing a set of criteria for evaluating the competing explanatory hypotheses reached by creative or selective abduction, also contemplating the experimental test. The combinatorial explosion of alternatives that has to be considered makes the task of finding the best explanation (as I have already said, in the sense of the provisionally most acceptable explanation) very costly. Peirce surely thinks abduction has to be *explanatory*, but also capable of experimental *verification* (that is evaluated inductively, cf. the model above), and *economic* (this includes the cost of verifying the hypothesis, its basic value, and other factors). Evaluation has a multi-dimensional and comparative character. Following Peirce the economics of abduction is driven in turn by three common factors: the cost of testing [Peirce, 1931-1958, 1.120], the intrinsic appeal of the hypothesis, e.g., its simplicity, [5.60 and 6.532], where simplicity seems to be a matter of naturalness [2.740]; and the consequences that a hypothesis might have for future research, especially if the hypothesis proposed were to break down [7.220].

Consilience [Thagard, 1988] can measure how much a hypothesis explains, so it can be used to determine whether one hypothesis explains more of the evidence (for instance, in diagnosis empirical or patient data) than another: thus, it deals with a form of corroboration. In this way a hypothesis is considered more consilient than another if it explains more "important" (as opposed to "trivial") data than the others do. In inferring the best explanation, the aim is not the sheer amount of data explained, but its relative significance. The assessment of relative importance presupposes that an inquirer has a rich background knowledge about the kinds of criteria that concern the data. The evaluation is strongly influenced by Ockham's razor: *simplicity* too can be highly relevant when discriminating between competing explanatory hypotheses; for example, it deals with the problem of the level of conceptual complexity of hypotheses when their consiliences are equal.

Explanatory criteria are needed because the rejection of a hypothesis requires demonstrating that a competing hypothesis provides a better explanation. Clearly, in some cases – for instance when choosing scientific hypotheses or theories, where the role of "explanation" is dominant – conclusions are reached according to rational criteria such as consilience or simplicity. In [Magnani, 2001b, chapter four] I have illustrated that in the case of selecting diagnostic hypotheses the epistemic reasons are dominant, whereas, in the case of selecting therapies, epistemic reasons are of course intertwined with pragmatic and ethical reasons, which will play a very important role. Hence, in reasoning to the best explanation, motivational, ethical or pragmatic criteria cannot be neglected. Indeed the context suggests that they are unavoidable: as we have just mentioned, this is for example true in some part of medical reasoning (in therapy planning), but scientists that must discriminate between competing scientific hypotheses or competing scientific theories have to recognize that sometimes they too are conditioned by motivationally biasing their inferences to the best explanation. Some epistemologists, like [Kuhn, 1962] and [Feyerabend, 1975], argued that in science these extra-rational motivation are unavoidable.

For example, the so-called theory of *explanatory coherence* [Thagard, 1989; Thagard, 1992] introduces seven ideal principles of plausibility that occurs in the acceptation of new hypotheses and theories in science;[31] the theory is susceptible to be treated at the computational level using a local connectionist network.

Josephson has stressed that evaluation in abductive reasoning has to be referred to the following criteria

1. How a hypothesis surpasses the alternatives.
2. How the hypothesis is good in itself.
3. Its confidence in the accuracy of the data.
4. How thorough was the search for alternative explanations [Josephson, 1998].

There is no agreement about which preference criteria to adopt. [Hendricks and Faye, 1999], speak, in the case of science, about correctness (concerning the world that it is investigating), empirical adequacy, simplicity (different kinds of), unification, consistency, practical usability, economy. [Poole and Rowen, 1990] list several criteria that have been proposed in the literature and it can be shown that some of these preference criteria are conflicting, i.e. in the same situation, they favor different conjectures. The problem is that all the proposed criteria do not work in all situations: they are in some sense *context dependent*. For instance, the (syntactic) criterion of minimality described by the sentential models of abduction (cf. the following section), is useless when the conjecture at hand is (syntactically) as simple as the conflicting conjectures.

We can also use mathematical probability to select among hypotheses evaluating them (Bayes's Theorem itself can be viewed as a modality for weighing alternative hypotheses [Krauss *et al.*, 1999], of course in case the appropriate knowledge of probabilities is present).[32]

The epistemological model (ST-model) I have previously illustrated should also be regarded as a very simple and schematic illustration of scientific theory change. In this case selective abduction is replaced by creative abduction and there exists a set of competing theories instead of diagnostic hypotheses. Furthermore the language of background scientific knowledge should be regarded as open: in the case of competing theories, as they are studied using the epistemology of theory change, we cannot – contrary to Popper's initial viewpoint [Popper, 1959] – reject a theory simply because it fails occasionally. If for example such a theory is simpler and

[31] The theory also fruitfully applies, with slight modifications, to many other fields like conceptual combination; adversarial problem-solving, when one has to infer an opponent's intentions; analogical reasoning; jury decisions in murder trials: contemporary debates about why the dinosaurs became extinct; psychological experiments on how beginning students learn physics; ethical deliberation; emotional decision.

[32] On the relationships between probabilism and explanationism and on the fact that probabilism is not appropriate to model abductive reasoning of actual individual human agents cf. [Gabbay and Woods, 2005, chapter six].

explains more significant data than its competitors, then it can be accepted as the best explanation.[33]

As already stressed, in accordance with the epistemological model previously illustrated, medical reasoning may be broken down into two different phases: first, patient data is abstracted and used *to select hypotheses*, that is hypothetical solutions of the patient's problem (selective abduction phase); second, these hypotheses provide the starting conditions for forecasts of expected consequences which should be compared to the patient's data in order *to evaluate* (corroborate or eliminate) those hypotheses which they come from (deduction-induction cycle).

If we consider the epistemological model as an illustration of medical diagnostic reasoning, the *modus tollens* is very efficacious because of the fixedness of language that expresses the background medical knowledge: a hypothesis that fails can nearly always be rejected immediately.

When Buchanan illustrates the old epistemological method of induction by elimination (and its computational meaning, evident if we add a "heuristic search", to limit the exhaustive enumeration of the derived hypotheses), first advanced by Bacon and Hooke and developed later on by John Stuart Mill, he is referring implicitly to my epistemological framework in terms of abduction, deduction and induction, as illustrative of medical diagnostic reasoning:

> The method of systematic exploration is [...] very like the old method of induction by elimination. Solutions to problems can be found and proved correct, in this view, by enumerating possible solutions and refuting all but one. Obviously the method is used frequently in contemporary science and medicine, and is as powerful as the generator of possibilities. According to Laudan, however, the method of proof by eliminative induction, advanced by Bacon and Hooke, was dropped after Condillac, Newton, and LeSage argued successfully that it is impossible to enumerate exhaustively all the hypotheses that could conceivably explain a set of events. The force of the refutation lies in the open-endedness of the language of science. Within a fixed language the method reduces to modus tollens [...]. The computational method known as heuristic search is in some sense a revival of those old ideas of induction by elimination, but with machine methods of generation and search substituted for exhaustive enumeration. Instead of enumerating all sentences in the language of science and trying each one in turn, a computer program can use heuristics enabling it to discard large classes of hypotheses and search only a small number of remaining possibilities [Buchanan, 1985, pp. 97–98].

[33] Rigourously speaking, [Gabbay and Woods, 2005, p. 137] usefully point out the following general result concerning abduction, which further stresses its ignorance preserving character: "If (H) is the conclusion of and abductive inference and H is subsequently shown to be false, this discredits neither the conclusion $C(H)$ nor the conclusion H^c. Corollary 5.12(a) This shows the importance of recognizing that the conclusions of abductions imbed (often implicitly) temporal parameters." [The meaning of the notations $C(H)$ and H^c is explained in the following chapter, subsection 2.1.1].

1.4 Sentential Abduction

Sentential abduction can be rendered in different ways. For example, in the syllogistic framework we have just described abduction is considered like something propositional and as a type of fallacious reasoning. If we want to model abduction in a computational logic-based system, the fundamental operation is *search* [Thagard, 1996]. When there is a problem to solve, we usually face several possibilities (hypotheses) and we have to select the suitable one (cf. selective abduction, above). Accomplishing the assigned task requires that we have to search through the space of possible solutions to find the desired one. In this situation we have to rely on heuristics, that are rules of thumb expressed in sentential terms that help in arriving at satisfactory choices without considering all the possibilities. An example of simple heuristic could be a rule like "Wear green socks with white pants but not with blue pants". The famous concept of *heuristic search*, which is at the basis of many computational systems based on propositional rules (cf. chapter two, section 2.7) can perform this kind of sentential abduction (selective). Of course other computational tools can be used to this aim, like neural and probabilistic networks, and frames-like representations.

Another important way of modeling abduction in a sentential way resorts to the development of suitable logical systems, that in turn are computationally exploitable in the area of the so-called logic programming (cf. section 1.4.1, below).

Many attempts have been made to model abduction by developing some formal tools in order to illustrate its computational properties and the relationships with the different forms of deductive reasoning (see, for example, [Bylander *et al.*, 1991; Console *et al.*, 1991; Console and Torasso, 1991; Coz and Pietrsykowski, 1986; Raedt and Bruynooghe, 1991; Jackson, 1989; Kakas *et al.*, 1993; Konolige, 1992; Josephson and Josephson, 1994; Levesque, 1989; O'Rorke, 1994; Poole, 1988; Reiter, 1987; Reiter and de Kleer, 1991; Shanahan, 1989]).

Some of the formal models of abductive reasoning, for instance [Boutilier and Becher, 1995], are based on the theory of the epistemic state of an agent [Alchourrón *et al.*, 1985; Gärdenfors, 1988; Gärdenfors, 1992], where the epistemic state of an individual is modeled as a consistent set of beliefs that can change by expansion and contraction (*belief revision framework*).[34]

Deductive models of abduction may be characterized as follows. An explanation for β relative to background theory T will be any α that, together with T, entails β (normally with the additional condition that $\alpha \cup T$ be consistent). Such theories are usually generalized in many directions: first of all by showing that explanations entail their conclusions only in a *defeasible* way (there are many potential explanations), thus joining the whole area of so-called nonmonotonic logic or of probabilistic treatments; second, trying to show how some of the explanations are relatively implausible, elaborating suitable technical tools (for example in terms of modal logic) able to capture the notion of preference among explanations (cf. Figure 1.4).

[34] Levi's theory of suppositional reasoning is also related to the problem of "belief change" [Levi, 1996].

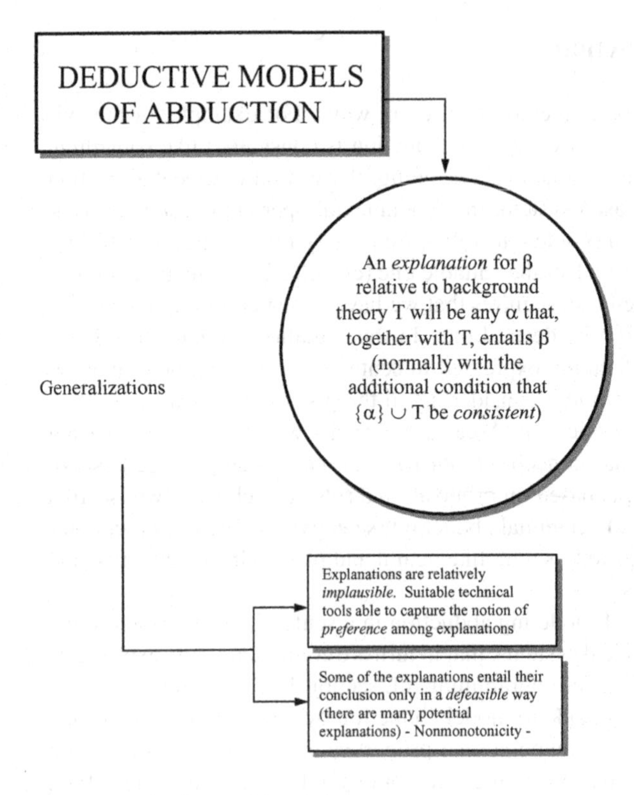

Fig. 1.4 Deductive models of abductive reasoning

Hence, we may require that an explanation makes the observation simply sufficiently probable [Pearl, 1988] or that the explanations that are more likely will be the "preferred" explanations: the involvement of a cat in breaking the glass can be considered less probable than the effect of wind. Finally, the deductive model of abduction does not authorize us to explain facts that are inconsistent with the background theory notwithstanding the fact that these explanations are very important and ubiquitous, for instance in diagnostic applications, where the facts to be explained contradict the expectation that the system involved is working according to specification.

[Boutilier and Becher, 1995] provide a formal account of the whole question in term of belief revision: if believing A is sufficient to induce belief in B, then A (epistemically) *explains* B; the situation can be semantically illustrated in terms of an ordering of plausibility or normality which is able to represent the epistemic state of an agent. The conflicting observations will require explanations that compel the agent to withdraw its beliefs (hypotheses), and the derived conditional logic is able to account for explanations of facts that conflict with the existing beliefs. The authors are able to reconstruct, within their framework, the two main paradigms of model-based diagnosis, abductive [Poole, 1988; Poole, 1991], and

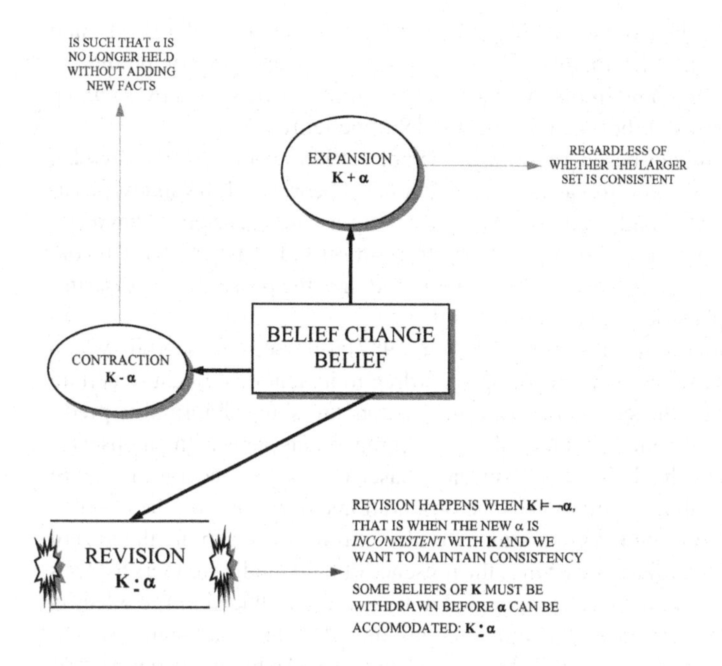

Fig. 1.5 Belief-revision

consistency-based [de Kleer *et al.*, 1990; Reiter, 1987], providing an alternative semantics for both in terms of a plausibility ordering over possible worlds.

Let us resume the kinds of change considered in the original belief revision framework (see Figure 1.5). The *expansion* of a set of beliefs K taken from some underlying language (considered to be the closure of some finite set of premise KB, or *knowledge base*, so $K = Cn(KB)$) by a piece of new information A is the belief set $K+A = Cn(K \cup A)$. The addition happens "regardless" of whether the larger set is *consistent*. The case of *revision* happens when $K \models \neg A$, that is when the new A is *inconsistent* with K and we want to maintain consistency: some beliefs in K must be withdrawn before A can be accommodated: $K \dot{-} A$. The problem is that it is difficult to detect which part of K has to be withdrawn. The least "entrenched" beliefs in K should be withdrawn and A added to the "contracted" set of beliefs. The loss of information has to be as small as possible so that "no belief is given up unnecessarily" [Gärdenfors, 1988]. Hence, *inconsistency resolution* in belief revision framework is captured by the concept of revision. Another way of belief change is the process of *contraction*. When a belief set K is contracted by A, the resulting belief set $K+A$ is such that A is no longer held, without adding any new fact.[35]

[35] [Aliseda, 1997; Aliseda, 2000] makes use of the belief revision framework to construct a theory of the epistemic transition between the states of doubt and belief able to account for many aspects of abductive reasoning. On the relationships between belief revision dynamics in data bases and abduction cf. [Aravindan and Dung, 1995].

After having explained the distinction between predictive explanations and "might" explanations, that merely allow an observation, and do not predict it, Boutilier and Becher show in the cited article how model-based diagnoses can be accounted for in terms of their new formal model of belief revision.

The *abductive* model-based reasoning[36] [Poole, 1988; Poole, 1991; Brewka, 1989] illustrated by some models, such as Poole's Theorist, allows many possible explanations, weak and predictive (so presenting a paraconsistent behavior: a non-predictive hypothesis can explain both a proposition and its negation). This old model, embedded in the new formal framework, acquires the possibility of discriminating certain explanations as preferred to others.

Reiter's *consistency-based* diagnosis [Reiter, 1987] is devoted to ascertain why a correctly designed system is not working according to its features. Because certain components may fail, the system description also contains some abnormality predicates (the absence of them will render the description inconsistent with an observation of an incorrect behavior). The consistency-based diagnosis concerns any set of components whose abnormality makes the observation consistent with the description of the system. A principle of parsimony is also introduced to capture the idea of preferred explanations/diagnoses. Since the presence of fault models renders Reiter's framework incorrect, new more complicated notions are introduced in [de Kleer *et al.*, 1990], where the presence of a complete fault model ensures that predictive explanations may be given for "every" abnormal observation. Without any description of correct behavior any observation is consistent with the assumption that the system works correctly. Hence, a complete model of correct behavior is necessary if we want the consistency-based diagnosis to be useful.

The idea of consistency that underlies some of the recent deductive consistency-based models of selective abduction (diagnostic reasoning) is the following: any inconsistency (anomalous observation) refers to an aberrant behavior that can usually be accounted for by finding some set of components of a system that, if behaving abnormally, will entail or justify the actual observation. The observation is anomalous because it contradicts the expectation that the system involved is working according to specification. This types of deductive model go beyond the mere treatment of selective abduction in terms of preferred explanations and include the role of those components whose abnormality makes the observation (no longer anomalous) consistent with the description of the system [Boutilier and Becher, 1995; Magnani, 2001a].

Without doubt the solution given by Boutilier and Becher furnishes a more satisfying qualitative account of the choice among competing explanations than Gärdenfors' in terms of "epistemic entrenchment"[37] which tries to capture the idea of an ordering of beliefs according to our willingness to withdraw them when

[36] Please distinguish here the technical use of the attribute *model-based* from the epistemological-cognitive one I am introducing in this chapter.

[37] Which of course may change over time or with the state of belief.

necessary. Moreover, the new formal account in terms of belief revision is very powerful in shedding new light on the old model-based accounts of diagnostic reasoning.

The framework of belief revision is sometimes called *coherence approach* [Doyle, 1992]. In this approach, it is important that the agent holds some beliefs just as long as they are consistent with the agent's remaining beliefs. Inconsistent beliefs do not describe any world, and so are unproductive; moreover, the changes must be epistemologically conservative in the sense that the agent maintains as many of its beliefs as possible when it adjusts its beliefs to the new information. It is contrasted to the *foundations approach*, according to which beliefs change as the agent adopts or abandons satisfactory reasons (or justifications). This approach is exemplified by the well-known "reason maintenance systems" (RMS) or "truth maintenance systems" (TMS) [Doyle, 1979], elaborated in the area of artificial intelligence to cooperate with an external problem solver. In this approach, the role of inconsistencies is concentrated on the negations able to invalidate justifications of beliefs; moreover, as there are many similarities between reasoning with incomplete information and acting with inconsistent information, the operations of RMS concerning revision directly involve logical consistency, seeking to solve a conflict among beliefs. The operations of *dependency-directed backtracking* (DDB) are devoted to this aim: RMS informs DDB whenever a contradiction node (for instance a set of beliefs) becomes believed, then DDB attempts to remove reasons and premises, only to defeat nonmonotonic assumptions: "If the argument for the contradiction node does not depend on any of these (i.e., it consists entirely of monotonic reasons), DDB leaves the contradiction node in place as a continuing belief" [Doyle, 1992, p. 36], so leaving the conflicting beliefs intact if they do not depend on defeasible assumptions, and presenting a paraconsistent behavior.

Both in the coherence and foundations approach the changes of state have to be epistemologically conservative: as already said above the agent maintains as many of its beliefs as possible when it adjusts its beliefs to the new information, thus following Quine's idea of "minimun mutilation" [Quine, 1979]. We have now to notice some limitations of the formal models in accounting for other kinds of inconsistencies embedded in many reasoning tasks.

This kind of *sentential frameworks* exclusively deals with selective abduction (diagnostic reasoning)[38] and relates to the idea of preserving *consistency*. Exclusively considering the sentential view of abduction does not enable us to say much about creative processes in science, and, therefore, about the nomological and most interesting creative aspects of abduction. It mainly refers to the *selective* (diagnostic) aspects of reasoning and to the idea that abduction is mainly an inference *to the best explanation* [Magnani, 2001b]: when used to express the creative events it is either empty or replicates the well-known *Gestalt* model of radical innovation. It is empty because the sentential view stops any attempt to analyze the creative processes: the

[38] As previously indicated, it is important to distinguish between *selective* (abduction that merely selects from an encyclopedia of pre-stored hypotheses), and *creative* abduction (abduction that generates new hypotheses).

event of creating something new is considered so radical and instantaneous that its irrationality is immediately involved.[39]

Already in the Peircean syllogistic and sentential initial conception of abduction – as the fallacy of affirming the consequent, we immediately see it is perfectly compatible with the *Gestalt* model of discovery. In the syllogistic model the event of creating something new (for example a new concept) is considered external to the logical process, so radical and instantaneous that its irrationality is immediately involved. In this case the process is not considered as algorithmic: "the abductive suggestion comes to us like a flash. It is an act of insight, although of extremely fallible insight" [Peirce, 1931-1958, 5.181]. Moreover, Peirce considers abduction as "a capacity of guessing right", and a "mysterious guessing power" common to all scientific research [Peirce, 1931-1958, 6.530].

Notwithstanding its non-algorithmic character it is well known that for Peirce abduction is an *inferential process* (for an explanation of the exact meaning of the word "inference" cf. below section 1.5). Hence abduction has to be considered as a kind of *ampliative* inference that, as already stressed, is not logical and truth preserving: indeed valid deduction does not yield any new information, for example new hypotheses previously unknown: abduction

> [...] is logical inference [...] having a perfectly definite logical form. [...] The form of inference, therefore, is this:
>
> The surprising fact, *C*, is observed;
> But if *A* were true, *C* would be a matter of course,
> Hence, there is reason to suspect that *A* is true [Peirce, 1931-1958, 5.188-189, 7.202].

C is true of the actual world and it is surprising, a kind of state of doubt we are not able to account for by using our available knowledge. *C* can be simply a *novel* phenomenon or may be in conflict with the background knowledge, that is *anomalous*.

To conclude, if we want to provide a suitable framework for analyzing the most interesting cases of conceptual change in science we do not have to limit ourselves

[39] Research into the logic of abduction has increased over recent years. I have edited special issues of journals and proceedings which present various articles on this subject. [Reyes *et al.*, 2006] propose a notion of abductive problems, *N*-abductive, which helps to provide an effective procedure for finding abductive solutions in first-order logic, by means of a modification of Beth's tableaux. [Carnielli, 2006] shows that the robust logics of formal inconsistency, a particular category of paraconsistent logics which permit the internalization of the concepts of consistency and inconsistency inside the object language, provide simple and yet powerful techniques for automatic abduction; moreover, the whole procedure is capable of automatization by means of the tableau proof-procedures available for such logics. [Inoue and Sakama, 2006] focus on the problem of identifying the equivalence of two abductive theories represented in first-order logic. [Bharathan and Josephson, 2006] suggest taking advantage of the specific structure of abductive reasoning to identify revision candidates among earlier beliefs, to propose specific revisions, to select among possible revisions, and to make the requisite changes to the system of beliefs; these adjustments are performed through meta-abductive processing over the recorded steps in an abductive agent's reasoning trace. Finally, on algorithms for diagnostic reasoning cf. [Luan *et al.*, 2006; Shangmin *et al.*, 2007], on abduction and counterfactuals and conditionals cf. [Pizzi, 2006; Pizzi, 2007], on abduction and fallacies cf. [Woods, 2007; Woods, 2010].

to the sentential view of theoretical abduction but we have to consider a broader *inferential* one which encompasses both sentential and what I call *model-based* sides of creative abduction (see, for details, section 1.5 below). The abductive inference includes all the operations whereby hypotheses and theories are constructed [Peirce, 1931-1958, 5.590] (see also [Hintikka, 1998]).

1.4.1 Abduction and Induction in Logic Programming

The syllogistic account of abduction we described above is the starting point of much research in AI and logic programming devoted to perform tasks such as diagnosis in medical reasoning and planning. In these logical and computational accounts abduction and induction are considered as separate forms of reasoning related to different tasks. Consequently, the distinction is very variable, context-dependent and different from the one we have seen operating in Peircean texts (where, as illustrated in subsection 1.3.2, abduction is especially – but not only - viewed as hypothesis generation and induction as a logic of hypothesis evaluation).

In a classical book edited by [Flach and Kakas, 2000b], many interesting contributions are dedicated to the analysis of the distinction between *abductive* and *inductive* reasoning.[40] Usually in these types of research abductive hypotheses are considered as providing explanations, and inductive hypotheses as providing generalizations: this explains, for example, why diagnosis is generally considered in AI like abductive and concept learning from examples inductive.[41] Usually abduction is regarded as reasoning from specific observations to their explanations, and induction as a Millian enumerative induction from samples to general statements.

In the case of the *abductive logic programming* (ALP), and assuming a common first-order language, possible abductive hypotheses are built from specific non-observable predicates Δ called *abducibles* (suitably distinguished from observable predicates). The problem is to be able to "select" among the so-called abductive extensions $T(\Delta p)$ of of T in which the given observation to be explained holds, by selecting the corresponding formula Δ. On the contrary, in the case of *inductive logic programming* (ILP) the problem is to "select" a *generalizing* hypothesis able to entail additional observable information on unobserved individuals (that is predictions), finding new individuals for which the addition of the hypothesis to our knowledge is necessary to derive some observable properties for them [Console and Saitta, 2000]. In the first case the abductive explanation Δ needs a given theory T, so it is "relative" to a "certain" theory T from which it is produced. In the case of induction the explanation does not depend on a particular theory: we can say that "all the beans from this bag are white" (Peirce's example, see above, subsection 1.3.1, this chapter), is an explanation for why the observed beans from the bag are white: this explanation does is in accordance with a particular model of the "world of beans" [Flach and Kakas, 2000a].

[40] Cf. also the clear article by [Console *et al.*, 1991].

[41] An overview on the relationships between abduction and induction is given in [Bessant, 2000].Cf. also the various contributions given by Abe; Christiansen; Inoue and Haneda; Mooney; Poole; Psillos; Sakama; Yamamoto, in [Flach and Kakas, 2000b].

Moreover, we can say that abductions explain a phenomenon by indicating enabling conditions like causes (this explains the fact that abduction needs a domain theory, often a causal theory, while induction does not):[42]

> If we want to explain, for instance, that the light appears in a bulb when we turn a switch on, an inductive explanation would say that this is because it happened hundreds of time before, whereas the abductive one can supply an explanation in terms of the electric current flowing into the bulb filament. If, at some moment, turning the switch on does not let the light bulbs starts burning, the inductive explanation just fails, whereas the abductive one can supply hints for understanding what happened and for suggesting remedies [Console and Saitta, 2000].

Hence, the two kinds of explanations are very different and distinct, induction firstly aims at providing generalizations, abduction explanations of particular observations. In Console and Saitta's terms, abductive reasoning extends the intension of known individuals (because abducible properties are rendered true for these individuals, for example by providing new *situated* "samples", which offer chances for further knowledge – cf. above the considerations I have illustrated at p. 14), without having a genuine generalization impact on the observables (it does not increase their extension).[43]

Another interpretation of the distinction between abduction and induction is given by [Josephson, 2000]. Following his point of view all inferences to the best explanation have to be considered as kinds of "smart" reasoning. In Josephson's terms induction and abduction are not distinct processes: the inductive generalization is a type of inference that points to some best explanation, so it can be considered as a kind of abduction (he agrees with the use of the term abduction according to the second meaning we previously illustrated, the one including generating – or selecting – and evaluating hypotheses). "Smart" inductive generalizations (or inductive hypotheses) do not explain "particular" observations but the frequencies with which the observations emerge, like in the well known AI case of "concept learning from examples": "An observed frequency is explained by giving a causal story that explains how the frequency came to be the way it was. This causal story typically

[42] The role of the causal-hypothetical reasoning is central in modern science: Galileo was already perfectly aware of this fact. He insists that interesting conclusions which reach far beyond experience can be derived from few experiments because "[...] the knowledge of a single fact acquired through a discovery of its cause prepares the mind to understand and ascertain other facts without need to recourse to experiment" [Galilei, 1914, p. 296]: in the case of his study of projectiles, once we know that their path is a parabola, we can derive using only pure mathematics, that their maximum range is 45°. Moreover, Newton says, using the ancient notion of "analysis": "By this way of analysis we may proceed from compounds to ingredients, and from motions to the forces producing them; and in general, from effects to their causes, and from particular causes to more general ones, [...] and the synthesis consists in assuming the causes discovered, and established as principles, and by them explaining the phenomena proceeding from them, and proving the explanations" [Newton, 1721, p. 380 ff.].

[43] Further results on the interaction and/or integration of abduction and induction in AI complex theory development tasks are given in [Michalski, 1993] – in terms of coexistence; [Dimopoulos and Kakas, 1996; Ourston and Mooney, 1994] – in terms of cooperation; [O'Rorke, 1994; Thompson and Mooney, 1994; Kakas and Riguzzi, 1997] – on the role of inducing and learning in abductive theories.

includes both the method of drawing the sample, and the population frequency in some reference class".

When inductive hypothesis are "smart" or "good" they are so because they are inferences to the best generalization-explanation of the sample frequencies, so they can be considered as a kind of abduction. As a consequence of his ideas on abduction and induction, Josephson concludes by arguing that the computational programs for inductive generalizations have to be constructed abductively.

1.5 Model-Based Creative Abduction

1.5.1 Conceptual Change and Creative Reasoning in Science

The sentential models of theoretical abduction are limited, because they do not capture various reasoning tasks [Magnani, 1999]:[44]

1. the role of statistical explanations, where what is explained follows only probabilistically and not deductively from the laws and other tools that do the explaining;
2. the sufficient conditions for explanation;
3. the fact that sometimes the explanations consist of the application of schemas that fit a phenomenon into a pattern without realizing a deductive inference;
4. the idea of the existence of high-level kinds of *creative* abductions;
5. the existence of model-based abductions (cf. the following section);
6. the fact that explanations usually are not complete but only furnish *partial* accounts of the pertinent evidence [Thagard and Shelley, 1997];
7. the fact that one of the most important virtues of a new scientific hypothesis (or of a scientific theory) is its power of explaining *new*, previously *unknown* facts: "[...] these facts will be [...] unknown at the time of the abduction, and even more so must the auxiliary data which help to explain them be unknown. Hence these future, so far unknown explananda, cannot be among the premises of an abductive inference" [Hintikka, 1998], observations become real and explainable only by means of new hypotheses and theories, once discovered by abduction.

We will see in the following subsection that it is in terms of *model-based abductions* (and not in terms of sentential abductions) that we have to think for example of the case of a successful synthesis of two earlier theoretical frameworks which might even have seemed incompatible. The old epistemological view sees Einstein's theory as an attempt to "explain" certain anomalies and facts such as the Michelson-Morley experiment: "The most instructive way of looking at Einstein's discovery is to see it as

[44] Important developments in the field of logical models of abduction – also touching some related problems in artificial intelligence (AI) and devoted to overcome the limitations above – are illustrated in [Flach and Kakas, 2000b] and in [Gabbay and Kruse, 2000; Gabbay and Woods, 2005; Gabbay and Woods, 2006]. Cf. also the recent papers contained in the collections [Magnani *et al.*, 2002a; Magnani, 2006c].

a way of reconciling Maxwell's electromagnetic theory with Newtonian mechanics [...] it would be ridiculous to say that Einstein's theory 'explains' Maxwell's theory any more than it 'explains' Newton's laws of motion" [Hintikka, 1998, p. 510]. This kind of abductive movement does not have that immediate explanatory effect illustrated by the sentential models of abduction: the new framework usually does not "explain" the previous ones but provides a very radical new perspective.

If we want to deal with the nomological and most interesting creative aspects of abduction we are first of all compelled to consider the whole field of the growth of scientific knowledge cited above.

We have anticipated that abduction has to be an inference permitting the derivations of *new* hypotheses and beliefs. Some explanations consist of certain facts (initial conditions) and universal generalizations (that is scientific laws) that deductively entail a given fact (observation), as showed by Hempel in his *law covering model* of scientific explanation [Hempel, 1966]: in this case the argument starts with the true premises and deduces the explained event. If T is a theory illustrating the background knowledge (a scientific or common sense *theory*) the sentence α explains the fact (observation) β just when $\alpha \cup T \models \beta$, it is difficult to govern the question involving nomological and causal aspects of abduction and explanation in the framework of the belief revision illustrated in the previous section: we would have to deal with a kind of belief revision that permits us to alter a theory with new conditionals.

We may also see belief change (cf. the above section 1.4) from the point of view of *conceptual change*, considering concepts either cognitively, like mental structures analogous to data structures in computers, or, epistemologically, like abstractions or representations that presuppose questions of justification. Belief revision is able to represent cases of conceptual change such as adding a new instance, adding a new weak rule, adding a new strong rule (see [Thagard, 1992], that is, cases of addition and deletion of beliefs, but fails to take into account cases such as adding a new part-relation, adding a new kind-relation, adding a new concept, collapsing part of a kind-hierarchy, reorganizing hierarchies by branch jumping and tree switching, in which there are reorganizations of concepts or redefinitions of the nature of a hierarchy.

Let us consider concepts as composite structures akin to frames of the following sort:

CONCEPT:
A kind of:
Subkinds:
A part of:
Parts:
Synonyms:
Antonyms:
Rules:
Instances:

It is important to emphasize (1) kind and part-whole relations that institute hierarchies, and (2) rules that express factual information more complex than simple slots. To understand the cases of conceptual revolutions we need to illustrate how concepts can fit together into conceptual systems and what is involved in the replacement of such systems. Conceptual systems can be viewed as ordered into kind-hierarchies and linked to each other by rules.

Adding new part-relations occurs when in the part-hierarchy new parts are discovered: an example is given by the introduction of new molecules, atoms, and subatomic particles. Thomson's discovery that the "indivisible" atom contains electrons was very sensational.

Adding new kind-relations occurs when it is added a new superordinate kind that combines two or more things previously taken to be distinct. In the nineteenth century scientists recognized that electricity and magnetism were the same and constructed the new concept of electromagnetism. Another case is shown by differentiation, that is the making of a new distinction that generates two kinds of things (heat and temperature were considered the same until the Black's intervention).

The last three types of conceptual change can be illustrated by the following examples. The Newtonian abandon of the Aristotelian distinction between natural and unnatural motion exemplifies the collapse of part of the kind-hierarchy. Branch jumping occurred when the Copernican revolution involved the recategorization of the earth as a kind of planet, when previously it had been considered special, but also when Darwin reclassified humans as a kind of animal. Finally, we have to say that Darwin not only reclassified humans as animals, he modified the meaning of the classification itself. This is a case of hierarchical tree redefinition:

Whereas before Darwin kind was a notion primarily of similarity, his theory made it a historical notion: being of common descent becomes at least as important to being in the same kind as surface similarity. Einstein's theory of relativity changed the nature of part-relations, by substituting ideas of space-time for everyday notions of space and time [Thagard, 1992, p. 36].

These last cases are the most evident changes occurring in many kinds of creative reasoning in science, when adopting a new conceptual system is more complex than mere belief revision. Related to some of these types of scientific conceptual change are different varieties of *model-based abductions*. In these cases the hypotheses "transcend" the vocabulary of the evidence language, as opposed to the cases of simple inductive generalizations: the most interesting case of creative abduction is called by [Hendricks and Faye, 1999] trans-paradigmatic abduction. This is the case where the fundamental ontological principles given by the background knowledge are violated, and the new discovered hypothesis transcends the immediate empirical agreement between the two paradigms, like for example in the well-known case of the abductive discovery of totally new physical concepts during the transition from classical mechanics to quantum mechanics.

To conclude, I have already said that, if we want to provide a suitable framework for analyzing the most interesting cases of conceptual change in science we do not have to limit ourselves to the sentential view of abduction but we have to consider a broader *inferential* one which encompasses both sentential and what I call *model-based* sides of creative abduction.

1.5.2 Model-Based Abduction and Its External Dimension

The last cases of creative reasoning in science we have just illustrated demonstrate the radical *conjectural* character of the new concepts and the incommensurability as regarding previous ones, that is the cases in which "revolutionary" changes happen and the most "counterinductive" acts can become visible. The analysis of model-based conceptual change helps us to study the revolutionary changes of science: different varieties of what I call *model-based* abduction are related to some of these types of conceptual change.

From Peirce's philosophical point of view, all thinking is in signs, and signs can be icons, indices or symbols. Moreover, all inference is a form of sign activity,[45] where the word sign includes "feeling, image, conception, and other representation" [Peirce, 1931-1958, 5.283], and, in Kantian words, all synthetic forms of cognition.[46] That is, a considerable part of the thinking activity is model-based. Of course model-based reasoning acquires its peculiar creative relevance when embedded in abductive processes, so that we can individuate a *model-based abduction*. Hence, we must think in terms of model-based abduction (and not in terms of sentential abduction) to explain complex processes like scientific conceptual change. Different varieties of *model-based abductions* [Magnani, 1999] are related to the high-level types of scientific conceptual change (see, for instance, [Thagard, 1992]).

For Peirce [Anderson, 1986] a Kantian keyword is synthesis, where the intellect constitutes in its forms and in a harmonic way all the material delivered by the senses. Surely Kant did not consider synthesis as a form of *inference* but, notwithstanding the obvious differences,[47] I think synthesis can be related to the Peircean concept of inference, and, consequently, of abduction. After all, when describing the ways the intellect follows to unify and constitute phenomena through imagination Kant himself makes use of the term *rule* "Thus we think a triangle as an object, in that we are conscious of the combination of the straight lines according to a rule by which such an intuition can always be represented" [Kant, 1929, A140, B179-180, p. 182], and also of the term *procedure* "This representation of a universal procedure of imagination in providing an image for a concept, I entitle the schema of this concept" [Kant, 1929, A140, B179-180, p. 182]. We know that rules and procedures represent the central features of the modern concept of inference. Moreover, according to Peirce, the central question of philosophy is "how synthetical reasoning is possible [...]. This is the lock upon the door of philosophy" [Peirce, 1931-1958, 5.348], and the mind presents a tendency to unify the aspects which are exhibited by phenomena: "the function of conception is to reduce the manifold of sensuous impressions to unity" [Peirce, 1931-1958, 1.545].

---

[45] Cf. also [Fischer, 2001].

[46] Also in the perspective of Thom's catastrophe theory, it is interesting to stress that signs are forms in space-time in its Euclidean validity, as the basic framework of all human experience. Consequently, "their spatio-temporal localization is one of the first factors to consider" [Thom, 1980, p. 270]. Cf. also this, book, chapter eight.

[47] For example Peirce considers space and time themselves as products of synthesis and not as forms of intuition [Davis, 1972].

Most of these forms of constitution of phenomena are creative and, moreover, characterized in a model-based way. Let me show some examples of model-based inferences. It is well known the importance Peirce ascribed to diagrammatic thinking, as shown by his discovery of the powerful system of predicate logic based on diagrams or "existential graphs". As we have already stressed, Peirce considers inferential any cognitive activity whatever, not only conscious abstract thought; he also includes perceptual knowledge and subconscious cognitive activity [Davis, 1972]. For instance in subconscious mental activities visual representations play an immediate role.

We may also see belief change from the point of view of *conceptual change*, considering concepts either cognitively, like mental structures analogous to data structures in computers, or, epistemologically, like abstractions or representations that presuppose questions of justification. Belief revision is able to represent cases of conceptual change such as adding a new instance, adding a new weak rule, adding a new strong rule [Thagard, 1992], that is, cases of addition and deletion of beliefs, but fails to take into account cases such as adding a new part-relation, adding a new kind-relation, adding a new concept, collapsing part of a kind-hierarchy, reorganizing hierarchies by branch jumping and tree switching, in which there are reorganizations of concepts or redefinitions of the nature of a hierarchy.

We should remember, as Peirce noted, that abduction plays a role even in relatively simple visual phenomena. *Visual (or iconic) abduction* [Magnani *et al.*, 1994; Magnani, 1996], a special form of non verbal abduction, occurs when hypotheses are instantly derived from a stored series of previous similar experiences. It covers a mental procedure that tapers into a non-inferential one, and falls into the category called "perception". Philosophically,[48] *visual perception* is viewed by Peirce as a fast and uncontrolled knowledge-production procedure. Perception, in this philosophical perspective, is a vehicle for the instantaneous retrieval of knowledge that was previously structured in our mind through more structured inferential processes. Peirce says: "Abductive inference shades into perceptual judgment without any sharp line of demarcation between them" [Peirce, 1955c, p. 304]. By perception, knowledge constructions are so instantly reorganized that they become habitual and diffuse and do not need any further testing: "[...] a fully accepted, simple, and interesting inference tends to obliterate all recognition of the uninteresting and complex premises from which it was derived" [Peirce, 1931-1958, 7.37]. Many visual stimuli – that can be considered the "premises" of the involved abduction – are ambiguous, yet people are adept at imposing order on them: "We readily form such hypotheses as that an obscurely seen face belongs to a friend of ours, because we can thereby explain what has been observed" [Thagard, 1988, p. 53]. This kind of image-based hypothesis formation can be considered as a form of what I have called *visual* [Magnani *et al.*, 1994; Magnani, 1996] (or *iconic*) *abduction*. Of course such subconscious visual abductions of everyday cognitive behavior are not of particular importance but we know that in science they may

[48] In philosophical tradition visual perception was viewed very often like a kind of inference [Kant, 1929; Fodor, 1983; Gregory, 1987; Josephson and Josephson, 1994]. On visual perception as abduction and its semi-encapsulated character cf. subsection 5.5.2, chapter five of this book.

be very significant and lead to interesting new discoveries [Magnani *et al.*, 1994; Shelley, 1996]. If perceptions are abductions they are withdrawable, just like the scientific hypotheses abductively found. They are "hypotheses" about data we can accept (sometimes this happens spontaneously) or carefully evaluate.

One more example is given by the fact that the perception of tone arises from the activity of the mind only after having noted the rapidity of the vibrations of the sound waves, but the possibility of individuating a tone happens only after having heard several of the sound impulses and after having judged their frequency. Consequently the sensation of pitch is made possible by previous experiences and cognitions stored in memory, so that one oscillation of the air would not produce a tone.

To conclude, for Peirce all knowing is *inferring* and inferring is not instantaneous, it happens in a process that needs an activity of comparisons involving many kinds of models in a more or less considerable lapse of time.[49] As I will illustrate in the first two sections 5.1 and 5.2 of chapter five this is not in contradiction with the fact that for Peirce the inferential and abductive character of creativity is based on the instinct (the mind is "in tune with nature") but does not have anything to do with irrationality and blind guessing. [Hanson, 1958, pp. 85-92] perfectly recognizes the model-based side of abductive reasoning, when he relates (and reduces) it to the activity of "interpretation" ("pattern of discovery") resorting to the well-known example of reversible perspective figures of *Gestalt* psychology. Unfortunately, this kind of analysis inhibits the possibility of gaining further knowledge about model-based reasoning. I think Hanson is inclined to consider the abductive event as instantaneous and not susceptible to further cognitive and epistemological examination.

All sensations or perceptions participate in the nature of a unifying hypothesis, that is, in abduction, in the case of emotions too:

> Thus the various sounds made by the instruments of the orchestra strike upon the ear, and the result is a peculiar musical emotion, quite distinct from the sounds themselves. This emotion is essentially the same thing as a hypothetic inference, and every hypothetic inference involved the formation of such an emotion [Peirce, 1931-1958, 2.643].

Also this example surely suggests that abductive movements have interesting extratheoretical effects (see the following chapter).[50] Human beings and animals have evolved in such a way that now they are able to recognize habitual and recurrent events and to "emotionally" deal with them, like in cases of fear, that appears to be a quick explanation that some events are dangerous. During the evolution such abductive types of recognition and explanation settled in their nervous systems: we can abduce "fear" as a reaction to a possible external danger, but also when affronting a different types of evidence, like in the case of "reading a thriller" [Oatley, 1996].

---

[49] This corresponds to Peirce's "philosophical" point of view, which delineates a very particular meaning of the word "inference", as illustrated above.

[50] Considering emotions as abductions, [Oatley and Johnson-Laird, 2002] have proposed a cognitive theory of emotions largely based on Peircean intuitions. A different aim is pursued by [O'Rorke and Ortony, 1992; O'Rorke, 1994]: using a computational tool implemented in PROLOG, AbMaL, and the situation calculus framework, they provide an abductive theory showing how it is possible to construct explanations of emotional states.

In all these examples Peirce is referring to a kind of hypothetical activity that is inferential but not verbal, where "models" of feeling, seeing, hearing, etc., are very efficacious when used to build both habitual abductions of everyday reasoning and creative abductions of intellectual and scientific life (see Figure 1.6).

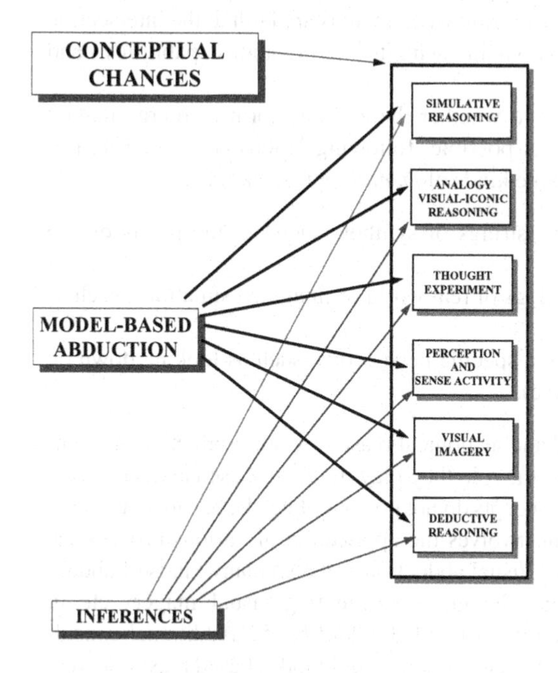

Fig. 1.6 Model-based abduction

Following Nersessian [1995; 1999b], the term "model-based reasoning" is used to indicate the construction and manipulation of various kinds of representations, not mainly sentential and/or formal, but mental (visual imagistic, analogical, etc.) and/or related to external mediators.[51] She proposes the so-called cognitive history and philosophy of science approach, which affords a reframing of the problem of

[51] See also the recent analysis of the role of models in science given by [Giere, 1988; Giere, 1999; Harris, 1999; Suarez, 1999]. For an account on the role of models in the history of recent philosophy of science cf. [Bailer-Jones, 1999]. [Zytkow, 1999; Winsberg, 1999] describe some aspects of model construction in automated computational systems aimed at reproducing scientific reasoning. On the mediating role of scientific models between theories and the real world cf. [Morgan and Morrison, 1999]. Further, differences in novice and expert reasoning skills in solving scientific problems (cf., e.g., [Chi *et al.*, 1981]) provide evidence that skills in modeling is something that develops with learning [Ippolito and Tweney, 1995]. More recent research can be found in [Magnani, 2006f; Magnani and Li, 2007]. Moreover, Nersessian relates model-based reasoning to some aspects of reasoning in terms of "mental models" described by [Johnson-Laird, 1988; Johnson-Laird, 1993], and recently enriched her perspective in the framework of distributed cognition cf. [Nersessian and Chandrasekharan, 2009; Nersessian and Patton, 2009].

conceptual formation and change in science that not only provides philosophical insights but also pays attention to the practices employed by real human agents in constructing, communicating and replacing representation of a domain. Common examples of model-based reasoning are constructing and manipulating visual representations, thought experiment, analogical reasoning, but also the so-called "tunnel effect" [Cornuéjols *et al.*, 2000], occurring when models are built at the intersection of some operational interpretation domain – with its interpretation capabilities – and a new ill-known domain.

Although controversy arises as to whether there is any form of representation other than strings of symbols, it is possible, following [Johnson-Laird, 1983] to assume the existence of at least three kinds of *mental* representations:

1. *propositional representations* (strings of symbols such as "the pot is on the table");
2. *mental models* (structural analogs of real world or imagined situations, such as a pot being on a table);
3. *images* (a mental model from a specific perspective, such as looking down on the pot on the table from above).

We have to remember that visual and analogical reasoning are productive in scientific concept formation too, where the role they play in model-based abductive reasoning is very evident; scientific concepts do not pop out of heads, but are elaborated in a problem-solving process that involves the application of various procedures: this process is a *reasoned process*. Visual abduction, but also many kinds of abductions involving analogies, diagrams, thought experimenting, visual imagery, etc. in scientific discovery processes, can be just called *model-based*. Additional considerations about the intersections between abduction and model-based reasoning (especially in experiment and thought experiment) are illustrated by [Gooding, 1990; Gooding, 2006]: the ability to integrate information from various sources is crucial to scientific inference and typical of all kinds of model-based reasoning also when models and representations are "external", like verbal accounts, drawings, various artifacts, narratives, etc.

We know that scientific concept formation has been ignored because of the accepted view that no "logic of discovery" – either deductive, inductive, or abductive algorithms for generating scientific knowledge – is possible.[52] The methods of discovery involve use of *heuristic* procedures (Peirce was talking of creative abduction

[52] It is well-known that Popper (and 3most of the philosophy of science tradition) confined scientific discovery to the realm of irrationality: "[...] there is not such thing as a logical method of having new ideas, or a logical reconstruction of this process. My view may be expressed by saying that every discovery contains 'an irrational element', or a 'creative intuition', in Bergson's sense" [Popper, 1959, p. 32]. This is also the case of the celebrated distinction between "context of discovery" and "context of justification" [Reichenbach, 1938] I have quoted at the beginning of this chapter. Rational analysis is only possible within the context of justification (verification, corroboration, falsification).

as the capacity and the "method" of making good conjectures);[53] cognitive psychology, artificial intelligence, and computational philosophy have established that heuristic procedures are reasoned (see the following section). Analogical reasoning is one such problem-solving procedure, and some reasoning from imagery is a form of analogical reasoning: [Holyoak and Thagard, 1995] elaborated an analysis of analogical reasoning that encompasses psychological, computational, and epistemological aspects. We have to remember that, among the various kinds of model-based reasoning, analogy received particular attention from the point of view of computational models designed to simulate aspects of human analogical thinking: for example, Thagard, et al. have developed ARCS (Analog Retrieval by Constraint Satisfaction; 1990) and ACME (Analogical Mapping by Constraint Satisfaction; Holyoak and Thagard, 1989), computational programs that are built on the basis of a multiconstraint theory.[54] [Holyoak and Thagard, 1995].

Finally, by recognizing the role of model-based abduction the analysis of conceptual change can overcome the negative issues that come from the reductionist theory of meaning and from the related incommensurability thesis, and illustrate the various grades of *commensurability* that can be found when dealing with the roles of model-based abduction in science. [Nersessian, 1998] exploits the representational and constructive virtues of model-based reasoning and makes use of Giere's general idea that "modeling is not at all ancillary to doing science, but central to constructing accounts of the natural world" [1999]: she illustrates how model-based abduction can explain that concept transformation and creation involves the construction of fluid and evolving frameworks that guarantee commensurability at many levels.

Manipulative abduction [Magnani, 2001b] – contrasted with theoretical abduction – happens when we are thinking through doing and not only, in a pragmatic sense, about doing. For instance, when we are creating geometry constructing and manipulating a triangle, like in the case given by Kant in the "Transcendental Doctrine of Method". So the idea of manipulative abduction (cf. Figure 1.7) goes beyond the well-known role of experiments as capable of forming new scientific laws by means of the results (nature's answers to the investigator's question) they present, or of merely playing a predictive role (in confirmation and in falsification). Manipulative abduction refers to an extra-theoretical behavior that aims at creating communicable accounts of new experiences to integrate them into previously existing systems of experimental and linguistic (theoretical) practices. As I said above, the existence of this kind of extra-theoretical cognitive behavior is also testified by the

[53] Analogy and abduction are separate types of reasoning practices, mutually independent both structurally and procedurally, but they are extremely useful in hypothesis-search in hypothesis selection tasks [Gabbay and Woods, 2005, p. 287].

[54] On analogy cf. also the contributions by [Kolodner, 1993] (analogy as a form of case-based reasoning in AI), [Davies and Goel, 2000; Nersessian *et al.*, 1997; Davies *et al.*, 2009] (visual analogy in AI); [Gentner, 1982; Gentner, 1983; Gentner *et al.*, 1997] (analogies and metaphors in cognitive science and history of science), [Shelley, 1999] (analogy in archaeology). Many theoretical and computational accounts of analogical reasoning have stressed the transfer of relational knowledge. Causal and functional relationships have been the focus of many theories [Holyoak and Thagard, 1995; Holyoak and Thagard, 1997; Bhatta and Goel, 1997; Falkenhainer, 1990; Winston, 1980] .

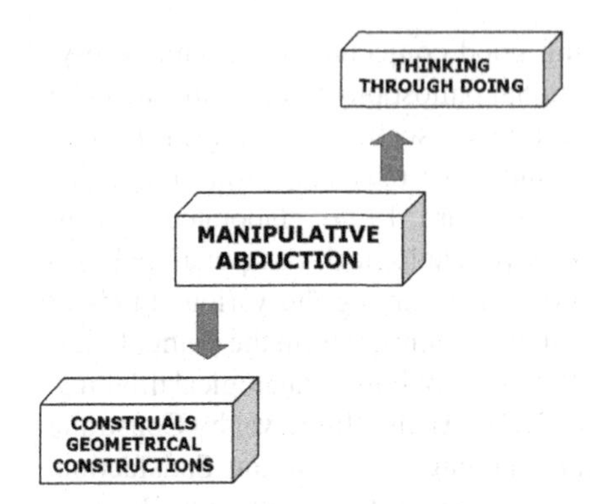

Fig. 1.7 Manipulative abduction

many everyday situations in which humans are perfectly able to perform very efficacious (and habitual) tasks without the immediate possibility of realizing their conceptual explanation.[55] In the following sections manipulative abduction will be considered from the perspective of the relationship between unexpressed knowledge and external representations.

I would like to reiterate that it is important to note that my epistemological distinction between theoretical and manipulative abduction is based on the possibility of separating the two aspects in actual cognitive processes, relying on the differentiation between off-line (theoretical, when only inner aspects are at stake) and on-line (manipulative, where the interplay between internal and external aspects is fundamental). As Wheeler has recently observed, some thinkers like Esther Thelen and Andy Clark have raised doubts about the on-line/off-line distinction "[...] on the grounds that no intelligent agent is (they claim) ever wholly on-line or wholly off-line" [Wheeler, 2004, p. 707, footnote 14]. I contend that, even if manipulative/on-line cases exist in great numbers, there are also cognitive processes that seem to fall into the class of off-line thinking, as we can simply introspectively recognize. Anyway, the distinction above is always rewarding from the epistemological perspective as a way of classifying and analyzing different cognitive levels, and it is endowed with an indisputable conceptual and explanatory usefulness.[56]

[55] [Rivera and Rossi Becker, 2007] have recently applied the ST-model above and especially my concept of manipulative abduction in the analysis of abductive/inductive reasoning of preservice elementary majors on patterns that consist of figural and numerical cues. Cf. also above footnote 24 at p. 14. Model-based reasoning and abduction in Felix Klein's heuristics are described in [Glas, 2009].

[56] An extreme case in which we see cognitive processes that seem to occur "completely" outside is that of a PC that performs a sophisticated AI program. In chapter two (section 3.7) I call this kind of artifact "mimetic mind"; in this case we must not forget that such an artifact still represents a cognitive "prosthesis" for the human brain, and so its cognitive performance still operates at the level of an on-line environment that also includes a human agent. In this cognitive sense the artifact is, in principle, no different from the mere use of a notebook for memorizing or of a hammer for building some piece of furniture.

1.6 Manipulative Abduction

1.6.1 Unexpressed Knowledge, Knowledge Creation, and External Mediators

The power of model-based abduction mainly depends on its ability to supply a certain amount of important information, unexpressed at the level of available data from the propositional point of view. It also has a fundamental role in the process of transformation of knowledge from its *tacit* to its *explicit* forms, and in the subsequent elicitation and use of knowledge. Let us describe how this happens.

As pointed out by Polanyi in his epistemological investigation, a large part of knowledge is not explicit, but tacit: we know more than we can tell and we can know nothing without relying upon those things which we may not be able to tell [1966]. Polanyi's concept of knowledge is based on three main theses: first, discovery cannot be accounted for by a set of articulated rules or algorithms; second, knowledge is public and also to a very great extent personal (i.e. it is constructed by humans and therefore contains emotions, "passions"); third, an important part of knowledge is tacit.

Hence, two levels of knowledge, mutually exclusive but complementary, as they interact in creative tasks, underlie every activity: there is a kind of knowledge we can call *focal*, that is the knowledge about the object or phenomenon in focus; and another kind of knowledge, masked under the first one, and often used as a tool to handle or improve what is being focused, we can call *tacit*. The first one is the knowledge that is transmissible through any systematic language, since it can be relatively easily formulated by means of symbols and it can be digitalized. Tacit knowledge, on the other hand, is characterized by the fact that it is personal, context specific, usually characterized as derived from direct experience, and therefore hard to elicit and communicate. It is a "non-codified, disembodied know-how that is acquired via the informal take-up of learned behavior and procedures" [Howells, 1996, p. 92].

[Fleck, 1996, p. 119] describes this form of knowledge as "a subtle level of understanding often difficult to put into words, a trained recognition and perception". Tacit knowledge is wholly embodied in the individual, rooted in practice and experience, expressed through skillful execution, and can become useful by means of watching and doing forms of learning and exploitation.

As Polanyi contends, human beings acquire and use knowledge by actively creating and organizing their own experience: tacit knowledge is the practical knowledge used to perform a task. The existence of this kind of not merely theoretical knowing behavior is also testified by the many everyday situations in which humans are perfectly able to perform very efficacious (and habitual) tasks without the immediate possibility of realizing their conceptual explanation. In some cases the conceptual account for doing these things was at one point present in memory, but now has deteriorated, and it is necessary to reproduce it, in other cases the account has to be constructed for the first time, like in creative experimental settings in science.

[Hutchins, 1995] illustrates the case of a navigation instructor that performed an automatized task for 3 years involving a complicated set of plotting manipulations and procedures. The insight concerning the conceptual relationships between relative and geographic motion came to him suddenly "as lay in his bunk one night". This example explains that many forms of learning can be represented as the result of the capability of giving conceptual and theoretical details to already automatized manipulative executions. The instructor does not discover anything new from the point of view of the objective knowledge about the involved skill, however, we can say that his conceptual awareness is new from the local perspective of his individuality.

We can find a similar situation also in the process of scientific creativity. In the cognitive view of science, it has been too often underlined that conceptual change just involves a *theoretical* and "internal" replacement of the main concepts. But usually researchers forget that a large part of these processes are instead due to *practical* and "external" *manipulations* of some kind, prerequisite to the subsequent work of theoretical arrangement and knowledge creation. When these processes are creative we can speak of manipulative abduction (cf. above). Scientists sometimes need a first "rough" and concrete experience of the world to develop their systems, as a *cognitive-historical* analysis of scientific change [Nersessian, 1992; Gooding, 1990] has carefully shown.

The prevailing perspective among philosophers is that the processes of discovery and the consequent new incoming scientific representations are too mysterious to be understood. This view receives support from numerous stories of genius' discoveries, such as Archimedean eureka-experiences. Such accounts neglect periods of intense and often arduous thinking activity, often performed by means of experiments and *manipulative* activity on external objects; these are periods that prepare such "instantaneous" discoveries. It is also important to understand that the scientific process is *complex* and *dynamic*: new representations do not emerge completely codified from the heads of scientists, but are constructed in response to specific problems by the systematic use of heuristic procedures – as pointed out by Herbert Simon's view on the "problem-solving process" [Simon, 1977].

Traditional examinations of how problem-solving heuristics create new representations in science have analyzed the frequent use of analogical reasoning, imagistic reasoning, and thought experiment, from an internal point of view. However, attention has not been focalized on those particular kinds of heuristics that resort to the existence of *extra-theoretical* ways of thinking – *thinking through doing* [Magnani, 2002b]. Indeed many cognitive processes are centered on *external representations*, as a means to create communicable accounts of new experiences ready to be integrated into previously existing systems of experimental and linguistic (theoretical) practices (cf. chapter three of this book).

Interesting insights can arise regarding these problems studying them from a different contrasting approach, which moves away from Simon's paradigm, but which can offer a rational solution to the problem of creativity and conceptual change in terms of mathematical models: the *dynamic* approach [Port and van Gelder, 1995]. The traditional computational view treats cognition as a process that computes internal symbolic representations of the external world. But this approach is considered too reductive, since it is based on the functionalist hypothesis (which cannot render the *external dimension* of cognition), and on a computation of static entities. It is useful to integrate it with a dynamical modeling of cognition, which is able to describe abductive processes as *dynamical entities* "unfolding" in real time (we can also gain a better cognitive-historical perspective) [Magnani and Piazza, 2005]. From this point of view it is possible to model the terms (objects or propositions) that constitute abduction by considering the *attractors* in a dynamical system. This can be achieved by topologically specifying the semantic content of the inferential process through the spatial relations between its defining attractors. We can therefore consider the process of progressive development of "new" concepts and replacement of old ones in terms of temporal evolving patterns defined by interactions between topological configurations of attractors.[57]

Moreover, a central point in the dynamical approach is the importance assigned to the "whole" cognitive system: cognitive activity is in fact the result of a complex interplay and simultaneous coevolution, in time, of the states of mind, body, and external environment. Even if, of course, a large portion of the complex environment of a thinking agent is internal, and consists in the proper software composed of the knowledge base and of the inferential expertise of the individual, nevertheless a "real" cognitive system is composed by "distributed cognition" among people and some "external" objects and technical artifacts [Hutchins, 1995; Norman, 1993].[58]

A recent special issue of the journal *Pragmatics & Cognition*, devoted to the theme of "distributed cognition", addresses many of the puzzling theoretical problems still open to debate. In particular the article by [Sutton, 2006] usefully emphasizes how distributed cognition is related to the "extended mind hypothesis" [Clark and Chalmers, 1998] and other similar approaches in terms of embodied, embedded, situated, and dynamical cognition and active and vehicle externalism, which of course present subtle nuances, that I nevertheless cannot account for here. Sutton nicely presents a taxonomy of the distributed resources that are studied in these fields of research: 1) external cultural tools, artifacts, and symbol systems; 2) natural environment features suitably endowed with cognitive value; 3) interpersonal and social distribution or cognitive "scaffolding"; 4) embodied capacities and skills interwoven in complex ways with our use of technological, natural, and social resources of the previous cases; 5) internalized cognitive artifacts. For Sutton the last two cases concern the analysis of the complex wholes made up when embodied brains couple with "cognition" amplifiers like objects – technologies for example, and other people, through a process I will mention in chapter three (subsection 3.6.4) "re-embodiment of the mind", as a kind of neural recapitulation of cognitive features – for example linguistic and model-based – found and distributed outside.

[57] On abduction, dynamic systems theory, and morphogenetical models cf. chapter eight of this book.

[58] [Skagestad, 1993] stresses the role of this coevolution in cognition in the framework of an analysis of Popperian writing on evolutionary epistemology and Peircean semiotics.

The external distributed resources have culturally specific different degrees of stability and so various chances to be re-internalized, varying from the very stable and reliable, like words and phrases of "natural" language and certain symbols, to others which are more evanescent and transient. Only when they are stable can we properly speak of the establishment of an "extended mind", like Wilson and Clark contend. Finally, internal and external resources (that is neural and environmental) are not identical but complementary – in this sense human beings can be appropriately considered "cyborgs" [Clark, 2003]. By showing their interaction from an epistemological and cognitive perspective I will illustrate in chapter three of this book various aspects of their interplay taking advantage of the concept of abduction.[59]

In the recent [Clark, 2008, p. 13], a deep analysis of various aspects of embodiment, environmental embedding, and of the so-called "extended mind hypothesis" is presented: Clark definitely contends that mind "leaches into body and world" (p. 29): "Inner neural processes [...] are often productively entangled with the gross bodily and extra-bodily processes of storage, representation, materialization, and manipulation" (p. 169). Embodiment, action, and situation are fundamental in human thought and behavior. The first chapter of Clark's new book also focuses on the so-called Principle of Ecological Assembly (PEA), which states that the cognizer tends to recruit "whatever mix of problem-solving resources will yield and acceptable result with a minimum effort". This recruitment process does not make a special distinction between neural, bodily, and environmental resources except insofar as these somehow affect the whole effort involved. The operation is not operated "in the neural system alone, but in the whole embodied system located in the world. [...] the embodied agent is empowered to use active sensing and perceptual coupling in ways that simplify neural problem-solving by making the most of environmental opportunities and information freely available in the optic array" (p.̃14).

In the case of the construction and examination of diagrams in geometrical reasoning, specific experiments serve as states and the implied operators are the manipulations and observations that transform one state into another. The geometrical outcome depends upon practices and specific sensorimotor activities[60] performed on a non-symbolic object, which acts as a dedicated external representational medium supporting the various operators at work. There is a kind of an epistemic negotiation between the sensory framework of the geometer and the external reality of the diagram [Magnani, 2002a]. This process involves an external representation consisting of written symbols and figures that for example are manipulated "by hand". The

[59] The role of external representations and resources – I call *moral mediators* – in ethics is described in my recent [Magnani, 2007d]. A detailed treatment of the theoretical and cognitive controversies concerning the extended mind hypothesis and the role of external representations is given in the recent collection [Schantz, 2004] and in [Clark, 2008, chapters five and six].

[60] "The agent's control architecture (e.g. nervous system) attends to and processes streams of sensory stimulation, and ultimately generates sequences of motor actions which in turn guide the further production and selection of sensory information. [In this way] 'information structuring' by motor activity and 'information processing' by the neural system are continuously linked to each other through sensorimotor loops" [Lungarella and Sporns, 2005, p. 25].

cognitive system is not merely the mind-brain of the person performing the geometrical task, but the system consisting of the whole body (cognition is *embodied*) of the person plus the external physical representation. In geometrical discovery the whole activity of cognition is located in the system consisting of a human together with diagrams.[61]

An external representation can modify the kind of computation that a human agent uses to reason about a problem: the Roman numeration system eliminates, by means of the external signs, some of the hardest parts of the addition, whereas the Arabic system does the same in the case of the difficult computations in multiplication. The capacity for inner reasoning and thought results from the internalization of the originally external forms of representation. In the case of the external representations we can have various objectified knowledge and structures (like physical symbols – e.g. written symbols, and objects – e.g. three-dimensional models, shapes and dimensions), but also external rules, relations, and constraints incorporated in physical situations (spatial relations of written digits, physical constraints in geometrical diagrams and abacuses) [Zhang, 1997]. The external representations are contrasted with the internal representations that consist in the knowledge and the structure in memory, as propositions, productions, schemas, models, prototypes, images.

The external representations are not merely memory aids: they can give people access to knowledge and skills that are unavailable to internal representations, help researchers to easily identify aspects and to make further inferences, they constrain the range of possible cognitive outcomes in a way that some actions are allowed and others forbidden. The mind is limited because of the restricted range of information processing, the limited power of working memory and attention, the limited speed of some learning and reasoning operations; on the other hand the environment is intricate, because of the huge amount of data, real time requirement, uncertainty factors. Consequently, we have to consider the whole system, consisting of both internal and external representations, and their role in optimizing the whole cognitive performance of the distribution of the various subtasks. In this case humans are not "just bodily and sensorily but also cognitively permeable agents" [Clark, 2008, p. 40]. It is well-known that in the history of geometry many researchers used internal mental imagery and mental representations of diagrams, but also self-generated diagrams (external) to help their thinking.

1.6.2 External Representations and Epistemic Mediators

I have illustrated above the notion of tacit knowledge and I have proposed an extension of that concept. From the perspective of a more adequate and updated account

[61] [Elveton, 2005] provides a survey concerning the problem of embodiment considered as addressing a kind of practical intelligence in contrast to a disembodied, symbol manipulating intelligence [Brooks, 1991]; a comparison of the recent cognitive perspectives with robotics and the classical philosophical insights given by Cassirer, Husserl and Heidegger is also illustrated. [Dourish, 2001] usefully demonstrates the importance of the concept of embodiment in human-computer interaction (*HCI*) and in the design of computational tools, technologies, and systems.

of cognition surely there is something more important beyond the tacit knowledge "internal" to the subject – considered by Polanyi as personal, embodied and context specific. We can also speak of a sort of tacit information "embodied" into the whole relationship between our mind-body system and suitable external representations. An information we can extract, explicitly develop, and transform in knowledge contents, to solve problems, as it was already manifest, for instance, in the geometrical problem contained in the *Meno* [Plato, 1977], even if philosophers know perfectly that Plato considered this activity to be just the result of reminiscence and not of discovery [Magnani, 2001b, chapter 1].

As I have already stressed, Peirce considers inferential any cognitive activity whatever, not only conscious abstract thought; he also includes perceptual knowledge and subconscious cognitive activity. For instance in subconscious mental activities visual representations play an immediate role. Peirce gives an interesting example of model-based abduction related to sense activity: "A man can distinguish different textures of cloth by feeling: but not immediately, for he requires to move fingers over the cloth, which shows that he is obliged to compare sensations of one instant with those of another" [Peirce, 1931-1958, 5.221]. This surely suggests that abductive movements have also interesting extra-theoretical characters and that there is a role in abductive reasoning for various kinds of manipulations of external objects. I would like to reiterate that for Peirce *all* knowing is *inferring* and inferring is not instantaneous, it happens in a process that needs an activity of comparisons involving many kinds of models in a more or less considerable lapse of time.

All these considerations suggest, then, that there exist a creative form of thinking through doing,[62] fundamental as much as the theoretical one. It is what I have called *manipulative abduction* (cf. above). As already said *manipulative* abduction happens when we are thinking *through* doing and not only, in a pragmatic sense, about doing. Of course the study of this kind of reasoning is important not only in delineating the actual practice of abduction, but also in the development of programs computationally adequate to rediscover, or discover for the first time, for example, scientific hypotheses and mathematical theorems or laws.

Various *templates* of manipulative behavior exhibit some regularities. The activity of manipulating external things and representations is highly conjectural and neither immediately explanatory nor necessarily immediately non-explanatory and instrumental[63] these templates are "hypotheses of behavior" (creative or already cognitively present in the scientist's mind-body system, and sometimes already applied) that abductively enable a kind of epistemic "doing". Hence, some templates of action and manipulation can be selected in the set of the ones available and pre-stored, others have to be created for the first time to perform the most interesting creative cognitive accomplishments of manipulative abduction.

---

[62] In this way the cognitive task is achieved on *external* representations used in lieu of internal ones. Here action performs an *epistemic* and not a merely performative role, relevant to abductive reasoning.

[63] I will illustrate these two non Peircean sorts of abductive cognition in the following chapter.

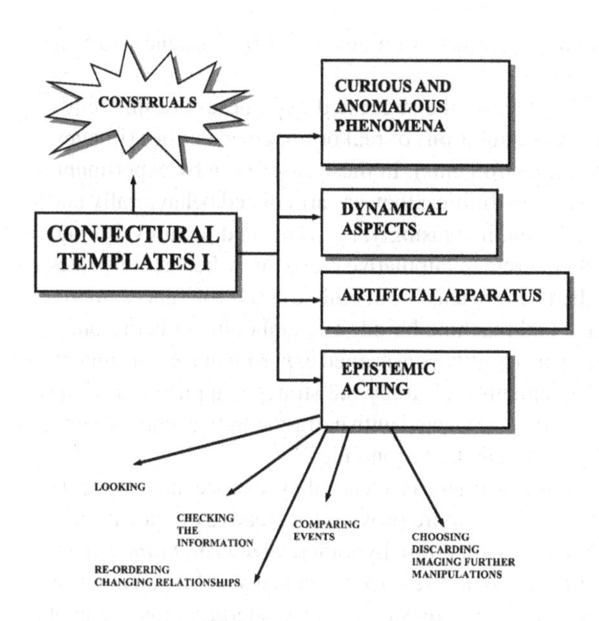

Fig. 1.8 Conjectural templates I

Some common features of the tacit templates of manipulative abduction (cf. Figure 1.8), that enable us to manipulate things and experiments in science are related to: 1. sensibility towards the aspects of the phenomenon which can be regarded as *curious* or *anomalous*; manipulations have to be able to introduce potential inconsistencies in the received knowledge (Oersted's report of his experiment about electromagnetism is devoted to describe some anomalous aspects that did not depend on any particular theory of the nature of electricity and magnetism); 2. preliminary sensibility towards the *dynamical* character of the phenomenon, and not to entities and their properties, common aim of manipulations is to practically reorder the dynamic sequence of events into a static spatial one that should promote a subsequent bird's-eye view (narrative or visual-diagrammatic); 3. referral to experimental manipulations that exploit *artificial apparatus* to free new possibly stable and repeatable sources of information about hidden knowledge and constraints (Davy well-known set-up in terms of an artifactual tower of needles showed that magnetization was related to orientation and does not require physical contact). Of course this information is not artificially made by us: the fact that phenomena are made and manipulated does not render them to be idealistically and subjectively determined; 4. various contingent ways of epistemic acting: *looking* from different perspectives, *checking* the different information available, *comparing* subsequent events, *choosing*, *discarding*, *imaging* further manipulations, *re-ordering* and *changing relationships* in the world by implicitly *evaluating* the usefulness of a new order (for instance, to help memory).[64]

In this kind of *action-based* abduction the suggested hypotheses are inherently ambiguous until articulated into configurations of real or imagined entities (images, models or concrete apparatus and instruments). In these cases only by experimenting, can we discriminate between possibilities: they are articulated behaviorally and concretely by manipulations and then, increasingly, by words and pictures. [Gooding, 1990] refers to this kind of concrete manipulative reasoning when he illustrates the role in science of the so-called "construals" that embody tacit inferences in procedures that are often apparatus and machine based. The embodiment is of course an expert manipulation of objects in a highly constrained experimental environment, and is directed by abductive movements that imply the strategic application of old and new *templates* of behavior mainly connected with extra-theoretical components, for instance emotional, esthetical, ethical, and economic.[65]

The hypothetical character of construals is clear: they can be developed to examine further chances, or discarded, they are provisional creative organization of experience and some of them become in their turn hypothetical *interpretations* of experience, that is more theory-oriented, their reference is gradually stabilized in terms of established observational practices. Step by step the new interpretation – that at the beginning is completely "practice-laden" – relates to more "theoretical" modes of understanding (narrative, visual, diagrammatic, symbolic, conceptual, simulative), closer to the constructive effects of theoretical abduction. When the reference is stabilized the effects of incommensurability with other stabilized observations can become evident. But it is just the construal of certain phenomena that can be

---

[64] The problem of manipulative abduction and of its tacit features is strongly related to the whole area of recent research on embodied reasoning (cf. [Anderson, 2003; Elveton, 2005]), but also to the studies on external representations and situated robotics (cf. [Clancey, 2002; Agree and Chapman, 1990; Brooks and Stein, 1994]). The role of manipulative abduction in ethical reasoning is illustrated in [Magnani, 2007d]. Further aspects of experiment design and its relationship with the problem of communication in science during the transition from the personal to the public domain are given in [Gooding and Addis, 1999]: only a small subset of many observations and measurements performed by individuals of research teams acquire the status of real and public phenomena. Moreover, additional properties of the agent in a scientific experimental setting are described: 1. ability to discriminate between observed results, 2. ability to make judgments about the likelihood of the occurrence of a result, 3. flexibility of the agent's change in perception of the world and his consequent capacity to respond to new information, 4. degrees of competence to build an experiment and observe the results, from novices to experts.

[65] [Tweney, 2006] has recently emphasized the importance of externalized cognitive artifacts used in the service of the "seeing" of scientists. In turn they are distributed "in the strong sense that not all of the agentive movement of thought is localized solely within an individual skin". I think a further light on the role of construals is shed by Franklin who usefully analyzes the so-called "exploratory experiments" that prior to theorizing investigate the world "without premature reflection of any great subtlety, like Bacon says [2000, p. 210], and where there is no particular hypothesis being pursued. They serve "[. . .] to find interesting patterns of activity from which the scientists could later generate a hypothesis" [Franklin, 2005, p. 894].

shared by the sustainers of rival theories.[66] [Gooding, 1990] shows how Davy and Faraday could see the same attractive and repulsive actions at work in the phenomena they respectively produced; their discourse and practice as to the role of their construals of phenomena clearly demonstrate they did not inhabit different, incommensurable worlds in some cases. Moreover, the experience is constructed, reconstructed, and distributed across a social network[67] of negotiations among the different scientists by means of construals.[68]

Gooding introduces the so called *experimental maps*[69] that are the epistemological two-dimensional tools that we can adopt to illustrate the conjecturing (abductive) role of actions from which scientists "talk and think" about the world. They are particularly useful to stress the attention to the interaction of hand, eye, and mind inside the actual four-dimensional scientific cognitive process. The various procedures for manipulating objects, instruments and experiences will be in their turn reinterpreted in terms of procedures for manipulating concepts, models, propositions, and formalisms. Scientists' activity in a material environment first of all enables a rich perceptual experience that has to be reported mainly as a visual experience by means of the constructive and hypothesizing role of the experimental narratives.

It is indeed interesting to note that in mathematics model-based and manipulative abductions are present. For example, I will illustrate in the following chapter and in chapter three that it is clear that in geometrical construction all these requirements are fulfilled. Geometrical constructions present situations that are curious and "at the limit". These are constitutively dynamic, artificial, and offer various contingent ways of epistemic acting, like looking from different perspectives, comparing subsequent appearances, discarding, choosing, re-ordering, and evaluating. Moreover, they present some of the features indicated below, typical of all abductive epistemic mediators, not only of the ones which play a scientific role in manipulative reasoning: simplification of the task and the capacity to get visual information otherwise unavailable.

---

[66] The theory of manipulative abduction can support Thagard's statement that oxygen and phlogiston proponents could recognize experiments done by each others [Thagard, 1992]: the assertion is exhibited as an indispensable requisite for his coherence-based epistemological and computational theory of comparability at the level of intertheoretic relations and for the whole problem of the creative abductive reasoning to the best explanation cited in the previous chapter.

[67] Cf. [Minski, 1985; Thagard and Shelley, 1997].

[68] [Gooding and Addis, 2008] further analyze the role of various kinds of experiments – ranging from the idealized crucial ones to those that are exploratory and/or controversial – like mediating models in the framework of an agent-based approach. Every agent or actor can investigate a world of experiments and other agents, in a setting where eventually scientists invent and negotiate ways of representing aspects of the world they are investigating. The process is "adaptive" and "inherently social" and inference is seen as a continuous activity of belief-revision where the distributed and collaborative aspects are acknowledged. In this perspective experiments are seen as mediating between at least four sets of objects: hypotheses, procedures, physical setups and observable outcomes.

[69] Circles denote concepts (mentally represented) that can be communicated, squares denote things in the material world (bits of apparatus, observable phenomena) that can be manipulated – lines denote actions.

These construals aim at arriving to a shared understanding overcoming all conceptual conflicts. As I said above they constitute a provisional creative organization of experience: when they become in their turn hypothetical interpretations of experience, that is more theory-oriented, their reference is gradually stabilized in terms of established and shared observational practices that also exhibit a cumulative character. It is in this way that scientists are able to communicate the new and unexpected information acquired by experiment and action.

To illustrate this process – from manipulations, to narratives, to possible theoretical models (visual, diagrammatic, symbolic, mathematical) – we need to consider some observational techniques and representations made by Faraday, Davy, and Biot concerning Oersted's experiment about electromagnetism. They were able to create consensus because of their conjectural representations that enabled them to resolve phenomena into stable perceptual experiences. Some of these narratives are very interesting. For example, Faraday observes: "[...] it is easy to see how any individual part of the wire may be made attractive or repulsive of either pole of the magnetic needle by mere change of position [...]. I have been more earnest in my endeavors to explain this simple but important point of position, because I have met with a great number of persons who have found it difficult to comprehend". Davy comments: "It was perfectly evident from these experiments, that as many polar arrangements may be formed as chord can be drawn in circles surroundings the wire". Expressions like "easy to see" or "it was perfectly evident" are textual indicators inside the experimental narratives of the stability of the forthcoming interpretations. Biot, in his turn, provides a three-dimensional representation of the effect by giving a verbal account that enables us to visualize the setup: "suppose that a conjunctive wire is extended horizontally from north to south, in the very direction of the magnetic direction in which the needle reposed, and let the north extremity be attached to the copper pole of the trough, the other being fixed to the zinc pole [...]" and then describes what will happen by illustrating a sequence of step in a geometrical way:

> Imagine also that the person who makes the experiment looks northward, and consequently towards the copper or negative pole. In this position of things, when the wire is paced above the needles, the north pole of the magnet moves towards the west; when the wire is placed underneath, the north pole moves towards the east; and if we carry the wire to the right or the left, the needle has no longer any lateral deviation, but is loses its horizontality. If the wire be placed to the right hand, the north pole rises; to the left, its north pole dips [...].[70]

It is clear that the possibility of "seeing" interesting things through the experiment depends from the manipulative ability to get the correct information and to create the possibility of a new interpretation (for example a simple mathematical form) of electromagnetic natural phenomena, so joining the theoretical side of abduction. Step by step, we proceed until Faraday's account in terms of magnetic lines and curves.

---

[70] The quotations are from [Faraday, 1821-1822, p. 199], [Davy, 821, pp. 282–283] and [Biot, 1821, p. 282-283], cited by [Gooding, 1990, pp. 35–37].

The whole activity of manipulation is in fact devoted to building various external *epistemic mediators*.[71] Therefore, manipulative abduction represents a kind of redistribution of the epistemic and cognitive effort to manage objects and information that cannot be immediately represented or found internally (for example exploiting the resources of visual imagery).[72]

It is difficult to establish a list of invariant behaviors that are able to illustrate manipulative abduction in science. As illustrated above, certainly the expert manipulation of objects in a highly constrained experimental environment implies the application of old and new *templates* of behavior that exhibit some regularities. The activity of building construals is highly conjectural and not immediately or necessarily explanatory: these templates are hypotheses of behavior (creative or already cognitively present in the scientist's mind-body system, and sometimes already applied) that abductively enable a kind of epistemic "doing". Hence, some templates of action and manipulation can be *selected* in the set of the ones available and prestored, others have to be *created* for the first time to perform the most interesting creative cognitive accomplishments of manipulative abduction.

Moreover, I think that a better understanding of manipulative abduction at the level of scientific experiment could improve our knowledge of induction, and its distinction from abduction: manipulative abduction could be considered as a kind of basis for further meaningful inductive generalizations. Different generated construals can give rise to different inductive generalizations.

If we see scientific discovery like a kind of opportunistic ability of integrating[73] information from many kinds of simultaneous constraints to produce explanatory hypotheses that account for them all, then manipulative abduction will play the role of eliciting possible hidden constraints by building external suitable experimental structures.

From the point of view of everyday situations manipulative abductive reasoning and epistemic mediators exhibit other very interesting templates (we can find the first three in geometrical constructions) (cf. Figure 1.9): 5. action elaborates a *simplification* of the reasoning task and a redistribution of effort across time [Hutchins, 1995], when we need to manipulate concrete things in order to understand structures which are otherwise too abstract [Piaget, 1974], or when we are in presence of

[71] This expression, I have introduced in [Magnani, 2001b], is derived from the cognitive anthropologist Hutchins, who coined the expression "mediating structure" to refer to various external tools that can be built to cognitively help the activity of navigating in modern but also in "primitive" settings. Any written procedure is a simple example of a cognitive "mediating structure" with possible cognitive aims, so mathematical symbols and diagrams: "Language, cultural knowledge, mental models, arithmetic procedures, and rules of logic are all mediating structures too. So are traffic lights, supermarkets layouts, and the contexts we arrange for one another's behavior. Mediating structures can be embodied in artifacts, in ideas, in systems of social interactions [...]" [Hutchins, 1995, pp. 290–291] that function as an enormous new source of information and knowledge. [Sterelny, 2004, p. 249] maintains that "epistemic tools support open-ended and counterfactually robust dispositions to succeed" and further stresses their social character.

[72] It is difficult to preserve precise spatial and geometrical relationships using mental imagery, in many situations, especially when one set of them has to be moved relative to another.

[73] On the role of opportunistic reasoning in design cf. [Simina and Kolodner, 1995].

redundant and unmanageable information; 6. action can be useful in presence of *incomplete* or *inconsistent* information – not only from the "perceptual" point of view – or of a diminished capacity to act upon the world: it is used to get more data to restore coherence and to improve deficient knowledge; 7. action enables us to build *external artifactual models* of task mechanisms instead of the corresponding internal ones, that are adequate to adapt the environment to the agent's needs: experimental manipulations exploit *artificial apparatus* to free new possible stable and repeatable sources of information about hidden knowledge and constraints. 8. action as a *control of sense data* illustrates how we can change the position of our body (and/or of the external objects) and how to exploit various kinds of prostheses (Galileo's telescope, technological instruments and interfaces) to get various new kinds of stimulation: action provides some tactile and visual information (e.g., in surgery), otherwise unavailable.

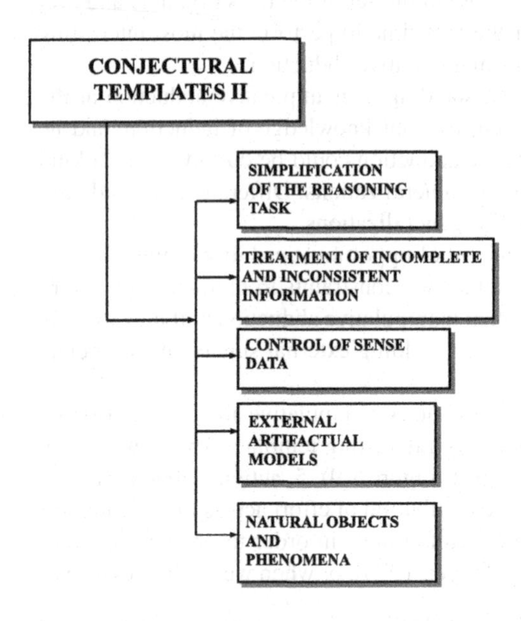

Fig. 1.9 Conjectural templates II

Also natural phenomena can play the role of external artifactual models: under Micronesians' manipulations of their images, the stars acquire a structure that "becomes one of the most important structured representational media of the Micronesian system" [Hutchins, 1995, p. 172]. The external artifactual models are endowed with functional properties as components of a memory system crossing the boundary between person and environment (for example they are able to transform the tasks involved in allowing simple manipulations that promote further visual inferences at the level of model-based abduction). The cognitive process is *distributed* between a person (or a group of people) and external representation(s), and so obviously *embedded* and *situated* in a society and in a historical culture.[74]

So external well-built structures (Biot's construals for example) and their contents in terms of new information and knowledge, will be projected onto internal structures (for instance models, or symbolic – mathematical – frameworks) so joining the constructive effect of theoretical abduction. The interplay consists of a superimposition of internal and external, where the elements of the external structures gain new meanings and relationships to one another, thanks to the constructive explanatory inner activity (for instance Faraday's new meanings in terms of curves and lines of force). This interplay expresses the fact that both internal and external processes are part of the same epistemic ecology.[75]

Not all epistemic and cognitive mediators are preserved, saved, and improved, as in the case of the ones created by Galileo at the beginning of modern science (see the following subsection). For example, in certain non epistemological everyday emergency situations some skillful mediators are elaborated to face possible dangers, but, because of the rarity of this kind of events, they are not saved and stabilized. [Hutchins, 1995, pp. 317–351] describes the interesting case of the failure of an electrical device, the gyrocompass, crucial for navigation, and the subsequent creation of substitutive contingent cognitive mediators. These cognitive mediators consisted of additional computations, redistributions of cognitive roles, and finally, of the discovery of a new shared mediating artifact in terms of divisions of labor – the so-called modular sum that is able to face the situation.

Finally, we have to observe that many external things that usually are cognitively inert can be transformed into epistemic or cognitive mediators. For example we can use our body: we can talk with ourselves, exploiting in this case the self-regulatory character of this action, we can use fingers and hands for counting.[76] We can also use external "tools" like writing, narratives, others persons' information,[77] concrete models and diagrams, various kinds of pertinent artifacts. Hence, not all of the cognitive tools are inside the head, sometimes it is useful to use external objects and structures as cognitive or epistemic devices. We indicated above that Micronesian navigator's stars, that are natural objects, become very complicated epistemic

[74] Modeling mechanisms of manipulative abduction is also related to the possibility of improving technological interfaces that provide restricted access to controlled systems, so that humans have to compensate by reasoning with and constructing internal models. New interfaces resources for action, related to task-transforming representations, can contribute to overcome these reasoning obstacles [Kirlik, 1998]. Further details on this issue will be provided in chapter six of this book.

[75] It is [Hutchins, 1995, p. 114] that uses the expression "cognitive ecology" when explaining the role of internal and external cognitive navigation tools. More suggestions on manipulative abduction can be derived by the contributions collected in [Morgan and Morrison, 1999], dealing with the mediating role of scientific models between theory and the "real world".

[76] Another example is given by the gestures that are also activated in talking, sometimes sequentially, sometimes in an overlapping fashion. On this problem cf. the updated critical survey given by [Clark, 2008, chapter six].

[77] The results of an empirical research that show the importance of collaborative discovery in scientific creative abduction and in explanatory activities are given in [Okada and Simon, 1997].

artifacts, when inserted in the various cognitive manipulations (of seeing them) related to navigation.

1.6.3 Segregated Knowledge and the "World of Paper"

I said that in the last part of the XX century the problem of the incommensurability of meaning has distracted the epistemologists from the procedural, extra-sentential and extra-theoretical aspects of scientific practice (cf. subsection 1.2.1 above). This is surprising especially if we consider that the emphasis on concrete manipulative reasoning in case of "construals", that embody tacit inferences in procedures that are often apparatus and machine based, is already clearly granted at the beginnings of modern science.

It is a very common philosophical view to assert that modern science uses experiment to get new information about the world, even if it is not always completely clear the manipulative character of this activity. The new world of the new knowledge has to be totally different from the one merely "of paper" of the Aristotelian tradition. An unbelievable amount of knowledge that was *segregated* had to be released. Accentuating the role of observational manipulations Galileo says:

> The anatomist showed that the great trunk of nerves, leaving the brain and passing through the nape, extended on down the spine and then branched out through the whole body, and that only a single strand as fine as a thread arrived at the heart [Galilei, 1989, p. 63].

Manipulating the cadaver, the anatomist is able to get new, not speculative, information that the Peripatetic philosopher immediately refuses:

> The philosopher, after considering for awhile, answered: "You have made me see this matter so plainly and palpably that if Aristotle's text were not contrary to it, stating clearly that the nerves originate in the heart, I should be forced to admit it to be true" (*ibid.*).

Ipse dixit: no room for the experience. Galileo-Salviati begs of Simplicius: "So put forward the arguments and demonstrations, Simplicius, [...] but not just texts and bare authorities, because our discourses must relate to the sensible world and not the one of paper" [Galilei, 1989, p. 68].

Manipulating observations to get new data, and "actively" building experiments, like the famous one from the leaning tower, sometimes with the help of artifacts, is the essence of the new way of knowing. Galileo says: "All these facts were discovered and observed by me many days ago with the aid of a spyglass which I devised, after first being illuminated by divine grace. Perhaps other things, still more remarkable, will in time be discovered by me or by other observers with the aid of such an instrument" [Galilei, 1957, p. 28]. Attaching a scale marked with equally spaced horizontal and vertical lines to his telescope, and manipulating objects "idealizing" them and not considering interesting and non influential factors, Galileo was able to record the daily histories of the four "starlets" accompanying Jupiter and to show that the data was consistent with the abduction that the starlets were indeed moons orbiting Jupiter with a constant period.

With Galileo's achievements, we observe that human scientific thinking is related to the manipulation of a material and experimental environment that is no longer natural. Knowledge is finally seen as something cognitively distributed across scientists, their internal "minds", and external artifacts and instruments. Experiments and instruments embody in their turn external crystallization of knowledge and practice. Modern science is made by this interplay of internal and external. Bacon too was very clear about this distribution of epistemic tasks:

> Those who handled sciences have been either men of experiment or men of dogmas. The men of experiment are like the ant, they only collect and use; the reasoners resemble spiders, who make cobwebs out of their own substance. But the bee takes a middle course: it gathers its material from the flowers of the garden and of the field, but transforms and digests it by a power of its own [Bacon, 2000, p. 52].

An immediate consequence of Galileo's and Bacon's ideas is the critique of the authority, that advocated the knowledge relevance of a "world of paper". Gooding observes: "It is ironical that while many philosophers admire science because it is empirical as well as rational, philosophical practice confines it to the literary view that Galileo rejected" [Gooding, 1990, p. xii]. Galileo's "book of nature" and his systematic use of the telescope are the revolutionary *epistemic mediators*[78] that characterize the cognitive power of the new way of producing intelligibility.

Changes in the modalities of distributing epistemic assignments are never without costs. We can just remark, even if well-known, that Galileo's new management of information and knowledge by means of inventing and stabilizing these mediators was not without individual and violent social costs. Because of the new knowledge provided, *Dialogue* was prohibited, and he was sentenced (1633 – the admonition is of 1616) to life imprisonment by the Holy Office with the added task of having to recite once a week for three years the seven penitential psalms. He read his abjuration and was released to the custody of the Archibishop of Siena; his daughter, Sister Maria Celeste was given permission to recite the psalms in his stead [van Helden, 1989]. The deterioration of the scientific climate and the decline of telescopic astronomy in Italy were the obvious immediate consequences. Notwithstanding the problems, these new epistemic mediators, that are at the roots of the tradition of scientific knowledge, were preserved, saved, and subsequently improved.

It is well known that recent philosophy of science has paid great attention to the so-called theory-ladenness of scientific facts (Hanson, Popper, Lakatos, Kuhn): in this light the formulation of observation statements presupposes significant knowledge, and the search for important observable facts in science is guided by that knowledge. It is absolutely true that theory is able to lead us to abduce new facts, but we cannot forget that a lot of new information is reached by observations and experiments, as fruit of various kinds of artifactual manipulations. Robert Hooke, using microscope to look at small insects, with practical interventions illuminated

[78] Together with the exploitation of mathematical models.

his specimens from different directions to establish which features remained invariant under such changes and discovered that some disagreements about data were apparent [Chalmers, 1999, p. 22]. Galileo did not have a theory about Jupiter's moons to test when he used his telescope, but the manipulations of the new technology offered a lot of new information. In these cases it is only later that theory is able to contribute new meanings to experimental results.

Following the so-called "new experimentalism" [Ackermann, 1989], we can say that "experiment" has a "life of its own" [Hacking, 1983], independent of theory. Hacking declares:

> Experimental work provides the strongest evidence for scientific realism. This is not because we test hypotheses about entities. It is because entities that in principle cannot be "observed" are regularly manipulated to produce new phenomena and to investigate other aspects of nature. They are tools, instruments not for thinking, but for doing (p. 262).

We are even able to manipulate the old "philosopher's favorite theoretical entity", the electron, and it is only in the early stages of our discovery of that entity, that we may merely test the hypothesis that it exists. We already said that a great part of the recent philosophy of science is theory-dominated: data is always considered as theory-laden. Many histories of scientific facts are written, in this light, to emphasize theory and disregard the experimental and technological aspects of research: experiments do not have an autonomous significance and the explanation of their characteristics, aims and results is made in terms of theoretical issues unknown to the experimenter. For instance: the experiment is considered significant only as a means to test a theory under scrutiny. Hacking provides an interesting analysis of Lakatos' treatment of Michelson's experiment: Hacking's description of this experiment tells us that it does not pursue any programme Lakatos writes about and it has a relative autonomy as regards theory. Classic positivism, pragmatism and kantism, the philosophies of science of Carnap, Popper, Lakatos, Feyerabend, Putnam, van Fraassen and others are characterized by a "single-minded obsession with representation and thinking and theory, at the expense of intervention and action and experiment" (p. 131).

Contrarily to a great part of the recent epistemological tradition, we have to follow Hacking and stress the attention on manipulative abduction and epistemic mediators also from the cognitive point of view. Creating an external cognitive support is very important to increase the possibility to get new information, to extend scientific knowledge, but also to improve and simplify many kinds of reasoning. Scientific thinking, like everyday thinking, has not to be viewed only like an internal speculative cognitive process, which occurs in a detached contemplation.

Hacking considers also the problem of realism by analyzing what we can use to intervene in the world to affect something else, or what the world can use to affect us. He shows, with the help of many interesting and sophisticated laboratory examples – some of them full of historical interest – that the significance of experiments sometimes has little to do with theory and representation. Entities whose causal powers are well understood are used as tools to investigate (and to intervene in) nature:

> Understanding some causal properties of electrons, you guess how to build a very ingenious complex device that enables you to line up the electrons the way you want, in order to see what will happen to something else. [...] Electrons are no longer ways of organizing our thoughts or saving the phenomena that have been observed. They are ways of creating phenomena in some other domain of nature. Electrons are tools (p. 263).

Concepts become tools endowed with absolutely unexpected outcomes. The experimentalists use various strategies for establishing the experimental effects without any recourse to theory. These strategies correspond to the expert manipulation of objects in a highly constrained experimental environment, we said directed by abductive movements that imply the application of old and new extra-theoretical *templates* of expert behavior. As possible creative organizations of experience some of them become in their turn hypothetical *interpretations* of experience, that is more theory-oriented, their reference is gradually stabilized in terms of established observational practices. Step by step the new interpretation relates to more "theoretical" modes of abductively understanding (visual, diagrammatic, symbolic, conceptual, simulative).

In this light it is not surprising that [Mayo, 1996], in her defense of experimentalism, has stressed attention to the possibility of delineating progress in science in terms of accumulation of experimental knowledge and expertise. She adds more arguments to the thesis of autonomy of experimental results illustrating many examples where the experiments are shown not as merely related to confirmation and falsification. In some cases they not only serve as a falsification of the assertion, but also to delineate new effects and ideas not previously known; moreover, they can bear on the comparison of radically different theories:[79] to resume, they can trigger revolutionary creative abductions, enabling us to learn from errors. To exemplify the positive role played by errors Mayo illustrates the famous case of the observation of the questionable features of Uranus's orbit that created problems for Newtonian theory: the detection of the source of this difficulty led to the discovery of Neptune.

1.7 Mirroring Hidden Properties through Optical Diagrams

It is well-known that in the whole history of geometry many researchers used internal mental imagery and mental representations of diagrams, but also self-generated diagrams (external) to help their thinking [Otte and Panza, 1999]. For example, it is clear that in geometrical construction many of the requirements indicated by the manipulative templates (cf. above subsection 1.6.2) are fulfilled. Indeed iconic geometrical constructions present situations that are curious and "at the limit". Because of their iconicity, they are constitutively dynamic, artificial, and offer various contingent ways of epistemic acting, like looking from different perspectives, comparing subsequent appearances, discarding, choosing, re-ordering, and evaluating. Moreover, they present the features typical of manipulative reasoning illustrated above,

[79] I already stressed at the beginning of this subsection the role played by construals of phenomena to overcome the problem of incommensurability.

such as the simplification of the task and the capacity to get visual information otherwise unavailable.

We have seen that manipulative abduction is a kind of abduction, usually model-based and so intrinsically "iconic", that exploits external models endowed with delegated (and often implicit) cognitive and semiotic roles and attributes. We can say that 1) the model (diagram) is *external* and the strategy that organizes the manipulations is unknown *a priori*; 2) the result achieved is *new* (if we, for instance, refer to the constructions of the first creators of geometry), and adds properties not contained before in the concept (the Kantian to "pass beyond" or "advance beyond" the given concept, [Kant, 1929, A154-B193/194, p. 192] I will describe in the following chapter, section 2.8).[80]

Hence, in the construction of mathematical concepts many external representations are exploited, both in terms of diagrams and of symbols. I am interested in my research in the diagrams which play various iconic roles: an *optical* role – microscopes (that look at the infinitesimally small details), telescopes (that look at infinity), windows (that look at particularly situation), a *mirror* role (to externalize rough mental models), and an *unveiling* role (to help to create new and interesting mathematical concepts, theories, and structures).[81] I also describe them as the *epistemic mediators* (cf. above) able to perform various abductive tasks (discovery of new properties or new propositions/hypotheses, provision of suitable sequences of models able to convincingly verifying theorems, etc.).[82]

An interesting epistemological situation I have recently studied is the one concerning the cognitive role played by some special epistemic mediators in the field of non-standard analysis, an "alternative calculus" invented by Abraham Robinson [1966], based on infinitesimal numbers in the spirit of Leibniz method.[83] It is a kind of calculus that uses an extension of the real numbers system $\mathbb{R}$ to the system $\mathbb{R}^*$ containing infinitesimals smaller in the absolute value than any positive real number. I maintain that in mathematics diagrams play various roles in a typical abductive way. Two of them are central:

- they provide an intuitive and mathematical abductive *explanation* facilitating the understanding of concepts difficult to grasp, that appear hidden, obscure, and/or epistemologically unjustified, or that are *not expressible* from an intuitive point of view;
- they help *create* new previously unknown concepts, playing a non-explanatory abductive role, as I will illustrate in the following chapter of this book.

[80] Other interesting applications of the concept of abduction in mathematical discovery and in the manipulation of symbols are illustrated in [Heeffer, 2007; Heeffer, 2009]. On the Cardano's abductive discovery of negative numbers and negative solution to a linear problem cf. [Heeffer, 2007].

[81] The epistemic and cognitive role of mirror and unveiling diagrams in the discovery of non-Euclidean geometry is also illustrated in [Magnani, 2002a]. Cf. also the following chapter, sections 2.9 and 2.11.

[82] Elsewhere I have presented some details concerning the role of optical diagrams in the calculus [Magnani and Dossena, 2005; Dossena and Magnani, 2007].

[83] Further updated details concerning Leibniz's mathematics and philosophy of infinitesimals are illustrated in [Mancosu, 1996].

Optical diagrams play a fundamental explanatory (and didactic) role in removing obstacles and obscurities and in enhancing mathematical knowledge of critical situations. They facilitate new internal representations and new symbolic-propositional achievements. In the example I have studied in the area of the calculus, the extraordinary role of the optical diagrams in the interplay standard/non-standard analysis is emphasized. In the case of our non-standard analysis examples, some new diagrams (microscopes within microscopes) provide new mental representations of the concept of tangent line at the infinitesimally small regions. Hence, external representations which play an "optical" role can be used to provide us with a better understanding of many critical mathematical situations and, in some cases, to more easily discover (or rediscover) sophisticated properties.

The role of an "optical microscope" that shows the behavior of a tangent line is illuminating. In standard analysis, the change dy in y along the tangent line is only an approximation of the change Δy in y along the curve. But through an optical microscope, that shows infinitesimal details, we can see that $dy = \Delta y$ and then the quotient $\Delta y/\Delta x$ is the same of dy/dx when $dx = \Delta x$ is infinitesimal (see Figure 1.10 and, for more details, [Magnani and Dossena, 2005]). This removes some difficulties of the representation of the tangent line as limit of secants, and introduces a more intuitive conceptualization: the tangent line "merges" with the curve in an infinitesimal neighborhood of the contact point.

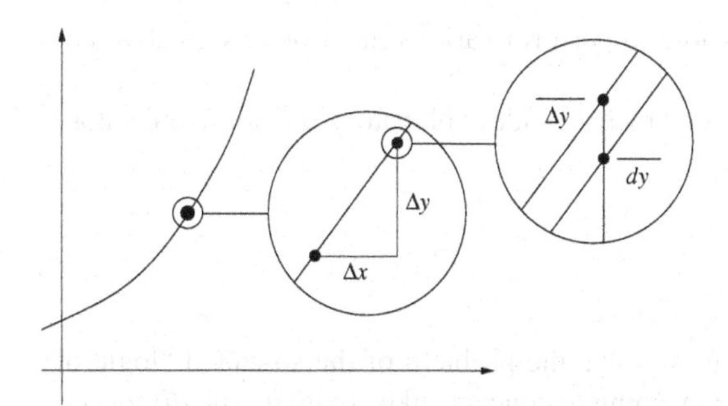

Fig. 1.10 An optical diagram shows an infinitesimal neighborhood of the graph of a real function

Only through a second more powerful optical microscope "within" the first (I call this kind of epistemic mediators *microscopes within microscopes*) (again, see Figure 1.10), we can see the difference between the tangent line and the curve. Under the first diagram, the curve looks like the graph of

$$f'(a)x,$$

i.e., a straight line with the same slope of its tangent line;[84] under the second, the curve looks like

[84] This is mathematically justified in [Magnani and Dossena, 2005].

$$f'(a)x - \frac{1}{2}f''(a).$$

This suggests nice new mental representations of the concept of tangent lines: through the optical lens, the tangent line can be seen as the curve, but through a more powerful optical lens the graph of the function and the graph of the tangent are distinct, straight, and parallel lines. The fact that one line is either below or above the other, depends on the sign of $f''(a)$, in accordance with the standard real theory: if $f''(x)$ is positive (or negative) in a neighborhood, then f is convex (or concave) here and the tangent line is below (or above) the graph of the function.

However, this easily mirrors a sophisticated *hidden* property. Let f be a two times differentiable function and let a be a flex point of it. Then $f''(a) = 0$ and so the second microscope shows again the curve as the same straight line: this means that the curve is "very straight" in its flex point a. Of course, we already know this property – the curvature in a flex point of a differentiable two times function is null – which comes from standard analysis, but through optical diagrams we can find it immediately and more easily (the standard concept of curvature is not immediate).

To conclude, I have already noted that some diagrams could also play an unveiling role, providing new light on mathematical structures: it can be hypothesized that these diagrams can lead to further interesting creative results.

I stated that in mathematics diagrams play various roles in a typical abductive way. We can add that:

- they are *epistemic mediators* able to perform various more or less creative abductive tasks in so far as
- they are *external representations* which provide explanatory and non-explanatory abductive results.

Summary

In this chapter we have seen that, to solve the problem of the so-called "logic of discovery", we need to clarify the meaning of concepts like *creativity* and *discovery*. Following Peircean ideas, I have stressed that recent computational modeling is very useful in a strictly pragmatical sense. We can produce and implement actual and then possible, *rational* models of creative reasoning and scientific discovery. In this intellectual framework a new paradigm, aimed at unifying the different perspectives, is provided by the fundamental concept of *abduction*. Many "working" abductive processes can be found and studied that are rational, unambiguous and perfectly communicable. I have maintained that the concepts of *sentential*, *model-based* and *manipulative* abduction are important not only in delineating the actual practice of abduction but also in the development of computationally adequate programs used to rediscover, or discover for the first time, for example, scientific hypotheses and mathematical theorems or laws.

It is clear that the manipulation of external objects helps human beings in their creative tasks. I have illustrated the strategic role played by the so-called traditional concept of "implicit knowledge" in terms of the cognitive and epistemological concept of manipulative abduction, considered as a particular kind of abduction that exploits external models endowed with delegated (sometimes implicit) cognitive roles and attributes. Abductive manipulations operate on models that are external and the strategy that organizes the manipulations is unknown and *a priori*. In the case of creative manipulations of course the result achieved is also *new* and adds properties not previously contained in the premises of reasoning. For example, in scientific practice there are many procedural, extra-sentential and extra-theoretical aspects indispensable to providing knowledge and information which are otherwise hard to grasp. By making them explicit we can rationally and positively integrate the previously existing scientific encyclopaedia. Enhancement of analysis of these important human skills can increase knowledge on inferences involving creative, analogical, spatial and simulative aspects, both in science and everyday situations, thereby extending epistemological, computational, and psychological theory. It is from this point of view that I have also described what I have called *epistemic mediators* and the *templates* of epistemic doing. These are able to illustrate the first features of the performance of various abductive tasks and refer to various external tools (and their manipulation) that can be built to cognitively help the activity of scientists.

At the end of the chapter I have also described some results in the specific domain of calculus, were diagrams which play an optical role such as microscopes, "microscopes within microscopes", telescopes, and windows, a mirror role (to externalize rough mental models) or an unveiling role (to help create new and interesting mathematical concepts, theories, and structures) are studied. They play the role of epistemic mediators able to perform the explanatory abductive task of endowing difficult mathematical concepts with new "intuitive" meanings and of providing a better understanding of the calculus through a non-standard model of analysis. I also maintain that they can be used in many other different epistemological and cognitive situations (other examples in the field of mathematical discovery will be given in the following chapter, section 2.8. [85]

[85] Another interesting application of the concept of manipulative abduction I have studied is in the area of chance discovery [Magnani *et al.*, 2002b]: concrete manipulations of the external world constitute a fundamental passage in chance discovery. By a process of manipulative abduction it is possible to build prostheses that furnish a kind of embodied and unexpressed knowledge that plays a key role in the subsequent processes of scientific comprehension and discovery but also in the extraction of the "unexpected" in ethical thinking and in moral deliberation.).

Chapter 2
Non-explanatory and Instrumental Abduction
Plausibility, Implausibility, Ignorance Preservation

In chapter one I have illustrated the basic distinction between theoretical and manipulative abduction and the other main features of abductive cognition. Further important cognitive and logico-epistemological considerations have to be added. First of all the fact that abduction is a procedure in which something that lacks classical explanatory epistemic virtue can be accepted because it has virtue of another kind: [Gabbay and Woods, 2005] contend that abduction presents an *ignorance preserving* (but also an *ignorance mitigating*) character. From this perspective abductive reasoning is a *response* to an ignorance-problem; through abduction the basic ignorance – that does not have to be considered a total "ignorance" – is neither solved nor left intact. Abductive reasoning is an ignorance-preserving accommodation of the problem at hand.

The chapter also introduces the basic distinction between the schematic representation of abduction which I have illustrated in chapter one (section 1.3), that [Gabbay and Woods, 2005] call *AKM*-schema, and the *GW*-schema, they propose in their recent book on abduction. The analysis and criticism of the *GW*-schema provides an opportunity to illustrate *non-explanatory* and [radical] *instrumentalist* aspects of abductive cognition. Examples of the non-explanatory features of abduction are present in logic and mathematical reasoning. The chapter gives an analysis of how the importance of non-explanatory abduction in logical and mathematical reasoning is clearly even if implicitly envisaged by Gödel. Furthermore, physics often aims at discovering physical dependencies which can be considered explanatorily undetermined. In this case abduction exhibits an *instrumental* aspect. In section 2.6 I contend that this character is sometimes related to the conventional nature of the involved hypotheses.

The new non-explanatory and instrumentalist aspects of abduction in turn lead to reconsideration, in a broader sense, of the role of *plausibility* in abductive cognition. Plausibility which occurs at the level of basic inner inferences of the real human agent is related to considerations of relevance or how characteristic a certain behavior is and can be called [Gabbay and Woods, 2005, p. 209] "propositional". However, there is also a "strategic" sense of plausibility that has to be taken into account, the one which occurs in the case of instrumental abduction, where

L. Magnani: Abductive Cognition, COSMOS 3, pp. 63–143.
springerlink.com

plausibility is no longer linked to standard characteristicness. For example, in scientific reasoning, an abductive hypothesis can be highly implausible from the "propositional" point of view and nevertheless it can be adopted for its instrumental virtues, such as in the Newtonian case of action-at-a-distance. Highly implausible hypotheses from the "propositional" point of view can be conjectured because of their high "instrumental" plausibility, where a different role of characteristicness is at stake.

The chapter further examines interesting cases of abduction that can be usefully labeled non-explanatory and/or instrumentalist, showing how often the abductive procedures are characterized by a mixture of explanatory, non-explanatory and instrumental aspects. The analysis of these cases is also an opportunity to demonstrate that contradictions and inconsistencies are fundamental in abductive reasoning, especially in science, and that abductive reasoning is appropriate for "governing" inconsistencies, both at the empirical and theoretical/conceptual level. The importance of the role of the so-called "preinventive forms" is also addressed.

Hence, contradiction is fundamental in abductive reasoning and it has a preference for strong hypotheses which are more easily falsified than weak ones. Moreover, hard hypotheses may be more easily weakened than weak ones, which subsequently prove more difficult to strengthen. Hypotheses may however be *unfalsifiable*, such as in the case of hypotheses which are fruit of "radical" instrumentalist abduction. In this case, it is impossible to find a contradiction from the empirical point of view but also from the theoretical point of view, in some area of the related conceptual systems. Notwithstanding this fact, it is sometimes necessary to construct methods for rejecting the unfalsifiable hypothesis at hand by resorting to some external form of negation. It would have to be an "external" criterion in order to avoid any arbitrary and subjective elimination, which could be rationally or epistemologically unjustified.

I contend that this problem of unfalsifiable hypotheses is strictly linked to the issue of non-explanatory and instrumentalist abduction, taking advantage of the analysis of hypothesis withdrawal in Freudian analytic reasoning and in Poincaré's conventionalism of the principles of physics. From the point of view of instrumentalist abduction my example concerning the conventional principles of physics shows a cognitive situation where such hypotheses are not subject to discharge except for their instrumental value, since the abduced hypothesis, fruit of a *radically instrumentalist* abduction, fails all tests that would reveal it as having the requisite classical epistemic value.

The last part of the chapter is devoted to illustrating the problem of the extra-theoretical dimension of cognition from the perspective of the famous discovery of non-Euclidean geometries. This case study is particularly appropriate to the present chapter because it shows relevant aspects of diagrammatic abduction, which involve intertwined processes of both explanatory and non-explanatory abduction acting at the model-based level in what I call *mirror* and *unveiling* diagrams. Finally, the section 2.7 also deals with the epistemologically very interesting computational AI applications that involve the abductive processes in scientific discovery and mathematical reasoning and creativity: some programs expressly devoted to the simulation of geometrical reasoning are illustrated.

2.1 Is Abduction an Ignorance-Preserving Cognition?

2.1.1 The Ignorance Preserving Character of Abduction

It is clear that "[...] abduction is a procedure in which something that lacks epistemic virtue is accepted because it has virtue of another kind" [Gabbay and Woods, 2005, p. 62]. For example: "Let S be the standard that you are not able to meet (e.g., that of mathematical proof). It is possible that there is a lesser epistemic standard S' (e.g., having reason to believe) that you do meet" [Woods, 2010, chapter eight]. Focusing attention on this cognitive aspect of abduction [Gabbay and Woods, 2005] contend that abduction (basically seen as a *scant-resource* strategy, which proceeds in absence of knowledge) presents an *ignorance preserving* (but also an *ignorance mitigating*) character. Of course "[...] it is not at all necessary, or frequent, that the abducer be wholly in the dark, that his ignorance be total. It needs not be the case, and typically isn't, that the abducer's choice of a hypothesis is a blind guess, or that nothing positive can be said of it beyond the role it plays in the subjunctive attainment of the abducer's original target (although sometimes this is precisely so)" (cit.). Abductive reasoning is a *response* to an ignorance-problem: one has an ignorance-problem when one has a cognitive target that cannot be attained on the basis of what one currently knows. Ignorance problems trigger one or other of three responses. In the first case, one overcomes one's ignorance by attaining some additional knowledge (subduance). In the second instance, one yields to one's ignorance (at least for the time being) (surrender). In the third instance, one abduces [Woods, 2010, chapter eight] and so has a positive and reasoned basis for new action even if in the presence of the constitutive ignorance.

From this perspective the general form of an abductive inference can be rendered as follows. Let α be a proposition with respect to which you have an ignorance problem. Putting T for the agent's epistemic target with respect to the proposition α at any given time, K for his knowledge-base at that time, K^* for an immediate accessible successor-base of K that lies within the agent's means to produce in a timely way,[1] R as the attainment relation for T, $\rightsquigarrow$ as the *subjunctive* conditional relation, H as the agent's hypothesis, $K(H)$ as the revision of K upon the addition of H, $C(H)$ denotes the conjecture of H and H^c its activation. The general structure of abduction can be illustrated as follows (GW-schema, cf. below subsection 2.1.3):

[1] "K^* is an accessible successor to K to the degree that an agent has the know-how to construct it in a timely way; i.e., in ways that are of service in the attainment of targets linked to K. For example if I want to know how to spell 'accommodate', and have forgotten, then my target can't be hit on the basis of K, what I now know. But I might go to my study and consult the dictionary. This is K^*. It solves a problem originally linked to K" [Woods, 2010, chapter eight].

1. $T!\alpha$	[setting T is an epistemic target with respect to a proposition α]
2. $\neg(R(K,T)$	[fact]
3. $\neg(R(K^*,T)$	[fact]
4. $H \notin K$	[fact]
5. $H \notin K^*$	[fact]
6. $\neg R(H,T)$	[fact]
7. $\neg R(K(H),T)$	[fact]
8. If H were true then it would be the case that $R(K(H),T)$	[fact]
9. H meets further conditions $S_1,....S_n$	[fact]
10. Therefore, $C(H)$	[sub-conclusion, 1-9]
11. Therefore, H^c	[conclusion, 1-10]

It is easy to see that the distinctive epistemic feature of abduction is captured by the schema. It is a given that H is not in the agent's knowledge-set. Nor is it in its immediate successor. Since H is not in K, then the revision of K by H is not a knowledge-successor set to K. Even so, $H \rightsquigarrow (K(H),T)$. So we have an ignorance-preservation, as required (cf. [Woods, 2010, chapter eight]).

[*Note*: Basically, line 9. indicates that H has no more plausible or relevant rival constituting a greater degree of subjunctive attainment. Characterizing the S_i is the most difficult problem for abductive cognition, given the fact that in general there are many possible candidate hypotheses. It involves for instance the consistency and minimality constraints (lines 4 and 5 of the standard schema below in subsection 2.1.3, p. 70). I will illustrate below (cf. subsection 2.3.1) that, in the case of inner processes in organic agents, this process is largely implicit, and so also linked to unconscious ways of inferring, or, in Peircean terms, to the activity of the instinct [Peirce, 1931-1958, 8.223] and of what Galileo called the *lume naturale* [Peirce, 1931-1958, 6.477], that is the innate fair for guessing right.

However, in more hybrid and multimodal (not merely inner) abductive processes (cf. chapter four, section 4.1), such as in the case of manipulative abduction, the assessment is reached – and constrained – taking advantage of the gradual acquisition of further external information with respect to future interrogation and control, even if not due to actual experimental tests. At least *three* kinds of actions are involved in these abductive processes (and we would have to also take into account the motoric aspect of inner "thoughts" too, cf. chapter four, section 4.4, p. 233). In this interplay the cognitive agent further triggers internal *thoughts* "while" modifying the environment and so (*i*) acting on it (thinking through doing). In this case the "motor actions" directed to the environment have to be intended as part and parcel of the whole embodied abductive inference, and so have to be distinguished from the final (*ii*) "actions" as fruit of the reached abductive result. In this perspective the proper experimental test involved in the Peircean evaluation phase, which for many researchers reflects in the most acceptable way the idea of abduction as inference to the best explanation, just constitutes a *special* subclass of the processes – which involve another kind (*iii*) of actions – of adoption of abductive hypotheses, and should be considered ancillary to the nature of abductive cognition, and inductive in its essence.

Finally, $C(H)$ is read "It is justified (or reasonable) to conjecture that H" and H^c is its activation, as the basis for *planned* "actions", in the sense I have just illustrated.]

In sum, T cannot be attained on the basis of K. Neither can it be attained on the basis of any successor K^* of K that the agent knows then and there how to construct. H is not in K: H is a hypothesis that when reconciled to K produces an updated $K(H)$. H is such that if it were true, then $K(H)$ would attain T. The problem is that H is *only hypothesized*, so that the truth is not assured. Accordingly Gabbay and Woods contend that $K(H)$ *presumptively* attains T. That is, having hypothesized that H, the agent just "presumes" that his target is now attained. Given the fact that presumptive attainment is not attainment, the agent's abduction must be considered as preserving the ignorance that already gave rise to her (or its, in the case for example of a machine) initial ignorance-problem. Accordingly, abduction does not have to be considered the "solution" of an ignorance problem, but rather a response to it, in which the agent reaches presumptive attainment rather than actual attainment. $C(H)$ expresses the conclusion that it follows from the facts of the schema that H is a worthy object of conjecture. It is important to note that in order to solve a problem it is not necessary that an agent actually conjectures a hypothesis, but it is necessary that she states that the hypothesis is *worthy of conjecture*.

It is remarkable that in the above schema

> [...] $R(K(H),T)$ is false and yet that $H \rightsquigarrow (K(H),T)$ is true. Let us examine a case. Suppose that your target T is to know whether α is true. Suppose that, given your present resources, you are unable to attain that target. In other words, neither your K nor your K^* enables you to meet your target. Let H be another proposition that you don't know. So $K(H)$ is not a knowledge-set for you. On the principle that you can't get to know whether α on the basis of what you don't know, $K(H)$ won't enable you to attain T either. This is a point of some subtlety. Pages ago, weren't we insisting that there are contexts – autoepistemic contexts – in which not knowing something is a way of getting to know something else? No, we said that not knowing something was a way of getting to presume something else. But just to be clear, let us point out that in the GW-schema α and H are not candidates for the autoepistemic inference of α from H or $K(H)$. So $R(K(H),T)$ is false. $H \rightsquigarrow (K(H),T)$ is different. It says, subjunctively, that if H were true, then the result of adding H to K would attain T. Clearly this can be true while, for the same H, K and T, $R(K(H),T)$ is false [Woods, 2010, chapter eight].

Finally, considering H justified to conjecture is not equivalent to considering it justified to send H to trial. H^c denotes the *decision* to release H for further premissory work in the domain of enquiry in which the original ignorance-problem arose, that is the activation of H as a positive basis for action. Woods usefully observes:

> There are lots of cases in which abduction stops at line 10, that is, with the conjecture of the hypothesis in question but not its activation. When this happens, the reasoning that generates the conjecture does not constitute a positive basis for new action, that is, for acting *on* that hypothesis. Call these abductions *partial* as opposed to full. Peirce has drawn our attention to an important subclass of partial abductions. These are cases in which the conjecture of H is followed by a decision to submit it to experimental test.

> Now, to be sure, doing this is an action. It is an action *involving* H but it is not a case of acting *on* it. In a full abduction, H is activated by being released for inferential work in the domain of enquiry within which the ignorance-problem arose in the first place. In the Peircean cases, what counts is that H is withheld from such work. Of course, if H goes on to test favourably, it may then be released for subsequent inferential engagement [Woods, 2009].

We have to remember that this last process is not abductive, it is inductive, as Peirce contended and I illustrated in the previous chapter. Woods adds: "Now it is quite true that epistemologists of a certain risk-averse bent might be drawn to the admonition that partial abduction is as good as abduction ever gets and that complete abduction, inference-activation and all, is a mistake that leaves any action prompted by it without an adequate rational grounding. This is not an unserious objection, but I have no time to give it its due here. Suffice it to say that there are real-life contexts of reasoning in which such conservatism is given short shrift, in fact is ignored altogether. One of these contexts is the criminal trial at common law" [Woods, 2009].

Here it cannot be said that testability is intrinsic to abduction, such as in the case of the ST-model illustrated above in subsection 1.4 of chapter one, and in the case of some passages of Peirce's writings.[2] This action, which in turn involves degrees of risk proportioned to the strength of the conjecture, is strictly cognitive/epistemic and inductive in itself, for example an experimental test, is an intermediate step to release the abduced hypothesis for inferential work in the domain of enquiry within which the ignorance-problem arose in the first place.

Through abduction the basic ignorance – that does not have to be considered total "ignorance" – is neither solved nor left intact: it is an ignorance-preserving accommodation of the problem at hand. As I have already stressed, in a defeasible way, further action can be triggered either to find further abductions or to "solve" the ignorance problem, possibly leading to what it is usually called the inference to the best explanation. It is clear that in this framework the inference to the best explanation – if considered as a truth conferring achievement – cannot be a case of abduction, because abductive inference is constitutively ignorance preserving. In this perspective the inference to the best explanation also involves – for example – the generalizing and evaluating role of *induction*. Of course it can be said that the requests of originary thinking are related to the depth of the abducer's ignorance.

2.1.2 Truth Preserving and Ignorance Preserving Inferences

From an agent-based perspective on logic similar to one adopted by Gabbay and Woods,[3] which pays attention to psychological/subjective aspects and strongly stresses the ignorance preserving character of the logic of abduction, Jean Yves Girard's annotations about the logics which model abduction can acquire further clarification. Girard says "'Epistemic' logics are supposed to illustrate 'abductive'

[2] When abduction stops at line 10., the agent is not prepared to accept $K(H)$, because of supposed adverse consequences.

[3] On this agent-based perspective cf. also chapter seven, this book.

principles of reasoning: from the fact that we don't know, one deduces something. [...] Epistemic logics are based upon the identification between 'not to know' and 'to know not'. If such an identification would be possible, it would suffice to add relevant axioms. One sees that this is impossible, since then, G being not provable, this fact would be provable (what is expressed by G), contradiction" [Girard, 2006, pp. 104–105]. Girard refers to the inconsistency of these systems as a corollary of Gödel theorem, and further stresses the problems created by the nonmonotonic logics of abduction, which, he says, without any hesitation, add a principle of the sort if $\neg A$ is not provable, then A is provable". The conclusion is strong:

> Granted adequate precautions, these systems are consistent and complete, so what do you complain about? They are simply non-deductive, because there is no way to activate the additional principle. One is no longer dealing with a formal system, since there is no way to know that something is not provable (this is already wrong in a deductive system, so in such a doohickey, good luck!). By the way, let us directly refute the algorithm analog of "non-monotonicity". One wants to answer any query and the "solution" is as follows: if the algorithm yields an answer, say "yes", answer "yes", if it says "no", answer "no", if it keeps silent answer whatever you like, "yes" or "no", nay "I don't know". This is impossible, because Turing's undecidability of the halting problem precisely tells us that there is no algorithmic way of knowing that one does not know [Girard, 2006, p. 105].

In sum, it seems that entering nonmonotonic logic of abduction means leaving deduction and demonstrative reasoning, at least in Girard's rigid sense. The ignorance preserving approach to abduction and its logics de facto acknowledges Girard's conclusion regarding the entire old-fashioned perspective on deductive logic: "What we logicians manipulate under the name true" is but an empty shell. A last word: one should not forget either that Gödel formula, this over-ornate artifact, before meaning 'I am not provable', says 'I mean nothing'" [Girard, 2006, pp. 106]. Indeed, the ignorance preserving approach permits us to further appreciate Girard's view. Following Girard, we can say that, in a sense, abductive logics like the nonmonotonic ones, certainly formalize ignorance preserving inferences, but the truth preserving character of deductive logic is achieved at the cost of skipping the problem of the multifaceted ignorance of cognitive agents. Truth preserving inferences involve a kind of ideal *intersubjectivity*, that is the fact that certain propositions $A, B, \ldots$ "share the same point of view" [Girard, 2007, p. 64], and it is at this price that deductive logic can say that the truth of A and of $A \rightarrow B$ implies the truth of B.

"In terms of intersubjectivity, logic still testimonies the caprices of 'common knowledge': no doubts about the fact that at this level the constitution of the subject is not admitted, everything is already available, and it is only necessary to select dispersed pieces of information" [Girard, 2007, pp. 64–65]. Nevertheless, the subjective side of truth implies that a theorem can also be true or false depending on the point of view. The idea of making cooperative A and B implies that they share something at the level of sense. This is basically (inter)subjective and expressed by a shared perspective on truth: it is with respect to this common point of view that truth is preserved. This subtle relativity to a "point of view" can be extended to the case of abduction: if deductive logic preserves truth, abduction preserves ignorance, but in both cases the preservation depends on some common contingent perspectives respectively on truth and ignorance.

2.1.3 AKM and GW Schemas of Abduction

The schematic representation of abduction I have illustrated in chapter one (section 1.3) expresses what [Gabbay and Woods, 2005] call *AKM*-schema,[4] which is contrasted to their own (*GW*-schema), which I have basically explained in the subsection 2.1.1 above.

The *AKM* can be illustrated as follows:

1. E
2. $K \not\hookrightarrow E$
3. $H \not\hookrightarrow E$
4. $K(H)$ is consistent
5. $K(H)$ is minimal
6. $K(H) \hookrightarrow E$
7. Therefore, H.

[Gabbay and Woods, 2005, pp. 48–49].

where of course the conclusion operator $\hookrightarrow$ cannot be classically interpreted.

Consequentialism and *explanationism* are the two main characters of this schema, which certainly grasps fundamental aspects of abduction. The target has to be an explanation and $K(H)$ bears R^{pres} [that is the relation of presumptive attainment] to T only if there is a proposition V and a consequence relation $\hookrightarrow$ such that $K(H) \hookrightarrow V$, where V represents a *payoff proposition* for T. In turn, in this schema explanations are interpreted in consequentialist terms. If E is an explanans and E' an explanandum the first explains the second only if (some authors further contend if and only if) the first implies the second. It is obvious to add that the *AKM* schema embeds a D-N (deductive-nomological) interpretation of explanation, as I have already stressed in [Magnani, 2001b, p. 39]. Moreover, the fact that abduction is a procedure in which something that lacks traditional epistemic virtue (at least in the sense epistemic virtues are intended in the epistemological tradition) is accepted because it has virtue of another kind, that is its ignorance preservation, (cf. above subsection 2.1.1) is tacitly recognized in the *AKM* model (and also in Peirce's philosophy), but only explicitly illustrated in the *GW* one.

I have said in the first chapter of this book that the *AKM* schema embeds the D-N model (the Hempel's *law covering model* of scientific explanation [Hempel, 1966]): an additional remark has to be added. The D-N model presents a classical logical structure and consequently the idea of explanation it involves is not in itself abductive, as clearly pointed out by Gabbay and Woods

[4] For A they refer to Aliseda [Aliseda, 1997; Aliseda, 2006], for K to Kowalski [Kowalski, 1979], Kuipers [Kuipers, 1999], and Kakas *et al.* [Kakas *et al.*, 1993], for M to Magnani [Magnani, 2001b] and Meheus [Meheus *et al.*, 2002].

> If explanation is taken in the D-N sense, then no successful explanationist abduction can embody the D-N notion of explanation unless the explanationism in question is subjunctive. [...] The explanations it captures also stand or fall independently of the epistemic states or interests of any agent (so the epistemic version is not intrinsic to the model). But when abduction enters the picture, it does so with the requisite structure and the necessary impairments or omissions of the agent's K-set. The requirements necessary for the explanationist hitting of an abductive target T make it impossible for the abducer to produce a D-N explanation of the state of affairs embraced by this target. The best he can do in this regard is produce a subjunctive D-N explanation, putting it loosely [Gabbay and Woods, 2005, p. 93].

However, this requirement is not sufficient: "Even so, it is a fateful turn. [...] There are senses of explanation for which the constraints of subjunctivity are redundant. Another way of saying this is that there are conceptions of explanations that are themselves subjunctivist in character, and for which, therefore, the subjunctivizing consequences of abductive employment amount to the transportation of coal to Newcastle" [Gabbay and Woods, 2005, p. 93].

2.2 Non-explanatory Abduction

[Gabbay and Woods, 2005] contend – and I agree with them – that abduction *is not intrinsically explanationist*, like for example its description in terms of inference to the best explanation would suggest. Not only that, abduction can also be merely *instrumental*. This conviction constitutes the main reason for proposing the GW-schema, which offers a representation of abductive cases not captured by that of the AKM-schema. In the following sections and subsections I will describe some non-explanatory (and instrumental) aspects of abduction that are explicitly acknowledged by the GW schema. In my previous book on abduction [Magnani, 2001b] I made some examples of abductive reasoning that basically are non-explanatory and/or instrumentalist without clearly acknowledging it. The contribution of Gabbay and Woods to the analysis of abduction has the logical and epistemological merit of having clarified these basic aspects of abduction, until now disregarded in the literature. Their distinction between explanatory, non-explanatory and instrumental abduction is orthogonal to mine in terms of the theoretical and manipulative (including the subclasses of sentential and model-based) and further allows us to explore fundamental features of abductive cognition.

Hence, if we maintain that E explains E' *only if* the first implies the second, certainly the reverse does not hold. This means that various cases of abduction are consequentialist but not explanationist [other cases are neither consequentialist nor explanationist]:

> It merits emphasis that not all T's either *specify* or have *payoff* propositions. If, for example, the target is to justify a recondite principle of logic L, it may suffice to produce a derivation of some obvious proposition of arithmetic A in which that logical principle occurs non-redundantly as premisses. Following Russell [...] we might well take this as grounds on which to hypothesize that the recondite principle L is indeed

justified. But it is as well to note that nowhere in this scenario is there any question that the abduction requires (or permits) that L itself is a payoff proposition for T or that L is in the counterdomain of any consequence relation on display in the abduction. Let the proof that does not deliver the goods for A be schematized as

$$\begin{array}{c} P_1 \\ \vdots \\ P_n \\ \hline A \end{array}$$

Assume now that if L is added as premiss, the proof goes through. In other words, whereas $\{P_1, ..., P_n\}$ doesn't suffice for A, $\{P_1, ..., P_n, L\}$ does. For this to be so, there must be a consequence relation on $\langle\{P_1, ..., P_n, L\}, A\rangle$. But A is not the payoff for T. Rather $\{P_1, ..., P_n, K\}$ is. And this is itself neither a proposition nor the consequent of any consequence relation of which the abduction must take note. We repeat: sometimes T has a payoff proposition; sometimes it specifies this proposition; and sometimes this proposition is required to be in the counterdomain of a consequence relation the abduction must take note of. When these facts obtain, it is essential that the abductive enterprise take them into account. When they do not obtain, there is nothing to take into account; and no schema should posit them unduly [Gabbay and Woods, 2005, pp. 51-52].

Non-explanatory modes of abduction are clearly exploited in the "reverse mathematics" pioneered by Harvey Friedman and his colleagues, e.g., [Friedman and Simpson, 2000], where propositions can be taken as axioms because they support the axiomatic proofs of target theorems. The target of reverse mathematics is to answer this fundamental question: What are the appropriate axioms for mathematics? The problem is to discover which are the appropriate axioms for proving particular theorems in central mathematical areas such as algebra, analysis, and topology (cf. [Simpson, 1999]). The idea of reverse mathematics originates with Russell's notion of the regressive method in mathematics [Russell, 1973], and is also present in some remarks of [Gödel, 1944; Gödel, 1990a].[5] [Gabbay and Woods, 2005, p. 128] conclude, following Russell, that regressive abduction is both instrumental and non-explanatory, and quote a Gödel's passage, which confirms their statement:

[...] even disregarding the intrinsic necessity of some new axiom, and even in case it has no intrinsic necessity at all, a probable decision about its truth is possible also in another way, namely inductively by studying its "success". Success here means fruitfulness in consequences, in particular, "verifiable" consequences, i.e., consequences demonstrable without the new axioms, whose proofs with the help of the new axiom, however, are considerably simpler and easier to discover, and make it possible to contract into one proof many different proofs [Gödel, 1990a, pp. 476–477].[6]

[5] For more details about this, see [Irvine, 1989], who also compares Russell's regressive method to Peirce's abduction.

[6] I will further illustrate Gödel's implicit acknowledgment of the role of abduction in logic in the following subsection.

Furthermore, often in physics the target is the discovery of physical dependencies which [Gabbay and Woods, 2005, pp. 122–123] consider explanatorily undetermined. In this case abduction can exhibit an *instrumental* aspect.[7] I will contend in section 2.6 below that this character is sometimes related to the conventional nature of the involved hypotheses. Moreover, also in many AI approaches based on logic programming and belief revision (cf. above, chapter one, subsections 1.4 and 1.4.1) explanationism tends to disappear and abduction is mainly considered as proof theoretic and algorithmic: "On this view, an H is legitimately dischargeable to the extent to which it makes it possible to prove (or compute) from a database a formula not provable (or computable) from it as it is currently structured. This makes it natural to think of AI-abduction in terms of belief-revision theory, of which belief-revision according to explanatory force is only a part" [Gabbay and Woods, 2005, p. 88]. However, the explanatory character is subsumed in these AI approaches as a philosophical conception.

2.2.1 Gödel and Abduction

I aim at further illustrating that the importance of non-explanatory abduction in logic and mathematics is clearly envisaged by Gödel. In the two essays of 1944 and 1947 "Russell's Mathematical Logic" and "What is Cantor's Continuum Problem?" [Gödel, 1944; Gödel, 1990b] and in the "Supplement" to the Second Edition of the 1947 essay [Benacerraf and Putnam, 1964, pp. 258–273], Gödel's ontology reveals itself as a space for reflection which allows us to rescue Cantor's classical analysis and set theory and to account for a possible expansion of the latter.

Gödel's ontology is articulated into a conceptual strategy with three levels of analysis: 1) the ontological extent of formal systems and mathematics; 2) the analogy between mathematics and logic on the one hand and natural sciences on the other; 3) the elaboration of an original notion of mathematical intuition to be understood as an activity of the constitution of mathematical objects, which we can interpret as essentially abductive.

Gödel considers "[...] mathematical objects to exist independently of our constructions and our having an intuition of them individually" ("What is Cantor's...", in [Benacerraf and Putnam, 1964, p. 262]). Mathematical entities, the objects of mathematical reality, are those studied by that area of mathematics called mathematical logic, i.e. "[...] classes, relations, combinations of symbols, etc., instead of numbers, functions, geometric figures, etc." ("Russell's Mathematical...", in [Benacerraf and Putnam, 1964, p. 211]. Moreover, all those objects from other areas of mathematics which one tries to derive "[...] actually from a very few logical concepts and axioms" (*ibid.*, p. 212) are entities too, such as for example, those of the *Principia* and in general those contained in formal systems meant to represent the entities themselves, that is the objects of mathematical reality.

[7] Other authors, and myself in [Magnani, 2001b, p. 17], disregarded this aspect and rather thought in this case we deal with a different kind of explanation, i.e simple non causal explanation.

We can, therefore, maintain that logical and mathematical objects transcend their own representation. Gödel is obviously interested in spotting the ontological constitution of the entities of mathematical logic: this is, by the way, of such great importance because not only does mathematical logic aim at achieving a representation and a foundation of mathematics, but "[...] it is a science prior to all others, which contains the ideas and principles underlying all sciences" (*ibid.*, p. 211).

Seeking a theory of knowledge adequate to address the issue of a Platonist-realist ontology, Gödel develops an analogy between logic and natural sciences already found in Russell's *Introduction to Mathematical Philosophy* [Russell, 1919]. Gödel finds in Russell the comparison between the axioms of logic and mathematics and the laws of nature and consequently the comparison between logical evidence (which [Wang, 1987, p. 303], calls "semi-perception"; for Gödel, in fact, logical evidence is "something like a perception") and sense perception. In this way, axioms do not appear to receive their evidence immediately themselves but, rather, their justification lies (exactly as is the case of hypotheses in physics) in the fact that they render possible an inference of more elementary "logical evidence", i.e., somehow relatable to sense perceptions. We can now make explicit that this inference is a kind of non-explanatory and instrumental abduction (plausibility at play is a mixed one, propositional and strategic, see the following section). Gödel's concept of the axiomatic method certainly goes beyond the concept of formal systems, since he regards classical physics, for instance, as an axiomatic system.

Logical axioms are, therefore, considered to be similar to physical hypotheses. Their hypothetical make-up indicates that they are meant to discover a world of mathematical objects and at the same time to make of these objects an original intelligibility and an original representability. I contend that the activity which is thought by Gödel to be operating in logic (and in mathematics) is to be considered as an abductive activity of the construction of a knowledge of mathematical objectivity. This is shown within an inferential process that allows to guess the axioms from which an argument can be traced back to the "logical evidence" (as similar to "sense perception"), which is similar to the process occurring in physics and natural sciences, though with a different degree of plausibility.

For Gödel, an example of this process is the case in which, in order to solve some arithmetical problems, it is deemed necessary to use very general principles, "[...] assumptions essentially transcending arithmetic, i.e., the domain of the kind of elementary indisputable evidence that may be most fittingly compared with sense perception" [Benacerraf and Putnam, 1964, p. 213]. This is also the case when, in order to solve set theory problems (which are, so to speak, far from the 'evident" level of arithmetic), it is necessary to enrich the available theories using "[...] new axioms based on some hitherto unknown idea" (*ibid.*). In my opinion the exploration of the analogy between mathematics and logic on the one hand and natural sciences on the other, does not superimpose a trivial theory of knowledge of a hypothetical-deductive type on the basic Platonist ontology, but rather serves to complicate it further with Kantian considerations, as we shall see below.

In the "Supplement" to the Second Edition of "What is Cantor's Continuum Problem?", Gödel clearly and profoundly points out the actual philosophical structure of the gnoseological medium which integrates with the basic Platonist ontological conception. Following through the analogy between logic and natural sciences, Gödel specifies the relationship between logical objects and the elementary level of "evidence", which, as we have seen above, had been held to be analogous with sense experience and sense perception in physics: "[...] despite their remoteness from sense experience, we do have something like a perception also of the objects of set theory, as is seen from the fact axioms force themselves upon us as being true" ('Supplement' to the Second Edition of "What is Cantor's...", [Benacerraf and Putnam, 1964, p. 271].

Sense perceptions, in the specific meaning they acquire in the case of logic and mathematics, trigger off a process of theory germination, just like perceptions themselves that, in the specific meaning they acquire in the case of physics and natural sciences, cause an abductive movement of hypothesis generation (Gödel, who does not refer to the concept of abduction, calls this movement inductive). More precision is needed: the abductive movement from "this type of perception" toward the generality of theories and logical concepts encounters, in its turn, the categorizing effect of another element which Gödel, after some hesitation, clearly indicates as "mathematical intuition".

Mathematical intuition is actually an activity of the constitution of mathematical objects themselves: "It should be noted that mathematical intuition need not be conceived of as a faculty giving an *immediate* knowledge of the objects concerned. Rather, it seems that, as in the case of physical experience, we *form* our ideas also of those objects on the basis of something else which *is* immediately given" (*ibid.*). Mathematical intuition does not therefore provide an immediate knowledge of its own object: it forms itself upon a basis, an original datum that "it" "is immediately given". And, moreover, a very important consideration, this datum "is *not*, or not primarily, the sensations" (*ibid*).

Sense perceptions in logic and mathematics, as already observed, contain something different from physical ones. Gödel explains himself very precisely: "That something besides the sensations actually is immediately given follows [...] from the fact that even our ideas referring to physical objects contain constituents qualitatively different from sensations or mere combinations of sensations, e.g., the idea of concept itself, whereas, on the other hand, by our thinking we cannot create any qualitatively new elements, but only reproduce and combine those that are given" (*ibid.*, pp. 271-272). These "constituents" are the basis/premise of the abductive inference.

With the statement concerning thinking, which reminds us of Kant's assertion regarding the incapability of concept constituting the object itself, it is possible to define the specific feature of that sense perception level ("evidence", "given") which operates in mathematical and logical knowledge or, more generally, the set of abstract aspects already present at the empirical level; in fact Gödel is once again very clear: "[...] the 'given' underlying mathematics is closely related to the abstract elements contained in our empirical ideas" (*ibid.*, p. 272). And again: "It by no means follows, however, that the data of this second kind, because they cannot be associated with actions of certain things upon our sense organs, are something purely subjective, as Kant asserted. Rather they, too, may represent an aspect of objective reality" (*ibid.*). At this point, the reasoning comes to an end: these abstract elements, besides representing an aspect of the objective reality, "[...] as opposed to the sensations [...] may be due to another kind of relationship between ourselves and reality" (*ibid*).

According to Gödel's philosophy, mathematical entities are therefore formed, and so abductively guessed, on the basis of something else which is *immediately given*, which is both their starting point and point of arrival. The gnoseological framework is given by an intuition as an abductive constitution of the mathematical object, but – it is important to note – not from a fixed and rigid perspective, but according to the image of a *continuous* and *progressive* activity (e.g., by means of the continuous introduction of new axioms and of the actual operations of constructivist proofs) which, starting from a given basis (the constituents qualitatively different from sensations), returns, having in the meantime produced the categorical and ontological representation of the true entities.

Gödel recognizes the continuing expansion of the results of our intuition and provides an illustration of this dynamic and progressive activity referring to the concept of set, that is "of synthetic nature", "[...] there is a close relationship between the concept of set [...] and the categories of pure understanding in Kant's sense. Namely, the function of both is 'synthesis', i.e., the generating of unities out of manifolds" (*ibid.*, footnote 40). The immediately given, therefore, does not consist of a merely sensuous manifold, but a manifold complicated by the presence of abstract elements and therefore different from data typical of sense perception, which are conceivable in the gnoseological process of the constitution of the physical object.[8]

Kant's considerations are for Gödel the ideal place from which to draw suggestions for a gnoseology which is adaptable to the requirement of a philosophy of logic. We can say that knowledge which represents mathematical and logical entities thus becomes the product of an abductive activity which, as seen above, is dynamic, progressive and iterative. But we cannot say that the abduced axioms have a clear explanatory function. If, therefore, the systems which are elaborated are many (and consequently the proofs change according to the systems in which they are activated), the fundamental feature of logic with regard to mathematics becomes relative, without falling into conventionalism or pragmatism.

[8] [Ladrière, 1981, pp. 296–297] provides an interpretation which I share, using analogies to Kant's *Critique of Pure Reason*: the original datum we spoke about coincides with transcendental schematism, where schemas are spotted, not as the data of an intuition, but as something intermediate which, on the one hand, result as being homogeneous with the category and, on the other, with the phenomenon, in order to render possible the application of the former to the latter. The schemas, therefore, are considered as deriving from a sort of vision that is neither merely conceptual nor merely sensible, but which is precisely transcendental imagination. This interpretation proves to be more satisfactory than the one provided by [Wang, 1987, pp. 295–304] who, not recognizing the different status of the given underlying logical and mathematical objects as a level which can be linked to transcendental schemas, undervalues the exclusively Kantian nature of Gödel's argument and must use the less appropriate notion of Husserl's *Wesenschau*. I have proposed a new interpretation in terms of abduction of some aspects of Kantian schematism in my book [Magnani, 2001c, chapter two].

2.3 Instrumental Abduction

Gabbay and Woods maintain we can face a kind of abduction that, basically,

- is not plausibilist

at least in the sense we have considered in the first chapter.

They say: "It is not uncommon for philosophers to speak of the contribution made by the hypothesis of action-at-a-distance as one of explaining otherwise unexplainable observational data. [...] Like numerous instances of D-N explanation, Newtonian explanations need convey no elucidation of their explicanda. They need confer no jot of further intelligibility to them. The action at-a-distance equation serves Newton's theory in a wholly instrumental sense. It allows the gravitational theory to predict observations that it would not otherwise be able to predict" [Gabbay and Woods, 2005, pp. 118-119]. In this case Newtonian explanations are seen as epistemically agnostic conjectures, that is they lack the classical epistemic virtues envisaged by the neopositivistic tradition. These abductions are secured by instrumental considerations and accepted because doing so enables one's target to be hit. They cannot be discharged because of their possible implausibility, for example on the basis of empirical disconfirmation.

We have to note that in some sense all abductions embed instrumental factors. In the general case, one accepts because doing so enables ones target to be attained, notwithstanding that lacks the relevant epistemic virtue. However, in cases such as Newtons, is selected notwithstanding that it is considered to be epistemically hopeless. [Gabbay and Woods, 2005, p. 119] call this extreme kind of abduction radically instrumental).

2.3.1 On Propositional and Strategic Plausibility and Abduction

Abductive reasoning occurs in a situation of "scant resources in quest of comparatively modest targets" [Gabbay and Woods, 2005, p. 58]. I have illustrated above that presumptive abductive hypotheses have to be relevant, plausible, and an economical substitute for any kind of potential exhaustive exploration, cheaper than the acquisition of relevant new knowledge. Hence, plausibility (which is traditionally considered similar to "reasonableness") is a central issue concerning hypothesis generation, choice, and selection in biological abducers like for example human beings and animals but also in artefactual abducers like ideal logical, probabilistic and computational agents.[9] In the case of organic agents plausibility processes are of course largely implicit, and so also linked to unconscious ways of inferring, or, in Peircean terms, to the activity of the instinct [Peirce, 1931-1958, 8.223] and of what Galileo called the *lume naturale* [Peirce, 1931-1958, 6.477], that is the innate fair for guessing right.[10]

[9] The different character of abduction in human and in logical and computational agents is illustrated in chapter six of this book.

[10] On the relationships between instinct and abduction in the framework of other related Peircean themes cf. chapter five, sections 5.1 and 5.2.

In the case of theoretical abduction, that is in the abductive processes basically performed thanks to the *inner* cognitive resources of human agents faced with some problems, some aspects concerning the role of plausibility have to be clearly stressed, taking advantage of the remarks made by [Gabbay and Woods, 2005, chapter seven]. In this case a key role is played by the following aspects, which are intertwined with the role of the *candidate spaces* and of the *resolution procedures* for filtrating them:

- surprise,
- relevance,
- characteristicness,
- plausibility,
- resolution.

Surprise refers to unexpected and uncharacteristic events that create problems which demand a solution (often anomalies, or just events/problems that urge explanations or solutions). These events are marked by both an epistemic disadvantage and an emotional rating.[11] Through a "filtration structure" [Gabbay and Woods, 2005, p. 212], human agents *internally* form a space of candidates, usually very small, found by means of an automatic inner exploration of the information and knowledge stored and available in memory: in this process of successive filtering a reduction from sets of potential explainer/solutions to the considerable actual ones is already at play. *Relevance*[12] dominates at this level, and the task is reached by applying – through relevance filters – further, helpful inner information to information which is considered important with respect to the problem at hand (abduction always aims – we have said – at reaching the best state of ignorance). Then *plausibility* enters the process, in which some candidates are rejected (judged *less plausible*, or clearly *implausible*, and thus *irrelevant*), often based on information that licenses defaults (generalizations for example)[13] for the agent. Resolution is the elimination of all candidates but one.

Looking for a plausibility logic able to illuminate the whole process [Gabbay and Woods, 2005, p. 202] stress the role of the "Auto Rule", which they illustrate in the following way, when speaking of a simple abductive situation involving the married couple Sarah and Harry: "*The Auto Rule*. To the extent possible, favours the option that has an element of autoepistemic backing. For example, in the case we are investigating, the generality claims about Sarah's never coming home early, is likely to be underwritten by two factors of autoepistemic significance. One is that if

[11] [Thagard, 2002b] already clearly stressed the central role of the emotion of surprise in finding problems and anomalies in scientific reasoning (and of the emotion of satisfaction caused by a discovery!).

[12] A further useful distinction between topical, full-use, irredundancy and probabilistic relevance is introduced in [Gabbay and Woods, 2005, pp. 239–250].

[13] Defaults are generalizations that tolerate exceptions and so are complicatedly linked to genericity and normalcy, and, in turn, to characteristicness, given the fact that generic claims also concern what is characteristic. Further details about the relationships between genericity, normalcy, default, characteristicness, common knowledge, and presumption – which in various ways qualify plausibility – are laid out in [Gabbay and Woods, 2005, pp. 213–238].

it were indeed true that Sarah never comes home early, this is something that Harry would know. And if today were to be an exception to that rule, this too is something that Harry may well have knowledge of." The rule is presumptive "Given that a candidate hypothesis is not known to be true, it is presumed to be untrue" (*ibid.*). The Auto Rule favors the most plausible hypothesis going beyond its characteristicness, which is often unconvincing. Characteristicness can fail because human beings obey to the conservative Quinean principle of minimal mutilation, so that a hypothesis should not be too strange. Cognitive psychologists have carried out some interesting empirical work on these issues but I think they are far from reaching general results.[14]

Until now I have described aspects of plausibility which are occurring at the level of basically inner inferences of the real human agent, and which are related to considerations of relevance and characteristicness: this kind of plausibility is called "propositional" by [Gabbay and Woods, 2005, p. 209]. There is also a "strategic" sense of plausibility that has to be taken into account, the one which is occurring in the case of instrumental and radically instrumental abduction (cf. above in this section), where plausibility is no longer linked to characteristicness. To make an example in the case of scientific reasoning, an abductive hypothesis can be highly implausible from the "propositional" point of view and nevertheless it can be adopted for its instrumental virtues, such as in the Newtonian case of action-at-a-distance. Highly implausible hypotheses from the "propositional" point of view can be conjectured because of their high *instrumental* plausibility, where a different role of characteristicness is at stake. I will illustrate below (subsection 2.6.2)[15] the case of the principle of physics (and of their conventional character): they are conjectured and adopted, beyond their (classical) epistemic plausibility – indeed they are *a priori* unfalsifiable – in so far as they are endowed with an instrumental epistemic value. It is "characteristic" in physics to adopt unifying and simple principles, and this is plausible in a "strategic" sense. Another striking example is the following, illustrated by Gabbay and Woods, where propositional plausibility is low and strategic plausibility high:

> Planck conjectured quanta for their contribution to the unification of the laws of black body radiation. He did so notwithstanding the extreme propositional implausibility of the existence of quanta. Even so, Planck thought it reasonable to proceed against the grain of this implausibility. Quanta were nothing like anything then known to physics; so they were uncharacteristic of what physics quantified over in 1900. Planck's was a conjecture grounded in its instrumental yield. It was, we say, a strategically plausible conjecture to make. Why would this be so? It would be so, as we saw earlier, because it is characteristic of the laws of physics to admit unification under the appropriate conditions. [...] Planck reasoned that black body radiation is such that it should be expected that it is subject to unified laws, and because such unifications are characteristic of physics, he made a conjecture that would achieve it [Gabbay and Woods, 2005, p. 218].

[14] In hypothesis search and selection tasks the role of analogy is also very important, cf. above subsection 1.5 of chapter one.

[15] Cf. also [Magnani, 2001b, chapter seven].

We have to remember that, in the case of real human agents plausibility assessment is of course largely inner and implicit, and it is unmediated by explicit and conscious reflections in terms of plausibility, characteristicness, and relevance, such as the ones that can be illustrated in the ideal epistemological and logico-cognitive reconstruction or in the functions of automatic abductive agents.[16]

In more hybrid (multimodal) abductive processes (cf. chapter four, section 4.1), such as in the case of manipulative abduction, of course the whole propositional plausibility assessment is reached – and constrained – taking advantage of the gradual acquisition of new external information with respect to future interrogation and control. In this interplay the cognitive agent further triggers internal plausibility *thoughts* "while" actively modifying the environment (thinking through doing). Indeed, as I have already noted in subsection 1.3.2 of the previous chapter, already at the generation/selection phase of genuine abduction many embodied processes – so to say – of a kind of confrontation (if not exactly of evaluation) with something external to the individual brain, even if not due to experimental tests, can be present, so that we can say that abduction considered as a way of forming hypotheses is of course immediately a generation/selection of "plausible" hypotheses. The role of these continuous stages of confrontation/coordination with external constrained (and fertile) cognitive offerings has been pointed out thanks to the concept of manipulative abduction, which takes into account the external dimension of abductive reasoning. Let me reiterate that in this perspective the proper experimental test involved in the Peircean evaluation phase, which for many researchers reflects in the most acceptable way the idea of abduction as inference to the best explanation, just represents a *particular* subclass of the processes of adoption of abductive hypotheses, and should be considered ancillary to the nature of abductive cognition, and inductive in its essence.

Some researchers could not be as convinced as Gabbay and Woods about the GW characterization of abduction, especially where it is not intrinsically explanatory. They advert to the observation by Gödel and others that putative mathematical axioms are sometimes endorsed because they help to prove conclusions that are accepted on other grounds. It could be maintained that such axioms are still explanatory in the broad sense that they are defeasible and must compare well with the alternatives. 1. The fact that the conclusions are accepted on other grounds suggests the plausibility of the putative axiom. 2. When constructing a proof, does a mathematician not try to select the best concepts or axioms to apply, among a set of alternatives? If so, then there is a kind of inference to the best proof occurring. 3. Old proofs and mathematical ideas are sometimes revisited and revised by later mathematicians. Some are found to be invalid. No proof is necessarily above revision, suggesting their defeasibility in practice. 4. If old proofs or ideas are revised, then they are, in effect, in competition with alternatives, suggesting that there is a kind of inference to the best proof at work in the overall process of mathematics (or logic).

[16] Automatic abductive agents are described below in section 2.7.

I have to note that – in these last cases – plausibility considerations are certainly still at play, but they range from various degrees that involve less "propositional" and more "strategical" and instrumental aspects, so that propositional plausibility is lower and strategic plausibility higher. These cases are far from the clear ones of explanatory abduction that are for example occurring in science and in various kinds of diagnosis. In non-explanatory abduction the cognitive virtues can be more strategical than epistemological, at least if we attribute to the word "epistemological" the standard meaning that neopositivistic tradition established through the specific "explanatory" tone of the D-N law-covering model of scientific explanation. To avoid misunderstandings it has to be stressed that also strategic and instrumental considerations "can" have other epistemological virtues related to scientific rationality, like it is in the case of the action at distance (cf. p. 64) or in the other cases (for example, the unifying principles of physics) I will describe in the following sections.

2.4 Governing Inconsistencies in Science through Explanatory, Non-explanatory, and Instrumental Abduction

In chapters six and seven of my previous book on abduction [Magnani, 2001b] I have illustrated other interesting cases of abductions that can be usefully and basically labeled non-explanatory and instrumentalist. Taking advantage of the role of inconsistencies in abductive reasoning, in the following I am summarizing my argument, showing how often the complex abductive procedures are characterized by a mixture of explanatory, non-explanatory, and instrumental aspects.

We have seen that for Peirce abduction is an *inferential process* in a very particular and wide semiotic sense (cf. chapter one, section 1.5). The aspect of *surprise* I have described in the previous subsection is central for Peirce: abduction is logical inference [...] having a perfectly definite logical form. [...] The form of inference, therefore, is this:

- The surprising fact, C, is observed;
- But if A were true, C would be a matter of course,
- Hence, there is reason to suspect that A is true [Peirce, 1931-1958, 5.188-189,7202].

C is true of the actual world and it is surprising, a kind of state of doubt we are not able to account for using our available knowledge. Philosophers of science in the last century have illustrated that inconsistencies and anomalies often play this role of surprise in the growth of scientific knowledge.

Hence, contradictions and inconsistencies are fundamental in abductive reasoning, and abductive reasoning is appropriate for "governing" inconsistencies: this section illustrates abductive reasoning in order to classify and analyze the different roles played by *inconsistencies* in different reasoning tasks and in scientific discovery. The special sensitivity to anomalies and inconsistencies, which is an undoubted endowment of human beings and many animals, becomes institutionalized in scientific mentality, and an important part of the scientific method. The aim of this section is to identify aspects of inconsistencies not covered by certain formalisms and to suggest extensions to present thinking, but also to delineate the first features of a broader constructive framework able to include abduction and to provide constructive solutions to some of the limitations of its formal models. There are many ways of "governing" inconsistencies: from the methods activated in diagnostic settings and consistency-based models (cf. chapter one, section 1.4) to the typical ones embedded in some forms of creative reasoning, from the interpretations in terms of conflicts and competitions to the actions performed on empirical and conceptual anomalies, from the question of generating inconsistencies by radical innovation to the connectionist treatment of coherence. The conclusions presented here aim at representing a step forward in the understanding of the use of inconsistencies in creative abductive reasoning both in scientific and practical settings.

In different theoretical changes we witness different kinds of discovery processes operating. Discovery methods are *data-driven* (generalizations from observation and from experiments) (induction), *explanation-driven* (explanatory abduction), and *coherence-driven* (formed to overwhelm contradictions) [Thagard, 1992]. Sometimes there is a mixture of such methods: for example, a hypothesis devoted to overcome a contradiction is found by explanatory abduction. The detection of an anomaly usually demonstrates that an explanation is needed.[17] The next move of the process of explanation is to obtain a possible explanation. Therefore, contradiction and its reconciliation play an important role in philosophy, in scientific theories and in all kinds of problem-solving. It is the driving force underlying change (thesis, antithesis and synthesis) in the Hegelian dialectic and the main tool for advancing knowledge (conjectures and refutations – [Popper, 1963] – and proofs and counter-examples – [Lakatos, 1976] – in the Popperian philosophy of science and mathematics).[18]

Following Quine's line of argument against the distinction between necessary and contingent truths [Quine, 1979], when in science a contradiction arises, consistency can be restored by rejecting or modifying any assumption which contributes to the derivation of contradiction: no hypothesis is immune from possible alteration. Of course there are epistemological and pragmatic limitations: some hypotheses contribute to the derivation of useful consequences more often than others, and some participate more often in the derivation of contradictions than others. For example, when faced with abduced hypotheses which we have decided to release for further premissory work (H^c) (cf. section 2.1.1), it might be useful to abandon the hypotheses which contribute least to the derivation of useful consequences leading to (empirical or theoretical) contradictions. If contradictions continue and the assessed utility of the hypotheses changes, it may be necessary to backtrack, reinstate a previously abandoned hypothesis and abandon another.

[17] In the last sections of this chapter I will describe the paradigmatic case of the anomaly of the postulate of parallels, which brings together explanatory and non-explanatory aspects.

[18] Also psychoanalysis relates creative thinking to something contradictory: creative expression is explained in terms of sublimation of unconscious *conflicts*, as Freud demonstrated in his famous analysis of the symbolic meanings of the works of Leonardo da Vinci [Freud, 1916].

Hence, the derivation of inconsistency contributes to the search for alternative, and possibly new, hypotheses: for each assumption which contributes to the derivation of a contradiction there exists at least one alternative new system obtained by abandoning or modifying the assumption.

Anomalies result not only from direct conflicts between inputs and system knowledge but also from conflicts between their ramifications: "[...] noticing a particular anomaly may require building long inference chains tracing ramifications until a contradiction is found" [Leake, 1992, p. xiii]. Any explanation must be suitably plausible and able to dominate the situation in terms of reasonable hypotheses. Moreover, the explanation has to be relevant to the anomaly, and resolve the underlying conflict. Finally, in some cases of everyday (and practical) anomaly-driven reasoning the explanation has to be useful, so it needs information that will point to the specific faults that need repair (on the role of plausibility and abduction in inner rehearsal of human agents cf. the previous subsection).

The classical example of a theoretical system that is opposed by a contradiction is the case in which the report of an empirical observation or experiment contradicts a scientific theory. Whether it is more beneficial to reject the report or the statement of the theory depends on the whole effect on the theoretical system. It is also possible that many alternatives might lead to non-comparable, equally viable, but mutually incompatible, systems.[19]

Why were the photographic plates in Röntgen laboratory continually blackened? Why does the perihelion of the Mercury planet advance? Why is the height of the barometer lower at the high altitudes than at the low ones? These are examples of problems that come from observation, but they are problematic in light of some theory, that is unexpected and anomalous. The first was problematic because it was tacitly supposed at that time that no radiation or emanation existed able to penetrate the container of the photographic plates; the second because it conflicted with the Newtonian theory; the third was problematic for the supporters of Galileo's theories because it contradicted the belief in the "force of vacuum" that was adopted as an explanation of why the mercury does not fall from a barometer tube [Chalmers, 1999].

Dealing with the problem of withdrawing scientific *paradigms* Kuhn writes:

> Discovery commences with the awareness of anomaly: i.e., with the recognition that nature has somehow violated the paradigm-induced expectations that govern normal science. It then continues with a more or less extended exploration of the area of anomaly. And it closes only when the paradigm theory has been adjusted so that the anomalous has become the expected. Assimilating a new sort of fact demands a more than additive adjustment of theory, and until that adjustment is completed – until the scientist has learned to see nature in a different way – the new fact is not quite a scientific fact at all. [Kuhn, 1962, p. 53, second edition].

[19] Thagard proposes a very interesting computational account of scientific controversies in terms of so-called *explanatory coherence* [Thagard, 1992] (cf. also chapter one, subsection 1.3.2, this book), which improves on Lakatos' classic one [1970], by explaining various aspects dealing with the comparison of scientific theories.

It is well-known that the recent falsificationist tradition in epistemology has focused attention on the role of anomalies (that can give rise to falsifications) establishing a sort of "received view" on the growth of scientific knowledge characterized by the fundamental role played by anomalies: Newton's theory is able to explain phenomena not touched on by Aristotle's theory, such as correlations between the tides and the location of the moon, and the variation in the force of gravity with respect to height above sea level; in turn, Einstein was able to do the same with respect to the Newtonian theory and its anomalies and falsifications.

As Lakatos argues, in a mature theory with a history of useful consequences, it is generally better to reject an anomalous conflicting report than it is to abandon the theory as a whole. The cases in which we have to abandon a whole theory are very rare: a theory may be considered as a complex information system in which there is a collection of cooperating individual statements some of which are useful and more firmly held than others; propositions that belong to the central core of a theory are more firmly held than those which are located closer to the border, where instead rival hypotheses may coexist as mutually incompatible alternatives. Accumulating reports of empirical observations can help in deciding in favor of one alternative over another.

I have to remember that even without restoring consistency, an inconsistent system can still produce useful information. Of course from the point of view of classical logic we are compelled to derive any conclusion from inconsistent premises, but in practice efficient proof procedures infer only "relevant" conclusions with varying degrees of accessibility, as reverberated by the criteria of non-classical *relevant entailment* [Anderson and Belnap, 1975].

We may conclude by asserting that

1. contradiction, far from damaging a system, helps to indicate regions in which it can be changed (and improved): it typically furnishes chances of "explanatory" abductive reasoning, it becomes possible to resolve/explain the inconsistency;
2. we have to remember that not all the configurations of new concepts are incoherence-driven and related to the highest case of creative abduction and creative analogical reasoning, ubiquitous in science, and more constructive than associative. For example, in conceptual combination in everyday reasoning, many new concepts are formed in a coherence-driven way, where a kind of reconciliation of associations and thematic relations operates. Thagard presents the case of the construction of the concept of "computational philosopher" where in order to understand the concept people need to make coherent sense of how a "modifier" such as "computational" can apply to a "head" such as "philosopher" [Thagard, 1997a]. In science hypotheses are abductively formed also in absence of triggering anomalies and inconsistencies, and in this case it is frequent to witness cases of non-explanatory (like for example in guessing new axioms in mathematics – cf. below section 2.9.1) and of instrumental abduction (like in the case of guessing conventions in physics – cf. below section 2.6.2); there is also the intermediate case of guessing hypotheses that have an explanatory descent but, in so far as they are not falsifiable, their instrumental character

tends to overwhelm the explanatory one, such as in the case of "constructions" during the Freudian psychoanalytic treatment: – cf. below 2.6;
3. of course contradiction is also a way of falsifying established hypotheses (H^c): it has a preference for strong hypotheses which are more easily falsified than weak ones; and moreover, hard hypotheses may more easily weakened than weak ones, which prove difficult subsequently to strengthen. It is always better to produce mistakes and then correct them than to make no progress at all.

We can see abductive inferences "[...] as answers to the inquirer's explicit or (usually) tacit questions put to some definite source of answers (information)" [Hintikka, 1998, p. 519] stressing the interrogative features of this kind of reasoning. If abduction is the making of a set of possible answers, the choice of the possible questions is also decisive (and this choice of course is not indifferent as regards the further process of finding answers). As already illustrated in chapter one (section 1.4)[20] we may see belief change from the point of view of *conceptual change*, considering concepts either cognitively, like mental structures analogous to data structures in computers, or, epistemologically, like abstractions or representations that presuppose questions of justification. Belief revision – even if extended by formal accounts such as illustrated above in chapter one[21] – is able to represent cases of conceptual change such as adding a new instance, adding a new weak rule, adding a new strong rule (see [Thagard, 1992, pp. 34–39]), that is, cases of addition and deletion of beliefs, but fails to take into account cases such as adding a new part-relation, adding a new kind-relation, adding a new concept, collapsing part of a kind-hierarchy, reorganizing hierarchies by branch jumping and tree switching, in which there are reorganizations of concepts or redefinitions of the nature of a hierarchy. These last cases are the most evident changes occurring in many kinds of creative reasoning, for example in science. Related to some of these types of conceptual change are different varieties of inconsistencies (see Figure 2.1), as explained in the following sections.

2.4.1 Empirical Anomalies and Explanatory Abduction

In chapter one (section 1.5) I argued that various logical accounts of abduction certainly illustrate much of what is important in abductive reasoning, especially the objective of selecting a set of hypotheses (diagnoses, causes) that are able to dispense good (preferred) explanations of data (observations), but tend to fail in accounting for many cases of explanations occurring in science or in everyday reasoning. For example they do not capture 1) the role of statistical explanations, where what is explained follows only probabilistically and not deductively from the laws and other tools that do the explaining; 2) the sufficient conditions for explanation; 3) the fact that sometimes the explanations consist of the application of schemas that fit a phenomenon into a pattern without realizing a deductive inference; 4) the idea of the

[20] Cf. also [Magnani, 1999].

[21] Or developed by others, see for example, [Katsuno and Mendelzon, 1992; Cross and Thomason, 1992].

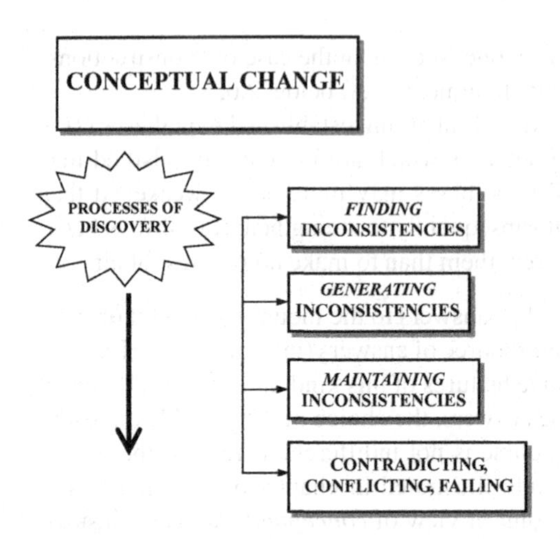

Fig. 2.1 Conceptual change and inconsistencies

existence of high-level kinds of *creative* abductions; 5) the existence of model-based abductions (for instance visual and diagrammatic); 6) the fact that explanations usually are not complete but only furnish *partial* accounts of the pertinent evidence (see [Thagard and Shelley, 1997];) 7) the fact that one of the most important virtues of a new scientific hypothesis (or of a scientific theory) is its power of explaining *new*, previously *unknown* facts.

Moreover, the recent logical accounts of abduction certainly elucidate many kinds of inconsistency government, which nevertheless reduce to the act of finding contradictions able to generate the withdrawal of some hypotheses, beliefs, reasons, etc.: these contradictions always emerge at the level of data (observations), and consistency is restored at the theoretical level.[22] This view may distract from important aspects of other kinds of reasoning that involve intelligent abductive performances.

For example, *empirical anomalies* result from data that cannot currently be fully explained by a theory. They often derive from predictions that fail, which implies some element of incorrectness in the theory. In general terms, many theoretical constituents may be involved in accounting for a given domain item (anomaly) and hence they are potential points for modification. The detection of these points involves defining which theoretical constituents are employed in the production of the anomaly. Thus, the problem is to investigate all the relationships in the explanatory area. In science, first and foremost, empirical anomaly resolution involves the localization of the problem at hand within one or more constituents of the theory. It is then necessary to produce one or more new hypotheses to account for the anomaly, and finally, these hypotheses need to be evaluated so as to establish which one best satisfies the criteria for theory justification. Hence, anomalies require a change in the theory, yet once the change is successfully made, anomalies are no longer

[22] We have to remember that the logical models in some cases exhibit a sort of paraconsistent behavior.

anomalous but in fact are now resolved. This can involve both transforming the domain knowledge and learning or discovering new schemas or rules endowed with explanatory power. Also in the tradition of the machine discovery programs (cf. below, section 2.7) failed predictions drive the mechanism which selects new experiments to guess new hypotheses [Zytkow, 1997]. The process is goal-driven. Of course an explainer will use different information for the objective of predicting a situation than for repairing or preventing it with a good explanation.

General strategies for anomaly resolution, as well as for producing new ideas and for assessing theories, have been studied by [Darden, 1991] in her book on reasoning strategies from Mendelian genetics. Anomaly resolution presents four aspects: 1) confirmation that the anomaly exists, 2) localization of the problem (not considering the cases where the anomaly either is an uninteresting monster or is outside the scope of the theory), 3) generation of one or more new hypotheses to account for the anomaly, that is, conceptual changes in the theory, 4) evaluation and assessment of the hypotheses chosen. Abductive steps are present at the third level (that normally is activated when step 2 fails to manage the anomaly): in this case we are trying to eliminate the already confirmed anomaly (step 1) in a creative way. At this level there are various kinds of conceptual changes: from the simple ones related to the possible alteration (deletion, generalization, simplification, complication, "slight" changes, proposal of the opposite, etc.), or addition of a new component of the theory, to the construction and discovery of a new "theoretical component" [Darden, 1991, pp. 269–275].

Let us relate this taxonomy to the cases of scientific conceptual change illustrated above and in the previous chapter. Darden's alterations and additions can be assimilated to the cases of conceptual change such as adding a new instance, adding a new weak rule, adding a new strong rule (that is, cases of addition and deletion of beliefs), adding a new part-relation, adding a new kind-relation, adding a new concept. On the contrary, Darden's case about the discovery of a new "theoretical component" relates to changes where there is collapse of part of a kind-hierarchy, the reorganization of hierarchies by branch jumping and tree switching, in which there are reorganizations of concepts or redefinitions of the nature of a hierarchy. We have seen that these last cases are the most evident changes occurring in many kinds of creative reasoning in science, when adopting a new conceptual system is more complex than mere belief revision: different varieties of *model-based abductions* are related to some of these types of scientific conceptual change.

Darden reminds to us that geneticists sometimes abandoned hypotheses on the basis of falsifying evidence (this is another way – unsuccessful resolution, unresolved anomaly – of seeing the problem of anomalies, already indicated at the step "alter a component-deletion"). And, of course, as clearly illustrated by Lakatos (see section 2.4 above) an anomalous observation statement can be rejected and the theory which it clashes retained. This is for example the case of the Copernicus' theory that was retained and the naked-eye observation of the sizes of Venus and Mars, which were in contradiction with that theory, eliminated.

Moreover, as taught by the recent epistemological tradition, and especially by Lakatos' falsificationism [1970], in science new creative abduced hypotheses (or new theories), originating from a case of anomaly resolution, have to lead to novel predictions. Galileo reported that the moon was not a smooth sphere but full of craters and mountains. His Aristotelian opponent had to admit that observation that nevertheless configured a terrible anomaly for the notion, common to many Aristotelians and coming from the ancient times, that the celestial bodies are perfect spheres. The hypothesis created by the Aristotelian to explain this anomaly is *ad hoc*: he advocated that there is an invisible substance on the moon, which fills the craters and covers the mountains so that the moon's aspect is absolutely spherical. Unfortunately, this modified theory of the moon did not lead to new testable consequences, and thus is not scientifically acceptable, following the falsificationist point of view.[23]

2.4.2 Conceptual Anomalies, Explanatory, and Non-explanatory Abduction

Empirical anomalies are not alone in generating impasses, there are also the so-called *conceptual anomalies*. To illustrate more features of a theory of the role of inconsistencies in (especially "model-based") abduction we can present the case of conceptual problems as triggers for hypotheses. The so-called conceptual problems represent a particular form of anomaly. In addition, resolving conceptual problems may involve satisfactorily answering questions about the nature of theoretical entities. Nevertheless such conceptual problems do not arise directly from data, but from the nature of the claims in the principles or in the hypotheses of the theory. It is far from simple to identify a conceptual problem that requires a resolution, since, for example, a conceptual problem concerns the adequacy or the ambiguity of a theory, and yet also its incompleteness or (lack of) evidence.

The formal sciences are especially concerned with conceptual problems. Let's consider an example deriving from the well-known case of the non-Euclidean revolution, which plays a remarkable role in illustrating some actual transformations in rational conceptual systems. The discovery of non-Euclidean geometries involves some interesting cases of *visual abductive reasoning*. It demonstrates a kind of visual abduction, as a strategy for anomaly resolution related to an interplay between explanatory and productive visual thinking, but also to the active role of *non-explanatory* abduction (cf. below sections 2.9 and 2.11).

Since ancient times the fifth postulate has been held to be not evident. This "conceptual problem" (just an anomaly) has caused much suspicion about the reliability of the whole theory of parallels, consisting of the theorems that can be only derived with the help of the fifth postulate. The recognition of this anomaly was fundamental to the development of the great non-Euclidean revolution. Two thousand years of attempts to resolve the anomaly have generated many more-or-less fallacious

[23] On the relationship between falsificationism (Popperian and Lakatosian) and conventionalism (at least in the ingenious case of Poincaré), cf. below section 2.6.2.

demonstrations of the fifth postulate (for example, a typical attempt was that of trying to prove the fifth postulate from the others), until the discovery of non-Euclidean geometries [Greenberg, 1974].

At the end of this chapter I will present some details derived from the historical discovery of non-Euclidean geometries which illustrate the relationships between strategies for anomaly resolution and visual thinking: I consider how Lobachevsky's strategy for resolving the anomaly of the fifth postulate was to manipulate the symbols, rebuild the principles, and then to derive new proofs and provide a new mathematical apparatus. The failure of the demonstrations of his predecessors induced Lobachevsky to believe that the difficulties that had to be overcome were due to causes other than those which had until then been focused on. I will show how some of the hypotheses created by Lobachevsky were mostly image-based trying to demonstrate that visual abduction is relevant to hypothesis formation in mathematical discovery, in an interplay between explanatory and non-explanatory steps.

The fact that inconsistencies may occur also at the theoretical level is further emphasized if we consider that in science or in legal reasoning [Thagard and Shelley, 1997], hypotheses are mainly *layered*, contrarily to the case of diagnostic reasoning, where we have a set of data that can be explained by a given set of diseases (that is with the explanation consisting of a mapping from the latter to the former). Hence, the organization of hypotheses is more complex than the one illustrated in formal models, and abduction is not only a matter of mapping from sets of hypotheses to a set of data.

In many "explanatory" abductive settings there are hypotheses that explain other hypotheses so that the selection or creation of explanations is related to these relationships.[24] In this case the plausibility of the hypothesis comes not only from what it explains, but also from it itself being explained. The Darwinian hypothesis stating that "Species of organic beings have evolved" gains plausibility from the many pieces of evidence it helps to explain. Moreover, it receives plausibility from above, from being explained by the hypothesis of natural selection, in its turn explained by the hypothesis concerning the struggle for existence. The principle of special relativity and the principle of the constancy of the speed of light explain (in this case the explanatory relation is "deductive") the Lorentz transformation, which explains the negative result of the Michelson-Morley experiment, but also they explain the convertibility of mass and energy which explains the nuclear transmutations detected by Rutherford in 1919. Hence the two principles explain the two experiments above by means of the intermediate layered hypotheses of Lorentz transformation and mass/energy conversion, but we also know the two principles directly explain the Fizeau experiment concerning the speed of light in a flowing fluid [Einstein, 1961]. In the tradition of machine discovery programs the question of layered hypotheses could be related to the one of postulating hidden structures where some hidden hypotheses can trigger discovery of other hypotheses at a higher level.

[24] This kind of hierarchical explanations has also been studied in the area of probabilistic belief revision [Pearl, 1988].

2.4.3 Generating Inconsistencies by Radical Innovation

The case of conceptual change such as adding a new part-relation, adding a new kind-relation, adding a new concept, collapsing part of a kind-hierarchy, reorganizing hierarchies by branch jumping and tree switching, in which there are reorganizations of concepts or redefinitions of the nature of a hierarchy are the most evident changes occurring in many kinds of *creative abduction*, for instance in the growth of scientific knowledge.

In *Against Method* [Feyerabend, 1975], Feyerabend attributes a great importance to the role of contradiction. He establishes a "counterrule" which is the opposite of the neopositivistic one that it is "experience", or "experimental results" which measures the success of our theories, a rule that constitutes an important part of all theories of corroboration and confirmation. The counterrule "[...] advises us to introduce and elaborate hypotheses which are inconsistent with well-established theories and/or well-established facts. It advises us to proceed counterinductively" [Feyerabend, 1975, p. 20]. Counterinduction is seen more reasonable than induction, because appropriate to the needs of creative reasoning in science: "we need a dream-world in order to discover the features of the real world we think we inhabit" (p. 29). We know that counterinduction, that is the act of introducing, inventing, and generating new inconsistencies and anomalies, together with new points of view incommensurable with the old ones, is congruous with the aim of inventing "alternatives" (Feyerabend contends that "proliferation of theories is beneficial for science"), is very important in all kinds of creative abductive reasoning.

When a scientist introduces a new hypothesis, especially in the field of natural sciences, he is interested in the potential rejection of an old theory or of an old knowledge domain. Consistency requirements in the framework of deductive models of abduction, governing hypothesis withdrawal in various ways, would arrest further developments of the new abduced hypothesis. In the scientist's case there is not the deletion of the old concepts, but rather the *coexistence* of two rival and competing views.

Consequently we have to consider this competition as a form of epistemological, and non logical inconsistency. For instance two scientific theories are conflicting because they compete in explaining shared evidence.

The problem has been studied in Bayesian terms but also in connectionist ones, using the so-called theory of explanatory coherence ([Thagard, 1992], cf. also footnote 19, above), which deals with the epistemological reasons for accepting a whole set of explanatory hypotheses conflicting with another one. In some cognitive settings, such as the task of comparing a set of hypotheses and beliefs incorporated in a scientific theory with the one of a competing theory, we have to consider a very complex set of criteria (to ascertain which composes the best explanation), that goes beyond the mere simplicity or explanatory power. The minimality criteria included in some of the formal accounts of abduction, or the idea of the choice among preferred models cited in section 1.4 of chapter one, are not sufficient to illustrate more complicated cognitive situations.

2.4.4 *Maintaining Inconsistencies: Static and Dynamic Aspects*

As noted above, when we create or produce a new concept or belief that competes with another one, we are compelled to maintain the derived inconsistency until the possibility of rejecting one of the two becomes feasible. We cannot simply eliminate a hypothesis and then substitute it with one inconsistent with it, because until the new hypothesis comes in competition with the old one, there is no reason to eliminate the old one. Other cognitive and epistemological situations present a sort of paraconsistent behavior: a typical kind of *inconsistency maintenance* is the well-known case of scientific theories that face anomalies. As noted above, explanations are usually not complete but only furnish partial accounts of the pertinent evidence: not everything has to be explained.

Newtonian mechanics is forced to cohabit with the anomaly of perihelion of Mercury until the development of the theory of relativity, but it also has to stay with its false prediction about the motion of Uranus. In diagnostic reasoning too, it is necessary to make a diagnosis even if many symptoms are not explained or remain mysterious. In this situation we again find the similarity between reasoning in the presence of inconsistencies and reasoning with incomplete information already stressed. Sometimes scientists may generate the so-called auxiliary hypotheses [Lakatos, 1970], justified by the necessity of overcoming these kinds of inconsistencies: it is well-known that the auxiliary hypotheses are more acceptable if able to predict or explain something new (the making of the hypothesis of the existence of another planet, Neptune, was a successful way – not an *ad hoc* maneuver – of eliminating the anomaly of the cited false prediction).

To delineate the first features of a constructive cognitive and formal framework that can handle the coexistence of inconsistent theories (and unify many of the themes concerning the limitations of formal models of abductions previously illustrated) we have first of all to be able to deal with the treatment of non sentential representations (that is model-based representations).

Moreover, I think that the problem of coexistence of inconsistent scientific theories and of reasoning from inconsistencies in scientific creative processes leads to analyze the characters of what I call the *best possible information* of a situation. It is also necessary to distinguish between the dynamic and the static sides of the best possible information. If we stress the *sequential* (dynamic) aspects we are more oriented to analyze anomalies as triggers for hypotheses: as illustrated by the traditional deductive models of abduction, the problem concerns the abductive steps of the sequential comprehension and integration of data into a hypothetical structure that represents the best explanation for them. Analogously, as we will see in the case of conceptual anomalies in geometry (this chapter, sections 2.8, 2.9, and 2.11), the "impasse" can also be a trigger for a whole process of model-based abduction. On the contrary, if we consider the *holistic* (static) aspects we are more interested in the coexistence of inconsistencies as potential sources of different reasoned creative processes. In this last case we have to deal with explanatory model-based abduction and its possible formal treatment; some suggestions can be derived from the area of paraconsistent and adaptive logic [Meheus, 1999; Meheus *et al.*, 2002;

Meheus and Batens, 2006], for instance handling hierarchies of inconsistent models of a given representation.[25]

When the holistic representation concerns the relationship between two competing theories containing some inconsistencies, a formal framework can be given by the connectionist tradition using a computational reconstruction of the epistemological concept of coherence, as already stated (see also the following section).

2.4.5 *Contradicting, Conflicting, Failing, and Instrumental Abduction*

Considering the *coherence* of a conceptual system as a matter of the simultaneous satisfaction of a set of positive and negative constraints leads to the *connectionist* models (also in computational terms) of coherence [Thagard, 2002a]. In this light logical inconsistency becomes a relation that furnishes a *negative* constraint and entailment becomes a relation that provides a *positive* constraint. For example, as already noted, some hypotheses are inconsistent when they simply compete, when there are some pragmatic incompatibility relations, when there are incompatible ways of combining images, etc. [Thagard and Shelley, 1997; Thagard and Verbeurgt, 1998].

From the viewpoint of the connectionist model of coherence, it spontaneously allows the situations in which there is a set of accepted concepts containing an inconsistency, for example in the case of anomalies: the system at hand may at any rate have a maximized coherence, when compared to another system. Moreover, another interesting case is the relation between quantum theory and general relativity, which individually have enormous explanatory coherence. According to the eminent mathematical physicist Edward Witten "[...] the basic problem in modern physics is that these two pillars are incompatible". Quantum theory and general relativity may be incompatible, but it would be premature given their independent evidential support to suppose that one must be rejected [Thagard, 1992, p. 223].

A situation that is specular to inconsistency maintenance (cf. previous section) is given when two theories are not intertranslatable but *observationally equivalent*, as illustrated by the epistemology of conventionalist tradition. In these cases they are unconcerned by inconsistencies (and therefore by crucial experiments, they are unfalsifiable) but have to be seen as rivals. The incommensurability thesis shows interesting relationships with the moderate and extreme conventionalism. If theories that are not intertranslatable, that is incommensurable, function in certain respects as do observationally equivalent theories (and they are unconcerned by crucial experiments), the role of observational and formal-structural invariants in providing comparability is central: it is impossible to find a contradiction in some areas of the conceptual systems they express. I think that it is necessary to study in general the reasons able to model the demise of such observationally equivalent "conventional" theories, showing how they can be motivationally abandoned. This problem

[25] See also the analysis of the relationships between inconsistency, generic modeling, and conceptual change given in [Nersessian, 1999a].

has been frequently stressed from the beginnings of research in automated discovery (cf. section 2.7): if many hypothetical patterns are discovered, all justified by their observational consequences, we are looking for the reasons to claim that one of them is the best [Zytkow and Fischer, 1996].

Moreover these theories can be seen as rivals in some sense not imagined in traditional philosophy of science. We already stressed that in these cases the role of observational and formal-structural invariants in providing comparability is central: it is impossible to find a contradiction in some area of the conceptual systems they express.

I have already said that contradiction has a preference for strong hypotheses which are more easily falsified than weak ones. Moreover, hard hypotheses may be more easily weakened than weak ones, which subsequently prove difficult to strengthen. Some hypotheses may however be *unfalsifiable*: they exhibit an *instrumental* abductive force and present various degrees of strategic plausibility (cf. above subsection 2.3.1). In this case, it is impossible to find a contradiction from the empirical, but also theoretical point of view, in the conceptual systems in which they are incorporated. Notwithstanding this fact, it is sometimes necessary to construct ways of rejecting the unfalsifiable hypothesis at hand by resorting to some external forms of negation, (external because we want to avoid any arbitrary and subjective elimination), which would be rationally or epistemologically unjustified. As I have already anticipated, in the section 2.6 I will consider a kind of abduced *instrumental* "weak" hypothesis that is hard to negate and the ways to make this easy. I will explore whether *negation as failure* can be employed to model hypothesis withdrawal in Freudian analytic reasoning, where "constructions" are hypotheses which oscillate between explanatory and instrumental roles, and in Poincaré's conventionalism of the principles of physics, where the abduced hypotheses (called "conventions") are essentially instrumental.

2.5 A Note on Preinventive Forms, Disconfirming Evidence, Unexpected Findings

I have said that intuitively an anomaly is something surprising, as Peirce already knew "The breaking of a belief can only be due to some *novel* experience" [Peirce, 1931-1958, 5.524] or "[...] until we find ourselves confronted with some experience contrary to those expectations" [Peirce, 1955b, 7.36] (cf. this chapter, section 2.4).[26] I have said that many biological organisms are very sensitive to anomalies, therefore it is not strange that something anomalous can be found in those kinds of structures the cognitive psychologists call *preinventive*. Cognitive psychologists have described many kinds of preinventive structures and described their desirable properties, that constitute particularly interesting ways of "irritating" the mind and stimulating creativity.

[26] Classical cognitive considerations on inconsistencies in reasoning can be found in [Schank, 1982; Schank and Abelson, 1987].

Preinventive structures are very important from the point of view of creative abduction, because of the propulsive role they play. [Finke *et al.*, 1992] list the following preinventive cognitive structures:[27] *visual patterns* and *objects forms* (one can generate two dimensional patterns resulting in creative products such as new types of symbols and artistic design or three-dimensional forms resulting in new inventions and spatial analogies); *mental blend* (two distinct entities are fused to create something new, one might imagine combining a lion with an ostrich to create a type of animal); exemplars of *unusual* or *hypothetical categories* (they show emergent features that lead to new and unexpected discoveries, for example, in attempting to construct a member of the category "alien creatures that inhabit a planet different from the earth", one might imagine a creature that resemble earth creatures in some respect but not others); *mental models* that represent various mechanical or physical systems (sometimes incomplete, unstable, and even unscientific), as well as conceptual systems; various kinds of *verbal combinations* (they can lead to poetic and other literary and narrative explorations, cf. subsection 7.8.4 of chapter seven). Moreover, some musical forms or actions schemas can be identified, as well as also other possibilities.

Some particular attributes of these structures are very important in contributing to discovery: *novelty*, *ambiguity* (ambiguous visual patterns are often interpreted in various creative ways), *implicit meaningfulness* (they seem to have hidden meanings: "a general perceived sense of 'meaning' in the structure [...] potential for inspiring or eliciting new and unexpected interpretations" [Finke *et al.*, 1992, p. 23], *emergence* (referred to the extension in the preinventive structures of unexpected relations and features), *incongruity* (that refers to conflict or contrast among elements).[28] Examples of creative reinterpretations of the preinventive form in experiments on spanning the object categories are given in Figures 2.2 and 2.3.

Fig. 2.2 A preinventive form in experiments on spanning the object categories, constructed using the bracket, hook, and half-sphere. (From [Finke *et al.*, 1992, p. 84], ©1992 Massachusetts Institute of Technology, by permission of The MIT Press).

[27] Cf. also [Finke, 1990].

[28] Already exploited by Koestler's theory of bisociation [1964], *divergence* refers to the possibility of finding various uses and meanings in the same structure, like in the case of a hammer, an unambiguous form that can be used in many ways. As can be easily seen, all these properties can be considered from the single theoretical point of view of the presence of something anomalous. All properties refer to a kind of detected surprise that can open the abductive exploratory processes of creativity.

Fig. 2.3 Possible reinterpretations of the preinventive form given in Figure 2.2, spanning eight object categories (left to right): lawn lounger (furniture), global earrings (personal items), water weigher, (scientific instruments) portable agitator (appliances), water sled (transportation), rotating masher (tool and utensils), ring spinner (toys and games), and slasher basher (weapons). (From [Finke *et al.*, 1992, p. 85], ©1992 Massachusetts Institute of Technology, by permission of The MIT Press).

[Koslowski, 1996] studies scientific reasoning observing the principles and strategies people use in generating and testing hypotheses in every day situations. In some situations subjects reason in a scientific way to a greater extent than considered in the existing literature. She illustrates experiments with subjects dealing with *hypothesis-testing* and examines how hypotheses (possible explanations) are generated and how they vary in credibility as a function of various sorts of evidence about the considered phenomena. Moreover, the experiments concern the ways in which humans deal with evidence or information "[...] that disconfirms or is anomalous to or at least unanticipated by an explanation" to focus on situations where there is the opportunity to engage hypothesis revision (or hypothesis withdrawal).

The results are quite interesting and show that theoretical concerns play a prominent role in recognizing the importance of the *anomalies*: subjects manage hypothesis revision as theory dependent in a scientific legitimate way; when the alternative explanatory hypotheses involved are in terms of causal (theoretical) mechanisms

– and not simply stated in terms of covariation with data – subjects are able to treat them as defeasible and to modify them in ways that are theoretically motivated [Klahr and Dunbar, 1988], not simply using *ad hoc* maneuvers. Decisions about whether to revise or reject as well as decisions about type of revisions that would considered as justified are highly theory-dependent and involve much more that merely information about covariation. In turn, ignoring the importance of the theoretical component (or mechanism), can underestimate subjects' willingness to reject hypotheses when rejection would be appropriate (this is the reason why these subjects have sometimes been regarded as poor scientists, like in the case described in Wason's task, [Wason, 1960].[29]

Finally, the empirical research by [Dunbar, 1995; Dunbar, 1999], in many molecular biology and immunology laboratory in US, Canada and Italy, has demonstrated the central role of the *unexpected* in creative abductive reasoning. Scientists expect the unexpected. By experimentally looking at the so-called "in vivo science" Dunbar analyzes three activities that are seen as the most important in scientific model building: analogical reasoning, attention given to unexpected findings (that is anomalies, errors, inconsistencies), experimental design, and distributed reasoning.

First of all scientists frequently use analogy where there is not a simple answer to a particular problem, and distant analogies are not so widespread as supposed, they are primarily used to explain concepts to others, but not in creative scientific reasoning.

Secondly, it is well know that the recent discoveries of naked DNA and Buckey balls, but also the old well-known of penicillin, nylon, and gravitation are charged to the unexpected: in the "in vivo" science we can see the unexpected is very common, for example it is a regular occurrence that the outcome of an experiment does not match the scientists' prediction. The scientists have to evaluate which findings are caused by methodological errors, faulty assumptions, and chance events. At the local level of experimentation in real scientific laboratories this research constitutes a kind of confirmation of the Popperian ideas on hypotheses falsification, made in that case at the macro-level of the whole growth of scientific knowledge. The hypotheses are activated to deal with such problematic findings, usually local analogies and model-based abductions, which can give rise to generalizations, causal explanation, visualizations, etc., for finding the common features of the unexpected findings, and possibly discover more general and deep explanations.

Third, experimental design is shown to have interesting cognitive components, illustrating the fact that sometimes the experiments are locally built independently of the hypotheses being tested. The problem is related to the role of manipulative abduction I described in chapter one, showing how we can find methods of constructivity based on external models and action-based reasoning in scientific and everyday reasoning, like the one embedded in experimental activity. Dunbar says

[29] On the role of conceptual change in childhood and in "intuitive" theories see [Carey, 1985; Carey *et al.*, 1996]. Analogies and differences between scientific and ordinary thought are illustrated in [Kuhn, 1991; Kuhn, 1996]. [Ram *et al.*, 1995] argue that a creative outcome is not an outcome of extraordinary mental processes, but of mechanisms that are on a continuum with those used in ordinary thinking.

scientists aim firstly at ensuring a robust internal structure of the experiment, optimizing the likelihood experiments will work, performing cost/benefits analysis on possible design components, ensuring acceptance of results in case of negotiation with other scientists of the community involved, and, finally, preferring experiments that have both conditions and control conditions [Dunbar, 1999, p. 95].

Finally, we have to remember that science happens, particularly at the "critical" moments, in a situation of distributed reasoning (see also [Thagard, 1997b]) by a group of scientists and not individual scientists. Abductive reasoning (to produce multiple hypotheses) and generalization are the main cognitive events that occur during social interactions among scientists. As I have already stressed in the previous chapter (section 1.6) and I will more clearly illustrate in the following chapters, real people (and so scientists) are some kinds of cognitive-epistemic "mediating structures" incorporating possible objective cognitive aims: epistemic structures can be embodied in artifacts, in ideas, but also in systems of social interactions.

2.6 Withdrawing Unfalsifiable Hypotheses Found through Explanatory and Instrumental Abduction

In the previous sections I have illustrated that contradiction is fundamental in abductive reasoning and that it has a preference for strong hypotheses which are more easily falsified than weak ones. Moreover, hard hypotheses may be more easily weakened than weak ones, which prove difficult subsequently to strengthen. Unfortunately, abductive hypotheses may be *unfalsifiable* and basically instrumental, such as in the case of hypotheses which are fruit of a radical *instrumentalist* abduction. In this case, it is impossible to find a contradiction from the empirical point of view but also from the theoretical point of view, in some area of the related conceptual systems. Notwithstanding this fact, it is sometimes necessary to construct ways of rejecting the unfalsifiable hypothesis at hand by resorting to some external forms of negation, external because we want to avoid any arbitrary and subjective elimination, which would be rationally or epistemologically unjustified.

In the following sections I will consider a kind of "weak" hypothesis in science that is hard to negate and the ways for making it easy. In these cases, the subject(s) can rationally decide to withdraw his hypotheses, and to activate abductive reasoning, even in contexts where it is impossible to find "explicit" contradictions; moreover, thanks to the new information reached simply by finding this kind of negation, the subject is free to abduce new hypotheses. I will explore whether *negation as failure* can be employed to model hypothesis withdrawal in Freudian analytic reasoning and in Poincaré's conventionalism of the principles of physics. The first case shows how conventions can be motivationally abandoned, the second one explains how the questioned problem of the probative value of clinical findings in psychoanalysis can be solved.

2.6.1 Negation as Failure in Query Evaluation

Computer and AI scientists have suggested an interesting technique for negating hypotheses and accessing new ones: negation as failure. The objective of this section is to consider how the use of *negation as failure* may be relevant to hypothesis withdrawal. There has been little research into the weak kinds of negating hypotheses, despite abundant reports that hypothesis withdrawal is crucial in everyday life and also in certain kinds of diagnostic or epistemological settings, such as medical reasoning and scientific discovery [Magnani, 2001b, chapters two and four].

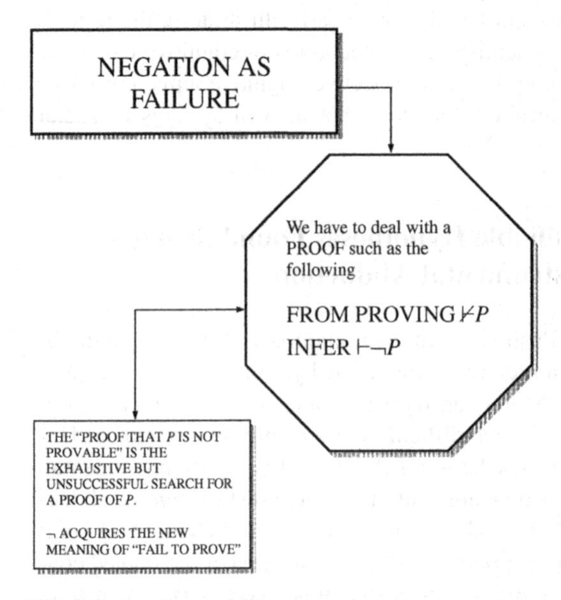

Fig. 2.4 Negation as failure

In the cases of conceptual change I describe, inferences are made using this kind of negation as a fundamental tool for advancing knowledge: new conclusions are issued on the basis of the data responsible for the failure of the previous ones. I plan to explore whether this kind of negation can be employed to model hypothesis withdrawal in Poincaré's conventionalism of the principles of physics and in Freudian analytic reasoning.

I consider this kind of logical account of negation, studied by researchers into logic programming, to be very important also from the epistemological point of view. Negation as failure is active as a "rational" process of withdrawing previously-abduced hypotheses in everyday life, but also in certain subtle kinds of diagnostic (analytic interpretations in psychoanalysis) and other epistemological settings. Contrasted with classical negation, with the double negation of intuitionistic logic, and with the philosophical concept of *Aufhebung*,[30] negation as failure shows how a subject can decide to withdraw his hypotheses, while maintaining the "rationality" of his argumentations, in contexts where it is impossible to find contradictions.

The statements of a logical database are a set of Horn clauses which take the form:

$$R(t_1,...,t_n) \leftarrow L_1 \wedge L_2 \wedge ... \wedge L_m$$

($m \geq 0, n \geq 0$, where $R(t_1,...,t_n)$ – conclusion – is the distinguished positive literal[31] and $L_1 \wedge L_2 \wedge ... \wedge L_m$ – conditions – are all literals, and each free variable is implicitly universally quantified over the entire implication). In more conventional notation this would be written as the disjunction

$$R(t_1,...,t_n) \vee \neg L_1 \vee \neg L_2 \vee ... \vee \neg L_m$$

where any other positive literal of the disjunctive form would appear as a negated precondition of the previous implication.

Let us consider a special query evaluation process for a logical database that involves the so-called negation as failure inference rule [Clark, 1978]. We can build a Horn clause theorem prover augmented with this special inference rule, such that we are able to infer $\neg P$ when every possible proof of P *fails*.

We know that a relational database only contains information about *true* instances of relations. Even so, many queries involve negation and we can answer them by showing that certain instances are *false*. For example, let's consider this simple case: to answer a request for the name of a student not taking a particular course, C, we need to find a student, S, such that the instance (atomic formula) $Takes(S,C)$ is false. For a logical database, where an atomic formula which is not explicitly given may still be implied by a general rule, the assumption is that an atomic formula is false if we *fail* to prove that it is true. To prove that an atomic formula P is *false* we do an exhaustive search for a proof of P. If *every* possible proof of P fails, we can infer $\neg P$. The well-known PROLOG programming language [Roussel, 1987] uses this method of manipulating negation.

We have to deal with a proof such as the following:

$$\text{from proving} \nvdash P \text{ infer} \vdash \neg P$$

where the "proof that P is not provable" [Clark, 1978, p. 120] is the *exhaustive* but *unsuccessful search* for a proof of P. Here the logical symbol $\neg$ acquires the new meaning of "fail to prove" (Figure 2.4).

Clark proposes a query evaluation algorithm based essentially on ordered linear resolution for Horn clauses (SLD) augmented by the negation as failure inference rule "$\neg P$ may be inferred if every possible proof of P fails" (SLDNF).[32]

[30] [Toth, 1991] models negation exploiting this Hegelian concept which he considers very significant for explaining non-Euclidean revolution.

[31] A *literal* is an atomic formula or the negation of an atomic formula.

[32] The links between negation as failure, completed databases [Clark, 1978, p. 120], and the closed world assumption have been classically studied in great detail. A survey can be found in [Lloyd, 1987].

What is the semantic significance of this kind of negation? Can we interpret a failed proof of P as a *valid* first order inference that P is false? Clark's response resorts to reconciling negation as failure with its truth functional semantics: if we can demonstrate that every failed attempt to prove P using the database of clauses B, is in effect a proof of $\neg P$ using the completed[33] database $C(B)$, then "negation as failure" is a derived inference rule for deductions from $C(B)$: the explicit axioms of equality and completion laws are therefore necessary at the object level in order to simulate failure of the matching algorithm at the meta-level. A negated literal $\neg P$ will be evaluated by recursively entering the algorithmic query evaluator (as an ordered linear resolution proof procedure, as stated above) with the query P. If every possible path for P ends in failure (failure proofs that can be nested to any depth), we return with $\neg P$ evaluated as true.

[Clark, 1978] has shown that for every meta-language proof of $\neg P$ obtained by a Horn clause theorem prover (query evaluation) augmented with negation as failure there exists a structurally similar object-language proof of $\neg P$. He has proved that a query evaluation with the addition of negation as failure will only produce results that are implied by first order inference from the completed database, that is, the evaluation of a query should be viewed as a "deduction" from the completed database (correctness of query evaluation). Consequently negation as failure is a sound rule for deductions from a completed database.

Although the query evaluation with negation as failure process is in general not complete, its main advantage is the efficiency of its implementation. There are many examples in which the attempt to prove neither succeeds nor fails, because it goes into a loop. To overcome these limitations it is sufficient to impose constraints on the logical database and its queries, and add loop detectors to the Horn clause problem solver: by this method the query evaluation process is guaranteed to find each and every solution to a query. However, because of the undecidability of logic, no query evaluator can identify all cases in which a goal is unsolvable. A best theorem prover does not exist and there are no limitations on the extent to which a problem solver can improve its ability to detect loops and to establish negation as failure.

2.6.2 Withdrawing Conventions and Instrumental Abduction

We will now consider some aspects dealing with Poincaré's famous conventionalism of the principles of physics and the possibility of negating conventions. From the point of view of radical instrumentalist abduction this example is striking because it shows how these abduced principles fail all tests that would reveal them as having a traditional epistemic value, so that they are not subject to discharge except for their instrumental value. An extension of Poincaré's so-called *geometric conventionalism*, according to which the choice of a geometry is only justified by considerations of simplicity, in a psychological and pragmatic sense ("commodisme"), is the *generalized conventionalism*, expressing the conventional character of the principles of physics:

[33] The notion of *database completion* can be found in [Clark, 1978], and in all textbooks on logic for computer science.

> The principles of mathematical physics (for example, the principle of conservation of energy, Hamilton's principle in geometrical optics and in dynamics, etc.) systematize experimental results usually achieved on the basis of two (or more) rival theories, such as the emission and the undulation theory of light, or Fresnel's and Neumann's wave theories, or Fresnel's optics and Maxwell's electromagnetic theory, etc. They express the common empirical content as well as (at least part of) the mathematical structure of such rival theories and, therefore, can (but need not) be given alternative theoretical interpretations [Giedymin, 1982, pp. 27–28].

From the epistemological point of view it is important to stress that the conventional principles usually survive the demise of theories and are therefore responsible for the continuity of scientific progress: in a sense they show a radical instrumental character Moreover, they are not empirically falsifiable; as stated by Poincaré in *Science and Hypothesis*:

> The principles of mechanics are therefore presented to us under two different aspects. On the one hand, they are truths founded on experiment, and verified approximately as far as almost isolated systems are concerned; on the other hand they are postulates applicable to the whole of the universe and regarded as rigorously true. If these postulates possess a generality and a certainty which the experimental truths from which they were deduced lack, it is because they reduce in final analysis to a simple convention that we have a right to make, because we are certain beforehand that no experiment can contradict it. This convention, however, is not absolutely arbitrary; it is not the child of our caprice. We admit it because certain experiments have shown us that it will be convenient, and thus is explained how experiment has built up the principles of mechanics, and why, moreover, it cannot reverse them [Poincaré, 1902, pp. 135–136].

Following Poincaré we can say that conventional principles of mechanics derive abductively from experience, as regards their "genesis", but cannot be falsified by experience because they contribute to "constitute" the experience itself, in a proper Kantian sense. The experience has only suggested their adoption because they are *convenient*: there is a precise analogy with the well-known case of geometrical conventions, but also many differences, which pertain the "objects" studied.[34]

Poincaré seeks also to stress that geometry is more abstract than physics, as is revealed by the following speculations about the difficulty of "tracing artificial frontiers between the sciences":

[34] The conventional principles of mechanics should not be confused with geometrical conventions: "The experiments which have led us to adopt as more convenient the fundamental conventions of mechanics refer to bodies which have nothing in common with those that are studied by geometry. They refer to the properties of solid bodies and to the propagation of light in a straight line. These are mechanical, optical experiments" [Poincaré, 1902, pp. 136–137], they are not, Poincaré immediately declares, "*des expériences de géométrie*" (*ibid.*): "And even the probable reason why our geometry seems convenient to us is, that our bodies, our hands, and our limbs enjoy the properties of solid bodies. Our fundamental experiments are pre-eminently physiological experiments which refer, not to the space which is the object that geometry must study, but to our body - that is to say, to the instrument which we use for that study. On the other hand, the fundamental conventions of mechanics and experiments which prove to us that they are convenient, certainly refer to the same objects or to analogous objects. Conventional and general principles are the natural and direct generalisations of experimental and particular principles" (*ibid.*)

> Let it not be said that I am thus tracing artificial frontiers between the sciences; that I am separating by a barrier geometry properly so called from the study of solid bodies. I might just as well raise a barrier between experimental mechanics and the conventional mechanics of general principles. Who does not see, in fact, that separating these two sciences we mutilate both, and that what will remain of the conventional mechanics when it is isolated will be but very little, and can in no way be compared with that grand body of doctrine which is called geometry [Poincaré, 1902, pp. 137–138].

I believe that the meaning of this passage refers primarily to the fact that physics cannot be considered completely conventional because we know that the conventional "principles" are derived from the "experimental laws" of "experimental mechanics", and then absolutized by the "mind". Second, Poincaré wants to demonstrate how geometry is more abstract than physics: geometry does not require a rich experimental reference as physics does, geometry only requires that experience regarding its genesis and as far as demonstrating that it is the most convenient is concerned. Here we are very close to Kant's famous passage about the *synthetical a priori* character of the judgments of (Euclidean) geometry, and of the whole of mathematics: "The science of mathematics presents the most splendid example of the extension of the sphere of pure reason without the help of the experience" [Kant, 1929, A712-B740, p. 576].

Even when separated from the reference to solid bodies, Euclidean geometry maintains all its conceptual pregnancy, as a convention that, in a proper Kantian sense, "constitutes" the ideal solid bodies themselves. This is not the case of the conventional principles of mechanics when separated from experimental mechanics: "[...] what will remain of the conventional mechanics [...] will be very little" if compared "[...] with that grand body of doctrine which is called geometry".

Poincaré continues:

> Principles are conventions and definitions in disguise. They are, however, derived from experimental laws, and these laws have, so to speak, been erected into principles to which our mind attributes an absolute value. Some philosophers have generalized far too much. They have thought that the principles were the whole of science, and therefore that the whole of science was conventional. This paradoxical doctrine, which is called nominalism, cannot stand examination. How can a law become a principle? [Poincaré, 1902, p. 138].

If the experimental laws of experimental physics are the source of the conventional principles themselves, conventionalism escapes nominalism.

As stated at the beginning of this section, conventional principles survive the demise (falsification) of theories in such a way that they underlie the incessant spectacle of scientific revolutions: "It is the mathematical physics of our fathers which has familiarized us little by little with these various principles; which has habituated us to recognize them under the different vestments in which they disguise themselves" [Poincaré, 1905, p. 95]. Underlying revolutions of physics, conventional principles guarantee the historicity and the growth of science itself. Moreover, the conventional principles surely imply "[...] *firstly*, that there has been a *growing tendency* in modern physics to *formulate and solve* physical problems *within powerful, and more abstract, mathematical systems of assumptions* [...]; *secondly*, the role of conventional principles has been growing and *our ability to discriminate experimentally between alternative abstract systems* which, with a great approximation, save the phenomena *has been diminishing* (by comparison to the testing of simple conjunctions of empirical generalizations)" [Giedymin, 1982, p. 28].

Moreover, as stated above, they are not empirically falsifiable: "The principles of mechanics [...] reduce in final analysis to a simple convention that we have a right to make, because we are certain beforehand that no experiment can contradict it" [Poincaré, 1902, p. 136].

Up to now I have considered in details how the conventional principles guarantee the revolutionary changes of physics and why they cannot be considered arbitrary, being motivated by – and abduced from – the *experimental laws* of the "experimental physics", that is by experience. Although arbitrary and conventional, the conventional principles too can be substituted by others. This is the main problem treated by Poincaré in the last passages of Chapter IX, "The Future of Mathematical Physics", in *The Value of Science*. Already the simple case of "linguistic" changes in science "[...] suffices to reveal generalizations not before suspected" [Poincaré, 1905, p. 78]. By means of the new discoveries, scientists arrive at a point where they are able to "[...] admire the delicate harmony of numbers and forms; they marvel when a new discovery opens to them an unexpected perspective" [Poincaré, 1905, p. 76], a new perspective that is always provisional, fallible, open to further confirmations or falsifications when compared to rival perspectives.

We have seen how the conventional principles of physics guarantee this continuous extension of experience thanks to the various perspectives and forms expressed by experimental physics. However, because conventional, "no experiment can contradict them". The experience only suggested the principles, and they, since absolute, have become constitutive just of the empirical horizon common to rival experimental theories.

Poincaré observes:

> Have you not written, you might say if you wished to seek a quarrel with me – have you not written that the principles, though of experimental origin, are now unassailable by experiment because they have become conventions? And now you have just told us that the most recent conquests of experiment put these principles in danger. Well, formerly I was right and today I am not wrong. Formerly I was right, and what is now happening is a new proof of it [Poincaré, 1905, p. 109].

Poincaré appeals to a form of weak negation, just as Freud did when dealing with the problem of withdrawing constructions in the analytic setting (cf. the following subsection). Let us follow the text. To pursue his point, Poincaré illustrates the attempts to reconcile the "calorimetric experiment of Curie" with the "principle of conservation of energy":

> This has been attempted in many ways; but there is among them one I should like you to notice; this is not the explanation which tends to-day to prevail, but it is one of those which have been proposed. It has been conjectured that radium was only an intermediary, that it only stored radiations of unknown nature which flashed through

> space in every direction, traversing all bodies, save radium, without being altered by this passage and without exercising any action upon them. Radium alone took from them a little of their energy and afterward gave it out to us in various forms [Poincaré, 1905, pp. 109–110].

At this point Poincaré resolutely asserts: "What an advantageous explanation, and how convenient! First, it is unverifiable and thus irrefutable. Then again it will serve to account for any derogation whatever to Mayer's principle; it answers in advance not only the objection of Curie, but all the objections that future experimenters might accumulate. This new and unknown energy would serve for everything" (p. 110). Now Poincaré can show how this *ad hoc* hypothesis can be identified with the non-falsifiability of the conventional principle of the conservation of energy:

> This is just what I said, and therewith we are shown that our principle is unassailable by experiment. But then, what have we gained by this stroke? The principle is intact, but thenceforth of what use is it? It enabled us to foresee that in such and such circumstance we could count on such total quantity of energy; it limited us; but now that this indefinite provision of new energy is placed at our disposal, we are no longer limited by anything [Poincaré, 1905, p. 110].

Finally, Poincaré's argumentation ends by affirming negation as failure: "[...] and, as I have written in 'Science and Hypothesis', if a principle ceases to be fecund, experiment without contradicting it directly will nevertheless have condemned it" (*ibid.*) (cf. Figure 2.5).

Let us now analyze this situation from the epistemological point of view: the conventional principle has to be withdrawn when it "ceases to be fecund" and so because it is no longer endowed with an acceptable degree of *strategical plausibility*, or when it seems that we have failed to prove it. It is clear that the principle exhibits in this case a kind of strategic, rather than propositional, plausibility, as I have described in subsection 2.3.1. Remember that for a logic database the assumption is that an atomic formula is false if we *fail* to prove that it is true. More clearly: as stated above, every conventional principle, suitably underlying some experimental laws, generates *expectations* with regard to the subsequent evidences of nature. I analogously consider as the proof of a conventional principle the fact that we can increasingly *extend* and complete the experimental laws related to it, adding the new (expected) evidence that "emerges" from the experimental research. If, after a finite period of time, nature does not provide this new "evidence" that is able to increase the fecundity of the conventional principle, this *failure* leads to its withdrawal: "[...] experiment without contradicting it directly will nevertheless have condemned it". Analogously to the Freudian case of constructions I will illustrate in the following subsection the "proof that a principle is not provable"[35] is the unsuccessful search for a proof of the principle itself. Here too, the logical symbol $\neg$ acquires the new meaning of "fail to prove" in the empirical sense.

Let us resume: if the old conventional principle does not produce new experimental "evidence" to underpin it, it is legitimate to abandon the principle, when

[35] Please keep in mind I am making an analogy between "not provable" and "not empirically fecund".

convenient: the opportunity to reject the old principle will happen just by exploiting the experimental evidence which, even if not suitable for contradicting it (Poincaré says, it is "unassailable by experiment"), is nevertheless suitable as a basis for conceiving a new alternative principle, generated by new creative abductions.

We can now interpret Popper's ideas about conventionalism in a different way. Popper writes: "Thus, according to the conventionalist view, it is not possible to divide systems of theories into falsifiable and non-falsifiable ones; or rather, such a distinction could be ambiguous. As a consequence, our criterion of falsifiability must turn out to be useless as a criterion of demarcation" [Popper, 1959, p. 81]. In the light of Poincaré's theory of the principles of physics that we have just illustrated, the nominalistic interpretation of conventionalism given by Popper (see also [Popper, 1963]) appears to be very reductive. Moreover, Popper's tendency to identify conventions with *ad hoc* hypotheses (a very bad kind of auxiliary hypotheses) is shown to be decidedly unilateral, since, as is demonstrated by the passages, immediately above, the *adhocness* is achieved only in a very special case, when the conventional principle is epistemologically exhausted.

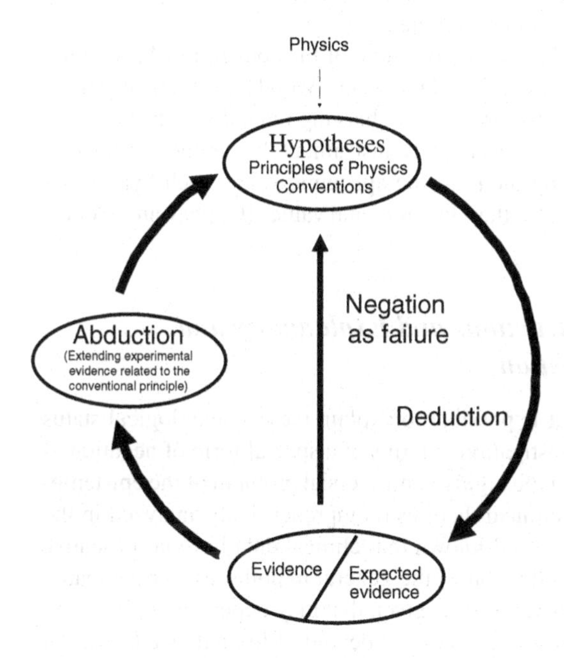

Fig. 2.5 Withdrawing conventions

In some sense Poincaré was already aware of the following fact, subsequently clearly acknowledged by Popper: the introduction of auxiliary hypotheses must not diminish the degree of falsifiability or testability (that will have to be performed by means of "severe tests") of the scientific theory in question, but, on the contrary, should increase it [Popper, 1959, p. 83]. The example of an unsatisfactory (*ad hoc*) auxiliary hypothesis given by Popper is the "[. . .] contraction hypothesis of Fitzgerald and Lorentz which had no falsifiable consequences but merely served to restore the agreement between theory and experiment – mainly the findings of Michelson and Morley" [Popper, 1959, p. 83].

In turn Lakatos' revision of falsificationism in terms of the theory of research programs [Lakatos, 1970] has definitely established that modified hypotheses (by means of auxiliary assumptions) have to be more falsifiable than the original versions, they have to lead to new testable consequences (and, moreover, independently testable, to use Popper's phrase – [Popper, 1959, p. 193]; progress in "scientific programs" is heavily related to the existence of novel predictions: one program is superior to another insofar as it is a more successful predictor of novel phenomena.

Something analogous – but in a more indirect way – operates in the case of the conventional principles described by Poincaré: it seems that conventionalism, at least in the Poincaré's case, does not treat all hypotheses like "stratagems", as maintained by Popper. Hypothetical conventional principles are unfalsifiable and should be withdrawn only when exhausted, when their indirect production of "novel" evidence is finished. Consequently, Poincaré's conventionalism is not simply a theory of *adhocness*, in the nominalistic Popperian sense.

As regards instrumentalist abduction my example of the conventional principles of physics shows a cognitive situation that Gabbay and Woods synthetically illustrate in the following way: "Proposition 5.7 (Discharging radically instrumentalist hypotheses) Since the hypothesis of a radically instrumentalist abduction fails all tests that would reveal it as having the requisite epistemic value, such hypotheses are not subject to discharge except for their instrumental value" [Gabbay and Woods, 2005, p. 120].

2.6.3 Withdrawing Constructions and Explanatory and Instrumental Abduction

First of all I will illustrate how it is possible to explain the epistemological status of Freud's method of clinical investigation in terms of a special form of negation as failure. I am not dealing here with the highly controversial problem of the epistemological status of psychoanalytic clinical theories (comprehensively analyzed in the classical [Grünbaum, 1984]): it is well-known that clinical data have no probative value for the confirmation or falsification of the general hypotheses of psychoanalytic clinical theories of personality, because, given that they depend completely on the specific nature of the clinical setting, they are devoid of the independence that characterizes observations endowed with scientific value.

Furthermore, because of the lack of probative value in the patient's clinical data with regard to the analyst's interpretations, any therapeutic gains from analysis may be considered to have been caused not by true insightful self-discovery but rather by placebo effects induced by the analyst's powers of suggestion. If the probative value of the analysand's responses is negated, then Freudian therapy might reasonably be considered to function as an emotional corrective (performed by a positive "transference" effect) and not because it enables the analysand to acquire self-knowledge;

instead he or she capitulates to proselytizing *suggestion*, which operates the more insidiously since under the pretense that analysis is nondirective. Suggestion is indeed responsible for the so-called epistemical contamination of the patient's responses.

Freud asks the patient to believe in the analyst's theoretical retrodictions of significant events in his early life and these theoretical retrodictions are communicated to him as *constructions* – actually they are *explanatory* abductions derived on the basis of some present evidence furnished by the patient:

> The analyst finishes a piece of construction and communicates it to the subject of the analysis so that it may work upon it; he then constructs a further piece out of the fresh material pouring in upon him, deals with it in the same way and proceeds in this alternating fashion until the end [Freud, 1953-1974, vol, 23, 1937, pp. 260–261].

The aim is to provoke the previously-cited true insightful self-discovery that guarantees the cure [Freud, 1953-1974, vol. 18, 1920, p. 18]. A single construction is built as a "sequence" of the interpretations (that have an obvious abductive character) that issue from clinical data found in the clinical setting, epistemologically characterized by "transference" and "countertransference":

> "Interpretation" applies to something that one does to some single element of the material, such as an association or a parapraxis. But it is a "construction" when one lays before the subject of the analysis a piece of his early history that he has forgotten, in some such way as this: "Up to your *n*th year you regarded yourself as the sole and unlimited possessor of your mother; then came another baby and brought you grave disillusionment. Your mother left you for some time, and even after her reappearance she was never again devoted to you exclusively. Your feelings towards your mother became ambivalent, your father gained a new importance for you," and so on [Freud, 1953-1974, vol. 23, 1937, p. 261].

A construction can be considered as a kind of "explanatory" "history" or "narrative" abductively obtained of the analysand's significant early life events, which is never complete, but that can be rendered more and more comprehensive by adding new interpretations. This abductive process, I call selective, presents a constitutive uncertainty due to its *nonmonotonicity* (cf. chapter one, section 1.3, this book), the analyst may always withdraw his or her interpretations (constructions) when new evidence arises. Every construction is generated by a "double" abduction: first of all the analyst has to select a suitable general psychoanalytic hypothesis, apply it to some "single element of the material" to produce an interpretation, then he/she has to select each of these general hypotheses in such a way that the sequence of the generated interpretations can give rise to a significant and consistent construction. Every "abduced" construction, suitably connected with some other clinical psychoanalytical hypotheses, generates expectations with regard to the analysand's subsequent responses and remarks.

Let us remember that Habermas considers therapy as due to a sort of Hegelian causality of fate: the analyst applies what Habermas calls "general interpretations"[36] [Habermas, 1968, p. 279] to the analysand's clinical data. This application generates

[36] They correspond to general "schemes" of possible constructions.

particular interpretations that combine into a "narrative" (Freud's "construction"). Within the scientophobic framework of Habermas's philosophy this application is regarded as "hermeneutic", because the constructions are presumed to be expressed in the "intentional" and motivational language of desires, affects, fantasies, sensations, memories, etc. We can more easily consider them abduction without resorting to the hermeneutical lexicon.

Of course the analyst aims at building *the most complete* construction. The problem here is the analyst cannot propose to the analysand any construction he wants, without some form of external testing. As stated above, the objection most often raised against psychoanalysis is that "[...] therapeutic success is *non*probative because it is achieved *not* by imparting veridical insight but rather by the persuasive suggestion of fanciful pseudoinsights that merely ring verisimilar to the docile patient" [Grünbaum, 1984, p. 138]. In one of his last papers, "Constructions in analysis" [Freud, 1953-1974, vol, 23, 1937, pp. 257–269], Freud reports that "a certain well-known man of science" had been "at once derogatory and unjust" because

> He said that in giving interpretations to a patient we treat him upon the famous principle of "Heads I win, tails you lose" [In English in the original]. That is to say, if the patient agrees with us, then the interpretation is right, but if he contradicts us, that is only a sign of his resistance, which again shows that we are right. In this way we are always in the right against the poor helpless wretch whom we are analysing, no matter how he may respond to what we put forward [Freud, 1953-1974, vol, 23, 1937, p. 257].

Freud looks for a criterion for justifying, in the clinical setting, the abandonment of constructions that have been shown to be inadequate (it is interesting to note that in the cited article Freud emphasizes the provisional role of constructions referring to them also as "hypotheses" or "conjectures"). This is the fundamental epistemological problem of the method of clinical investigation: Freud is clear in saying that therapeutic success will occur only if incorrect analytic constructions, spuriously confirmed by "contaminated" responses from the patient, are discarded in favor of new correct constructions (that are constitutively *provisional*) derived from clinical data not distorted by the patient's compliance with the analyst's communicated expectations.

Freud then proceeds "[...] to give a detailed account of how we are accustomed to arrive at an assessment of the 'Yes' or 'No' [considered as "direct evidences"] of our patients during analytic treatment – of their expression of agreement or of denial" (p. 257).

Analytic constructions cannot be falsified by dissent from the patient because "[...] it is in fact true that a 'No' from one of our patients is not as a rule enough to make us abandon an interpretation as incorrect" (p. 257). It might seem to Freud that patient dissent from an interpretation can be always discounted as inspired by neurotic resistance. It is only "in some rare cases" that dissent "turns out to be the expression of legitimate dissent" (p. 262). A "patient's 'No' is no evidence of the correctness of a construction, though it is perfectly compatible with it" (p. 263). Rather, a patient's 'No' might be more adequately related to the "incompleteness" of the proposed constructions: "[...] the only safe interpretation of his 'No' is that it points to incompleteness" (p. 263).

Even if a patient's verbal assent may result from genuine recognition that the analyst's construction is true, it may nevertheless be spurious because it derives from neurotic resistance, as already seen in his or her dissent. Assent is "hypocritical" when it serves "[...] to prolong the concealment of a truth that has not been discovered" (p. 262). On the other hand, assent is genuine and not hypocritical when patient's verbal assent will be followed and accompanied by new memories: "The 'Yes' has no value unless it is followed by indirect confirmations, unless the patient, immediately after his 'Yes', produces new memories which complete and extend the construction" (p. 262) .

Since "Yes" and "No" do not have any importance to test a construction it is necessary to see other facts, such as "the material" that has "come to light" after having proposed a construction to the patient:

> [...] what in fact occurs [...] is rather that the patient remains as though he were untouched by what has been said and reacts to it with neither a "Yes" nor a "No". This may possibly mean no more than that his reaction is postponed; but if nothing further develops we may conclude that we have made a mistake and we shall admit as much to the patient at some suitable opportunity without sacrificing any of our authority (pp. 261–262).

Let us now analyze this situation from the epistemological point of view: the analyst has to withdraw the construction (a narrative complex hypothesis) when he has failed to prove it. Remember that for a logic database the assumption is that an atomic formula is false if we *fail* to prove that it is true. More precisely: as stated above, every construction, suitably connected with some other clinical psychoanalytical hypotheses, generates expectations with regard to the analysand's subsequent responses and remarks. We consider the fact that we can continuously extend and complete a construction by adding the new (*expected*) material that "has come to light" from the patient as proof of the construction validity. If the patient does not provide this new "material" which is able to extend the proposed construction, this *failure* leads to the withdrawal of the construction itself. So the "proof that a construction is not provable" is the *unsuccessful search* for a proof of the construction itself. Here the logical symbol $\neg$ acquires the new meaning of "fail to prove" in the empirical sense.[37]

Let us resume: if the patient does not provide new "material" which extends the proposed construction, "if", as Freud declares, "[...] nothing further develops we may conclude that we have made a mistake and we shall admit as much to the patient at some suitable opportunity without sacrificing any of our authority". The "opportunity" of rejecting the proposed construction "will arise" just

> [...] when some new material has come to light which allows us to make a better construction and so to correct our error. In this way the false construction drops out, as if it has never been made; and indeed, we often get an impression as though, to borrow the words of Polonius, our bait of falsehood had taken a carp of truth (p. 262).

[37] Again, please keep in mind in this case I am making an analogy between "not provable" and "not empirically fecund".

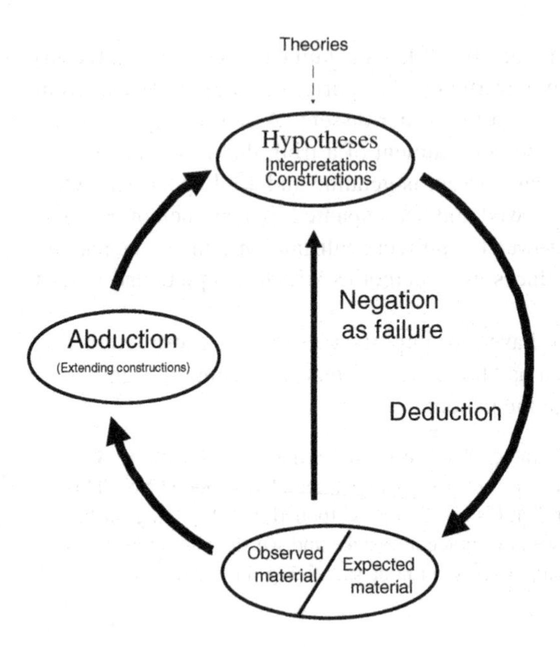

Fig. 2.6 Withdrawing constructions

A new cycle very similar to the one previously started with the assumption of the first construction takes place: a new construction (derived by applying new clinical psychoanalytical hypotheses and schemes) is provisionally conceived on the basis of the new material that came to light when the analyst was seeking to extend the old one ("we often get an impression as though [...] our bait of falsehood had taken a carp of truth") (cf. Figure 2.6). The inferential process is clearly *nonmonotonic*: in an initial phase we have some material coming from the patient and which provides the background for an initial abduced construction; in a second phase we have to add to the initial construction the material which emerges after having communicated to the patient the first construction. If in this second phase the new material is not suitable for extending the first construction, the negation as failure compels the analyst to withdraw and reject the construction. The whole process is nonmonotonic because the increase of material does not generate an increase in (the number of) constructions: the old construction is abandoned ("the false construction drops out, as if it has never been made").

I should stress that the epistemological role of what Freud calls "*indirect confirmations*" or *disconfirmations* of analytic constructions is in my opinion negligible. These patient responses, other than verbal assent or dissent, Freud declares, "are in every respect trustworthy" (p. 263). Examples are when a patient has a mental association whose content is similar to that of the construction, or when a patient commits a parapraxis as part of a direct denial. Moreover, when a masochistic patient is averse to "receiving help from the analyst", an incorrect construction will not affect his symptoms, but a correct one will produce "[...] an unmistakable aggravation of his symptoms and of his general condition" (p. 265). The indirect confirmations, in Freud's opinion, provide a "[. . .] valuable basis for judging whether the construction is likely to be [further] confirmed in the course of analysis" (p. 264).

We should not confuse the kind of transformations of "No" in "Yes" (or vice versa) that pertains to the "indirect confirmations" or disconfirmations, with the above-described extension of constructions toward the most complete one on the basis of new (expected) material that emerges. In conclusion, a patient's "Yes" or "No", whether direct or indirect, has no role to play in withdrawing constructions. I think that these cases do not have any function in the process of abandoning a construction because they always keep open the possibility of extending it: moreover, the indirect confirmations or disconfirmations do not increase the acceptability of constructions. In my opinion, we should not consider Freud as an inductivist, despite his emphasis on these kind of indirect evidence.

Moreover, as stated by Grünbaum, who tends to consider Freudian description of clinical method as inductive, this presumption of the consilience of clinical inductions is "spurious" because

> [. . .] the *independence* of the inferentially concurring pieces of evidence is grievously jeopardized by a *shared* contaminant: the analyst's influence. For *each* of *seemingly* independent clinical data may well be more or less alike confounded by the analyst's suggestion as to conform to his construction, at the cost of their epistemic reliability or probative value. For example a "confirming" early memory may be compliantly produced by the patient on the heels of giving docile assent to an interpretation [Grünbaum, 1984, p. 277].

The second section of "Constructions in analysis" concludes with a very explicit affirmation of nonmonotonicity:

> Only the further course of the analysis enables us to decide whether our constructions are correct or unserviceable. [. . .] an individual construction is anything more than a conjecture which awaits examination, confirmation or rejection. We claim no authority for it, we require no direct agreement from the patient, nor do we argue with him if at first he denies it. In short, we conduct ourselves on the model of a familiar figure in one of Nestroy's farces – the manservant who has a single answer on his lips to every question or objection: "It will all become clear in the course of future developments" (p. 265).

But I have shown that Freud considers important only the rejection achieved by negation as failure. The epistemological aim is not to validate a construction by extensions provided by new material or by indirect confirmations or disconfirmations. Freud aims to reject it. Perhaps in Freud's considerations there are some ambiguities in perceiving the asymmetry between falsification and confirmation, but it would seem that my interpretation of Freud as a special falsificationist can be maintained without fear of distorting his methodological intention. Freud is a special type of "falsificationist" because negation as failure guarantees the possibility of freely withdrawing an abduced construction and substituting it with a rival and better one. In the computational case, negation as failure is achieved by suitable algorithms related to the knowledge that is handled (see above, subsection 2.6.1).

In the human and not computational case, negation as failure is played out in the midst of the analyst-analysand interaction, where transference and countertransference are the human epistemological operators and "reagents". Negation as failure is therefore a limitation on the dogmatic and autosuggestive exaggerations of (pathological) countertransference.

We can say again, from the point of view of instrumentalist abduction that my example shows a cognitive situation that Gabbay and Woods synthetically illustrate in the following way and that we have already quoted: "Proposition 5.7 (Discharging radically instrumentalist hypotheses) Since the hypothesis of a radically instrumentalist abduction fails all tests that would reveal it as having the requisite epistemic value, such hypotheses are not subject to discharge except for their instrumental value" [Gabbay and Woods, 2005, p. 120]. In this Freudian case the explanatory aspects of hypothetical interpretations and constructions shades in their final radical instrumentalist/strategic character. They are only indirectly activated for further explanatory premissory work; so to say, they are always constitutively taken on epistemological explanatory sufferance because no other chances are practicable, and their main epistemic (instrumental, strategic) virtue – which guarantees they are withdrawable – is stimulating the emergence of new material able to enrich the construction.[38]

2.7 Automatic Abductive Scientists

Paul Thagard [1988] illustrates four kinds of abduction that have been implemented in PI, a system devoted to explaining in computational terms the main problems of the traditional philosophy of science, such as scientific discovery, explanation, evaluation, etc. He distinguishes between simple, existential, rule-forming, and analogical abduction. Simple abduction generates hypotheses about individual objects. Existential abduction postulates the existence of previously unknown objects, such as new planets. Rule-forming abduction generates rules that explain laws. Analogical abduction uses past cases of hypothesis formation to construct hypotheses similar to existing ones. If the pure philosophical task is to state correct rules of reasoning in an abstract and objective way, the use of computer modeling may be a rare tool to investigate abduction in science because of its rational correctness. The increase in knowledge provided by this intellectual interaction is manifest.

Early works on *machine scientific discovery*, such as the well-known Logic Theorist [Newell *et al.*, 1957], DENDRAL, in chemistry [Lindsay *et al.*, 1980], and AM, in mathematics, [Lenat, 1982], have shown that *heuristic search* in combinatorial spaces is an advantageous and general framework for automating scientific

[38] Other interesting examples of instrumentalist abduction in science are illustrated in [Gabbay and Woods, 2005, p. 120]: the axiom of choice – which has an abductive role in proofs of Löwenheim-Skolem theorem, Planck's discovery of the quantum hypothesis (cf. above p. 79), and the hypothesis of *gravitons*.

discovery[39]. In these programs abduction is mainly rendered in a sentential way, using rules and heuristics.

There are many ways for identifying a commonality in computational scientific discovery programs that will take a next step beyond the acknowledged general – but weak – framework of heuristic search (cf. also [Tweney, 1990]. For example, [Valdés-Pérez, 1999], characterizes discovery in science as the generation of novel, interesting, plausible, and intelligible knowledge about the objects of study. Looking for a common general pattern he analyzes four machine discovery programs that match those requirements in different ways:

1. MECHEM, which hypothesizes reaction mechanisms in chemistry based on the available experimental evidence [Zeigarnik *et al.*, 1997]
2. ARROSMITH, which notices connections between drugs or dietary factors and diseases in medicine [Swanson and Smalheiser, 1997]
3. GRAFFITI, which makes conjectures in graph theory and other similar mathematical fields [Fajtlowicz, 1988]
4. MDP/KINSHIP, which delineates the classes within a classification in linguistics [Pericliev and Valdés-Pérez, 1998].

In turn [Boden, 1992] expecially stresses the distinction between classical programs able to *re-produce* historical cases of scientific discovery in physics (BACON systems and GLAUBER, [Langley *et al.*, 1987]), and systems able to perform *new* discoveries (DENDRAL and AM, cited above). Other authors (for example, [Shunn and Klahr, 1995], who constructed the program ACT-R) emphasize the distinction between computational systems that address the process of hypothesis formation and evaluation (BACON; PHINEAS, [Falkenhainer, 1990]; AbE, [O'Rorke *et al.*, 1990]; ECHO, [Thagard, 1989; Thagard, 1992]; TETRAD, [Glymour *et al.*, 1987]; MECHEM), those that address the process of experiment (like DEED, [Rajamoney, 1993]; DIDO, [Scott and Markovitch, 1993]), and, finally, those that address both the processes (like KEKADA, [Kulkarni and Simon, 1988]; SDDS, [Klahr and Dunbar, 1988], LIVE, [Shen, 1993], and others).

All these AI systems explicitly or implicitly perform epistemological tasks. From the point of view of the task of abduction it is interesting to note that some of them model a kind of sentential creative abduction, others are dealing with model-based creative abduction, and there are also the ones related to model the activity of experiment (that relate to the problem of what I call "manipulative abduction").

Sentential creative abduction. In the first case we have to note that systems like BACON, GLAUBER, built in terms of heuristic search, notwithstanding they

[39] Already in 1995 the AAAI Society organized a Spring Symposium on "Systematic Methods of Scientific Discovery" and in 1997 the Journal *Artificial Intelligence* devoted a special issue to "Machine Discovery" (91, 1997, – [Okada and Simon, 1997]). Classical books where the reader can find the description of the most interesting research and the description of historical machine discovery programs are [Langley *et al.*, 1987], and [Shrager and Langley, 1990]. Cf. also [Zytkow, 1992] (Proceedings of MD-92 Workshop on "Machine Discovery"), and [Colton, 1999] (Proceedings of AISB'99). In 1990 AAAI Society already organized a Spring Symposium on the problem of "automated abduction", devoted to the illustration of many computational programs able to perform various abductive tasks.

perform outputs that can be presented as a fruit of the creative abductive task of reproducing well-known past discovery of physics, they actually execute a selective abduction: starting from given data, they just have to "select" among a pre-stored encyclopedia of mathematical equations capable of explaining the data. Consequently they are similar, because of the epistemology of their architecture, to the computational programs devoted to perform diagnostic reasoning in medicine (cf. [Magnani, 2001b, chapter four]).

Model-based creative abduction. In the second case the programs are capable of performing model-based abductions: for example by providing causal and analogical reasoning, like the previously cited AbE (theory revision in science), CHARADE (discovery of the causes of scurvy) [Corruble and Ganascia, 1997], CDP (discovery of urea cycle, [Grasshoff and May, 1995]), GALATEA (explanation tasks) [Davies and Goel, 2000], PROTEUS [Davies *et al.*, 2009] (analogical reasoning), PHINEAS, that exploits the representational resources of qualitative physics [Forbus, 1984; Forbus, 1986]. to perform analogical reasoning in liquid flow.[40] AbE and PHINEAS explicitly and directly refer to abductive tasks, other programs employ the word induction, even if they are achieving a more complicated task than mere generalization from data.[41] A system that explicitly addresses model-based abduction (the so-called generic modeling) in science is TORQUE [Nersessian *et al.*, 1997], devoted to perform tasks of visualizations able to account for various cases of discovery in science (Faraday, Maxwell).[42]

More recent AI programs have been built to simulate abduction in mathematical and geometrical reasoning. This is the case of ARCHIMEDES ([Lindsay, 1994; Lindsay, 1998; Lindsay, 2000b; Lindsay, 2000a], cf. also the last section of this chapter, section 2.12), which realizes cases of manipulative diagrammatic abduction in elementary geometry, and HR [Colton, 1999; Pease *et al.*, 2005], which creates new concepts in the field of algebra also taking advantage of Lakatosian epistemology of formal reasoning. [Trickett and Trafton, 2007] recently stressed the role of conceptual simulation in the so called "what if" reasoning (mental experiment, thought experiment, inceptions, mental simulations) in scientists' strategies to resolve informational uncertainty. Actually the analysis illustrates the abductive role played by a wide range of model-based and manipulative ways of discovering in science. For a recent survey of the relationships between computation and the problem

[40] A system that aims at constructing causal hypotheses is TETRAD [Glymour *et al.*, 1987], but it manipulates numeric data – and not model-based types of reasoning – and is deeply entrenched in a probabilistic framework.

[41] On the ambiguities and relationships between abduction and induction cf. chapter one, subsection 1.4.1 and chapter seven, section 7.4.

[42] Other tools that could be proven useful in the area of abduction and machine discovery come from the field of genetic algorithms and evolving neural networks (cf. [Pennock, 1999; Pennock, 2000]), where creative reasoning is studied improving Darwinian mechanisms described by evolutionary theories, and may be from the very recent so-called DNA computers [Boneh *et al.*, 1996].

of scientific explanation and discovery in philosophy of science cf. [Thagard and Litt, 2008]..[43]

Manipulative abduction. In the third case, when dealing with the simulation of experiment, the computational programs join the area of manipulative abduction. An interesting and neglected point of contention about human reasoning is whether or not concrete manipulations of external objects influence the generation of hypotheses, for example in science: in the following chapter I will delineate the first features of what I call *manipulative abduction* showing how we can find methods of constructivity in scientific and everyday reasoning based on external models and "epistemic mediators". Manipulation of external objects in scientific experiments realizes a kind of epistemic mediation, also exploiting the cognitive resources of human body and its performances. The discovery programs that address the process of experiment constitute the first attempt to automatize these abilities, that could further extend the interest of machine discovery in science also to the whole area of robotics.[44]

It is well known that epistemology and logic are not alone in investigating reasoning. Reasoning is also a major subject of investigation in AI cognitive psychology, and the whole area of cognitive science. Epistemological (and logical) theories of reasoning, when implemented in a computer, become AI programs. The theories and the programs are, quite literally, two different ways of expressing the same thing.

[43] Recent workshops and conferences that cover computational AI applications that involve abductive processes have been organized worldwide. Here a list of the more recent ones: AAAI Symposium on Automated Scientific Discovery, Stanford, November 2008; International Joint Workshop on Computational Creativity (IJWCC2008), Madrid, Spain, September, 2008; Fourth Joint Workshop on Computational Creativity (ECAI2006), London, UK, June 2007; Workshop Abduction and Induction in AI and Scientific Modeling (ECAI2006), Riva del Garda, Italy, August 2006; Third Joint Workshop on Computational Creativity (ECAI2006), Riva del Garda, Italy, August 2006; Fourth International Workshop on Computational Models of Scientific Reasoning and Applications (CMSRA-IV) Lisbon, Portugal, September 2005; The Second Joint International Workshop on Computational Creativity (IJCAI'05, International Joint Conference of Artificial Intelligence), UK, August 2005; Workshop on Chance Discovery: from Data Interaction to Scenario Creation, The 22nd International Conference on Machine Learning (ICML 2005), Bonn, Germany, August 2005; Computational Creativity 2004, at the Seventh European Conference on Case-Based Reasoning (ECCBR-04), Madrid, Spain, 2004; The 6th International Conference on Discovery Science, Sapporo, Japan, 2003; The Third International Workshop on Computational Models of Scientific Reasoning and Applications (III CMSRA), Buenos Aires, Argentina, 2003. To update knowledge about recent research in the interesting field of AI the reader is directed to the web pages related to the events above. Of particular interest to the problem of manipulative abduction are the studies on natural language and visual interpretation in the context of human-robot interaction as a mathematical abductive process, where every potential interpretation has an associated set of relevant manipulative actions that an agent should perform in every reasoning and epistemic decision step and decision about of reasoning.

[44] [Clark, 2008, p. 202] is rather optimistic about the possibility that in the near future classical artificial intelligence and robotics will be able to include aspects related to embodiment, action, and situatedness: "The increasingly popular image of functional, computational, and information-processing approaches to mind as flesh-eating demons is thus subtly misplaced. For rather than necessarily ignoring the body, such approaches may instead help target larger organizational wholes in ways that help reveal where, why, how, and even how much [...] embodiment and environmental embedding really matter for the construction of mind and experience".

After all, theories of reasoning are about rules for reasoning and these are rules telling us to do certain things in certain circumstances. Writing a program allows us to state such rules precisely.

Some philosophers might insist that, between epistemology (and logic) and cognitive psychology, there is little, if any connection. The basis for such claims is that epistemology and logic are normative while psychology is descriptive. That is, psychology is concerned with how scientists do reason, whereas epistemology and logic with how scientists ought to reason. One of the central dogmas of philosophy is that you cannot derive an ought from an is.[45]

Nevertheless, this kind of ought might be called a "procedural ought". The apparent normativity of epistemology and logic is just a reflection of the fact that epistemology and logic are concerned with rules for how to do something. It would be considerably unreasonable to design a computational model of scientific discovery and reasoning without taking into account how scientists actually reason, what scientists know, and what data scientists can acquire. Nevertheless, the general goal is not the complete simulation of scientists themselves, but rather the achievement of discoveries about the world, using methods that extend human cognitive capacities. The goal is to build prosthetic scientists: just as telescopes are designed to extend the sensory capacity of humans, computational models of scientific discovery and reasoning are designed to extend our cognitive capacities. This cooperation should prove very fruitful from an educational perspective too: reciprocally clarifying both philosophical and AI theories of reasoning will provide new and very interesting didactic tools.

2.8 Geometrical Construction Is a Kind of Manipulative Abduction

Let's quote an interesting passage by Peirce about constructions. Peirce says that mathematical and geometrical reasoning "[...] consists in constructing a diagram according to a general precept, in observing certain relations between parts of that diagram not explicitly required by the precept, showing that these relations will hold for all such diagrams, and in formulating this conclusion in general terms. All valid necessary reasoning is in fact thus diagrammatic" [Peirce, 1931-1958, 1.54]. Not dissimilarly Kant says that in geometrical construction "[...] I must not restrict my attention to what I am actually thinking in my concept of a triangle (this is nothing more than the mere definition); I must pass beyond it to properties which are not contained in this concept, but yet belong to it" [Kant, 1929, A718-B746, p. 580].

We have seen that manipulative abduction is a kind of, usually model-based, abduction that exploits external models endowed with delegated (and often implicit) cognitive roles and attributes. 1. The model (diagram) is external and the strategy that organizes the manipulations is unknown a priori. 2. The result achieved is new (if we, for instance, refer to the constructions of the first creators of geometry), and

[45] Chapter seven of this book analyzes in details this gap between "ideal" or "institutional" models of reasoning and concrete inferences in "beings-like-us".

adds properties not contained before in the concept (the Kantian to "pass beyond" or "advance beyond" the given concept [Kant, 1929, A154-B194, p. 192]).[46]

Humans and other animals make a great use of perceptual reasoning and kinesthetic and motor abilities. We can catch a thrown ball, cross a busy street, read a musical score, go through a passage by imaging if we can contort out bodies to the way required, evaluate shape by touch, recognize that an obscurely seen face belongs to a friend of ours, etc. Usually the "computations" required to achieve these tasks are not accessible to a conscious description. Mathematical reasoning uses language explanations, but also non-linguistic notational devices and models. Geometrical constructions represent an example of this kind of extra-linguistic machinery we know as characterized in a model-based and manipulative – abductive – way. Certainly a considerable part of the complicated environment of a thinking agent is internal, and consists of the proper software composed of the knowledge base and of the inferential expertise of that individual. Nevertheless, I have already pointed out, any cognitive system consists of a "distributed cognition" among people and "external" technical artifacts [Hutchins, 1995; Zhang, 1997].

In the case of the construction and examination of diagrams in geometry, a sort of specific "experiments" serve as states and the implied operators are the manipulations and observations that transform one state into another. The mathematical outcome is dependent upon practices and specific sensorimotor activities performed on a non-symbolic object, which acts as a dedicated external representational medium supporting the various operators at work. There is a kind of an epistemic negotiation between the sensory framework of the mathematician and the external reality of the diagram. This process involves an external representation consisting of written symbols and figures that are manipulated "by hand". The cognitive system is not merely the mind-brain of the person performing the mathematical task, but the system consisting of the whole body (cognition is *embodied*) of the person plus the external physical representation. For example, in geometrical discovery the whole activity of cognition is located in the system consisting of a human together with diagrams.

An external representation can modify the kind of computation that a human agent uses to reason about a problem: the Roman numeration system eliminates, by means of the external signs, some of the hardest parts of the addition, whereas the Arabic system does the same in the case of the difficult computations in multiplication [Zhang, 1997]. All external representations, if not too complex, can be transformed in internal representations by memorization. But this is not always necessary if the external representations are easily available. Internal representations can be transformed in external representations by externalization, that can be productive "[...] if the benefit of using external representations can offset the cost associated with the externalization process" (*ibid.*, p. 181). Hence, contrarily to the old view in cognitive science, not all cognitive processes happen in an internal model

[46] Of course in the case we are using diagrams to demonstrate already known theorems (for instance in didactic settings), the strategy of manipulations is already available and the result is not new. Further details on this issue are illustrated in chapter three.

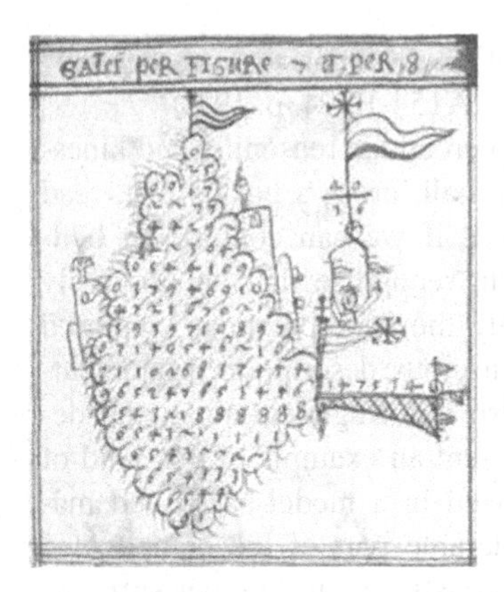

Fig. 2.7 Galley division, XVI Century, from an unpublished manuscript of a Venetian monk. The title of the work is Opus Artimetica D. Honorati veneti monachj coenobij S. Lauretij.

of the external environment. The information present in the external world can be directly picked out without the mediation of memory, deliberation, etc. Moreover, various different external devices can determine different internal ways of reasoning and cognitively solve the problems, as is well-known. Even a simple arithmetic task can completely change in presence of an external tool and representation. In the Figure 2.7 an ancient external tool for division is represented.

Following the approach in cognitive science related to the studies in distributed cognition, I contend that in the construction of mathematical concepts many external representations are exploited, both in terms of diagrams and of symbols. I have been interested in my research in diagrams which play an *optical* role[47]– microscopes (that look at the infinitesimally small details), telescopes (that look at infinity), windows (that look at a particular situation), a *mirror* role (to externalize rough mental models), and an *unveiling* role (to help create new and interesting mathematical concepts, theories, and structures).

Moreover optical diagrams play a fundamental explanatory (and didactic) role in removing obstacles and obscurities (for example the ambiguities of the concept of infinitesimal)[48] and in enhancing mathematical knowledge of critical situations (for example the problem of parallel lines, cf. the following sections). They facilitate new internal representations and new symbolic-propositional achievements. The mirror and unveiling diagrammatic representation of mathematical structures activates *perceptual operations* (for example identifying the interplay between conflicting structures: for example how the parallel lines behave to infinity). These perceptual operations provide mirror and unveiling diagrammatic representations of mathematical structures.

To summarize we can say mathematics diagrams play various roles in a typical abductive way; moreover, they are external representations which, in the cases I

---

47 This method of visualization was invented by [Stroyan, 2005] and improved by [Tall, 2001].

48 Chapter one, this book, section 1.7.

will present in the following sections, are devoted to provide explanatory and non-explanatory abductive results. Two of them are central:

- they provide an intuitive and mathematical *explanation* able to help the understanding of concepts difficult to grasp or that appear obscure and/or epistemologically unjustified. I will present in the following section some mirror diagrams which provided new mental representations of the concept of parallel lines.
- they help abductively *create* new previously unknown concepts that are *non-explanatory*, as illustrated in the case of the discovery of the non-Euclidean geometry.

2.9 Mirror Diagrams: Externalizing Mental Models to Represent Imaginary Entities

I have already said that empirical anomalies result from data that cannot currently be fully explained by a theory. They often derive from predictions that fail, which implies some element of incorrectness in the theory. In general terms, many theoretical constituents may be involved in accounting for a given domain item (anomaly) and hence they are potential points for modification. The detection of these points involves defining which theoretical constituents are employed in the explanation of the anomaly. Thus, the problem is to investigate all the relationships in the explanatory area.

As illustrated in section 2.4, first and foremost, anomaly resolution involves the localization of the problem at hand within one or more constituents of the theory, it is then necessary to produce one or more new hypotheses to account for the anomaly, and, finally, these hypotheses need to be evaluated so as to establish which one best satisfies the criteria for theory justification. Hence, anomalies require a change in the theory. We know that empirical anomalies are not alone in generating impasses. The so-called *conceptual problems* represent a particular form of anomaly (cf. above in this chapter). Resolving conceptual problems may involve satisfactorily answering questions about the status of theoretical entities: conceptual problems arise from the nature of the claims in the principles or in the hypotheses of the theory. Usually it is necessary to identify the conceptual problem that needs a resolution, for example by delineating how it can concern the adequacy or the ambiguity of a theory, and yet also its incompleteness or (lack of) evidence.

I have also already illustrated that formal sciences are especially concerned with conceptual problems. The discovery of non-Euclidean geometries presents an interesting case of visual/spatial abductive reasoning, where both *explanatory* and *non-explanatory* aspects are intertwined. First of all it demonstrates a kind of *visual/spatial abduction*, as a strategy for anomaly resolution connected to a form of explanatory and productive visual thinking. Since ancient times the fifth postulate has been held to be not evident. This "conceptual problem" has generated many difficulties about the reliability of the theory of parallels, consisting of the theorems that can be only derived with the help of the fifth postulate. The recognition of this anomaly was crucial to the development of the non-Euclidean revolution. Two thousand years of attempts to resolve the anomaly have produced many fallacious demonstrations of the fifth postulate: a typical attempt was that of trying to prove the fifth postulate from the others. Nevertheless, these attempts have also provided much theoretical speculation about the unicity of Euclidean geometry and about the status of its principles.

Here, I am primarily interested in showing how the anomaly is recognizable. A postulate that is equivalent to the fifth postulate states that for every line l and every point P that does not lie on l, there exists a unique line m through P that is parallel to l. If we consider its model-based (diagrammatic) counterpart (cf. Figure 2.8), the postulate may seem "evident" to the reader, but this is because we have been conditioned to think in terms of Euclidean geometry. The definition above represents the most obvious level at which ancient Euclidean geometry was developed as a formal science – a level composed of *symbols* and *propositions*.

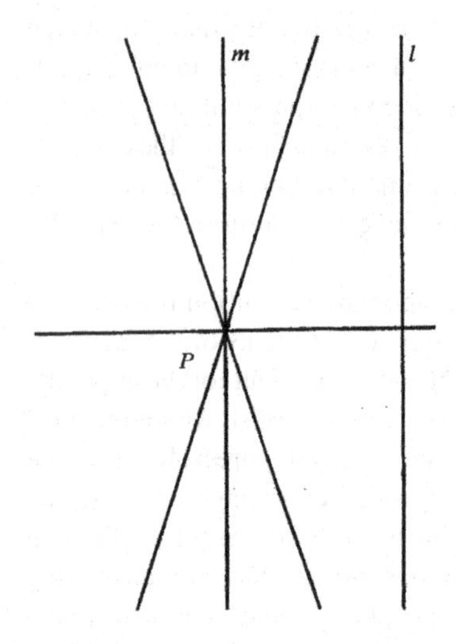

Fig. 2.8

Furthermore, when we also consider the other fundamental level, where model-based aspects (diagrammatic) are at play, we can immediately detect a difference between this postulate and the other four if we regard the first principles of geometry as abstractions from experience that we can in turn represent by drawing figures on a blackboard or on a sheet of paper or on our "visual buffer" [Kosslyn and Koenig, 1992] in the mind. We have consequently a *double passage* from the sensorial experience to the abstraction (expressed by symbols and propositions) and from this abstraction to the experience (sensorial and/or mental).

We immediately discover that the first two postulates are abstractions from our experiences drawing with a straightedge, the third postulate derives from our experiences drawing with a compass. The fourth postulate is less evident as an abstraction, nevertheless it derives from our measuring angles with a protractor (where the sum of supplementary angles is 180°, so that if supplementary angles are congruent to each other, they must each measure 90°) [Greenberg, 1974, p. 17].

In the case of the fifth postulate we are faced with the following serious problems: 1) we cannot verify empirically whether two lines meet, since we can draw only segments, not lines. Extending the segments further and further to find if they meet is not useful, and in fact we cannot continue indefinitely. We are forced to verify parallels indirectly, by using criteria other than the definition; 2) the same holds with regard to the representation in the "limited" visual buffer. The "experience" localizes a problem to solve, an ambiguity, only in the fifth case: in the first four cases our "experience" *verifies* without difficulty the abstraction (propositional and symbolic) itself. In the fifth case the formed images (mental or not) are the images that are able to *explain* the "concept" expressed by the definition of the fifth postulate as problematic (an anomaly): we cannot draw or "imagine" the two lines at infinity, since we can draw and imagine only segments, not the lines themselves.

The *selected* visual/spatial image or imagery derived from the propositional and symbolic level of the definition is nothing more than the explanation of the anomaly of the definition itself. As stated above, the image demonstrates a kind of visual abduction, as a strategy for anomaly localization related to a form of explanatory visual/spatial thinking.

Once the anomaly is detected, the way to anomaly resolution is opened up – in our case, this means that it becomes possible to discover non-Euclidean geometries. That Euclid himself did not fully trust the fifth postulate is revealed by the fact that he postponed using it in a proof for as long as possible – until the twenty-ninth proposition. As is well-known, Proclus tried to solve the anomaly by proving the fifth postulate from the other four. If we were able to prove the postulate in this way, it would become a theorem in a geometry which does not require that postulate (the future "absolute geometry") and which would contain all of Euclid's geometry.

Without showing all the passages of Proclus's argument [Greenberg, 1974, p. 119-121] we need only remember that the argument seemed correct because it was proved using a diagram. Yet we are not allowed to use that diagram to justify a step in a proof. Each step must be proved from stated axioms or previously proven theorems. We may visualize parallel lines as railroad tracks, everywhere equidistant from each other, and the ties of the tracks as being perpendicular to both parallels. Yet this imagery is valid only in Euclidean geometry. In the absence of the parallel postulate we can only consider two lines as "parallel" when, by the definition of "parallel", they do not possess any points in common. It is not possible implicitly to assume that they are equidistant; nor can it be assumed that they have a common perpendicular. This is an example in which a *selected* abduced image is capable of compelling you to make a mistake, and in this way it was used as a means of evaluation in a proof: we have already stated that in this case it is not possible to use that image or imagery to justify a step in a proof because it is not possible to use that image or imagery that attributes to experience more than the experience itself can deliver.

For over two thousand years some of the greatest mathematicians tried to prove Euclid's fifth postulate. For example, Saccheri's strategy for anomaly resolution in the XVIII century was to abduce two opposite hypotheses.[49] of the principle, that is, to negate the fifth postulate and derive, using new logical tools coming from non-geometrical sources of knowledge, all theorems from the two alternative hypotheses by trying to detect a contradiction. The aim was indeed that of demonstrating/explaining that the anomaly is simply apparent. We are faced with a kind of explanatory abduction. New axioms are hypothesized and adopted in looking for outcomes which can possibly help in explaining how the fifth postulate is unique and so not anomalous. At a first sight this case is similar to the case of non-explanatory abduction pointed out at p. 72, speaking of reverse mathematics, but the similarity is only structural (i.e. guessing "new axioms"): in the case of reverse mathematics axioms are hypothesized to account for already existing mathematical theories and do not aim at explanatory results.

The contradiction in the elliptic case ("hypothesis of obtuse angle", to use the Saccheri's term designing one of the two future elementary non-Euclidean geometries) was found, but the contradiction in the hyperbolic case ("hypothesis of the acute angle") was not so easily discovered: having derived several conclusions that are now well-known propositions of non-Euclidean geometry, Saccheri was forced to resort to a metaphysical strategy for anomaly resolution: "Proposition XXXIII. The 'hypothesis' of acute angle [that is, the hyperbolic case] is absolutely false, because repugnant to the nature of the straight line" [Saccheri, 1920]. Saccheri chose to state this result with the help of the somewhat complicated imagery of infinitely distant points: two different straight lines cannot both meet another line perpendicularly at one point, if it is true that all right angles are equal (fourth postulate) and that two different straight lines cannot have a common segment. Saccheri did not ask himself whether everything that is true of ordinary points is necessarily true of an infinitely distant point. In Note II to proposition XXI some "physico-geometrical" experiments to confirm the fifth postulate are also given, invalidated unfortunately by the same incorrect use of imagery that we have observed in Proclus's case. In this way, the anomaly was resolved unsatisfactorily and Euclid was not freed of every fleck: nevertheless, although he did not recognize it, Saccheri had discovered many of the propositions of non-Euclidean geometry [Torretti, 1978, p. 48].

In the following sections I will illustrate the example of Lobachevsky's discovery of non-Euclidean geometry where we can see the abductive role played in a discovery process by new considerations concerning visual sense impressions and productive imagery representations.

2.9.1 Internal and External Representations

Lobachevsky was obliged first of all to rebuild the basic Principles and to this end, it was necessary to consider geometrical principles in a new way, as neither ideal nor *a priori*. New interrelations were created between two areas of knowledge: Euclidean

[49] On the strategies adopted in anomaly resolution cf. [Darden, 1991, pp. 272–275].

geometry and the philosophical tradition of empiricism/sensualism. In the following section I will describe in detail the type of abduction that was at play in this case. Lobachevsky's target is to perform a geometrical abductive process able to create the new and very abstract concept of non-Euclidean parallel lines. The whole epistemic process is mediated by interesting manipulations of external mirror diagrams.

I have already said that for over two thousand years some of the greatest mathematicians tried to prove Euclid's fifth postulate. Geometers were not content merely to construct proofs in order to discover new theorems and thereby to try to resolve the anomaly (represented by its lack of evidence) without trying to reflect upon the status of the symbols of the principles underlying Euclidean geometry represent. Lobachevsky's strategy for resolving the anomaly of the fifth postulate was first of all to manipulate the symbols, second to rebuild the principles, and then to derive new proofs and provide a new mathematical apparatus; of course his analysis depended on some of the previous mathematical attempts to demonstrate the fifth postulate. The failure of the demonstrations – of the fifth postulate from the other four – that was present to the attention of Lobachevsky, lead him to believe that the difficulties that had to be overcome were due to causes traceable at the level of the first principles of geometry.

We simply can assume that many of the internal visualizations of the working geometers of the past were spatial and imaginary because those mathematicians were precisely operating with diagrams and visualizations. By using internal representations Lobachevsky has to create new external visualizations and to adjust them tweaking and manipulating [Trafton *et al.*, 2005] the previous ones in some particular ways to generate appropriate spatial transformations (the so-called *geometrical constructions*).[50] In cognitive science many kinds of spatial transformations have been studied, like mental rotation and any other actions to improve and facilitate the understanding and simplification of the problem. It can be said that when a spatial transformation is performed on external visualizations, it is still generating or exploiting an internal representation.

Spatial transformations on external supports can be used to create and transform external diagrams and the resulting internal/mental representations may undergo further mental transformations. Lobachevsky mainly takes advantage of the transformation of external diagrams to create and modify the subsequent internal images. So mentally manipulating both external diagrams and internal representations is extremely important for the geometer that uses both the drawn geometrical figure and her own mental representation. An active role of these external representations, as *epistemic mediators* able to favor scientific discoveries – widespread during the ancient intuitive geometry based on diagrams – can be curiously seen at the beginning of modern mathematics, when new abstract, imaginary, and counterintuitive non-Euclidean entities are discovered and developed.

There are *in vivo* cognitive studies performed on human agents (astronomers and physicists) about the interconnection between mental representations and the external scientific visualizations. In these studies "pure" spatial transformations, that

[50] I maintain that in general spatial transformations are represented by a visual component and a spatial component [Glasgow and Papadias, 1992].

is transformations that are performed – and based – on the external visualizations dominate: the perceptual activity seems to be prevalent, and the mental representations are determined by the external ones. The researchers say that there is, in fact, some evidence for this hypothesis: when a scientist mentally manipulates a representation, 71% of the time the source is a visualization, and only 29% of the time it is a "pure" mental representation. Other experimental results show that some of the time scientists seem to create and interpret mental representations that are different from the images in the visual display: in this case it can be hypothesized that scientists use a comparison process to connect their internal representation with the external visualizations [Trafton *et al.*, 2005].

In general, during the comparison between internal and external representation the scientists are looking for discrepancies and anomalies, but also equivalences and coherent shapes (like in the case of geometers, as we will see below). The comparison between the transformations acted on external representations and their previously represented "internal" counterpart forces the geometer to merge or to compare the two sides (some aspects of the diagrams correspond to information already represented internally as symbolic-propositional).[51]

External geometrical diagrams activate perceptual operations, such as searching for objects that have a common shape and inspecting whether three objects lie on a straight line. They contain permanent and invariant geometrical information that can be immediately perceived and kept in memory without the mediation of deliberate inferences or computations, such as whether some configurations are spatially symmetrical to each other and whether one group of entities has the same number of entities as another one. Internal operations prompt other cognitive operations, such as making calculations to get or to envision a result. In turn, internal representations may have information that can be directly retrieved, such as the relative magnitude of angles or areas.

2.10 Mirror Diagrams and the Infinite

As previously illustrated the failure of the demonstrations (of the fifth postulate from the other four) of his predecessors induced Lobachevsky to believe that the difficulties that had to be overcome were due to causes other than those which had until then been focused on.

Lobachevsky was obliged first of all to rebuild the basic principles: to this end, it was necessary to consider geometrical principles in a new way, as neither ideal nor *a priori*. New interrelations were created between Euclidean geometry and some claims deriving from the philosophical tradition of empiricism/sensualism.

[51] Usually scientists try to determine *identity*, when they make a comparison to determine the individuality of one of the objects; *alignment*, when they are trying to determine an estimation of fit of one representation to another (e.g. visually inspecting the fit of a rough mental triangular shape to an external constructed triangle); and *feature comparison*, when they compare two things in terms of their relative features and measures (size, shape, color, etc.) [Trafton *et al.*, 2005].

2.10.1 Abducing First Principles through Bodily Contact

From this Lobachevskyan perspective the abductive attainment of the basic concepts of any science is in terms of senses: the basic concepts are always acquired through our *sense impressions*. Lobachevsky builds geometry upon the concepts of body and bodily contact, the latter being the only "property" common to all bodies that we ought to call geometrical. The well-known concepts of depthless surface, widthless line and dimensionless point were constructed considering different possible kinds of bodily contact and dispensing with, *per abstractionem*, everything but preserving the contact itself: these concepts "[...] exist only in our representation; whereas we actually measure surfaces and lines by means of bodies" for "[...] in nature there are neither straight lines nor curved lines, neither plane nor curved surfaces; we find in it only bodies, so that all the rest is created by our imagination and exists just in the realm of theory" [Lobachevsky, 1897, Introduction]. The only thing that we can know in nature is movement "[...] without which sense impressions are impossible. Consequently all other concepts, e.g. geometrical concepts, are generated artificially by our understanding, which derives them from the properties of movement; this is why space in itself and by itself does not exist for us" (*ibid.*).

It is clear that in this inferential process Lobachevsky performs a kind of *model-based* abduction, where the perceptual role of sense impressions and their experience with bodies and bodily contact is cardinal in the generation of new concepts. The geometrical concepts are "[...] generated artificially by our understanding, which derives them from the properties of movement". Are these abductive hypotheses explanatory or not? I am inclined to support their basic "explanatory" character: they furnish an explanation of our sensorial experience with bodies and bodily contact in ideal and abstract terms.

On the basis of these foundations Lobachevsky develops the so-called *absolute geometry*, which is independent of the fifth postulate: "Instead of commencing geometry with the plane and the straight line as we do ordinarily, I have preferred to commence it with the sphere and the circle, whose definitions are not subject to the reproach of being incomplete, since they contain the generation of the magnitudes which they define" [Lobachevsky, 1929, p. 361].)

This leads Lobachevsky to abduce a very remarkable and modern hypothesis – anticipatory of the future Einstein's theoretical atmosphere of general relativity – which I consider to be largely *image-based:* since geometry is not based on a perception of space, but constructs a concept of space from an experience of bodily movement produced by physical forces, there could be place in science for two or more geometries, governing different kinds of natural forces:

> To explain this idea, we assume that [...] attractive forces decrease because their effect is diffused upon a spherical surface. In ordinary Geometry the area of a spherical surface of radius r is equal to $4r^2$, so that the force must be inversely proportional to the square of the distance. In Imaginary Geometry I found that the surface of the sphere is
>
> $$(e^r - e^{-r})^2,$$

> and it could be that molecular forces have to follow that geometry [...]. After all, given this example, merely hypothetical, we will have to confirm it, finding other more convincing proofs. Nevertheless we cannot have any doubts about this: forces by themselves generate everything: movement, velocity, time, mass, matter, even distances and angles [Lobachevsky, 1897, p. 9].

Lobachevsky did not doubt that something, not yet observable with a microscope or analyzable with astronomical techniques, accounted for the reliability of the new non-Euclidean imaginary geometry. Moreover, the principles of geometry are held to be testable and it is possible to prepare an experiment to test the validity of the fifth postulate or of the new non-Euclidean geometry, the so-called *imaginary geometry*. He found that the defect of the triangle formed by Sirius, Rigel and Star No. 29 of Eridanus was equal to $3.727 + 10^{-6}$ seconds of arcs, a magnitude too small to be significant as a confirmation of imaginary geometry, given the range of observational error. Gauss too had claimed that the new geometry might be true on an astronomical scale. Lobachevsky says:

> Until now, it is well-known that, in Geometry, the theory of parallels had been incomplete. The fruitlessness of the attempts made, since Euclid's time, for the space of two thousand years, aroused in me the suspicion that the truth, which it was desired to prove, was not contained in the data themselves; that to establish it the aid of experiment would be needed, for example, of astronomical observations, as in the case of other laws of nature. When I had finally convinced myself of the justice of my conjecture and believed that I had completely solved this difficult question I wrote, in 1826, a memoir on this subject *Exposition succincte des principes de la Géométrie* [Lobachevsky, 1897, p. 5].

With the help of the explanatory abductive role played by the new sensualist considerations of the basic principles, by the empiricist view and by a very remarkable productive visual hypothesis, Lobachevsky had the possibility to proceed in discovering the new theorems. Following Lobachevsky's discovery the fifth postulate will no longer be considered in any way anomalous – we do not possess any proofs of the postulate, because this proof is *impossible*. Moreover, the new non-Euclidean hypothesis is reliable: indeed, to understand visual thinking we have also to capture its status of guaranteeing the reliability of a hypothesis. In order to prove the relative consistency of the new non-Euclidean geometries we should consider some very interesting visual and mathematical "models" proposed in the second half of XIX century (i.e. the Beltrami-Klein and Poincaré models), which involve new uses of visual images in theory assessment.

In summary, the abductive process of Lobachevsky's discovery can be characterized in the following way, taking advantage of the nomenclature introduced in the previous and in the present chapter:

1. the inferential process Lobachevsky performs to rebuild the first principles of geometry is prevalently a kind of *manipulative* and *model-based* abduction, endowed with an *explanatory* character: the new abduced principles furnish an explanation of our sensorial experience with bodies and bodily contact in ideal and abstract terms;

2. at the same time the new principles found offer the chance of further *multimodal*[52] and *distributed* abductive steps (that is based on both on both visual and sentential aspects, and on both internal and external representations) which are mainly *non-explanatory* and provide unexpected mathematical results. These further abductive processes:

    a. first of all have to provide a different multimodal way of describing parallelism (both from a diagrammatical and propositional perspective, cf. subsection 2.10.4 and Figure 2.11);
    b. second, on the basis of the new concept of parallelism it will be possible to derive new theorems of a new non-Euclidean geometrical system exempt from inconsistencies just like the Euclidean system. Of course this process show a moderately instrumental character, more or less present in all abductions (cf. below section 2.11).

Let us illustrate how Lobachevsky continues to develop the absolute geometry. The immediate further step is to define the concept of plane, which is defined as the geometrical locus of the intersections of equal spheres described around two fixed points as centers, and, immediately after, the concept of straight line (for example BB' in the mirror diagram of the Figure 2.9) as the geometrical locus of the intersections of equal circles, all situated in a single plane and described around two fixed points of this plane as centers. The straight line is so defined by means of "finite" parts (segments) of it: we can prolong it by imaging a repeatable movement of rotation around the fixed points (cf. Figure 2.9) [Lobachevsky, 1829-1830, 1835-1838, §25].

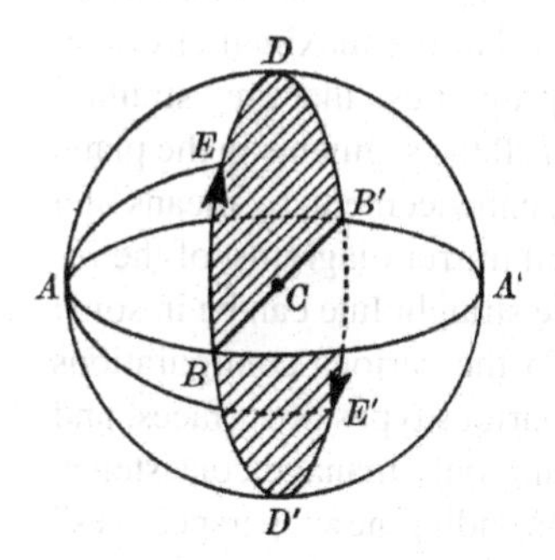

Fig. 2.9

Rectilinear angles (which express arcs of circle) and dihedral angles (which express spherical lunes) are then considered; and the solid angles too, as generic parts of spherical surfaces – and in particular the interesting spherical triangles. π means for Lobachevsky the length of a semicircumference, but also the solid angle that

[52] Cf. below chapter four, this book.

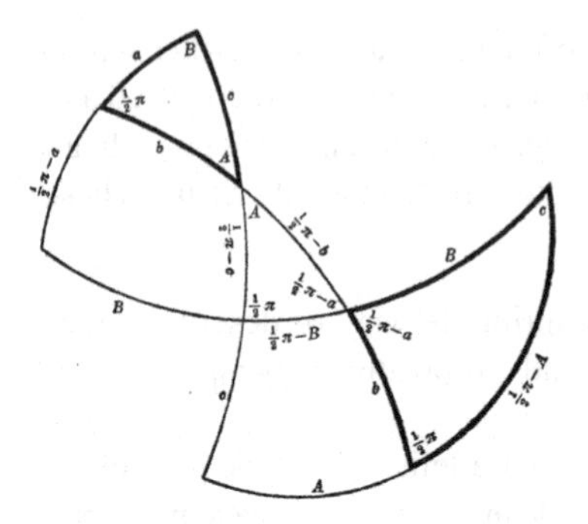

Fig. 2.10

corresponds to a semisphere (straight angle). The surface of the spherical triangles is always less than π, and, if π, coincides with the semisphere. The theorems about the perpendicular straight lines and planes also belong to absolute geometry.

2.10.2 *Expansion of Scope Strategy*

We have to note some general cognitive and epistemological aspects which characterize the development of this Lobachevskyan absolute geometry.

Spherical geometry is always treated together with the plane geometry: the definitions about the sphere are derived from the ones concerning the plane when we substitute the straight lines (geodetics in the plane) with the maximal circles (geodetics in the spherical surface). Lobachevsky says that the maximal circle on the sphere with respect to the other circles presents "properties" that are "similar" to the ones belonging to the straight line with respect to all the segments in the plane [Lobachevsky, 1829-1830, 1835-1838, §66]. This is an enhancement, by means of a kind of *analogical* reasoning, reinforced by the external mirror diagrams, of the internal representation of the concept of straight line. The straight line can be in some sense thought (because it is "seen" and "imagined" in the various configurations provided by the external diagrams) as "belonging" to various types of surfaces, and not only to the plane. Consequently, mirror diagrams not only manage consistency requirements, they can also act in an additive way, providing new "perspectives" and information on old entities and structures. The directly perceivable information strongly guides the discoverer's selections of moves by servicing the discovery strategy *expansion-of-the-scope* (of the concept of straight line). This possibility was not indeed available at the simple level of the internal representation. The Figure 2.10 [Lobachevsky, 1829-1830, 1835-1838, §79] is another example of the exploitation of the analogy plane/spherical surface by means of a diagram that exploits the perspective of the two-dimensional flat plane.

2.10.3 *Infinite/Finite Interplay*

In all the previous cases the external representations are constructions that have to respect the empirical attitude described above: because of the fact that the geometrical bodies are characterized by their "finiteness" the external representation is just a coherent mirror of finite internal images. The "infinite" can be perceived in the "finite" constructions because the infinite is considered only as something potential that can be just mentally and artificially thought: "defined artificially by our understanding". As the modern axiomatic method is absent, the geometer has to conceptualize infinite situations exploiting the finite resources offered by diagrams. In front of the question: "How is it that the finite human resources of internal representations of human mind can conceptualize and formalize abstract notions of infinity?" – notions such as the specific ones embedded in the non-Euclidean assumptions – the geometer is aware we perceive a finite world, act upon it, and think about it. Moreover, the geometer operates in "[...] a combination of perceptual *input*, physical *output*, and internal mental processes. All three are finite. But by thinking about the possibility of performing a process again and again, we can easily reach out towards the potential infinite" [Tall, 2001]. Lobachevsky states: "Which part of the lines we would have to disregard is arbitrary", and adds, "our senses are deficient" and it is only by means of the "artifice" consisting of the continuum "enhancement of the instruments" that we can overcome these limitations [Lobachevsky, 1829-1830, 1835-1838, §38]. Given this epistemological situation, it is easy to conclude saying that *instruments* are not just and only telescopes and laboratory tools, but also diagrams.

Let us continue to illustrate the geometer's inventions. In the Proposition 27 (a theorem already proved by Euler and Legendre) of the *Geometrical Researches of the Theory of Parallels*, published in 1840, [Lobachevsky, 1891], Lobachevsky states that if A, B, and C are the angles of a spherical triangle, the ratio of the area of the triangle to the area of the sphere to which it belongs will be equal to the ratio of

$$\frac{1}{2}(A+B+C-\pi)$$

to four right angles; that the sum of the three right angles of a rectilinear triangle can never surpass two right angles (Prop. 19), and that, if the sum is equal to two right angles in any triangle, it will be so in all (Prop. 20).

2.10.4 *Non-euclidean Parallelism: Coordination and Inconsistency Detection*

The basic unit is the manipulation of diagrams. Before the birth of the modern axiomatic method the geometers still and strongly have to exploit external diagrams, to enhance their thoughts. It is impossible to mental imaging and evaluating the alternative sequences of symbolic calculations being only helped by the analytic tools, such as various written equations and symbols and marks: it is impossible to do a complete anticipation of the possible outcomes, due to the limited power of working memory and attention. Hence, because of the complexity of the geometrical problem space and the limited power of working memory, complete mental search is impossible or difficult. Geometers may use perceptual external biases to make decisions. Moreover, in those cognitive settings, lacking in modern axiomatic theoretical awareness, certainly perceptual operations were epistemic mediators which need less attentional and working memory resources than internal operations. "The directly perceived information from external representations and the directly retrieved information from internal representation may elicit perceptual and cognitive biases, respectively, on the selections of actions. If the biases are inconsistent with the task, however, they can also misguide actions away from the goal. Learning effect can occur if a task is performed more than once. Thus, the decision on actions can also be affected by learned knowledge" [Zhang, 1997, p. 186].

The new external diagram proposed by Lobachevsky (the diagram of the drawn parallel lines of Figure 2.11) [Lobachevsky, 1891] is a kind of *analogous* both of the mental image we depict in the mental visual buffer and of the symbolic-propositional level of the postulate definition. It no longer plays the *explanatory* role of showing an anomaly, like it was in the case of the diagram of Figure 2.8 (and of other similar diagrams) during the previous centuries. I have already said I call this kind of external tool in the geometrical reasoning *mirror diagram*. In general this diagram mirrors the internal imagery and provides the possibility of detecting anomalies, like it was in the case of the similar diagram of Figure 2.8. The external representation of geometrical structures often activates direct perceptual operations (for example, identify the parallels and search for the limits) to elicit consistency or inconsistency routines. Sometimes the mirror diagram biases are inconsistent with the task and so they can make the task more difficult by misguiding actions away from the goal. If consistent, we have already said that they can make the task easier by instrumentally and non-explanatorily guiding actions toward the goal. In certain cases the mirror diagrams biases are irrelevant, they should have no effects on the decision of abductive actions, and play lower cognitive roles.

In the case of the diagram of the parallel lines of the similar Figure 2.8 it was used in the history of geometry to make both consistent and in-consistent the fifth Euclidean postulate and the new non-Euclidean perspective (more details on this epistemological situation are given in [Magnani, 2001c]).

I said that in some cases the mirror diagram plays a negative role and inhibits further creative abductive theoretical developments. As I have already indicated (p. 121), Proclus tried to solve the anomaly by proving the fifth postulate from the other four. If we were able to prove the postulate in this way, it would become a theorem in a geometry which does not require that postulate (the future "absolute geometry") and which would contain all of Euclid's geometry. We need only remember that the argument seemed correct because it was proved using a diagram. In this case the mirror diagram biases were consistent with the task of *justifying* Euclidean geometry and they made this task easier by guiding actions toward the goal, but they inhibited the discovery of non-Euclidean geometries [Greenberg, 1974, pp. 119–121]; cf. also [Magnani, 2001c, pp. 166–167].

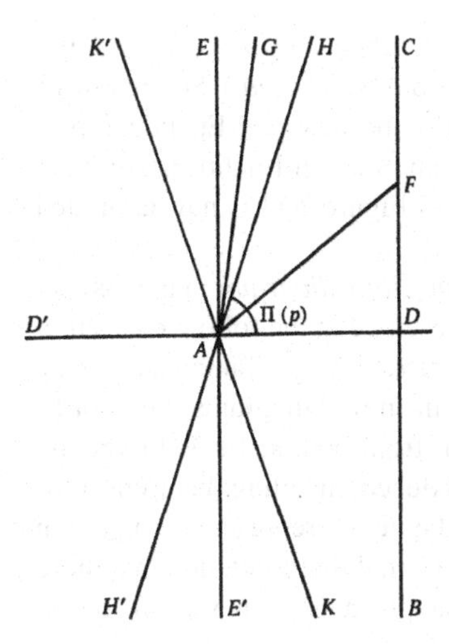

Fig. 2.11 Non-Euclidean parallel lines

In sum, contrarily to the diagram of Figure 2.8, the diagram of Figure 2.11 does not aim at explaining anything given, it is fruit of a non-explanatory and instrumental abduction, as I have anticipated at p. 127: the new related principle/concept of parallelism offers the chance of further *multimodal* and *distributed* abductive steps (based on both visual and sentential aspects, and on both internal and external representations) which are mainly *non-explanatory*. On the basis of the new concept of parallelism it will be possible to derive new theorems of a new non-Euclidean geometrical system exempt from inconsistencies just like the Euclidean system (cf. below section 2.11).

The diagram now favors the new definition of parallelism [Lobachevsky, 1891, Prop. 16], which introduces the non-Euclidean atmosphere: "All straight lines which in a plane go out from a point can, with reference to a given straight line in the same plane, be divided in two classes – into *cutting* and *not-cutting*. The boundary lines of the one and the other class of those lines will be called *parallel to the given lines*" (p. 13).

The external representation is easily constructed like in Figure 2.11 of [Lobachevsky, 1891, p. 13], where the angle HAD between the parallel HA and the perpendicular AD is called the angle of parallelism, designated by $\Pi(p)$ for $AD = p$. If $\Pi(p)$ is $< \frac{1}{2}\pi$, then upon the other side of AD, making the same angle $DAK = \Pi(p)$ will lie also a line AK, parallel to the prolongation DB of the line DC, so that under this assumption we must also make a distinction of sides in parallelisms. Because of the fact that the diagrams can contemplate only finite parts of straight lines it is easy to represent this new postulate in this mirror image: we cannot know what happens at the infinite neither in the internal representation (because of the limitations of visual buffer), nor in the external representation: "[...] in the uncertainty whether the perpendicular AE is the only line which does not meet DC, we will assume it may be possible that there are still other lines, for example *AG*, which do not cut *DC*, how far so ever they may be prolonged" (*ibid.*). So the mirror image in this case is seen as consistently supporting the new non-Euclidean perspective. The idea of constructing an external diagram of a non-Euclidean situation is considered normal and reasonable. The diagram of Figure 2.11 is now exploited to unveil new fruitful consequences.

A first analysis of the exploitation of what I call *unveiling diagrams* in the discovery of the notion of non-Euclidean parallelism is presented in the following section related to the exploitation of diagrams at the stereometric level.[53] Taking advantage of the Lobachevskyan case I have illustrated that in mirror diagrams the coordination between perception and cognition is central, from both static and dynamic (constructions) points of view; in the case of the abduced unveiling diagrams allocating and switching attention between internal and external representation govern the reasoning strategy, by integrating internal and external representation in a more dynamical and complicated way – essentially non-explanatory – as we will see in the following section.

2.11 Unveiling Diagrams in Lobachevsky's Discovery as Gateways to Imaginary Entities

2.11.1 Euclidean/Non-euclidean Model Matching Strategy

Lobachevsky's target is to perform a geometrical abductive process able to create new and very abstract entities: the whole epistemic process is mediated by interesting manipulations of external unveiling diagrams. The first step toward the exploitation of what I call *unveiling diagrams* is the use of the notion of non-Euclidean parallelism at the stereometric level, by establishing relationships between straight lines and planes and between planes: Proposition 27 (already proved by Lexell and Euler): "*A three-sided solid angle equals the half sum of surface angles less a right-angle*" (p. 24, Figure 2.12). Proposition 28 (directly derived from Prop. 27): "*If three planes cut each other in parallel lines, then the sum of the three surface angles equals two rights*" (p. 28), Figure 2.13. These achievements are absolutely important: it is established that for a certain geometrical configuration of the new geometry (the three planes cut each other in parallel lines that are parallel in Lobachevskyan sense) some properties of the ordinary geometry hold.

The important notions of *oricycle* and *orisphere* are now defined to search for a possible symbolic counterpart able to express a foreseen consistency (as a justification) of the non-Euclidean theory. This consistency is looked at from the point a view of a possible "analytic" solution, that is in terms of verbal-symbolic (not diagrammatic) results (equations).

[53] [Magnani and Dossena, 2005; Dossena and Magnani, 2007] illustrate that external representations like the ones I call *unveiling diagrams* can enhance the consistency of a cognitive process but also provide more radically creative suggestions for new useful information and discoveries.

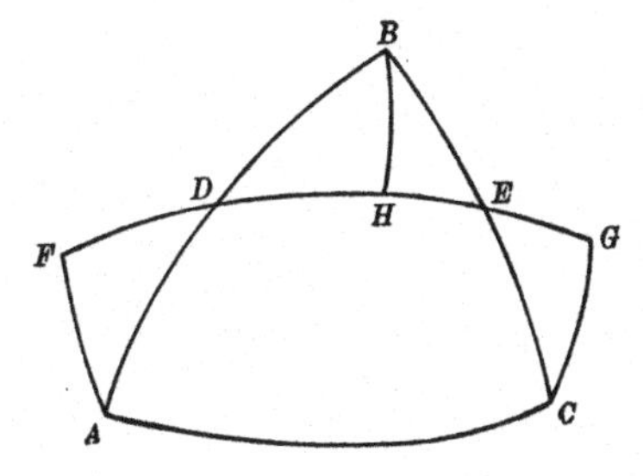

Fig. 2.12

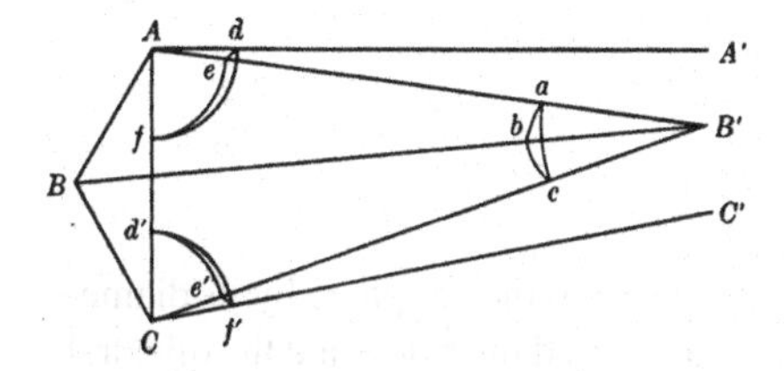

Fig. 2.13

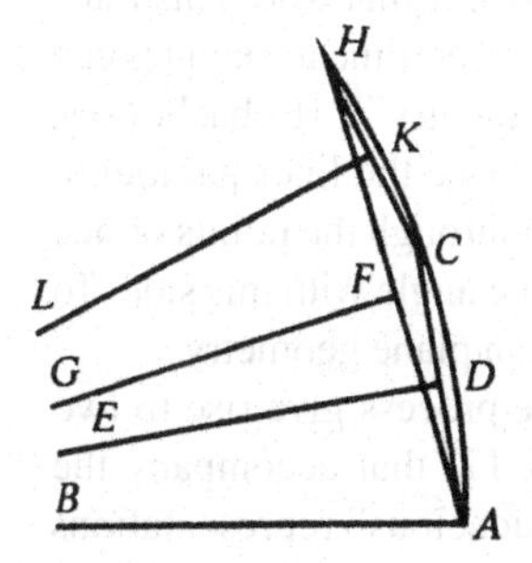

Fig. 2.14

Such is the case of the Proposition 31. "*We call boundary line (oricycle) that curve lying in a plane for which all perpendiculars erected at the mid-ponts of chords are parallel to each other.* [...] The perpendicular *DE* erected upon the chord *AC* at its mid-point *D* will be parallel to the line *AB*, which we call the *Axis of the boundary line*" (pp. 30-31), cf. Figure 2.14. Proposition 34. "*Boundary surface*[54] we call that surface which arises from the revolution of the boundary line about one of its axes, which, together with all other axes of the boundary-line, will be also an axis of the

[54] Also called limit sphere or orisphere.

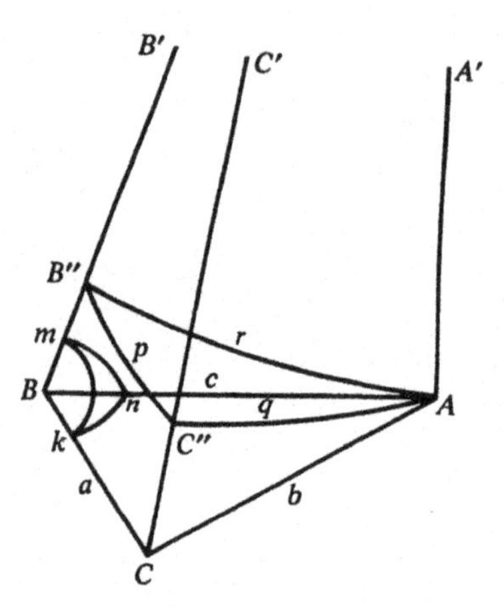

Fig. 2.15

boundary surface" (p. 33). Moreover, the intersections of the orisphere by its diametral planes are limit circles. The limit circle arcs are called the sides, and the dihedral angles between the planes of these arcs the angles of the "orisphere triangle".

A part of the surface of the orisphere bounded by three limit circle arcs will be called an orisphere triangle. From Prop. 28 follows that the sum of the angles of an orisphere triangle is always equal to two right angles: "everything that is demonstrated in the ordinary geometry of the proportionality of the sides or rectilinear triangles can therefore be demonstrated in the same manner in the pangeometry"[55] [Lobachevsky, 1929, p. 364] of the orisphere triangles if only we will replace the lines parallel to the sides of the rectilinear triangle by orisphere arcs drawn through the points of one of the sides of the orisphere triangle and all making the same angle with this side. To conclude, the orisphere is a "partial" *model* of the Euclidean plane geometry.

The last constructions of the Lobachevskyan abductive process give rise to two fundamental *unveiling diagrams* (cf. Figures 2.15 and 2.17) that accompany the remaining proofs. They are more abstract and exploit "audacious" representations in the perspective of three dimensional geometrical shapes.

The construction given in Figure 2.15 aims at diagrammatically "representing" a stereometric non-Euclidean form built on a *rectilinear right angled triangle ABC* to which the Theorem 28 above can be applied (indeed the parallels AA', BB', CC', which lie on the three planes are parallels in non-Euclidean sense), so that Lobachevsky is able to further apply symbolic identifications; the planes make with each other the angles $\Pi(a)$ at AA', a right angle at CC', and, consequently $\Pi(a')$ at BB'.[56] The diagram is enhanced by constructing a spherical triangle *mnk*, in which

[55] Lobachevsky called the new theory "imaginary geometry" but also "pangeometry".

[56] Given that Lobachevsky designates the size of a line by a letter with an accent added, e.g. x', in order to indicate this has a relation to that of another line, which is represented by the same letter without the accent x, "which relation is given by the equation $\Pi(x)+\Pi(x')=\pi$" (Prop. 35).

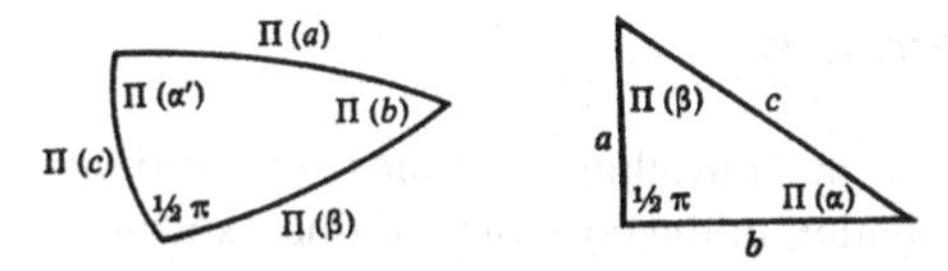

Fig. 2.16

the sides are $mn = \Pi(c)$, $kn = \Pi(\beta)$, $mk = \Pi(a)$ and the opposite angles are $\Pi(a)$, $\Pi(\alpha')$, $\frac{1}{2}\pi$ realizing that with the "existence" of a rectilinear triangle with the sides a, b, c (like in the case of the previous one) "we must admit" the existence of a related spherical triangle (cf. Figure 2.16), etc. Not only, a *boundary surface* (orisphere) can be constructed, that passes through the point A with AA' as axis, and those intersections with the planes the parallels form a boundary triangle (that is a triangle situated upon the given orisphere), whose sides are $B''C'' = p$, $C''A = q$, $B''A = r$, and the angles opposite to them $\Pi(\alpha)$, $\Pi(\alpha')$, $\frac{1}{2}\pi$ and where consequently (this follows from the Theorem 34):

$$p = r\sin\Pi(a), q = r\cos\Pi(a).$$

As I will illustrate in the following subsections in this way Lobachevsky is able to further apply symbolic identifications and to arrive to new equations which consistently (and at the same time) connect Euclidean and non-Euclidean perspectives. This kind of diagram strongly guides the geometer's selections of moves by eliciting what I call the *Euclidean-inside non-Euclidean* "model matching strategy".

Inside the perspective representations (given by the fundamental unveiling diagram of a *non-Euclidean structure*, cf. Figure 2.15), a *Euclidean spherical triangle* and the *orisphere* (and its boundary triangle where the Euclidean properties hold) are constructed. The directly perceivable information strongly guides the geometer's selections of moves by eliciting the Euclidean-inside non-Euclidean "model matching strategy" I have quoted above. This maneuver also constitutes an important step in the affirmation of the modern "scientific" concept of *model*. We have to note that other perceptions activated by the diagram are of course disregarded as irrelevant to the task, as it usually happens when exploiting external diagrammatic representations in reasoning processes. Because not everything in external representations is always relevant to a task, high level cognitive mechanisms need to use task knowledge (usually supplied by task instructions) to direct attention and perceptual processes to the relevant features of external representations.

The different selected representational system – that still uses Euclidean icons – determines in this case quite different possibilities of constructions, and thus different results from iconic experimenting. New results are derived in diagrammatic reasoning through modifying the representational systems, adding new meaning to them, or in reconstructing their systematic order.

2.11.2 Consistency-Searching Strategy

This external representation in terms of the unveiling diagram illustrated in Figure 2.15 activates a perceptual reorientation in the construction (that is identifies possible further constructions); in the meantime the consequent new generated internal representation of the external elements activates directly retrievable information (numerical values) that elicits the strategy of building further non-Euclidean structures together with their *analytic counterpart* (cf. below the non-Euclidean trigonometry equations). Moreover, the internal representation of the stereometric figures activates cognitive operations related to the *consistency-searching strategy*. In this process, new "imaginary" and strange mathematical entities, like the oricycle and the orisphere, are – non-explanatorily – abduced and *unveiled*, and related to ordinary and unsuspected perceptive entities.

Finally, it is easy to identify in the proof the differences between perceptual and other cognitive operations and the differences between *sequential* – the various steps of the constructed unveiling diagram – and *parallel* perceptual operations. Similarly, it is easy to distinguish between the forms that are directly perceptually inspected and the elements that are mentally computed or computed in external symbolic configurations.

To arrive to the second unveiling diagram the old diagram (cf. Figure 2.15) is further enhanced by a new construction, by breaking the connection of the three principal planes along the line BB', and by turning them out from each other so that they, together with all the lines lying in them, come to lie in one plane, where consequently the arcs p, q, r will unite to a single arc of a boundary-line (oricycle), which goes through the point A and has AA' as its axis, in such a manner that (Figure 2.17) on the one side will lie, the arcs q and p, the side b of the triangle, which is perpendicular to AA' at A, the axis CC' going from the end of b parallel to AA' and through C'' the union point of p and q, the side a perpendicular to CC' at the point C, and from the end-point of a the axis BB' parallel to AA' which goes

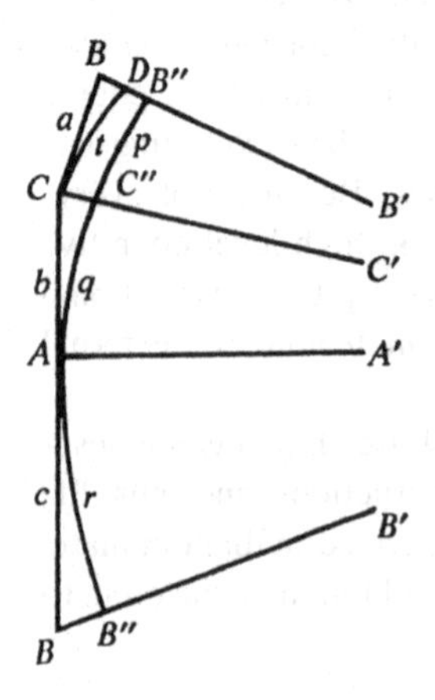

Fig. 2.17

through the end-point B'' of the arc p, etc. Finally, taking CC' as axis, a new boundary line (an arc of oricycle) from the point C to its intersection with the axis BB' is constructed. What happens?

2.11.3 Loosing Intuition

In this case we see that the external representation completely looses its spatial intuitive interest and/or its capacity to simulate internal spatial representations: it is not useful to represent it as an internal spatial model in order to enhance the problem solving activity. The diagram of Figure 2.17 does not have to depict internal forms coherent from the intuitive spatial point of view, it is just devoted to suitably "unveil" the possibility of further calculations by directly activating perceptual information that, in conjunction with the non-spatial information and cognitive operations provided by internal representations in memory, determine the subsequent problem solving behavior. This diagram does not have to prompt an internal "spatially" intuitively coherent model. Indeed *perception* often plays an autonomous and central role, it is not a peripheral device. In this case the end product of perception and motor operations coincides with the intermediate data highly analyzed, processed, and transformed, that is prepared for high-level cognitive mechanisms in terms of further *analytic* achievements (the equations).[57]

We have to note that of course it cannot be said that the external representation would work independently without the support of anything internal or mental. The mirror and unveiling diagrams have to be processed by perceptual mechanisms that are of course internal. And in this sense the end product of the perceptual mechanisms is also internal. But it is not an internal model of the external representation of the task: the internal representation is the knowledge and structure of the task in memory; and the external representation is the knowledge and structure of the task in the environment. The end product of perception is merely the situational information in working memory that usually only reflects a fraction (crucial) of the external representation [Zhang, 1997]. At this point it is clear that the perceptual operations generated by the external representations "mediated" by the unveiling diagrams are central as mechanisms of the whole geometrical abductive and manipulative process; they are not less fundamental than the cognitive operations activated by internal representations, in terms of images and/or symbolic-propositional. They constitute an extraordinary example of complex and perfect coordination between perceptual, motor, and other inner cognitive operations.

Let us conclude the survey on Lobachevsky's route to an acceptable assessment of its non-Euclidean theory. By means of further *symbolic/propositional* designations taken from both internal representations followed from previous results and "externalized" calculations, the reasoning path is constrained to find a general "analytic" counterpart for (some aspects of) the non-Euclidean geometry (we skip the exposition of this complicated passage – cf. [Lobachevsky, 1891]). Therefore we obtain the equations.

[57] In other problems solving cases, the end product of perception – directly picked-up – is the end product of the whole problem solving process.

$$\sin \Pi(c) = \sin \Pi(a) \sin \Pi(b)$$
$$\sin \Pi(\beta) = \cos \Pi(\alpha) \sin \Pi(a)$$

Hence we obtain, by mutation of the letters,

$$\sin \Pi(\alpha) = \cos \Pi(\beta) \sin \Pi(b)$$
$$\cos \Pi(b) = \cos \Pi(c) \cos \Pi(\alpha)$$
$$\cos \Pi(a) = \cos \Pi(c) \cos \Pi(\beta)$$

that express the mutual dependence of the sides and the angles of a non-Euclidean triangle. In these equations of plane non-Euclidean geometry we can pass over the equations for spherical triangles. If we designate in the right-angled spherical triangle (Figure 2.16) the sides $\Pi(c)$, $\Pi(\beta)$, $\Pi(a)$, with the opposite angles $\Pi(b)$, $\Pi(\alpha')$, by the letters a, b, c, A, B, then the obtained equations take of the form of those which we know as the equations of spherical trigonometry for the right-angled triangle

$$\sin(a) = \sin(c) \sin(A)$$
$$\sin(b) = \sin(c) \sin(B)$$
$$\cos(A) = \cos(A) \sin(B)$$
$$\cos(B) = \cos(B) \sin(A)$$
$$\cos(c) = \cos(a) \cos(b).$$

The equations are considered to "[...] attain for themselves a sufficient foundation for considering the assumption of imaginary geometry as possible" (p. 44). The new geometry is considered exempt from possible inconsistencies together with the acknowledgment of the reassuring fact that it presents a very complex system full of surprisingly harmonious conclusions. A new contradiction which could have emerged and which would have forced to reject the principles of the new geometry would have been already contained in the equations above. Of course this is not true from the point of view of modern deductive axiomatic systems and a satisfactory model of non-Euclidean geometry has not yet been built (as Beltrami and Klein will do with the so-called "Euclidean models of non-Euclidean geometry").[58] As for now the argument rests on a formal agreement between two sets of equations, one of which is derived from the new non-Euclidean geometry. Moreover, the other set of equations does not pertain to Euclidean geometry; rather they are the equations of spherical trigonometry that does not depend on the fifth postulate (as maintained by Lobachevsky himself). Nevertheless, we can conclude that Lobachevsky is not far from the modern idea of *model*.

We can say that geometrical diagrammatic thinking represented the capacity to extend *finite* perceptual experiences to give known (Euclidean) and infinite unknown (non-Euclidean) mathematical structures that appear consistent in themselves and that have quite different properties each other.

[58] On the limitations of the Lobachevskyan perspective cf. [Torretti, 1978] and [Rosenfeld, 1988].

Many commentators (and myself in [Magnani, 2001c]) contend that Kant did not imagine that non-Euclidean concepts could in some way be constructed in *intuition*[59](a Kantian expression which indicated our iconic external representation), through the mediation of a model, that is preparing and constructing a Euclidean model of a specific non-Euclidean concept (or group of concepts). Yet Kant also wrote that "[...] the use of geometry in natural philosophy would be insecure, unless the notion of space is originally given by the nature of the mind (so that if anyone tries to frame in his mind any relations different from those prescribed by space, he will labor in vain, for he will be compelled to use that very notion in support of his figment)" [Kant, 1968, Section 15E].

[Torretti, 2003, p. 160] observes:

> I find it impossible to make sense of the passage in parentheses unless it refers precisely to the activity of constructing Euclidean models of non-Euclidean geometries (in a broad sense). We now know that one such model (which we ought rather to call quasi-Euclidean, for it would represent plane Lobachevskian geometry on a sphere with radius $\sqrt{-1}$) is mentioned in the *Theorie der Parallellinien* that Kant's fellow Königsbergian Johann Heinrich Lambert [Lambert, 1786] wrote about 1766. There is no evidence that Kant ever saw this tract and the few extant pieces of his correspondence with Lambert do not contain any reference to the subject, but, in the light of the passage I have quoted, it is not unlikely that Kant did hear about it, either from Lambert himself, or from a shared acquaintance, and raised the said objection.

I agree with Torretti, Kant had a very wide perspective about the resources of "intuition", anticipating that a geometer would have been "compelled" to use the notion of space "given by nature", that is the one that is at the origins of our external representation, "in support of his figment", for instance the non-Euclidean Lobachevskyan abstract structures we have treated above in Figure 2.15, which exhibits the non-Euclidean through the Euclidean.

2.12 Mechanizing Manipulative Abduction

2.12.1 Automatic Geometrical Constructions as Extra-Theoretical Epistemic Mediators

A very interesting artificial intelligence computer program has been built, ARCHIMEDES [Lindsay, 1994; Lindsay, 1998], that represents geometrical diagrams (points, line segments, polygons, and circles) both as pixels arrays and as propositional statements.[60] For example a triangle will be represented from a propositional description as a set of marked pixels in an array, together with a set of data naming the given triangle and storing facts about it (for instance that it is right) and constraints upon it (perhaps that it remains right throughout this use of the diagram).

---

[59] We have seen how Lobachevsky did this by using the Figure 2.15.

[60] This approach in computer science, involving the use of diagram manipulations as forms of acceptable methods of reasoning, was opened by Gelernter's Geometry Machine [Gelertner, 1959], but the diagrams played a very secondary role.

Hence, a computational equivalent of a physical diagram is represented, plus some human propositional knowledge about it.

The program is able to manipulate and modify its own representations of diagrams, that is it is able to make geometrical constructions (called "simulation constructions"): adding parts or elements, moving components about, translating and rotating by preserving metric properties, of course subordinated to the given specific constraints and to the whole structure of the two-dimensional space. Some knowledge of algebra is added, and of the taxonomic hierarchy of geometric figures (all square are rectangles, etc.); moreover, additional knowledge is also included, like side-angle-side congruency theorem and the sum of the interior angles of a triangle, knowledge of problem solving strategies and heuristics, knowledge of logic (for example: a universal statement can be disproved by a single counterexample) [Lindsay, 2000a].

When the program manipulates the specific diagram, it records the new information that comes out, then it can for example detect sets of area equivalences, and so on: for example, it is able to verify that a demonstration of the Pythagorean Theorem is correct, mirroring its truth in terms of constructions and manipulations. To account for the universality of geometrical theorems and propositions many different methods for learning and "generalizing" the specific instance of the constructed diagram are exploited [Lindsay, 1998, pp. 260-264].

These methods come from a kind of predicative knowledge "exogenous" of course to the mere diagrammatic representation: generalization is not a possible product of the pure diagrammatic understanding. For instance, one suggestion is to break the problem into cases, to individuate a "representative" instance for each case, and to demonstrate that this conclusion holds for each of these instances.[61] Another one is to exploit the simulative aspects of constructions by "running experiments" that show how some parts of a diagram co-vary with changes in others: the observation of the interaction of the diagrams parts as one property is varied allows us to grasp and understand the "universal" value of some geometric relations and results (for example congruency theorem, asymptotic behaviors, periodic relations, and some symmetric relations). It is interesting to note that one of the construction manipulations proposed by the program to verify the Pythagorean Theorem intends to show that it is not true of (some examples of) non-right triangles.

An extension of the program (described in [Lindsay, 2000b]), aims at autonomously building constructions that can demonstrate a given proposition,[62] It accomplishes the further complex task of discovering conjectures that can lead to the constructions of demonstrations, illustrating the possible role of diagrammatic reasoning in creativity.[63]

[61] A method already suggested by Johnson-Laird to deal with generalization using the so-called "mental models" [Johnson-Laird, 1983].

[62] The program is able to "discover demonstrations", that is to find sequences of manipulations that achieve a particular end, rather than simply verify them.

[63] On the role of abduction in automated scientific discovery and machinery for generating hypotheses cf. the recent [Ray, 2007]. On the historical systems cf. the section "Automatic abductive scientists" above (section 2.7).

2.12.2 Automatic "Thinking through Doing"

Geometric constructions are certainly epistemic mediators that exploit the semantics of two dimensional diagrams (rather than the syntax of formal propositions) to perform various manipulative abductive tasks (discover a new property or new proposition/hypothesis, selecting suitable sequences of constructions as able to convincingly verifying theorems, etc).

Hence, geometrical construction, one of the most ancient exploitations of two-dimensional diagrams for both practical and mathematical problem solving, is "embodied" in a computational program, that is, finally, in a machine. From the epistemological point of view it is important to note that the program shows how it is possible to delineate the rules and the procedures that underlie the diagrams as models of propositions about space, that is able to capture the structure of space. The kind of reasoning described is very rich and takes advantage of almost all the resources of two-dimensional space (going beyond the simple use of topological properties like in the case of Euler/Venn diagrams). Moreover, the physical diagram necessarily (that is for the objective reasons of its materiality) preserves topological and geometric properties of two-dimensional space.[64]. The following passage is very clear:

> Geometric diagrams [...] are intended to reflect all structural properties of two dimensional space, although this requires that the observer ignore line width and so forth and deal with the intended idealization. Since they are recorded in space (on a piece of paper for example) the mapping is iconic. The intuitive structure of two- dimensional space is partially captured by Euclid's definitions, postulates and axioms. Although it is now known that this is not the veridical description of actual space, for objects of human scale it is adequate both for reasoning and for successful intercourse with the environment, just as Newtonian mechanics is adequate for modeling physical processes that occur at a human scale though it is inaccurate at very large and very small scale [...] [Lindsay, 2000a].

We know that the structure of intuitive space is also embraced by analytic geometry. Lindsay observes that, in general, it could be better to use an analytic representation because conventional digital computers are "a natural match for numerical representation". If we consider the various ways of representing geometrical diagrams and their behavior – the analytic representation is an example – it is important to point out their real cognitive nature. I think we have to agree with the following position: "This does not mean, however, that diagrams represented numerically are not really diagrams. What makes them diagrams is not bits or voltages or axioms or CCD signals. What makes them diagrams is that they capture the structure of space. This is another way of saying that they enforce constraints on the behavior of the representations that reflect restrictions on the behavior of objects in space" (*ibid.*).

The computational embodiment generates a kind of "squared" epistemic mediator: geometrical constructions, as epistemic mediators, are further mediated. A mental performance not simply reproduced, like in more traditional AI systems, but just

[64] This is not the case of the two-dimensional depictions of a three dimensional object (like in the clear cases illustrated by Escher's drawings) [Lindsay, 1998, p. 266].

a way of "thinking through doing", as illustrated by manipulative abduction. Humans can think using geometrical constructions also without "doing", for instance in the case of "thinking through drawing" at the level of imagination. Kant (and Proclus) were perfectly clear when they referred to the role of imagination as the condition of possibility of the empirical drawing itself [Magnani, 2001c].

It has been established that imagination is able to perform some tasks at intermediate levels of complexity. Humans, when working from an external physical display, are aided by actual pencil shading or tracing. When working from mental images humans have greater difficulty keeping track of what has been inspected. Moreover, in the case of imaginary geometrical representations (or in data structures in a computer) we can use whatever rules we want (for instance the rules that reflect the motion of rigid bodies but also the ones that violate them and do not happen – or are not yet seen to happen – in the physical world). This of course does not mean that the mind can imagine anything whatsoever, or everything with equal facility: "[...] human mind is indeed predisposed to handle certain types of imagery and simulations better than others, presumably those kinds for which evolution has best prepared us, especially the motion of rigid objects" [Lindsay, 1998, p. 267].

In the past some researchers emphasized the role of visualizations in geometrical reasoning. For example in 1847 Byrne prepared an amazing and attractive edition of the first six books (which range from the most elementary plane geometry to the theory of proportions) of the Elements of Euclid "in which coloured diagrams and symbols are used instead of letters for the greater ease of learners".[65] The aim was to try to use as little text – and in particular, labels – as possible. Byrne considered the use of colors in geometrical constructions a suitable tool to this aim. Interesting exploitations of the visual devices of the Internet are some Java applications/manipulations where it is possible to "construct by clicking".[66]

Summary

In this chapter we have seen that, to better understand how abductive cognition works, non-explanatory and instrumental aspects clearly have to be taken into account. I have stressed that Gabbay and Woods' *GW*-model is very useful to this end: it stimulates the analysis of many "working" abductive processes that have non-explanatory and pragmatic/instrumental importance. From this perspective the

[65] [Byrne, 1847]. See the web site http://sunsite.ubc. ca/DigitalMathArchive/ Euclid/byrne.html. The home page provides links to other web sites where it is possible to find Java editions of Euclid and other "visual" information. The web site devoted to illustrating a list of many proofs of Pythagoras' theorem in Java is particularly interesting, where the user can "construct" geometrical demonstrations by clicking at the figures presented, moving points and features of geometrical diagrams (http://sunsite.ubc.ca/DigitalMathArchive/Euclid/java/html/pythagoras. html).

[66] Cf. the Banchoff's web page http://www.geom.umn.edu / banchoff/, where it is possible to manipulate diagrams that correspond to mathematical problems. Joyce's appealing Java edition of Euclid where it is possible to "construct by clicking", is given at the web site http://aleph0.clarku.edu/ djoyce/java /elements/elements.html.

concept of plausibility, central to abductive reasoning, has been reshaped by stressing its "strategic" components, more pragmatically and in a less epistemologically qualified manner.

Finally, I would like to reiterate the importance of the sections devoted to illustrating the concrete examples taken from the history and practice of epistemology, science, and mathematics, seen from the perspective of the interconnection between explanatory, non-explanatory, and instrumental abduction. This analysis also allowed me to shed new light on traditional epistemological concepts, which are much better understood, such as in the case of falsification of the conventionalist first principles of physics and regarding the problem of falsification of construction in psychoanalytic interpretation. In these cases newly acquired knowledge about abductive cognition better accounts for the role of contradiction and inconsistency in creative reasoning. I also believe it can help us to overcome some consequences of the old-fashioned assumptions typical of the epistemological tradition which resorted to affirmation of the different status of inferential processes at work in natural and formal sciences.

The analysis of mirror and unveiling diagrams described at the end of the chapter, taking advantage of the cognitive-epistemological reconstruction of the discovery of non-Euclidean geometry, entails some general consequences concerning the epistemology of mathematics and formal sciences. The concept of mirror diagram plays a fundamental explanatory role in the epistemology of removing obstacles and obscurities related to the ambiguities of the problem of parallel lines and, in general, in enhancing mathematical knowledge regarding critical situations. In the case of the more instrumental unveiling diagrams, the allocating and switching of attention between internal and external representation better reveals how to govern the reasoning strategy at hand by integrating internal and external representation in a more dynamic and complicated way. This account in terms of mirror and unveiling diagrams seems empirically adequate to integrate findings from research on cognition and findings from historical-epistemological research into models of actual mathematical practices. I contended that the assessment of the fit between cognitive findings and historical-epistemological practices helps to elaborate richer and more realistic models of cognition and presents a significant advance over previous epistemological work on actual mathematical reasoning and practice.

I also think that: i) the role of optical diagrams in a geometry teaching environment could be relevant. Experimental research could be performed on high school geometry students devoted to detecting the didactic effects and learning improvements due to mirror and unveiling diagrams in producing abstract entities; ii) the activity of zooming (especially in the case of magnification) of optical diagrams could be studied in other areas of human abductive model-based reasoning, such as the ones involving creative, analogical, and spatial inferences, both in science and everyday situations so that this can extend the psychological theory.

Chapter 3
Semiotic Brains and Artificial Minds
How Brains Make Up Material Cognitive Systems

In chapter one the important role of external representations and epistemic mediators was stressed, when illustrating the concept of manipulative abduction, especially in the field of scientific reasoning. Further insight can be granted by some considerations that also take into account Turing's seminal ideas about human and machine intelligence, some paleoanthropological results, and a Peircean semiotic perspective.

Following Peirce's semiotics, the interplay between internal and external representation can be further depicted taking advantage of what I call *semiotic brains*. They are brains that make up a series of signs and that are engaged in making or manifesting or reacting to a series of signs. Through this semiotic activity they are at the same time engaged in "being minds" and thus in thinking intelligently. An important effect of this semiotic brain activity is a continuous process of disembodiment of mind that exhibits a new cognitive perspective on the mechanisms underlying the semiotic emergence of meaning processes. To illustrate this process I will take advantage of Turing's comparison between the so-called "unorganized" brains and "logical" and "practical" machines, and of some paleoanthropological results on the birth of material culture, that provide an evolutionary perspective on the origin of intelligent behaviors.

Then I will describe the centrality to semiotic cognitive information processes of the disembodiment of mind from the point of view of the cognitive interplay between internal and external representations, both *mimetic* and *creative*, where the problem of the continuous interaction between on-line – like in the case of manipulative abduction – and off-line (for example in inner rehearsal) intelligence can properly be addressed (section 3.6.5). I consider this interplay critical in analyzing the relation between meaningful semiotic internal resources and devices and their dynamical interactions with the externalized semiotic materiality already stored in the environment. This materiality plays a specific role in the interplay due to the fact that it exhibits (and operates through) its own cognitive constraints. Hence, minds are "extended" and artificial in themselves.

L. Magnani: Abductive Cognition, COSMOS 3, pp. 145–217.
springerlink.com

From this perspective Turing's "unorganized" brains can be seen as structures that organize themselves through a semiotic activity that is reified in the external environment and then re-projected and reinterpreted through new configurations of neural networks and chemical processes. I will show how the disembodiment of mind can nicely account for low-level semiotic processes of meaning creation, bringing up the question of how higher-level processes could be comprised and how they would interact with lower-level ones. To better explain these higher-level semiotic mechanisms I will return to analysis of the role of model-based and manipulative abduction and of external representations. The example of elementary geometry will also be examined, where many external things, usually inert from the cognitive/semiotic point of view, can be transformed into what I have called "epistemic mediators" (cf. [Magnani, 2001b] and the previous chapter of this book) that then give rise – for instance in the case of scientific reasoning – to new signs, new chances for "interpretants", and thus to new interpretations. In this interplay abduction can be seen as fully *multimodal*, in that both data and hypotheses can have a full range of verbal and sensory representations. Some basic aspects of this constitutive hybrid nature of abduction – involving words, sights, images, smells, etc. but also kinesthetic experiences and other feelings such as pain – will be investigated.[1]

Taking advantage of Turing's comparison between "unorganized" brains and "logical" and "practical" machines the concept of the *mimetic mind* is introduced. This sheds new cognitive and philosophical light on the role of Turing's machines and computational modeling, outlines the decline of the so-called Cartesian computationalism and emphasizes the possible impact of the construction of new types of universal "practical" machines, available over there, in the environment, as new tools underlying the emergence of meaning processes. Language itself can be seen as a mediating "ultimate artifact" (section 3.4.2): from this perspective the brain would merely be a pattern completing device while language would be considered an external resource/tool which is – through coevolution – obviously fitted to the human brain helping and supporting it to enhance its cognitive capacities.

Finally, the thesis of this chapter being that the externalization/disembodiment of mind is a significant cognitive perspective able to unveil some basic features of abduction and creative/hypothetical thinking, its success in explaining the semiotic interplay between internal and external representations (mimetic and creative) is evident. This is also clear at the level of some intellectual issues stressed by certain results of psychoanalytic research and therapy, such as in the case of creative meaning formation. To this aim I will focus on the epistemological status of the psychoanalytic concepts of projections and introjections, and of the mythologization of external observation. Also taking advantage of the concept of manipulative abduction, I will stress the role of some external artifacts (symbols, Mandala, ritual tools) in what Jung calls "psychic energy flow", where the mobility and disposability of psychic energy are seen as the secret of cultural development. I contend

[1] On the concept of multimodal abduction cf. also the following chapter, section 4.1.

these artifacts are tools which can be usefully represented as memory mediators that "mediate" and make available the story of their origin and the actions related to them, which can be learnt and/or re-activated when needed. From this wide perspective symbols in psychoanalysis can be seen as memory mediators which maximize abducibility, because they maximize recoverability, in so far as they are the best possible expression of something not yet grasped by consciousness.

In summary, a key issue of the chapter is "What kind of brain could or would use external representations as an aid or method of thinking?" The description of cognitive mediators in the light of multimodal and manipulative abduction and the introduction of the concept of maximization of abducibility try to furnish a first answer.

3.1 Turing Unorganized Machines

If we decide to increase knowledge on the semiotic character of high-level types of cognition it is first of all necessary to develop a model of creativity able to represent not only "novelty" and "unconventionality", but also some features commonly referred to as the entire creative process, such as the expert use of background knowledge and ontology (defining new concepts and their new meanings and searching heuristically among the old ones) and the modeling activity developed in the so called "incubation time" (generating and testing, transformations in the space of the hypotheses). The philosophical concept of *abduction* I have illustrated in the first two chapters is a candidate to solve this problem, and offers an approach to model creative processes of meaning generation in a completely explicit and formal way, which can fruitfully integrate the narrowness proper of a merely psychological approach, too experimentally human-oriented. I have already stressed that abductive reasoning is active in many scientific disciplines but also in everyday reasoning: it is essential in scientific discovery, medical and non medical diagnosis, generation of causal explanations, generations of explanations for the behaviors of others, minds interplay, when for example we attribute intentions to others, empathy, analogy, emotions, as an appraisal of a given situation endowed with an explanatory or instrumental power, etc. I have illustrated in chapter one that the concept of manipulative abduction can nicely account for the relationship between meaningful behavior and dynamical interactions with the environment. In the following sections we will see in detail that at the roots of the creation of new meanings there is a process of *disembodiment of mind* that exhibits a new cognitive description of the mechanisms underling the emergence of meaning processes through semiotic "delegations" to the environment.[2]

[2] To illustrate the importance of the semiotic processes in complex cognitive systems [Loula *et al.*, 2009] describe an interesting digital ecosystem in which the emergence of self-organized and adaptive symbol-based communication among distributed (and semiotic) artificial creatures is simulated.

3.1.1 Logical, Practical, Unorganized, and Paper Machines

Aiming at building intelligent machines Turing first of all provides an analogy between human brains and computational machines. In "Intelligent Machinery", written in 1948 [Turing, 1969] maintains that "[...] the potentialities of human intelligence can only be realized if suitable education is provided" (p. 3). The concept of *unorganized machine* is then introduced, and it is maintained that the infant human cortex is of this nature.[3] The argumentation is indeed related to showing how such machines can be educated by means of "rewards and punishments".

Unorganized machines are listed among different kinds of existent machineries:

- *(Universal) Logical Computing Machines (LCMs)*. A LCM is a kind of discrete machine Turing introduced in 1937 that has

> [...] an infinite memory capacity obtained in the form of an infinite tape marked out into squares on each of which a symbol could be printed. At any moment there is one symbol in the machine; it is called the scanned symbol. The machine can alter the scanned symbol and its behavior is in part described by that symbol, but the symbols on the tape elsewhere do not affect the behavior of the machine. However, the tape can be moved back and forth through the machine, this being one of the elementary operations of the machine. Any symbol on the tape may therefore eventually have innings [Turing, 1992, p. 6].

This machine is called Universal if it is "[...] such that if the standard description of some other LCM is imposed on the otherwise blank tape from outside, and the (universal) machine then set going it will carry out the operations of the particular machine whose description is given" (p. 7). The importance of this machine resorts to the fact that we do not need to have an infinity of different machines doing different jobs. A single one suffices: it is only necessary "to program" the universal machine to do these jobs.

- *(Universal) Practical Computing Machines (PCMs)*. PCMs are machines that put their stored information in a form very different from the tape form. Given the fact that in LCMs the number of steps involved tends to be enormous because of the arrangement of the memory along the tape, in the case of PCMs "[...] by means of a system that is reminiscent of a telephone exchange it is made possible to obtain a piece of information almost immediately by 'dialing' the position of this information in the store" (p. 8). Turing adds that "nearly" all the PCMs under construction have the fundamental properties of the Universal Logical Computing Machines: "[...] given any job which could have be done on an LCM one can also do it on one of these digital computers" (*ibid.*) so we can speak of Universal Practical computing Machines.

---

[3] I am taking advantage here of the concept of unorganized brain (and machine) to stress the historical/epistemological interest of Turing's discoveries. Of course the concept acquires a specific meaning in the context of those Turing's philosophical speculations that lead to the proposal of the new idea of universal logical computing machine and to the related computational- representational-understanding of the mind (CRUM). Therefore the concept is obviously unrelated to current neuroscience results and only presents a philosophical concern.

- *Unorganized Machines.* Machines that are largely random in their constructions are called "Unorganized Machines": "So far we have been considering machines which are designed for a definite purpose (though the universal machines are in a sense an exception). We might instead consider what happens when we make up a machine in a comparatively unsystematic way from some kind of standard components. [...] Machines which are largely random in their construction in this way will be called 'Unorganized Machines'. This does not pretend to be an accurate term. It is conceivable that the same machine might be regarded by one man as organized and by another as unorganized." (p. 9). They are machines made up from a large number of similar units. Each unit is endowed with two input terminals and has an output terminals that can be connected to the input terminals of 0 or more of other units. An example of the so-called unorganized A-type machine with all units connected to a synchronizing unit from which synchronizing pulses are emitted at more or less equal intervals of times is given in Figure 3.1 (the times when the pulses arrive are called moments and each unit is capable of having two states at each moment). The so-called A-type unorganized machines are considered very interesting because they are the simplest model of a nervous system with a *random arrangement of neurons* (cf. the following section 3.2, "Brains as unorganized machines").

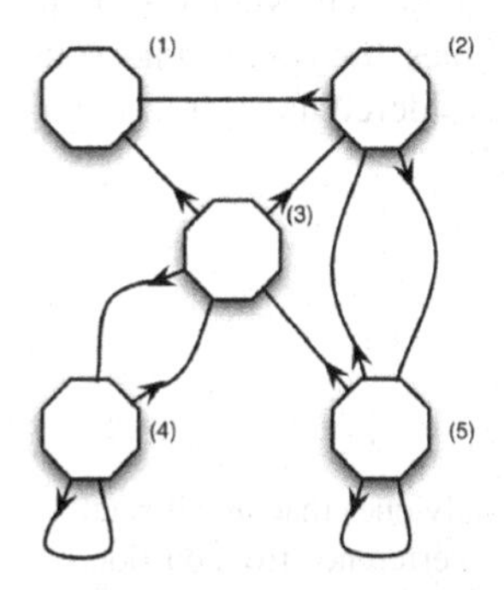

Fig. 3.1 An example of the so-called unorganized A-type machine

- *Paper Machines.* "It is possible to produce the effect of a computing machine by writing down a set of rules of procedure and asking a man to carry them out. [...] A man provided with paper, pencil and rubber, and subject to strict discipline, is in effect a universal machine" (p. 9). Turing calls this kind of machine "Paper Machine".

3.1.2 Continuous, Discrete, and Active Machines

The machines described above are all *discrete* machines because it is possible to describe their possible states as a discrete set, with the motion of the machines occurring by jumping from one state to another. However, Turing remarks that all machinery can be also regarded as continuous (where the states form a continuous manifold and the behavior of the machine is described by a curve on this manifold) but "[...] when it is possible to regard it as discrete it is usually best to do so. Moreover machineries are called "controlling" if they only deal with information, and "active" if aim at producing some definite physical effect. A bulldozer will be a continuous and active machine, a telephone continuous and controlling. But also brains can be considered machines and they are – Turing says "probably" – continuous and controlling but "very similar to much discrete machinery" (p. 5).

Brains very nearly fall into this class [discrete controlling machinery – when it is natural to describe its possible states as a discrete set] and there seems every reason to believe that they could have been made to fall genuinely into it without any change in their essential properties. However, the property of being "discrete" is only an advantage for the theoretical investigator, and serves no evolutionary purpose, so we could not expect Nature to assist us by producing truly "discrete brains" (p. 6).[4]

Brains can be treated as machines but they can also be considered discrete machines. The epistemological reason is clear: this is just an advantage for the "theoretical investigator" that aims at knowing what are intelligent machines, but certainly it would not be an evolutionary advantage. "Real" humans brains are of course continuous systems, only "theoretically" they can be treated as discrete.

Following Turing's perspective we have derived two new achievements about machines and intelligence: brains can be considered machines, the simplest nervous systems with a random arrangement of neurons can be considered unorganized machines, in both cases with the property of being "discrete".

3.1.3 Mimicking Human Education

Turing also says:

> The types of machine that we have considered so far are mainly ones that are allowed to continue in their own way for indefinite periods without interference from outside. The universal machines were an exception to this, in that from time to time one might change the description of the machine which is being imitated. We shall now consider machines in which such interference is the rule rather than the exception (p. 11).

Screwdriver interference is when parts of the machine are removed and replaced with others, giving rise to completely new machines. *Paper* interference is when mere communication of information to the machine modifies its behavior. It is clear that in the case of the universal machine, paper interference can be as useful as screwdriver interference: we are interested in this kind of interference. We can say that each time an interference occurs the machine is probably changed. It has to be noted that paper interference provides information that is both external and material.

Turing thought that the fact that human beings have already made machinery able to imitate various parts of a man was positive in order to believe in the possibility

[4] Further details on the so-called "discretization of knowledge" will be given in chapter eight, section 8.1.1.

of building thinking machinery: trivial examples are the microphone for the ear, and the television camera for the eye. What about the nervous system? We can copy the behavior of nerves with suitable electrical models and the electrical circuits which are used in electronic computing machinery seem to have essential properties of nerves because they are able to transmit information and to store it.

Education in human beings can model "education of machinery" "Mimicking education, we should hope to modify the machine until it could be relied on to produce definite reactions to certain commands" (p. 14). A graduate has had interactions with other human beings for twenty years or more and at the end of this period "[...] a large number of standard routines will have been superimposed on the original pattern of his brain" (*ibid.*).

Turing maintains that

1. in human beings the interaction is mainly with other men and the receiving of visual and other stimuli constitutes the main forms of interference;
2. it is only when a man is "concentrating" that he approximates a machine without interference;
3. even when a man is concentrating his behavior is mainly conditioned by previous interference.

3.2 Brains as Unorganized Machines

3.2.1 The Infant Cortex as an Unorganized Machine

In many unorganized machines when a configuration[5] is reached and possible interference suitably constrained, the machine behaves as one organized (and even universal) machine for a definite purpose. Turing provides the example of a B-type unorganized machine with sufficient units where we can find particular initial conditions able to make it a universal machine also endowed with a given storage capacity. The set up of these initial conditions is called "organizing the machine" that indeed is seen a kind of "modification" of a preexisting unorganized machine through external interference.

Infant brain can be considered an unorganized machine. Given the analogy previously established (cf. subsection above "Logical, Practical, Unorganized, and Paper Machines), what are the events that modify it in an organized universal brain/machine? "The cortex of an infant is an unorganized machinery, which can be organized by suitable interference training. The organization might result in the modification of the machine into a universal machine or something like it. [...] This picture of the cortex as an unorganized machinery is very satisfactory from the point of view of evolution and genetics." (p. 16). The presence of human cortex is not meaningful in itself: "[...] the possession of a human cortex (say) would be virtually useless if no attempt was made to organize it. Thus if a wolf by a mutation

[5] A configuration is a state of a discrete machinery.

acquired a human cortex there is little reason to believe that he would have any selective advantage" (*ibid.*). Indeed the exploitation of a big cortex (that is its possible organization) requires a suitable environment: "If however the mutation occurred in a milieu where speech had developed (parrot-like wolves), and if the mutation by chance had well permeated a small community, then some selective advantage might be felt. It would then be possible to pass information on from generation to generation" (*ibid.*).

Hence, organizing human brains into universal machines strongly relates to the presence of

1. *speech* (even if only at the level of rudimentary but meaningful parrot-like wolves);
2. and a *social setting* where some "techniques" are learnt "[...] the isolated man does not develop any intellectual power. It is necessary for him to be immersed in an environment of other men, whose techniques he absorbs during the first twenty years of his life. He may then perhaps do a little research of his own and make a very few discoveries which are passed on to other men. From this point of view the search for new techniques must be regarded as carried out by human community as a whole, rather than by individuals" (p. 23).

This means that a big cortex[6] can provide an evolutionary advantage only in presence of that massive storage of meaningful information and knowledge on external supports that only an already developed small community can possess. Turing himself considers this picture rather speculative but evidence from paleoanthropology can support it, as I will describe in the following section.

Moreover, the training of a human child depends on a system of rewards and punishments, that suggests that organization can occur only through two inputs. The example of an unorganized P-type machine, that can be regarded as a LCM without a tape and largely incompletely described, is given. Through suitable stimuli of pleasure and pain (and the provision of an external memory) the P-type machine can become an universal machine (p. 20).

When the infant brain is transformed in an intelligent one both discipline and initiative are acquired: "[...] to convert a brain or machine into a universal machine is the extremest form of discipline. [...] But discipline is certainly not enough in itself to produce intelligence. That which is required in addition we call initiative. [...] Our task is to discover the nature of this residue as it occurs in man, and try and copy it in machines" (p. 21).

Examples of problems requiring initiative are the following: "Find a number n such that ...", "see if you can find a way of calculating the function which will

[6] [Evans *et al.*, 2005] illustrate recent research in neurogenetics which shows the role of a gene, Microcephalin, which regulates brain size and which has evolved under strong positive selection in the evolution of primate lineage, leading to *Homo sapiens* and beyond. One genetic variant of it in modern humans, which arose ~37,000 years ago, increased in frequency too rapidly to be compatible with neutral drift, so supporting the ongoing brain evolutionary plasticity of the human brain.

enable us to obtain the values for arguments ....". The problem is equivalent to that of finding a program to put on the machine in question.

We have seen how a brain can be "organized", in Turing's sense, but how can that brain be a creative brain able to account for the emergence of interesting meaning processes?

3.3 From the Prehistoric Brains to the Universal Machines

I have said that what I call *semiotic brains* are brains that make up a series of signs and that are engaged in making or manifesting or reacting to a series of signs: through this semiotic activity they are at the same time engaged in "being minds" and so in thinking intelligently. In this section I will illustrate the process of "disembodiment of mind" as an important aspect of this semiotic activity of brains.

We have seen in the previous section that following Turing's point of view [1969] a big cortex can provide an evolutionary advantage only in presence of a massive storage of meaningful information and knowledge on external supports that only an already developed small community of human beings can possess. Evidence from paleoanthropology seems to support this perspective. Some research [Mithen, 1996; Mithen, 1999; Humphrey, 2002; Lewis-Williams, 2002] in cognitive paleoanthropology – even if rather speculative – teaches us that high level and reflective consciousness in terms of thoughts about our own thoughts and about our feelings (that is consciousness not merely considered as raw sensation) is intertwined with the development of *modern language* (speech) and *material culture*. After 250.000 years ago several hominid species had brains as large as ours today, but their behavior lacked any sign of art or symbolic behavior. If we consider high-level consciousness as related to a high-level organization – in Turing's sense – of human cortex, its origins can be related to the active role of environmental, social, linguistic, and cultural aspects.[7]

Handaxes were made by Early Humans and firstly appeared 1,4 million years ago, still made by some of the Neanderthals in Europe just 50.000 years ago. The making of handaxes is seen as strictly intertwined with the development of consciousness. Many needed capabilities constitute a part of an evolved psychology that appeared long before the first handaxes were manufactured. It seems humans were pre-adapted for some components required to make handaxes [Mithen, 1996; Mithen, 1999] (cf. Figure 3.2):

1. imposition of *symmetry* (already evolved through predators escape and social interaction). It has been an unintentional by-product of the bifacial knapping technique but also deliberately imposed in other cases. [Dennett, 1991] hypothesizes that the attention to symmetry may have developed through social interaction and predator escape, as it may allow one to recognize that one is being

[7] [Logan, 2006] further stresses the coevolution of brain and culture and the supposed limits of brain size due to obstetrical and mobility reasons and to the the expensiveness in terms of energy.

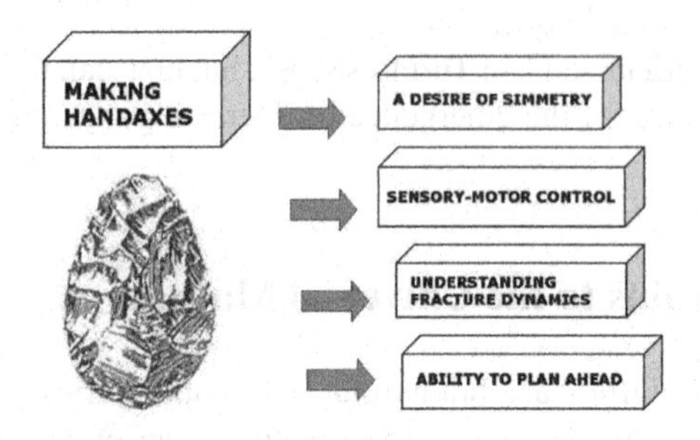

Fig. 3.2 From [Mithen, 1999, p. 286]. ©1999 Massachusetts Institute of Technology, by permission of The MIT Press.

directly stared at. It also seems that "Hominid handaxes makers may have been keying into this attraction to symmetry when producing tools to attract the attention of other hominids, especially those of the opposite sex" [Mithen, 1999, p. 287];

2. understanding *fracture dynamics* (for example evident from Oldowan tools and from nut cracking by chimpanzees today);
3. ability to *plan ahead* (modifying plans and reacting to contingencies, such unexpected flaws in the material and miss-hits), still evident in the minds of Oldowan tool makers and in chimpanzees;
4. high degree of *sensorimotor control*: "Nodules, preforms, and near finished artifacts must be struck at precisely the right angle with precisely the right degree of force if the desired flake is to be detached" [Mithen, 1999, p. 285]. The origin of this capability is usually tracked back to encephalization – the increased number of nerve tracts and of the integration between them allows for the firing of smaller muscle groups – and bipedalism – that requires a more complex integrated highly fractionated nervous system, which in turn presupposes a larger brain.

The combination of these four resources produced an important semiotic revolution: the birth of what Mithen calls technical intelligence of early human mind, that is consequently related to the construction of handaxes and their new semiotic values. Indeed they indicate high intelligence and good health. They cannot be compared to those artifacts made by animals, like honeycomb or spider web, deriving from the iteration of fixed actions which do not require plastic consciousness and intelligence.[8]

[8] I will illustrate animal artifacts due to more "plastic" abductive endowments in chapter five.

3.3.1 Private Speech and Fleeting Consciousness

Two central factors play a fundamental role in the combination of the four resources above:

- the exploitation of *private speech* (speaking to oneself) to trail between planning, fracture dynamic, motor control and symmetry (also in children there is a kind of private muttering which makes explicit what is implicit in the various abilities);
- a good degree of *fleeting consciousness* (thoughts about thoughts).

Of course they furnish a kind of blackboard where the four – previously distinct – resources can be exploited all together and in their dynamic interaction. In the meantime these two aspects obviously played a fundamental role in the development of consciousness and thought:

> So my argument is that when our ancestors made handaxes there were private mutterings accompanying the crack of stone against stone. Those private mutterings were instrumental in pulling the knowledge required for handaxes manufacture into an emergent consciousness. But what type of consciousness? I think probably one that was fleeting one: one that existed during the act of manufacture and that did not the endure. One quite unlike the consciousness about one's emotions, feelings, and desires that were associated with the social world and that probably were part of a completely separated cognitive domain, that of social intelligence, in the early human mind [Mithen, 1999, p. 288].

This use of private speech can be certainly considered a semiotic internal "tool" for organizing brains and so for manipulating, expanding, and exploring minds, a tool that probably coevolved with another: talking to each other.[9] Both private and public language act as tools for thought and play a fundamental role in the evolution "opening up our minds to ourselves" and so in the emergence of new meaning processes.

3.3.2 Material Culture as Distributed Cognition and Semiosis

Another semiotic tool appeared in the latter stages of human evolution, that played a great role in the evolution of primitive minds, that is in the organization of human brains. Handaxes also are at the birth of *material culture*, so as new cognitive chances can coevolve:

- the mind of some early humans, like the Neanderthals, were constituted by relatively isolated cognitive domains, [Mithen, 1999] calls *different intelligences*, probably endowed with different degrees of consciousness about the thoughts and knowledge within each domain (natural history intelligence, technical

[9] On natural languages as cognitive artifacts cf. [Carruthers, 2002a; Clark and Chalmers, 1998; Clark, 2003; Clark, 2005; Norman, 1993; Clowes and Morse, 2005], and below section 3.4.2.

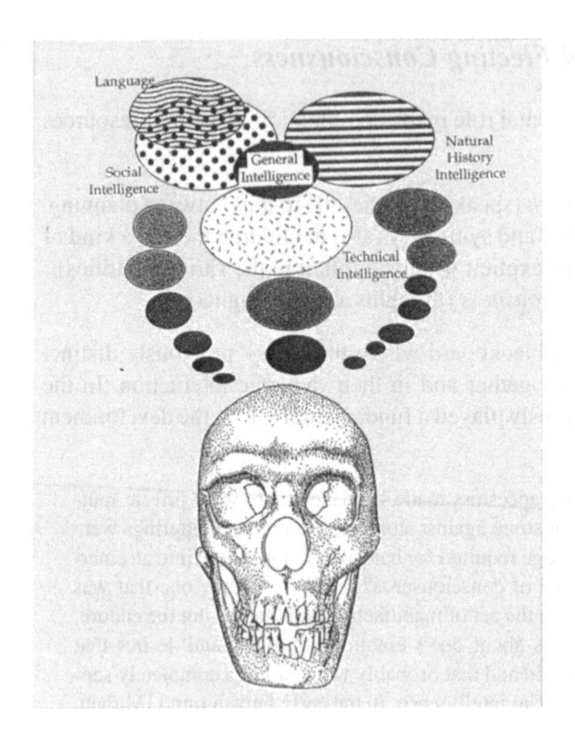

Fig. 3.3 From [Mithen, 1999, p. 290]. ©1999 Massachusetts Institute of Technology, by permission of The MIT Press.

intelligence, social intelligence). These isolated cognitive domains became integrated also taking advantage of the role of public language (cf. Figure 3.3.);[10]

- degrees of *high level consciousness* appear, human beings need thoughts about thoughts;
- *social intelligence* and *public language* arise.[11]

[10] In this book I will address the problem of the origin of human language from many current perspectives related to philosophy, cognitive science, biology, and paleoanthropology, addressing phylogenetic, ontogenetic, and learning issues. A general description of the various approaches to the problem of the emergence of natural language is provided by [Számadó and Szathmáry, 1997].

[11] Posed in the late 1980s [Whiten and Byrne, 1988; Whiten and Byrne, 1997; Byrne and Whiten, 1988], the "social brain hypothesis" (also called "Machiavellian intelligence hypothesis") holds that the relatively large brains of human beings and other primates reflect the computational demands of complex social systems and not only the need of processing information of ecological relevance: ability to manipulate information and not simply to remember it, to recognize visual signals to identify other individuals, sufficient memory for faces and to remember who has a relationship with whom, use of tactical deception, coalition, ability to understand intentions, to hold false beliefs, and "mind-read", known as "theory of mind" etc. Language itself would have at a certain point grooming as a way of creating social cohesion as the size and complexity of the social group increased (cf. also [Dunbar, 1998; Dunbar, 2003]).

It is extremely important to stress that material culture is not just the product of this massive cognitive chance but also cause of it. “The clever trick that humans learnt was to disembody their minds into the material world around them: a linguistic utterance might be considered as a disembodied thought. But such utterances last just for a few seconds. Material culture endures” [Mithen, 1999, p. 291].

In this perspective we acknowledge that material artifacts are tools for thoughts as is language: tools (and their related new “signs”) for exploring, expanding, and manipulating our own minds. In this regard the evolution of culture is inextricably linked with the evolution of consciousness and thought.[12]

Early human brain becomes a kind of universal “intelligent” machine, extremely flexible so that we did no longer need different “separated” intelligent machines doing different jobs. A single one will suffice. As the engineering problem of producing various machines for various jobs is replaced by the office work of “programming” the universal machine to do these jobs, so the different intelligences become integrated in a new universal device endowed with a high-level type of consciousness.[13]

From this perspective the semiotic expansion of the minds is in the meantime a continuous process of *disembodiment* of the minds themselves into the *material world* around them. In this regard the evolution of the mind is inextricably linked with the evolution of large, integrated, material cognitive semiotic systems. In the following sections I will illustrate this extraordinary interplay between human brains and the cognitive systems they make.

3.3.3 Semiotic Delegations through the Disembodiment of Mind

A wonderful example of meaning creation through disembodiment of mind is the carving of what most likely is the mythical being from the last ice age, 30.000 years ago, a half human/half lion figure carved from mammoth ivory found at Hohlenstein Stadel, Germany.

> An evolved mind is unlikely to have a natural home for this being, as such entities do not exist in the natural world, the mind needs new chances: so whereas evolved minds could think about humans by exploiting modules shaped by natural selection, and about lions by deploying content rich mental modules moulded by natural selection and about other lions by using other content rich modules from the natural history

[12] A further analysis of the domain specific mentality of Neanderthals, related to social behavior, natural world, and technology, contrasted with the cognitive “fluidity” of modern mind, is illustrated in [Mithen, 2007].

[13] On the relationship between material culture and the evolution of consciousness cf. [Donald, 1998; Donald, 2001; Dennett, 2003]. The fact that bigger brains found an evolutionary success has been recently also related to the change in diet, caused by the fact that in Pliocene the climate conditions began to change. Bigger brains (together with growth of body size and reduction in the dentition) were important because they helped to solve dietary problems and to arrive to higher quality diet: so the birth of social and material culture would be related to the problem of solving dietary problems, brains are also required to retain a mental map of plant and food supplies [Milton, 2006].

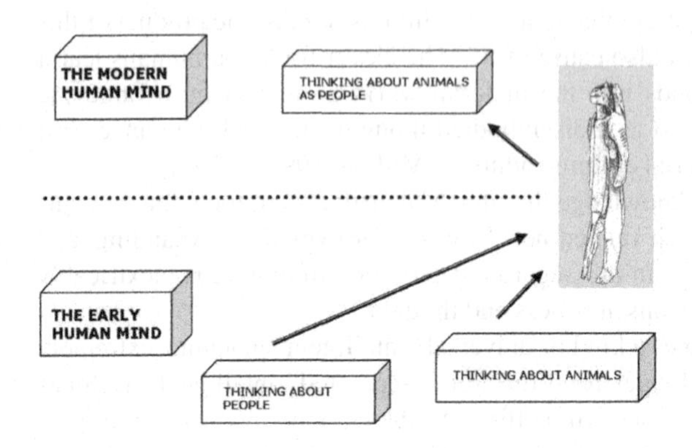

Fig. 3.4 From [Mithen, 1999, p. 293]. ©1999 Massachusetts Institute of Technology, by permission of The MIT Press.

> cognitive domain, how could one think about entities that were part human and part animal? Such entities had no home in the mind [Mithen, 1999, p. 291].

A mind consisting of different separated intelligences cannot come up with such entity (Figure 3.4). The only way is *to extend* the mind into the *material word*, building in the environment primitive *scenarios*[14] and exploiting in a semiotic way rocks, blackboards, paper, ivory, and writing, painting, and carving: "[...] artifacts such as this figure play the role of anchors for ideas and have no natural home within the mind; for ideas that take us beyond those that natural selection could enable us to possess" [Mithen, 1999, p. 291].

In the case of our figure we face with an anthropomorphic thinking created by the material representation serving to semiotically anchor the cognitive representation of supernatural being. In this case the material culture disembodies thoughts, that otherwise will soon disappear, without being transmitted to other human beings, and realizes a systematic semiotic delegation to the external environment. The early human mind possessed two separated intelligences for thinking about animals and people. Through the mediation of the material culture the modern human mind can arrive to *internally* think about the new concept of animal and people at the same time. But the new meaning occurred over there, in the external material world where the mind picked up it.

---

[14] Ohsawa proposed this term to explain some aspects of "chance discovery". Cf. for example [Nara and Ohsawa, 2004] that stress the attention on the activity of extracting chances through communication of scenarios: given that a scenario is a time series of events under a given context, during a communication scenario drawings are generated, externalized (that is disembodied) and subsequently analyzed in the abductive process of chance discovery. [Abe *et al.*, 2006] usefully apply abduction to dynamic risk management in nursing, where rare chances (risks or accidents) can be abduced as novel generated hypotheses. The set of nursing activities is regarded as a scenario where risks and accidents are considered as a scenario violation.

Artifacts as *external semiotic objects* allowed humans to loosen and cut those chains on our unorganized brains imposed by our evolutionary past. Chains that always limited the brains of other human beings, such as the Neanderthals. Loosing chains and securing ideas to external objects was also a way to creatively re-organize brains as universal machines for thinking.[15]

In the remaining part of this chapter I will describe the centrality to semiotic cognitive information processes of the disembodiment of mind from the point of view of the cognitive interplay between internal and external representations. I consider this interplay critical in analyzing the relation between meaningful semiotic internal resources and devices and their dynamical interactions with the externalized semiotic materiality already stocked in the environment. Hence, minds are "extended" and artificial in themselves.[16] With the aim of explaining these higher-level mechanisms I will provide a cognitive framework where model-based and manipulative abduction together with external representations and epistemic mediators play a central role.

3.4 Mimetic and Creative Representations

We have seen that unorganized brains organize themselves through a semiotic activity that is reified in the external environment and then re-projected and reinterpreted through new configurations of neural networks and chemical processes. I also think the disembodiment of mind can nicely account for low-level semiotic processes of meaning creation, bringing up the question of how could higher-level processes be comprised and how would they interact with lower-level ones.

3.4.1 External and Internal Representations

We have said that through the mediation of the material culture the modern human mind can arrive to *internally* think the new meaning of animals and people at the

[15] Actually, the brains of the humans who made those artifacts were highly organized, not unorganized. I have to reiterate that my argumentation is related to the concept of unorganized brain in its specific Turing's sense. We do tend to think of early humans as being like our children but they were perfectly viable adults.

[16] To correctly place this view in present, general cognitive debate it is useful to read the book by [Shapiro, 2004]. The author, speaking of the so-called "multiple realizability thesis" (that is: are mental states multiply realizable?) offers a full and critical treatment of the various current theories that deal with the concept of multiple realizable mind, also treating topics like "envatment" – mind as distinct from the body, embodied and extended mind – mind "beyond" and "outside" the body, etc. The perspectives range from those which only see some creatures which are different to us in their physical composition as possessing a mind to the ones stating that every suitable organized system, regardless of its physical composition, can have minds like ours. On the multirealizability thesis cf. also [Polger, 2006] and the recent discussion, also related to the so-called "separability thesis", given by [Clark, 2008, chapter nine]. On the frequent misunderstandings and confusions in current discussions of multiple realization, cf. the clear criticisms contained in [Polger, 2008].

same time. We can account for this process of disembodiment from an impressive cognitive point of view.

I maintain that representations are external and internal. We can say that

- *external representations* are formed by external materials that express (through reification) concepts and problems already stored in the brain or that do not have a *natural home* in it;
- *internalized representations* are internal re-projections, a kind of recapitulations (learning), of external representations in terms of neural patterns of activation in the brain. They can sometimes be "internally" manipulated like external objects and can originate new internal reconstructed representations through the neural activity of *transformation* and *integration*.

This process explains why human beings seem to perform both computations of a *connectionist* type[17] such as the ones involving representations as

- (I Level) *patterns of neural activation* that arise as the result of the interaction between body and environment (and suitably shaped by the evolution and the individual history): pattern completion or image recognition,[18]

and computations that use representations as

- (II Level) *derived combinatorial syntax and semantics* dynamically shaped by the various external representations and reasoning devices found or constructed in the environment (for example geometrical diagrams); they are neurologically represented contingently as pattern of neural activations that "sometimes" tend to become stabilized structures and to fix and so *to permanently belong to the I Level* above.

The I Level originates those *sensations* (they constitute a kind of "face" we think the world has), that provide room for the II Level to reflect the structure of the environment, and, most important, that can follow the computations suggested by these external structures. It is clear we can now conclude that the growth of the brain and especially the synaptic and dendritic growth are profoundly determined by the environment.

When the fixation is reached the patterns of neural activation no longer need a direct stimulus from the environment for their construction. In a certain sense they can be viewed as *fixed internal records* of *external structures* that *can exist* also in the absence of such external structures. These patterns of neural activation that

[17] Here the reference to the word "connectionism" is used on the plausible assumption that all mental representations are brain structures: verbal and the full range of sensory representations are neural structures endowed with their chemical functioning (neurotransmitters and hormones) and electrical activity (neurons fire and provide electrical inputs to other neurons). In this sense we can reconceptualize cognition neurologically: for example the solution of an explanatory problem can be seen as a process in which one neural structure representing an explanatory target generates another neural structure that constitutes a hypothesis for the solution.

[18] Andy Clark, adopting a connectionist perspective, maintains that human brain is essentially a device for pattern-association, pattern-completion, and pattern-manipulation.

constitute the I Level Representations always keep record of the experience that generated them and, thus, always carry the II Level Representation associated to them, even if in a different form, the form of *memory* and not the form of a vivid sensorial experience. Now, the human agent, via neural mechanisms, can retrieve these II Level Representations and use them as *internal* representations or use parts of them to construct new internal representations very different from the ones stored in memory (cf. also [Gatti and Magnani, 2005]).[19]

3.4.2 Language as the Ultimate Artifact

The example of recent cognitive theories concerning natural language is particularly useful to illustrate the interplay between external and internal representations. Following [Clark, 1997, p. 218] language is an "ultimate artifact". In this perspective brain is just a pattern completing device (as I have illustrated introducing the I level in the previous subsection), while language is an external resource/tool which is – along a process of coevolution – obviously fitted to the human brain, helping and supporting it to enhance its cognitive capacities [Wheeler, 2004]. Language is culturally passed from one generation to the next and is thus learnt again and again just through exposure to a sample of it, and then suitably generalized.[20] It is not only an important part of the *cognitive niche* built by human beings a long time ago, but it also formed a permanent artificial environment that in turn created a further selective pressure in evolution, in the coevolutionary interplay between genes and culture, as I will describe in chapter six: in a recent article [Clark, 2006, p. 370] himself acknowledges that "language is a self-constructed cognitive niche" consisting of structures that "combine with appropriate culturally transmitted practices to enhance problem-solving".

Exactly like hammers and PCs are fitted to the human brain and to the structure and capacities of human hands, language is a medium of communication and information and it "[...] alters the nature of the computational tasks involved in various kinds of problem solving" that affect human beings (and their brains) [Clark, 1997,

[19] The role of external representations has already been stressed in some central traditions of cognitive science and artificial intelligence, from the area of distributed and embodied cognition and of robotics [Brooks, 1991; Clark, 2003; Zhang, 1997] to the area of active vision and perception [Gibson, 1979; Thomas, 1999]). I also think this discussion about external and internal representations can be used to extend and enhance the Representational Redescription model introduced by [Karmiloff-Smith, 1992], which accounts for how these levels of representation are generated in the infant mind. [Sterelny, 2004] lists some of the most important results we can obtain thanks to external representations: they 1) ease memory burdens, 2) transform difficult cognitive problems into easier perceptual problems, 3) transform difficult perceptual problems into easier ones, 4) transform difficult learning problems into easier ones, 5) engineer workspaces to complete tasks more rapidly and reliably.

[20] On the importance of arbitrariness in natural languages cf. [Gasser, 2004]. Research acknowledging the fact that language understanding cannot be performed through the manipulations of arbitrary symbols alone, but has to be based on the body interaction with the environment is described in [Glenberg and Kaschak, 2003; Zwaan, 2004]. In this perspective language acquisition and meaning comprehension are partly achieved through the same simulative structures used to plan and guide action [Svensson and Ziemke, 2004].

p. 193]. It is said that language *scaffolds* cognition for the mind [Clowes and Morse, 2005]. Basically, language is for Clark a cognitive tool that facilitates thought and cognition through 1) memory augmentation, 2) environmental simplification, 3) coordination of activities through control of attention and resource allocation, 4) the activity of transcending path-dependent learning (the learning of linguistic organisms is not constrained by complicated cognitive paths that are circumvented thanks to language), 5) control loops (that act for our future behavior: for example writing plans difficult to keep in one's head), 6) data manipulation and representation [Bermúdez, 2003, p. 151]. From this perspective there is no innate domain-specific language processing system, – like for example the one maintained by [Chomsky, 1986] and language does not deeply alter the "basic modes of representation and computation" of the brain [Clark, 1997, p. 198].[21]

The acquisition of language is a kind of reprogramming of the computational resources of the human brain in such a way that "[...] our innate pattern-completing neural architecture comes to simulate a kind of logic-like serial processing device" [Wheeler, 2004, p. 696], without a substantial modification of the brain's processing architecture. Just like diagrams can help us in many cognitive tasks and especially in mathematical reasoning (cf. below sections 3.6.1, 3.6.2 and 3.6.3) language helps in various cognitive tasks, for instance as a sensory relay in human communication (and in other various simulations of basic psychic endowments), when writing in notebooks, building databases, organizing actions and plans, creating narratives and theories, etc. Moreover, language helps us in a more internal modality, such as in self-directed speech (silent, in auditory imagery, or aloud), when for example we repeat some instructions to ourselves.[22]

In Clark's words, "[...] exposure to, or rehearsal [of spoken and written language, through visual, auditory, and haptic sensorial systems] [...] always activates or otherwise exploits many other kinds of internal representational or cognitive resources" that are able "to provide a new kind of cognitive niche whose features and properties *complement*, but do not need to replicate the basic modes of operation and representation of the biological brain [Clark, 2006, pp. 370–371]. Various experiments provide evidence that the adoption of language (and symbols) would favor the de-coupling of the cognitive agent from the "immediate pull of the encountered scene" and would provide a "new realm of perceptible objects" which simplify certain kinds of attentional, reasoning, and learning tasks (*ibid.*). For example, in the case of the use of external linguistic tags or symbols (for example numbers), the brain is enabled – by re-presenting them when needed – to solve problems previously seen as puzzling. Studies on writing as thinking show how their coupling involves a kind of reciprocal influence, where inner and outer features

[21] On the received view on language, the so-called "language myth" cf. [Love, 2004], who discusses Clark's rejection of the idea that natural languages are codes and usefully analyzes some aspects of Saussure's and Harris' perspectives. A computational framework for studying the emergence of language and communication, which sees language as a heterogeneous set of artifacts implicated in cultural and cognitive activities is presented in [Cangelosi, 2007], also taking into account social, sensorimotor, and neural capabilities of cognitive agents.

[22] On the role of the so-called "inner rehearsal" cf. below subsection 3.6.5.

have a causal influence on one another which is occurring over time [Harris, 1989; Menary, 2007]: "The restructuring of thought which writing introduces depends upon prising open a conceptual gap between sentence and utterance. [...] Writing is crucial here because autoglottic inquiry presupposes the validity of unsponsored language. Utterances are automatically sponsored by those who utter them, even if they merely repeat what has been said before. Sentences by contrast, have no sponsors: they are autoglottic abstractions. The Aristotelian syllogism like the Buddhist *panchakarani*, presupposes writing" [Harris, 1989, p. 104].

Language would stabilize and discipline (or "anchor", Clark says) intrinsically fluid and context-sensitive modes of thought and reason:[23] one of the fruitful qualities of connectionist or artificial neural-network models is their capability and their need to be stabilized. Moreover, words act on mental off-line[24] *inner states* affecting not only other internally represented words but also many other model-based and sensorimotor representations and modes, between and within humans: "Words and sentences act as artificial input signals, often (as in self-directed inner speech) entirely self-generated, that nudge fluid natural systems of encoding and representation along reliable and useful trajectories", where a "a semi-anarchic parallel organization of competing elements" (a metaphor taken from [Dennett, 1991]) is at play and explains the origin of language. These elements take control at different times in a distributed structure informed by "[...] a wealth of options involving intermediate grades of intelligent and semi-intelligent orchestration, and of hierarchical and semi-hierarchical control" [Clark, 2006, p. 372].[25]

Quoting Clark, we have stressed above that "exposure to, or rehearsal of [spoken and written language, through visual, auditory, and haptic sensorial systems] [...] always activates or otherwise exploits many other kinds of internal representational or cognitive resources" that are able "[...] to provide a new kind of cognitive niche whose features and properties *complement*, but do not need to replicate the basic modes of operation and representation of the biological brain" [Clark, 2006, pp. 370–371].

[23] I agree with [Bermúdez, 2003, p. 155] who, speaking of Clark's approach says: "His view, I suspect, is that the environmental simplification that language provides applies to a perceived environment that is already parsed into objects or objects like entities" (on prelinguistic reification in animal cognition cf. chapter five, section 5.5, on the role of spatial cognition in reification cf. chapter four, section 4.7).

[24] On this concept cf. below subsection 3.6.5.

[25] Interesting considerations on the scaffolding role of language in mathematical thought are presented in [Clark, 2008, chapter, section four]. The never-ending problem of the role of language in "necessarily" rendering thoughts possible, and even its role in any form of conceptual thinking, or at least in the mere acquisition of thoughts or in scaffolding them and in making communication, consciousness and mind-reading possible, is extensively treated in [Carruthers, 2002a]. Coherently with Clark's contention which I have just described, the author maintains that language is the medium for "non-domain specific thinking", which fulfils the role of integrating the outputs of a variety of domain-specific conceptual faculties (or "central-cognitive quasi-modules"). The opinion that embodiment in cognitive science undervalues concepts such as conventions/norms, representations, and consciousness, as essential properties of language, is provided by [Zlatev, 2007].

The semantic approach to language can take advantage of this perspective in a more traditional framework that does not take into account the concept of cognitive niche, but is oriented towards a dynamic systems framework: [Logan, 2006, p. 153] nicely expresses an analogous consideration. A word is "[...] a strange attractor for all the percepts associated with the concept represented by that word", and a concept can be characterized like an "artificial or virtual percept". Instead of "[...] bringing the mountain or the percept of the mountain directly to the mind the word brings the mind to the mountain through the concept of the mountain" so accessing and capturing suitable memories. In the terms of dynamic systems approach [Logan, 2006, p. 155] "An attractor is a trajectory in phase space towards which all of the trajectories of a non-linear dynamic system are attracted. The meaning of the word [as an attractor] being uttered does not belong simply to the individual but to the community to which the individual belongs [...] and emerges in the context in which it is being used". The variability of the context explains that "The attractor is a strange attractor because the meaning of a word never exactly repeats itself" for instance because of the variability of the constraints imposed by the medium at hand.[26] According to the theory of dissipative systems [Prigogine and Stengers, 1984], spoken and syntaclilized language and abstract conceptual thinking can be seen as having emerged at exactly the same time as the "bifurcation" of the brain which shifted from the concrete percept-based thinking of prelingual hominids to that of the fully fledged human species, *Homo sapiens sapiens*, providing an example of both punctuated equilibrium and a new order coming out of a chaotic linear system. Of course *Homo sapiens sapiens* vestigially retains the perceptual-oriented features of hominid brains.

In tune with this dynamic approach to semantics are the considerations made in terms of the catastrophe theory: at the level of human individuals we can hypothesize that there exists an "[...] isomorphism between the mental mechanisms which ensure the stability of a concept Q, and the physical and material mechanisms which ensure the stability of the actual object K represented by Q" [Thom, 1980, p. 248]. Here the semantic depth of a concept is characterized by the time taken by the mental mechanisms of analysis to reduce this concept to its representative sign. The more complex the concept is, the more its stability needs regulator mechanisms, the greater is its semantic density to an actual object, as obviously happens in the case of nouns which refer to a substance: "The supreme prize is handed to animate beings, and most likely to man. An animal to live must periodically resort to a whole spectrum of activities: eating, sleeping, moving,... etc. To these fundamental physiological activities are added (for man) mental activities almost as indispensable to the meaning of being human: speaking, thinking, believing, ... etc., which constitute a form of regulation which superimposes itself at the beginning and on the presupposed" [Thom, 1980, p. 248].

[26] Details on the concept of attractor are given in chapter four, section 4.7.4 and in chapter eight. On the constraints imposed by the materiality of the external *medium* in the the process of abductive formation of new meanings and in meaning change cf. below section 3.6.7.

In the following section I will illustrate some fundamental aspects of the interplay above in the light of basic semiotic aspects of general abductive reasoning (cf. also section 1.3).

3.5 Model-Based Abduction and Semiosis beyond Peirce

I think there are two basic kinds of external representations active in the process of externalization of the mind: *creative* and *mimetic*. Mimetic external representations mirror concepts and problems that are already represented in the brain and need to be enhanced, solved, further complicated, etc. so they sometimes can creatively give rise to new concepts and meanings. In the examples I will illustrate in the following sections it will be clear how for instance a *mimetic geometric representation* can become *creative* and give rise to new meanings and ideas in the hybrid interplay between brains and suitable cognitive environments, as "cognitive niches"[27] that consequently are appropriately reshaped.

What exactly is model-based abduction from a philosophical point of view? I have already said that Peirce stated that all thinking is in signs, and signs can be icons, indices, or symbols and that all *inference* is a form of sign activity, where the word sign includes "feeling, image, conception, and other representation" [Peirce, 1931-1958, 5.283] (for details cf. [Kruijff, 2005]), and, in Kantian words, all synthetic forms of cognition. In this light it can be maintained that a considerable part of the creative meaning processes is *model-based*. Moreover, a considerable part of meaning creation processes (not only in science) occurs in the middle of a relationship between brains and external objects and tools that have received cognitive and/or epistemological delegations (cf. the previous section and the following subsection).

Following this Peircean perspective about inference I think it is extremely useful from a cognitive point of view to consider the concept of reasoning in a very broad way (cf. also [Brent, 2000, p. 8]). We have three cases:

1. reasoning can be fully conscious and typical of high-level worked-out ways of inferring, like in the case of scientists' and professionals' performances;
2. reasoning can be "acritical" [Peirce, 1931-1958, 5.108], which includes every day inferences in conversation and in various ordinary patterns of thinking;
3. reasoning can resort to "[...] operations of the mind which are logically analogous to inference excepting only that they are unconscious and therefore uncontrollable and therefore not subject to logical criticism" [Peirce, 1931-1958, 5.108].

Immediately Peirce adds a note to the third case: "But that makes all the difference in the world; for inference is essentially deliberate, and self-controlled. Any operation which cannot be controlled, any conclusion which is not abandoned, not merely as

[27] This expression, Clark used in the different framework of the cognitive analysis of language appears very appropriate also in this context [Pinker, 2003].

soon as criticism has pronounced against it, but in the very act of pronouncing that decree, is not of the nature of rational inference – is not reasoning" (*ibid.*).

As Colapietro clearly states [Colapietro, 2000, p. 140], it seems that for Peirce human beings semiotically involve unwitting trials and unconscious processes. Moreover, it seems clear that unconscious thought can be in some sense considered "inference", even if not rational; indeed, Peirce says, it is not reasoning. Peirce further indicates that there are in human beings multiple trains of thought at once but only a small fraction of them is conscious, nevertheless the prominence in consciousness of one train of thought is not to be interpreted an interruption of other ones.

In this Peircean perspective, which I adopt in this chapter, where inferential aspects of thinking dominate, there is no intuition, in an anti-Cartesian/anti-dualistic way. We know all important facts about ourselves in an *inferential* abductive way:

> [...] we first form a definite idea of ourselves as a hypothesis to provide a place in which our errors and other people's perceptions of us can happen. Furthermore, this hypothesis is constructed from our knowledge of "outward" physical facts, such things as the sounds we speak and the bodily movements we make, that Peirce calls signs [Brent, 2000, p. 8].

Recognizing in a series of *material*, physical events, that they make up a series of signs, is to know the existence of a "mind" (or of a group of minds) and to be absorbed in making, manifesting, or reacting to a series of signs is to be absorbed in "being a mind". "[...] all thinking is dialogic in form" [Peirce, 1931-1958, 6.338], both at the intrasubjective[28] and intersubjective level, so that we see ourselves exactly as others see us, or see them exactly as they see themselves, and we see ourselves through our own speech and other interpretable behaviors, just others see us and themselves in the same way, in the commonality of the whole process [Brent, 2000, p. 10].

As I will better explain later on in the following sections, in this perspective minds are material like brains, in so far as they consist in intertwined internal and external semiotic processes: Peirce clearly anticipated the "extended mind" hypothesis maintaining that "[...] the psychologists undertake to locate various mental powers in the brain; and above all consider it as quite certain that the faculty of language resides in a certain lobe; but I believe it comes decidedly nearer the truth (though not really true) that language resides in the tongue. In my opinion it is much more true that the thoughts of a living writer are in any printed copy of his book than they are in his brain" [Peirce, 1931-1958, 7.364].

---

[28] "One's thoughts are what he is 'saying to himself', that is saying to that other self that is just coming to life in the flow of time. When one reasons, is that critical self that one is trying to persuade: and all thought whatsoever is a sign, and is mostly in the nature of language" [Peirce, 1931-1958, 5.421].

3.5.1 Man Is an External Sign

Peirce's semiotic motto "man is an external sign" is very clear about the materiality of mind and about the fact that the conscious self is a cluster actively embodied of flowing intelligible signs:[29]

> It is sufficient to say that there is no element whatever of man's consciousness which has not something corresponding to it in the word; and the reason is obvious. It is that the word or sign which man uses *is* the man himself. For, as the fact that every thought is a sign, taken in conjunction with the fact that life is a train of thoughts, proves that man is a sign; so, that every thought is an *external sign*, proves that man is an external sign. That is to say, the man and the *external sign* are identical, in the same *sense* in which the words *homo* and *man* are identical. Thus my language is the sum total of myself; for the man is the thought [Peirce, 1931-1958, 5.314].

It is by way of signs that we ourselves *are* semiotic processes – for example a more or less coherent cluster of narratives. If all thinking is in signs it is not true that thoughts are in us because we are in thoughts.[30]

The systemic perspective of the catastrophe theory[31] also stresses the role of signs in their creation of semiotic brains. In the structure of signs (as potential messages for humans) there is always a kind of dynamic instability, which renders them less probable than naturally created forms: "The imprint of a finger on the sand, the tracing of a stylet on clay, are so many naturally fragile marks of man's deliberate acts" [Thom, 1980, p. 284]. Nevertheless, on being perceived by human organisms – which consequently also "become" semiotic processes – these unstable structures return to a normal stability, and in so doing they activate semantic values, generating – mentally – the content signified by the message.

I think it is at this point clearer what I meant in section 1.5 of chapter one, when I explained the concept of model-based abduction and said, adopting a Peircean perspective, that all thinking is in signs, and signs can be icons, indices, or symbols and that, moreover, all *inference* is a form of sign activity, where the word sign includes feeling, image, conception, and other representation. The model-based aspects of human cognition are central, given the central role played for example by signs like images and feeling in the inferential activity "[. . .] man is a sign developing according to the laws of inference. [. . .] the entire phenomenal manifestation of mind is a sign resulting from inference" [Peirce, 1931-1958, 5.312 and 5.313].

Moreover, the "person-sign" is future-conditional, that is not fully formed in the present but depending on the future destiny of the concrete semiotic activity (future thoughts and experience of the community) in which she will be involved. If Peirce maintains that when we think we appear as a sign [Peirce, 1931-1958, 5.283] and, moreover, that everything is present to us is a phenomenal manifestation of

[29] Consciousness arises as "a sort of public spirit among the nerve cells" [Peirce, 1931-1958, 1.354]. The contemporary researcher on consciousness Donald fully acknowledges the "materiality of mind" [Donald, 2001, pp. 96-99].

[30] It is similar to the situation of the dreamer who is so deeply involved in the dream (we say, "she is lost in her dreams") that she does not feel she is in the dream.

[31] See also chapter eight of this book.

ourselves, then feelings, images, diagrams, conceptions, schemata, and other representations are phenomenal manifestations that become available for interpretations and thus are guiding our actions in a positive or negative way. They become *signs* when we think and interpret them. It is well-known that for Peirce all semiotic experience – and thus abduction – is also providing a guide for action. Indeed the whole function of thought is to produce habits of action.[32]

Let us summarize some basic semiotic ideas that will be of help in the further clarification of the cognitive and computational features of model-based and manipulative abduction. One of the central property of signs is their reinterpretability. This occurs in a social process where signs are referred to *material objects*.

As is well-known, for Peirce *iconic* signs are based on similarity alone, the psychoanalytic patient who thought he was masturbating when piloting the plane interpreted the cloche as an extension of his body, and an iconic sign of the penis; an ape may serve as an icon of a human.[33] *Indexical* signs are based on contiguity and dynamic relation to the object, a sign which refers to an object that it denotes by virtue of being "really affected" by that object: a certain grimace indicates the presence of pain, the rise of the column of mercury in a thermometer is a sign of a rise in temperature, indexical signs are also the footprints in the sand or a rap on the door. Consequently we can say indexical signs "point". A *symbol* refers to an artificial or conventional ("by virtue of a law") interpretation of a sign, the sign ∞ used by mathematicians would be an example of Peirce's notion of symbol, almost all words in language, except for occasional onomatopoeic qualities, are symbols in this sense, associated with referents in a wholly arbitrary manner.[34]

We have to immediately note that from the semiotic point of view *feelings* too are signs that are subject to semiotic interpretations at different levels of complexity. Peirce considered feelings elementary phenomena of mind, comprising all that is immediately present, such as pain, sadness, cheerfulness. He believes that a feeling is a state of mind possessing its own living qualities independent of any other state of the mind. Neither icon, index, nor symbol actually functions as a sign until it is interpreted and recognized in a semiotic activity and code. To make an example, it is the evolutionary kinship that makes the ape an icon of the man, in itself the similarity of the two animals does not mean anything.

Where cognition is merely possible, sign action, or *semiosis*, is working. Knowledge is surely inferential as well as abduction, that like any inference requires three

---

[32] On this issue cf. for example the contributions contained in a recent special issue of the journal *Semiotica* devoted to abduction [Queiroz and Merrell, 2005].

[33] Iconic signs preserve the relational structure governing their objects. This fact does not always have to be interpreted as a mirror-like resemblance, it can be seen as a "relation of reason" [Peirce, 1931-1958, 1.369] with the object. Rather, the structural relation would be better and more generally grasped through the mathematical notion of homomorphism – between icons and icons and their referents, as already indicated by [Barwise and Etchemendy, 1990; Stenning, 2000], and recently stressed by [Ambrosio, 2007]. A general homomorphic relationship would also be more satisfactory to account for the case in which the manipulation of diagrams is able to creatively convey new information and chances, like in the case of algebraic representations which I will illustrate below in subsection 3.6.4.

[34] On the role of symbols in mathematical abduction cf. [Heeffer, 2007].

elements: a *sign*, the *object* signified, and the *interpretant*. Everywhere "*A* signifies *B* to *C*".

There is a continuous activity of interpretation and a considerable part of this activity – as we will see – is abductive. The Peircean notion of interpretant plays the role of explaining the activity of interpretation that is occurring in semiosis. The interpretant does not necessarily refer to an actual person or mind, an actual interpreter. For instance the communication to be found in a beehive[35] where the bees are able to communicate with the others by means of signs is an example of a kind of "mindless" triadic semiosis: indeed we recognize that a sign has been interpreted not because we have observed a mental action but by observing another material sign. To make another example, the person recognizing the thermometer as a thermometer is an interpretant, as she generates in her brain a thought. In this case the process is conscious, but also unconscious or emotional interpretants are widespread. Again, a person points (index) up at the sky and his companion looks up (interpretant) to see the object of the sign. Someone else might call out "What do you see up there?" that is also another interpretant of the original sign. As noted by Brent "For Peirce, any appropriate response to a sign is acting as another sign of the object originally signified. A sunflower following the sun across the sky with its face is also an interpretant. Peirce uses the word interpretant to stand for any such development of a given sign" [Brent, 2000, p. 12].

Semiosis is in itself a dynamic and interactive process that happens in time and presupposes the notions of environment and agents. As anything can be seen as a sign, the collection of potential signs may encompass virtually everything available within the agent, including all data gathered by its sensors. In the context of the science of complexity semiosis can be depicted as an emergent property of a semiotic system: emergent properties constitute a certain class of higher-level properties, related in a certain way to the microstructure of a class of system, that thus become able to produce, transmit, receive, compute, and interpret signs of different kinds. In this last sense they are more than simple reactive systems which in principle are not able to use something as a sign for something else [Gomes *et al.*, 200; Loula *et al.*, 2009]. It has to be stressed that semiotic systems are obviously materially embodied because they can be only realized through physical implementation.

Finally, an interpretant may be the thought of another person, but may as well be simply the further thought of the first person, for example in a soliloquy the succeeding thought is the interpretant of the preceding thought so that an interpretant is both the interpretant of the thought that precedes it and the object of the interpretant thought that succeeds it. In soliloquy sign, object, and interpretant are all present in the single train of thought.

Interpretants, mediating between signs[36] and their objects have three distinct levels in hierarchy: feelings, actions, and concepts or habits (that is various generalities

[35] This kind of communication is studied in [Monekosso *et al.*, 2004].

[36] It has to be noted that for Peirce no sign is so general that it cannot be amended, hence all general signs are to an extent incomplete. Consequently, a sign holds the chance of taking any particular feature previously unknown to its interpreters, many of these new features remaining inconsistent with other possibilities.

as responses to a sign). They are the effect of a sign process. The interpretant produced by the sign can lead to a feeling (*emotional* interpretant), or to a muscular or mental effort, that is to a kind of action – *energetic* interpretant (not only outward, bodily action, but also purely inward exertions like those "mental soliloquies strutting and fretting on the stage of imagination" – [Colapietro, 2000, p. 142]. Finally, when it is related to the abstract meaning of the sign, the interpretant is called *logical*,[37] as a generalization requiring the use of verbal symbols. It is a further development of semiosis in the hierarchy of iconic, enactive, and symbolic communication: in short, it is "an interpreting thought", related for instance not only to the intellectual activity but also to initiate the ethical action in so far as a "modification of a person's tendencies toward action" [Peirce, 1931-1958, 5.476].

The logical interpretants are able to translate percepts, emotions, unconscious needs, and experience needs, and so to mediate their meanings to arrive to provisional stabilities. They can lead to relatively stable cognitive or intellectual habits and belief changes as self-controlled achievements like many abductive conceptual results, that Peirce considers the most advanced form of semiosis and the ultimate outcome of a sign. Indeed abduction – hypothesis – is the first step toward the formation of cognitive habits: "[...] every concept, every general proposition of the great edifice of science, first came to us as a conjecture. These ideas are the *first logical interpretants* of the phenomena that suggested them, and which, as suggesting them, are signs" [Peirce, 1931-1958, 5.480].[38]

Orthogonal to the classification of interpretants as emotional, energetic, and logical is the alternate classification given by Peirce: M. E. Q. Gonzalez and W. F. G. Haselager interpretants can also be immediate, dynamic, and normal. Some interpreters consider this classification a different way of expressing the first one. It is sufficient to note this classification can be useful in studying the formation of a subclass of debilitating and facilitating psychic habits [Colapietro, 2000, pp. 144–146] and, I would add, of certain reasoning devices that are used by human agents.[39] Colapietro proposes the concept of quasi-final interpretants – as related to the Peircean normal interpretants – as "[...] effective in the minimal sense that they allow the conflict-ridden organism to escape being paralyzed agent: they permit the body-ego to continue its ongoing negotiations with these conflicting demands, even if only in a precarious and even debilitating manner. In brief, they permit the body-ego to

[37] The logical interpretant is not "logical" in the sense in which deductive reasoning is studied by a discipline called "logic", but rather because it attributes a further meaning to the emotion or to the mental effort that preceded it by providing a conceptual representation of that effort.

[38] Habits also appear in organic and inorganic matter: "Empirically, we find that some plants take habits. The stream of water that wears a bed for itself is forming a habit" [Peirce, 1931-1958, 5.492]. In human beings, it has to be stressed that Peirce's habit is not a purely mental, rational, or intellectual result of the semiotic process, but it is a mental representation that is always connected to the somatic and motor level, and thus constitutively embodied. On the abductive creative formation of habit as typical of self-organizing dynamic systems and processes, cf. [Gonzalez and Haselager, 2005].

[39] On the role of agency in distributed cognitive systems cf. also [Giere, 2006]. I have illustrated the role of these kinds of – more or less conscious – reasoning processes in real "human-agents", as contrasted with the abstract templates of thinking as crystallized and stabilized in the so-called "ideal logical agents", in [Magnani and Belli, 2006].

go on" [Colapietro, 2000, p. 146]. For instance there are some sedimented unconscious reactions of this type in immediate puzzling environments – later on useless and stultifying in wider settings – but there also is the recurrent reflective and – provisionally – productive use of fallacious ways of reasoning like hasty generalizations and other arguments [Woods, 2004].[40]

Some Peircean words about instinctual beliefs are very interesting and can be stressed to further comprehend the character of the unconscious reactions above: "[...] our indubitable beliefs refer to a somewhat primitive mode of life" [Peirce, 1931-1958, 5.511] but it seems their authority is limited to that domain "While they never become dubitable in so far as our mode or life remains that of somewhat primitive man, yet as we develop degrees of self-control unknown to that man, occasions of actions arise in relation to which the original beliefs, if stretched to cover them, have no sufficient authority" (*ibid.*).

3.5.2 Cultured Unconscious and External/Internal Representations

In the perspective of the disembodiment of mind I have illustrated above in section 3.3 we can also understand how both modern human beings and externalized culture contain within them "implicit" traces of each of the previous stages of cognitive evolution. The first case of externalized distributed culture is evident: remains, buildings, manuscripts, and so on, are fragments of ancient "cognitive niches" from which we can retrieve cultural knowledge.

In the second case it can be hypothesized that much of what Freud attributes to the unconscious is truly unconscious only in the cultural sense of the word, that is formed by "things that are not expressed or are repressed at the level of culture". It has to be acknowledged that in recent cognitive science, and in the sense I have attributed to it in the previous subsections on "man as an external sign" the unconscious is a solipsistic notion, not a cultural one and concerns a part of human mind that is a priori outside the reach of consciousness, a golem, an "automaton world of instincts and zombies", like Donald eloquently says. An example is object vision: "It serves up all the richness of the three-dimensional visual world of awareness, gratis and fully formed. But we can never gain access to the mysterious region of mind that delivers such images. It lies on the other side of cognition, permanently outside the purview of consciousness" [Donald, 2001, pp. 286-287].

In the case of psychoanalysis unconscious is constructed by drives, intuitions, and representations that are shaped by the brain/culture symbiosis and interplay and so are not *a priori* inaccessible to awareness. It is interesting to remember that Jung has also hypothesized the existence of a collective unconscious, that is that part of individual unconscious we would share with others humans, shaped by the evolution of the above interplay, which, so to say, "wired" in it archetypes, also very ancient, that still would act in our present behavior. An example can be the "scapegoat" mechanism, typical of ancient groups and societies, where a paroxysm of violence

[40] Further details are illustrated in chapter seven of this book.

would tend to focus on an arbitrary victim and a unanimous antipathy generated by "mimetic desire" (and the related envy) would grow against him. The brutal elimination of the victim would reduce the appetite for violence that possessed everyone a moment before, and leaves the group suddenly appeased and calm so granting the equilibrium of the related social organization (for us repugnant, but not less useful for those societies for this reason).[41]

Like Girard [1986] says, and many researchers maintain, this kind of archaic brutal behavior, fruit of a conscious (at that time) cultural religious invention of our ancestors is still present in civilized human conduct in rich countries, it is almost always implicit and unconscious, for example in spontaneous and incognizant racist and mobbing behaviors. Given the fact that these kinds of behavior are widespread and partially unconsciously performed it is easy to understand how they can be implicitly "learned" during infancy and then implicitly "wired" by the individual in that *cultured unconscious* we humans collectively share with others. The result is that they are there, available in our minds/brains, to be picked up and executed – paradoxically, given the fact we are often convinced we are meant to be civil modern human beings – as archaic forms of "social" behavior.

3.5.3 Duties, Abductions, and Habits

The Peircean theory of "habits" can help us understand duties as imposed on ourselves from a philosophical, evolutionary, and pragmatic viewpoint, a conception I consider to be in tune with the idea of abduction that I am proposing in this chapter: as I contended above, all semiotic experience – and thus abduction – also provides a guide for action. For example, the logical interpretant, as a hypothetical fruit of abductive thinking requiring the use of verbal symbols is in itself "an interpreting thought", related for instance not only to the intellectual activity but also to initiate the ethical action in so far as a "modification of a person's tendencies toward action" [Peirce, 1931-1958, 5.476]. Indeed the whole function of thought is to produce habits of action, Peirce says that "[...] conduct controlled by ethical reason tends toward fixing certain habits of conduct, the nature of which [...] does not depend upon any accidental circumstances, and *in that sense* may be said to be *destined*" [Peirce, 1931-1958, 5.430]. This philosophical attitude "[...] does not make the *summum bonum* to consist in action, but makes it to consist in that process of evolution whereby the existent comes more and more to embody those generals which [...] [are] *destined*, which is what we strive to express in calling them *reasonable*" [Peirce, 1931-1958, 5.433]. This process, Peirce adds, is related to our "capacity of learning": increasing our "knowledge" will occur through time and generations, "[...] by virtue of man's capacity of learning, and by experience continually pouring over him" [Peirce, 1931-1958, 5.402 n. 2]. It is in this process of anthroposemiosis that civilization moves toward clearer understanding and greater reason. It is in this process of anthroposemiosis, Peirce maintains, that we build highly beneficial habits that help us to acquire various "ethical propensities".

[41] On this archaic mechanism and its effect in the violence that characterizes modern societies cf. [Girard, 1977; Girard, 1986]. Cf. also subsection 8.6.2, chapter eight of this book.

Not only abductions, but also reiterations originate ethical habits as logical interpretants, and in this case the interplay between internal and external representations is still fundamental, related to the exercise of rational self-control and self-reproach guilt feelings which can be further strengthened by direct commands to oneself: "Reiterations in the inner world – fancied reiterations – if well-intensified by direct effort, produce habits, *just as do reiterations in the outer world*; and these habits will have power to influence actual behaviour in the outer world; especially, *if each reiteration be accompanied by a peculiar strong effort that is usually likened to issuing a command to one's future self*" [Peirce, 1931-1958, 5.487]. Moreover, reiterations originate habits both through imaginary and actual exertions[42] – for example repeated outward actions – but also in a hybrid way, in the suitable combination of the two (cf. above section 3.4, on the interplay between external and internal representations).

Moreover, it may be useful to recall here what Peirce says about instinctual beliefs, we have already quoted above: "our indubitable beliefs refer to a somewhat primitive mode of life" [Peirce, 1931-1958, 5.511], but their authority is limited to such a primitive sphere. "While they never become dubitable in so far as our mode of life remains that of somewhat primitive man, yet as we develop degrees of self-control unknown to that man, occasions of action arise in relation to which the original beliefs, if stretched to cover them, have no sufficient authority." (*ibid.*) The problem Peirce touches on here relates to the role of emotions in ethical reasoning: I agree with him that it is only in a constrained and educated – not primitive – way that emotions like love, compassion, and good will, for example, can guide us "morally.[43]

Anyway, a link between ethical rules and conventions and drives and instincts can be hypothesized at a more basic level, as Damasio contends in the framework of a neurological perspective: "Although such conventions and rules need be transmitted only through education and socialization, from generation to generation, I suspect that the neural representations of the wisdom they embody, and of the means to implement that wisdom, are inextricably linked to the neural representations of innate regulatory biological processes" [Damasio, 1994, p. 125]. Of course in this perspective drives and instincts have to be considered not only innate but also acquired, like in the case of educated emotions (cf. [Moorjani, 2000, p. 116]).

Natural entities exhibit different habits and various degrees, ways, and speeds with which they abandon old habits and adopt (or integrate the old ones with) new ones. Peirce says "The highest quality of mind involves greatest readiness to take habits, and a great readiness to lose them" [Peirce, 1931-1958, 6.613]. Colapietro observes that "[...] this capacity entails a measure of consciousness below that of the most acute sensations (e.g., intense pleasure or pain) but above that of our quasi-automatic reactions resulting from the unimpeded operation of effective habits in familiar circumstances" [Colapietro, 2000, p. 139]. In this sense inanimate matter is more reluctant than – for example – brains, to lose old habits and assume new ones,

[42] "[...] every sane person lives in a double world, the outer and the inner world, the world of percepts and the world of fancies" [Peirce, 1931-1958, 5.487].

[43] I defended this perspective in a recent book [Magnani, 2007d, chapter six].

but it is absolutely not exempt from habit-change. We must not forget that for Peirce there is a real cosmic tendency to acquire novel dispositions that is extremely strong in well encephalized human beings.[44]

The previous two sections have introduced to both the interplay between internal and external representations and to some basic semiotic aspects of abductive reasoning: the following sections will take advantage of this background. I will describe how the interplay of signs, objects, and interpretants is working in important aspects of abductive reasoning. Of course model-based cognition acquires its peculiar creative relevance when embedded in abductive processes. I will show some examples of model-based inferences. It is well known the importance Peirce ascribed to diagrammatic thinking (a kind of iconic thinking), as shown by his discovery of the powerful system of predicate logic based on diagrams or "existential graphs". As I have already stressed, Peirce considers inferential any cognitive activity whatever, not only conscious abstract thought; he also includes perceptual knowledge and subconscious cognitive activity. For instance in subconscious mental activities visual representations play an immediate role [Queiroz and Merrell, 2005].

3.6 Constructing Meaning through Mimetic and Creative External Objects

3.6.1 Constructing Meaning through Manipulative Abduction

Manipulative abduction occurs when many external things, usually inert from the semiotic point of view, can be transformed into what in the first chapter I have called, in the case of scientific reasoning, "epistemic mediators" (see also [Magnani, 2001b]) that give rise to new signs, new chances for interpretants, and new interpretations.

We can cognitively account for this process of externalization[45] taking advantage of the concept of *manipulative* abduction (cf. chapter one section 1.6 and Figure 1.6.2). It happens when we are thinking *through* doing and not only, in a pragmatic sense, about doing. It happens, for instance, when we are creating geometry constructing and manipulating an external suitably realized icon like a triangle looking for new meaningful features of it, like in the case given by Kant in the "Transcendental Doctrine of Method" (cf. [Magnani, 2001c], and the following subsection). It

[44] The idea of morality as "habit" – originated through the long negotiation between instinctual impulses and the inescapable pressure of cultural practices – is also supported by James Q. Wilson in a strict Darwinian framework: "I am not trying to discover 'facts' that will prove 'values'; I am endeavoring to uncover the evolutionary, developmental, and cultural origins of our moral habits and our moral sense." He also argues for a biological counterpart that would facilitate the formation of these habits. He continues "But in discovering these origins, I suspect we will encounter uniformities; and by revealing uniformities, I think that we can better appreciate what is general, non-arbitrary, and emotionally compelling about human nature" [Wilson, 1993, p. 26].

[45] I have illustrated above in this chapter a significant contribution to the comprehension of this process in terms of the so–called "disembodiment of the mind".

refers to an extra–theoretical behavior that aims at creating communicable accounts of new experiences to integrate them into previously existing systems of experimental and linguistic (semantic) practices. [Gooding, 1990] refers to this kind of concrete manipulative reasoning when he illustrates the role in science of the so-called "construals" that embody tacit inferences in procedures that are often apparatus and machine based. I have described them in chapter one, section 1.6.2.

It is difficult to establish a list of invariant behaviors that are able to describe manipulative abduction in science.[46] Even if abduction operates, like Peirce says, according to the aesthetic process of *musement*: "a certain agreeable occupation of the mind" [Peirce, 1992-1998, II, p. 436] which must follow "the very law of liberty" [Peirce, 1931-1958, 6.458], as I have already illustrated above, the expert manipulation of objects in a highly semiotically constrained experimental environment certainly implies the application of old and new *templates* of behavior that exhibit some regularities.[47] The activity of building construals is highly conjectural and not necessarily or immediately explanatory: these templates are hypotheses of behavior (creative or already cognitively present in the scientist's mind-body system, and sometimes already applied) that abductively enable a kind of epistemic "doing". Hence, some templates of action and manipulation can be *selected* in the set of the ones available and pre-stored, others have to be *created* for the first time to perform the most interesting creative cognitive accomplishments of manipulative abduction.

3.6.2 Manipulating Meanings through External Semiotic Anchors

If the structures of the environment play such an important role in shaping our semiotic representations and, hence, our cognitive processes, we can expect that physical manipulations of the environment receive a cognitive relevance.

Several authors have pointed out the role that physical actions can have at a cognitive level. In this sense Kirsh and Maglio [1994] distinguish actions into two categories, namely *pragmatic actions* and *epistemic actions*. Pragmatic actions are the actions that an agent performs in the environment in order to bring itself physically closer to a goal. In this case the action modifies the environment so that the latter acquires a configuration that helps the agent to reach a goal which is understood as physical, that is, as a desired state of affairs. Epistemic actions are the actions that an agent performs in a semiotic environment in order to discharge the mind of a cognitive load or to extract information that is hidden or that would be very hard to obtain only by internal computation.

---

[46] A list is provided in chapter one, section 1.6.2.

[47] It is simple to explain why abduction works according to musement. This is the general attitude we adopt when we are wondering about the beauty and the harmony of universes and their connections [Peirce, 1992-1998, II, p. 436]. I think that beauty plays a kind of exciting emotional role in abductive reasoning, very similar to the one played by anomalies and surprise. Cf. also [Maddalena, 2005, p. 247].

In this subsection I want to focus specifically on the relationship that can exist between manipulations of the environment and representations. In particular, I want to examine whether external manipulations can be considered as means to construct external representations.

If a manipulative action performed upon the environment is devoted to create a configuration of signs that carries relevant information, that action will well be able to be considered as a cognitive semiotic process and the configuration of elements it creates will well be able to be considered an external representation. In this case, we can really speak of an embodied cognitive process in which an action constructs an external representation by means of manipulation. We define *cognitive manipulating* as any manipulation of the environment devoted to construct external configurations that can count as representations.

An example of cognitive manipulating is the diagrammatic demonstration illustrated in Figure 3.5, taken from the field of elementary geometry. In this case a simple manipulation of the triangle in Figure 3.5(a) gives rise to an external configuration – Figure 3.5(b) – that carries relevant semiotic information about the internal angles of a triangle "anchoring" new meanings.

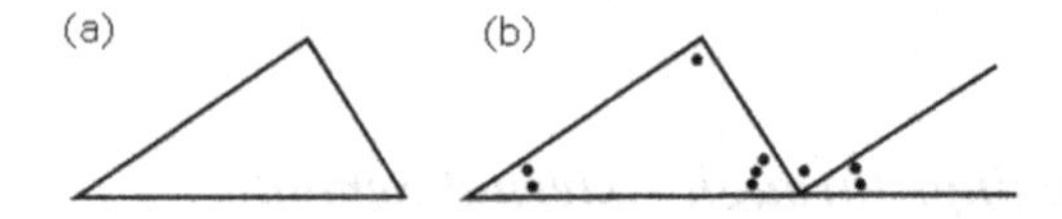

Fig. 3.5 Diagrammatic demonstration that the sum of the internal angles of any triangle is 180. (a) Triangle. (b) Diagrammatic manipulation/construction.

The entire process through which an agent arrives at a physical action that can count as cognitive manipulating can be understood by means of the concept of manipulative abduction. In this perspective manipulative abduction is a specific case of cognitive manipulating in which an agent, when faced with an external situation from which it is hard or impossible to extract new meaningful features of an object, selects or creates an action that structures the environment in such a way that it gives new information which would be otherwise unavailable and which is used specifically to infer explanatory hypotheses.

In this way the semiotic result is achieved on *external* representations used in lieu of the internal ones. Here action plays an *epistemic* and not merely performatory role, for example relevant to abductive reasoning. The process also illustrates a synthesis between a constructive procedure of motor origin (the putting the new segment end to end parallel to one side in the externally represented given triangle), followed by a sensory procedure, "visual" (calculation of the sizes of the now clearly – externally – "seen" angles).[48]

It is important to note that in this manipulative and "multimodal"[49] abductive case abduction and induction play a role similar to the one described in the area of logic programming: abductive reasoning extends the intension of known individuals (because abducible properties are rendered true for these individuals, for example by providing new situated "samples", as "anchors" which offer chances for further knowledge), without having a genuine generalization impact on the observables (it does not increase their extension). Abductively building new situated results through manipulation of the external diagram is in this case central to make possible an "induction" able to generate new general knowledge, not reachable through abduction (on the variety of the roles played by induction in its interaction with abduction cf. above the considerations I have illustrated at p. 14 and p. 30).

3.6.3 Geometrical Construction Is a Kind of Manipulatxive Abduction

Let's quote Peirce's passage about mathematical constructions. Peirce says that mathematical and geometrical reasoning "[...] consists in constructing a diagram according to a general precept, in observing certain relations between parts of that diagram not explicitly required by the precept, showing that these relations will hold for all such diagrams, and in formulating this conclusion in general terms. All valid necessary reasoning is in fact thus diagrammatic" [Peirce, 1931-1958, 1.54]. This passage clearly refers to a situation like the one I have illustrated in the previous subsection. This kind of reasoning is also called by Peirce "theorematic" and it is a kind of "deduction" necessary to derive significant theorems (Necessary Deduction]: "[...] is one which, having represented the conditions of the conclusion in a diagram, performs an ingenious experiment upon the diagram, and by observation of the diagram, so modified, ascertains the truth of the conclusion" [Peirce, 1931-1958, 2.267]. The experiment is performed with the help of "[...] imagination upon the image of the premiss in order from the result of such experiment to make corollarial deductions to the truth of the conclusion" [Peirce, 1976, IV, p. 38].

---

[48] "The essential step in the construction of the Euclidean space, has been the possibility of the *division* of a motor field; and here we come up against an evident physiological impossibility. Greek geometry resolved this problem of the division of a segment into equal segments by the discovery of Thales' Theorem: equidistant parallel lines cut two secants in proportional segments" [Thom, 1980, p. 134]. Furthermore, following Thom, I think this ancient Greek geometry example already represents the quintessence of the scientific approach, that is "[...] replacing a non-local operation (for example, taking the intersection of two lines in a plane) by a verbal description the formal analysis of which became the demonstration that it was virtually autonomous, that is, able to be rendered independent of the non-local intuitive approaches which described it" [Thom, 1980, p. 135]. The use of literary symbols, which are empty of sense, together with the axiomatic approach realizes the localization of the non-local intuition of the plane (and of space).

[49] Both propositional and model-based aspects are at play. On the concept of multimodal abduction cf. this book, chapter four, section 4.1.

The "corollarial" reasoning is mechanical (Peirce thinks it can be performed by a "logical machine") and not creative, "A Corollarial Deduction is one which represents the condition of the conclusion in a diagram and finds from the observation of this diagram, as it is, the truth of the conclusion" [Peirce, 1931-1958, 2.267] (cf. also [Hoffmann, 1999]).

In summary, the point of theorematic reasoning is the transformation of the problem by establishing an *unnoticed* point of view to get interesting – and possibly new – insights. The demonstrations of "new" theorems in mathematics are examples of theorematic deduction.

Not dissimilarly Kant says that in geometrical construction of external diagrams "[...] I must not restrict my attention to what I am actually thinking in my concept of a triangle (this is nothing more than the mere definition); I must pass beyond it to properties which are not contained in this concept, but yet belong to it" [Kant, 1929, A718-B746, p. 580].

Theorematic deduction can be easily interpreted in terms of manipulative abduction. We have seen that manipulative abduction is a kind of abduction, mainly model-based, that exploits external models endowed with delegated (and often implicit) cognitive and semiotic roles and attributes:

1. the model (diagram) is *external* and the strategy that organizes the manipulations is unknown *a priori*;
2. the result achieved is *new* (if we, for instance, refer to the constructions of the first creators of geometry), and adds properties not contained before in the concept (the Kantian to "pass beyond" or "advance beyond" the given concept [Kant, 1929, A154-B193/194, p. 192]).[50]

Iconicity in theorematic reasoning is central. Peirce, analogously to Kant, maintains that "[...] philosophical reasoning is reasoning with words; while theorematic reasoning, or mathematical reasoning is reasoning with specially constructed schemata" [Peirce, 1931-1958, 4.233]; moreover, he uses diagrammatic and schematic as synonyms, thus relating his considerations to the Kantian tradition where schemata mediate between intellect and phenomena.[51] The following is the famous related passage in the *Critique of Pure Reason* ("Transcendental Doctrine of Method"):

> Suppose a philosopher be given the concept of a triangle and he be left to find out, in his own way, what relation the sum of its angles bears to a right angle. He has nothing but the concept of a figure enclosed by three straight lines, and possessing three angles. However long he meditates on this concept, he will never produce anything new. He can analyse and clarify the concept of a straight line or of an angle or of the number three, but he can never arrive at any properties not already contained in these concepts.

[50] Of course in the case we are using diagrams to demonstrate already known theorems (for instance in didactic settings), the strategy of manipulations is not necessary unknown and the result is not new, like in the Peircean case of corollarial deduction.

[51] Schematism, a fruit of the imagination is, according to Kant, "[...] an art concealed in the depths of the human soul, whose real modes of activity nature is hardly likely ever to allow us to discover, and to have open to our gaze" [Kant, 1929, A141-B181, p. 183].

> Now let the geometrician take up these questions. He at once begins by constructing a triangle. Since he knows that the sum of two right angles is exactly equal to the sum of all the adjacent angles which can be constructed from a single point on a straight line, he prolongs one side of his triangle and obtains two adjacent angles, which together are equal to two right angles. He then divides the external angle by drawing a line parallel to the opposite side of the triangle, and observes that he has thus obtained an external adjacent angle which is equal to an internal angle – and so on.[52] In this fashion, through a chain of inferences guided throughout by intuition, he arrives at a fully evident and universally valid solution of the problem [Kant, 1929, A716-B744, pp. 578-579].

We can depict the situation of the philosopher described by Kant at the beginning of the previous passage taking advantage of some ideas coming from the catastrophe theory (cf. also this book, chapter eight). As a human being who is not able to produce anything new relating to the angles of the triangle, the philosopher experiences a feeling of frustration (just like the Kölher's monkey which cannot keep the banana out of reach). The bad affective experience "deforms" the organism's regulatory structure by complicating it and the cognitive process stops altogether. The geometer instead "at once constructs the triangle", that is, he makes an external representation of a triangle and acts on it with suitable manipulations. Thom thinks that this action is triggered by a "sleeping phase" generated by possible previous frustrations which then change the cognitive status of the geometer's available and correct internal idea of triangle (like the philosopher, he "has nothing but the concept of a figure enclosed by three straight lines, and possessing three angles", but his action is triggered by a sleeping phase). Here the idea of the triangle is no longer the occasion for "meditation", "analysis" and "clarification" of the "concepts" at play, like in the case of the "philosopher". Here the inner concept of triangle – symbolized as insufficient – is amplified and transformed thanks to the sleeping phase (a kind of Kantian imagination active through schematization) in a prosthetic triangle to be put outside, in some external support. The instrument (here an external diagram) becomes the extension of an organ:

> What is strictly speaking the end [...] [in our case, to find the sum of the internal angles of a triangle] must be set aside in order to concentrate on the means of getting there. Thus the problem arises, a sort of vague notion altogether suggested by the state of privation. [...] As a science, heuristics does not exist. There is only one possible explanation: the affective trauma of privation leads to a folding of the regulation figure. But if it is to be stabilized, there must be some exterior form to hold on to. So this anchorage problem remains whole and the above considerations provide no answer as to why the folding is stabilized in certain animals or certain human beings whilst in others (the majority of cases, needless to say!) it fails [Thom, 1988, pp. 63–64].[53]

---

[52] It is Euclid's Proposition XXXII, Book I, cf. above Figure 3.5.

[53] A full analysis of the Kölher's chimpanzee getting hold of a stick to knock a banana hanging out of reach in terms of the mathematical models of the perception and the capture catastrophes is given in [Thom, 1988, pp. 62–64]. On the role of emotions, for example frustration, in scientific discovery cf. [Thagard, 2002b].

As we have already said, for Peirce the whole mathematics consists in building diagrams that are "[...] (continuous in geometry and arrays of repeated signs/letters in algebra) according to general precepts and then [in] observing in the parts of these diagrams relations not explicitly required in the precepts" [Peirce, 1931-1958, 1.54]. Peirce contends that this diagrammatic nature is not clear if we only consider syllogistic reasoning "which may be produced by a machine" but becomes extremely clear in the case of the "logic of relatives, where any premise whatever will yield an endless series of conclusions, and attention has to be directed to the particular kind of conclusion desired" [Peirce, 1987, pp. 11–23].

In ordinary geometrical proofs auxiliary constructions are present in terms of "conveniently chosen" figures and diagrams where strategic moves are important aspects of deduction. The system of reasoning exhibits a dual character: deductive and "hypothetical". Also in other – for example logical – deductive frameworks there is room for strategic moves which play a fundamental role in the generations of proofs. These strategic moves correspond to particular forms of abductive reasoning.

We know that the kind of reasoned inference that is involved in creative abduction goes beyond the mere relationship that there is between premises and conclusions in valid deductions, where the truth of the premises guarantees the truth of the conclusions, but also beyond the relationship that there is in probabilistic reasoning, which renders the conclusion just more or less probable. On the contrary, we have to see creative abduction as formed by the application of *heuristic procedures* that involve all kinds of good and bad inferential actions, and not only the mechanical application of rules. It is only by means of these heuristic procedures that the acquisition of *new* truths is guaranteed. Also Peirce's mature view illustrated above on creative abduction as a kind of inference seems to stress the strategic component of reasoning.

Many researchers in the field of philosophy, logic, and cognitive science have maintained that deductive reasoning also consists in the employment of logical rules in a heuristic manner, even maintaining the truth preserving character: the application of the rules is organized in a way that is able to recommend a particular course of actions instead of another one. Moreover, very often the heuristic procedures of deductive reasoning are performed by means of model-based abductive steps where iconicity is central.

We have seen that the most common example of manipulative creative abduction is the usual experience people have of solving problems in geometry in a model-based way trying to devise proofs using diagrams and illustrations: of course the attribute of creativity we give to abduction in this case does not mean that it has never been performed before by anyone or that it is original in the history of some knowledge (they actually are cases of Peircean corollarial deduction).[54]

[54] We have to say that model-based abductions – which for example exploit iconicity – also operate in deductive reasoning. On the role of strategies and heuristics in deductive proofs cf. chapter seven, section 7.3.2.

3.6.3.1 Theoric Reasoning and Creativity

I think the previous considerations concerning manipulative abduction also hold for Peircean theorematic reasoning. Let us quote again the important Peirce's passage about theorematic reasoning:

> A Necessary Deduction is a method of producing Dicent Symbols[55] by the study of a diagram. It is either Corollarial or Theorematic. A Corollarial Deduction is one which represents the conditions of the conclusion in a diagram and finds from the observation of this diagram, as it is, the truth of the conclusion. A Theorematic Deduction is one which, having represented the conditions of the conclusion in a diagram, performs an ingenious experiment upon the diagram, and by the observation of the diagram, so modified, ascertains the truth of the conclusion [Peirce, 1931-1958, 2.267].

As I have already indicated Peirce further distinguished a "corollarial" and a "theoric" part within "theorematic reasoning", and connected theoric aspects to abduction [Hoffmann, 1999, p. 293]: "Thêoric reasoning [...] is very plainly allied to" what is normally called abduction [Peirce, 1966, 754, ISP, p. 8]. Indeed theoric reasoning is considered a kind of creative diagrammatic reasoning: following [Hoffmann, 2003, p. 167] we can say that it

> [...] is based on just the same idea, which is the idea of "the transformation of the problem, – or its statement,– due to viewing it from another point of view" [Peirce, 1966, 1907, 318; CSP, p. 68; ISP, p. 225]. Peirce takes the term "theoric" from the Greek θεωρία (theory) which he translates as 'the power of looking at facts from a novel point of view' [Peirce, 1966, 1907, 318; CSP, p. 50; ISP, p. 42]. For Peirce, the most important discoveries in mathematics are also based on reaching new perspectives, as he shows in his 1907 manuscript about "Pragmatism" with the example of the proof of the "ten points theorem".

In a passage of the manuscripts not contained in the microfilm edition, quoted by [Hoffmann, 2003, p. 293],[56] Peirce clearly states his own surprise at seeing retroduction (that is abduction) at work in a non empirical science, like mathematics, where *new* results are reached through an abductive reasoning that strangely leads to *indisputable* achievements.

> Further study, however, leads me to lop off/discard a corollarial part from/of the Theorematic Deductions, which follows that which originates a new point of view. I will call this part of the theorematic procedure, thêoric reasoning. It is very plainly allied to retroduction, from which it only differs, as far as I now see, in that it is "indisputable" [Peirce, 1966, 754; ISP, p. 8].

I think Peirce is referring here to the creative and non-explanatory side of what I have called manipulative abduction. Indeed he also said he "[...] would regard the great hypotheses of pure mathematics [...] as coming to us through retroduction

---

[55] A dicent symbol (such as a proposition or a description) is a sign which may be interpreted to refer to an actually existing object.

[56] Already indicated by [Levy, 1997, p. 106] and [Levy, 1997, p. 482].

from considering what for want of a better word I may call the facts of mathematics" [Peirce, 1966, 754; ISP, p. 3]. I wholeheartedly agree with Peirce's following observation, written about a century before the new perspective on mathematical reasoning offered by Lakatos's work [1976]: "It has long been a puzzle how it could be that, on the one hand, mathematics is purely deductive in nature, and draws its conclusions apodictically, while in the other hand, it presents as rich and apparently unending a series of surprising discoveries as any observational science" [Peirce, 1931-1958, 1885, 3.363].

Of course, as already stressed, we have to remember this abductive aspect of mathematical reasoning can be performed both in creative [theorematic] (to find new theorems and mathematical hypotheses) and non creative [corollarial] (merely "selective") ways, for example in the case that we are using diagrams to demonstrate already known theorems (for instance in didactic settings), where selecting the strategy of manipulations is among chances not necessarily unknown and where the result is not new. With respect to abduction in empirical science abduction in mathematics aims at hypothesizing ideal objects, which later we can then possibly insert into a deductive apodictic and truth preserving framework.[57]

The epistemological situation of this geometrical example is similar to the one I have illustrated in the last sections of chapter two, concerning the analysis in terms of abduction of the non-Euclidean discovery:

1. the inferential process is a kind of *manipulative* and *model-based* abduction (visual), first of all endowed with an *explanatory* character: the abduced mirror diagrams of the triangle furnishes a visual explanation/description of our internal representation of it;
2. at the same time the external image, and the construction based on it offer the chance of a further *multimodal* and *distributed* abductive step (based on both internal and external representations, and on both visual and sentential aspects) mainly *non-explanatory* and *instrumental*. This further abductive process makes possible to derive the new "indisputable" result/theorem of the sum of the internal angles of a triangle.

Finally, the example of diagrams in geometry furnishes a semiotic and epistemological example of the nature of the cognitive interplay between internal neuronal representations (and embodied "cognitive" kinesthetic and motor abilities) and external representations I have illustrated above: also for Peirce, more than a century before the new ideas derived from the field of distributed cognition, the two aspects are intertwined in the pragmatic and semiotic view, going beyond the rigidity of the Kantian approach in terms of schematism. Diagrams are icons that take material and semiotic form in an external environment endowed with

[57] In a recent analysis of further aspects of the relationship between abduction and the inference to the best explanation, [Minnameier, 2004, p. 85] contends theorematic deduction should be basically considered an inverse deduction, that is "[...] an inference not from the premises of ordinary deduction (which Peirce terms 'corollarial') to the conclusion, but *from the (prospective) conclusion to the premises* of the deductive argument from which the conclusion follows". I am inclined to think this interpretation underestimates the role played by the creative side (theoric) in theorematic reasoning, and its manipulative character.

- constraints depending on the specific cognitive delegation performed by human beings and
- the particular intrinsic constraints of the materiality at play.[58]

Concrete manipulations on them can be done for instance to get new data and cognitive information and/or to simplify the problem at issue (cf. the epistemic templates illustrated above in subsection 3.6.1).

3.6.4 The Semiosis of Re-embodiment and Its Sensorimotor Nature

Some interesting semiotic aspects of the above illustrated process can be nicely analyzed. Imagine that a suitable *fixed internal record* exists – deriving from the cognitive exploitation of the previous suitable interplay with *external structures* – at the level of neural activation and that for instance it embeds an abstract concept endowed with all its features, for example the concept of triangle. Now, the human agent, via neural mechanisms and bodily actions, can "re-embody" that concept by making an external perceivable *sign*, for instance available to the attention of other human or animal senses and brains. For instance that human agent can use what in semiotics is called a *symbol* (with its conventional character: *ABC*, for example), but also an *icon* of relations (a suitable diagram of a triangle), or a *hybrid representation* that will take advantage of both. In Peircean terms:

> A representation of an idea is nothing but a sign that calls up another idea. When one mind desires to communicate an idea to another, he embodies his idea by making an outward perceptible image which directly calls up a like idea; and another mind perceiving that image gets a like idea. Two persons may agree upon a conventional sign which shall call up to them an idea it would not call up to anybody else. But in framing the convention they must have resorted to the primitive diagrammatic method of embodying the idea in an outward form, a picture. Remembering what likeness consists in, namely, in the natural attraction of ideas apart from habitual outward associations, I call those signs which stand for their likeness to them *icons*.
>
> Accordingly, I say that the only way of directly communicating an idea is by mean of an icon; and every indirect method of communicating an idea must depend for its establishment upon the use of an icon [Peirce, 1966, 787, 26–28].

We have to note that for Peirce an idea "[...] is not properly a conception, because a *conception* is not an idea at all, but a *habit*. But the repeated occurrence of a general idea and the experience of its *utility*, results in the formation or strengthening of that habit which is the conception" [Peirce, 1931-1958, 7.498].

Habits, as beliefs and vehicles of cognition and at the same time anticipation of future chances for action, are usually considered bodily states in so far as they are, according to Peirce, comparable to "dispositions" [Peirce, 1931-1958, 5.440]. In the

[58] An admirable analysis of the constraints intrinsic to the external material exploited by the ancient Greek mathematicians (especially geometricians) is given in [Netz, 1999]. The book also illustrates the most important semiotic tools at work in ancient Greek mathematical reasoning.

light of the cognitive interplay internal/external I have described above it is better to interpret them as forms of interaction with the suitable circumstances involved in the related action, as [Määttänen, 2009] stresses. In this perspective perception and action are strictly intertwined simply as different degrees of interaction with the environment: in action "our modification of other things is more prominent that their reaction on us" as compared to perception "where their effect on us is overwhelmingly greater than our effect on them" [Peirce, 1931-1958, 1.324].

Of course what is external to the body is not necessarily external to the process realizing cognition and basically resorts to "sensorimotor" representations that are *created* or *re-activated* (if already formed and stable) during the interaction with the physical world.[59] Nevertheless, very often a considerable part of the cognitive process occurs outside, thanks to the suitable materiality endowed with contingent cognitive delegations. It is in this sense that the possible establishment of a habit is in itself also the institution of new meanings. As maintained by Peirce, acquired habits are in themselves meanings, being the result of interactions with the environment they are not literally only in the head but also intertwined with motor action: they are embodied. "Sensory inputs [we could add: that are abductively matched to a suitable habit] are associated not only with each other but also, and more importantly, with neural mechanisms controlling overt motor action" [Määttänen, 1997; Määttänen, 2009].

Peirce pays much attention to interactional play when dealing with consciousness and "Secondness":

> We are continually bumping up against hard fact. We expect one thing, or passively take it for granted, having the image of it in our minds, but experience then forces that idea into the background, and compels us to think quite differently. You get this kind of consciousness in some approach to purity when you put your shoulder against a door and try to force it open. You have a sense of resistance and at the same time a sense of effort. There can be no resistance without effort; there can be no effort without resistance. They are only two ways of describing the same experience. It is a double consciousness. We become aware of ourselves in becoming aware of the not-self. The waking state is a consciousness of reaction and, as consciousness itself is two-sided, it has two varieties: namely, action, where our modification of other things

[59] [Noë, 2005; Noë, 2006] and other researchers like K. O'Regan [O'Regan and Noë, 2001] and S. Hurley propose a sensorimotor theory of perception that challenges its merely traditional representational character: perceptual experience – and so visual system – is fundamentally structured (even if not caused) by our sensorimotor competence, which allows us to access the world and to act. In the first case [Fusaroli, 2007; Fusaroli and Vandi, 2009] contend that the traditional concept of "representation" can be either rejected or it can acquire a new operational status. In this case, perception is taken as direct access to the world and representations do not play any role; the world would serve as its own representation and as an external memory, as is similarly contended by researchers in the field of active vision (cf. above section 3.4.1); in the latter case – action – it is the "body" of the human being or of the animal that governs the entire cognitive process of externalization/re-embodiment and that furnishes the suitable final "meaning" and interpretation. On the sensorimotor theory of perception as contrasted to the explanation of perception as inner encoding cf. the deep and rich considerations given by [Clark, 2008, chapters seven and eight]: Clark eloquently label these approaches "strongly" sensorimotor models of perception (SSM).

> is more prominent than their reaction to us, and perception, where their effect on us is overwhelmingly greater than our effect on them. This notion, of being what other things make us, is such a prominent part of our life that we also conceive other things to exist by virtue of their reactions against each other. The idea of other, of not, becomes central to our thinking. To this element I give the name of Secondness [Peirce, 1931-1958, 1.324].

The process of re-embodiment concerns the formation of internal "mental" representations (as mere brain states) which are also strictly intertwined with motor aspects, at both the neural and the somatic level. From this perspective the world is experienced as giving various opportunities to carry out "habitual" actions, that is it provides what Gibson calls *affordances*: habits of action *abductively* "reveal" affordances.[60]

This is the correct sense in which we can say that cognition is "embodied", as "the sharing of neural mechanisms between sensorimotor processes and higher-level cognitive processes": many, if not all, higher-level cognitive processes seem body-based in the sense that "they make use of (partial) emulations or simulations of sensorimotor processes through the reactivation of neural circuitry that is also active in bodily perception and action" [Svensson and Ziemke, 2004, p. 1309], as already stressed by the theory of autopoiesis (self-organization) put forward by Maturana and Varela [Maturana and Varela, 1980; Varela *et al.*, 1991]. The traditional distinction between perception and action as well as between sensorimotor and cognitive processes has to be given up: the same neural structures that are at the basis of actions and/or perception would also be exploited in the performance of various cognitive tasks. Empirical data have suggested that perceptual and motor areas of the brain can be covertly activated either separately or in sequence, for example there are similarities between the neural structures activated during preparation and execution of an action and those employed in its mental simulation through what is called motor imagery, as well as in the case of perception and visual imagery (it is easy to suppose that in both cases the same representational formats are at play, so this does not need to resort to the computer metaphor of internal symbol manipulation). Simulating an action involves some simulator [Barsalou *et al.*, 2003; Decety, 1996; Decety and Grèzes, 2006; Frith and Dolan, 1996; Hesslow, 2002; Jeannerod, 2001] (or emulator [Grush, 2004a; Grush, 2007]) devices that abductively anticipate the perceptual feedback that would have occurred in the case of the executed action.[61]

Research on neurons located in the rostral part of the inferior premotor cortex (area F5) has demonstrated that they discharge and respond to goal directed actions such as grasping, holding, or tearing that have the same meaning. They are interpreted as internal representations of action, rather than motor or movement commands. Some of those neurons, which are called canonical neurons, discharge both

[60] A full treatment of the concept of affordance in its relationship with abduction is given in chapter six.

[61] [Pickering and Garrod, 2006] further stress that in the case of language, comprehenders use prediction and imitation to construct an "emulator" using the production system, and combine predictions with the input dynamically.

during the action they code and when an object affords that action, so that it can be said they express affordances coding for example the graspability of things from both the perceptual and the action perspectives; consequently they account for a basic link between an agent and its environment. Internally reactivated perceptions and actions are also at play 1) in the case of various inner planning and problem solving performances [Dagher *et al.*, 1999; Hesslow, 2002], and (2) in social cognition (especially emotive states based on social stimuli), where simulation of "bodily states" occurs: they are not conscious mediating knowledge structures, but they affect higher cognition [Barsalou *et al.*, 2003]. F5 also contains the well-known "mirror neurons" that become activated both when performing a specific action and when observing the same goal-directed hand (or mouth) movements of an experimenter.[62] They provide a clear example of sensorimotor brain parts performing a resonance task devoted to both performing and understanding actions.

It is well-known that for Peirce every picture is an icon and thus every diagram, even if it lacks a sensuous similarity with the object, but just exhibits an analogy between the relations of the part of it and of the object:

> All iconic signs, like diagrams, rarely function as pure icons: symbolic aspects are often at stake and background knowledge of such conventions is necessary to obtain the desired information. Particularly deserving of notice are icons in which the likeness is aided by conventional rules. Thus, an algebraic formula is an icon, rendered such by the rules of commutation, association, and distribution of the symbols; that it might as well, or better, be regarded as a compound conventional sign. It may seem at first glance that it is an arbitrary classification to call an algebraic expression an icon. But it is not so. For a great distinguishing property of the icon is that by direct observation of it other truths concerning its object can de discovered than those which suffice to determine its construction. Thus, by means of two photographs a map can be drawn, etc. Given a conventional or other general sign of an object, to deduce any other truth than which it explicitly signifies, it is necessary, in all cases, to replace that sign by an icon. This capacity of revealing unexpected truth is precisely that wherein the utility of algebraic formulae consists, so that the icon in character is the prevailing one [Peirce, 1966, 787, CSP 26–28].

Stressing the role of iconic dimensions of semiosis[63] in the meantime celebrates the virtues of analogy, as a kind of "association by resemblance", as contrasted to "association by contiguity".

[Stenning, 2000] provides an indepth philosophical and cognitive analysis of diagrammatic reasoning which further clarifies the Peircean observation above about the fact that algebraic formulas can be considered icons. He acknowledges that homomorphism – which resorts to "likeness" – between diagrams and their referents is what distinguishes diagrammatic semantics from sentential semantics. From a Peircean perspective languages are fundamentally symbolic and have indexical aspects, and diagrams – which are of course iconic – also contain symbolic elements such as words, often indexically related to their reference like labels by spatial

[62] On the role of mirror neurons in social cognition cf. chapter four, section 4.3.2.

[63] We have to remember that in this perspective any proposition is a diagram as well, because it represents a certain relation of symbols and indices.

deixis. Stenning proposes a new distinction that is intertwined with the Peircean one but avoids some of its problems: we have to distinguish between direct interpretation (for diagrammatic semantics) and indirect interpretation (for sentential semantics). In this last case the interpretation is indirect because between representation and the referenced world an abstract syntax based on concatenation is interposed: "The interpretation is indirect because the significance between two elements being spatially [in written language] or temporally [in spoken language] concatenated cannot be assessed without knowing what abstract syntactic relation holds between them" [Stenning, 2000, p. 136].

Some two-dimensional representation systems, like semantic networks and conceptual graphs, are close to language systems because they have a semantics that interposes an abstract syntax. For example Peirce clearly stated that syntactic relations in a language with an abstract syntax could be iconic, like in the case of algebraic relations that represent transitivity iconically (cf. the quotation above).

Some interesting differences between sentential and diagrammatic reasoning have to be pointed out. [Shin, 2002] has demonstrated that Peircean existential graphs (that lack *connectives*) can have multiple equivalents in sentential calculi with connectives, just like the fact that two sentences are translations of the same existential graph demonstrates they are logically equivalent. Moreover, sentential graphs are written from left to right while existential graphs do not have this rigid layout. Sentential representation systems use sentences *discursively* (when a conclusion is drawn from earlier sentences, it is rewritten on a new line of results), but diagrammatic systems are often used and represented *agglomeratively* (in diagrams all the icons are automatically related to each other by all the interpreted spatial relations and if a new assumption is made, it is represented in the particular existential graph at play). The inferences in the diagram are made by modifying the "individual" existential graph, in this way giving rise to a derivation. In this last case *history* is erased (history in sentential representation systems is very much needed because it guarantees the legitimacy of later inferences). [Of course the derivations can be expressed by drawing diagrams in a sequence or as a tree]. In the case of Euler diagrams they have to be used in an agglomerative mode that preserves history but they do not present inferences because of the lack of diagram combinations.

The interplay between indirect interpretation (discursive use) and direct interpretation (agglomerative use) can help exceed their relative limits, as is clearly seen in the case of some geometrical diagrammatic proof, that appears neither useless nor adequate. Proof without words can be justified in a static diagram unaccompanied by words and can be directly interpreted and agglomeratively used as in the case of the diagrammatic proof contained in a single diagram of the Pythagoras theorem. But what about the statement of the theorem?

> If we do not know what the Pythagoras' theorem is, we are unlikely to find it in this diagram. Even if we know that it is about triangles, there are still many triangles to choose from in the figure and no indication which is the topic of the theorem. To even state Pythagoras' theorem requires a representation system that can provide only part of the truth about right angled triangles. In particular, a representation system that can represent the square of the hypotenuse without representing the rest (especially the

> sum of squares on the other two sides) and v.v.; and can represent the equality of areas between these two entities. Only indirectly interpreted systems with their syntactic articulation can do this. On the other hand, proof without words can supply a kind of insight into proof which sentential proof finds it hard to match. Proof-without words can be sequentialised, after a fashion, by providing a kind of "comic-strip" animation of a sequence of constructions. This however raises problems about which inferences and constructions are to count as "formal" [...] It is intriguing that an interaction between diagrams and language appears to have been what gave rise to the first invention of Greek geometry [Stenning, 2000, pp. 147–148].[64]

[Clark, 2006, p. 371] has recently stressed the cognitive role in mathematics of "tokening" numerical expressions in some way, as symbol strings of our own public language so that human arithmetic thinking is de facto hybrid, made up of a combination of this tokening and the activation of some basic biological/neurological resources.[65]

Human beings delegate cognitive features to external representations through semiotic attributions because for example in many problem solving situations the internal computation would be impossible or it would involve a very great effort because of human mind's limited capacity. First a kind of "alienation" is performed, second a recapitulation is accomplished at the neuronal level by re-representing internally that which was "discovered" outside. Consequently only later on do we internally perform cognitive operations on the structure of data that synaptic patterns have "picked up" in an analogical way from the environment. We can maintain that internal representations used in cognitive processes like many events of *meaning creation* have a deeper origin in the experience lived in the semiotic environment. [Hutchins, 2005, p. 1575] further clarifies this process of recapitulation: "[...] when a material structure becomes very familiar, it may be possible to imagine the material structure when it is not present in the environment. It is even possible to imagine systematic transformations applied to such a representation. This happened historically with the development of mathematical and logical symbol systems in our own cultural tradition".

As we will see in chapter six, in this interplay of re-embodiment diagrams "afford" some actions as being possible and the embodied result can be considered as the establishment of a habit, in a Peircean sense, not only a theoretical result but also a kind of "know how": "We imagine cases, place mental diagrams before our mind's eye, and multiply these cases, until a habit is formed of expecting what always turns out the case, which has been seen to be the result in all the diagrams. To appeal to such a habit is a very different thing from appealing to any immediate instinct of rationality. That the process of forming a habit of reasoning by the use of

[64] On the interplay between spatial convention of order preference and various kinds of formal knowledge cf. [Landy and Goldstone, 2007b; Landy and Goldstone, 2007a]. This perspective further challenges the conception that human reasoning with formal systems exploits only the formal properties of symbolic notations: people also use other regularities, formal, visual, diagrammatic, rule-based, and statistical.

[65] On the role of external symbols and natural language in mathematical discovery and reasoning cf. also [Dehaene, 1997; Dehaene *et al.*, 1999].

diagrams is often performed there is no room for doubt" [Peirce, 1931-1958, 2.170]. D. Landy and R. L. Goldstone

I already illustrated in section 3.4 that I think there are two kinds of artifacts that play the role of *external objects* (representations) active in this process of externalization of the mind: *creative* and *mimetic*. Mimetic external representations mirror concepts and problems that are already represented in the brain and need to be enhanced, solved, further complicated, etc. so they sometimes can creatively give rise to new concepts and meanings.

Following my perspective it is at this point evident that the "mind" transcends the boundary of the individual and includes parts of that individual's environment. It is in this sense that the mind is constitutively semiotic and artificial.

3.6.5 On-line and Off-line Intelligence Intertwined: The Problem of Language and of Inner Rehearsal

I have said above in subsection 3.6.2 that the entire process through which an agent arrives at a physical action – that counts as a more or less creative cognitive manipulating – can be understood by means of the concept of manipulative abduction. In this case the agent, when faced with an external situation from which it is for example hard or impossible to extract new meaningful cognitive features of a concept, selects or creates an action that structures a mimetic representation referring to the concept in the environment in such a way that it can give information, which would be otherwise unavailable and which is used specifically to infer hypotheses. Moreover, as subsequently illustrated in the previous subsection, I have stressed the fact that in this interplay first a kind of "alienation" is performed, and second a recapitulation is accomplished at the neuronal level by re-representing internally that which was "discovered" outside. [Dartnall, 2005, p. 136] says that in this case the "world leaks into the mind".

Only later on the agent internally performs cognitive operations on the structure of data that synaptic patterns have "picked up" in an analogical way from the mimetic representation stored in the environment. I have already said that it is possible to conclude that internal representations exploited in cognitive processes like many events of *meaning creation* have a deep origin in the experience lived in the semiotic environment.[66]

Following Clark's perspective on language as an external tool [Clark, 1997] Wheeler qualifies the double process above speaking of *on-line* – like in the case of manipulative abduction – and *off-line* thinking (also called *inner rehearsal*). Language is inner: "[...] just so long as there are private thought processes which are

[66] The process can be accounted for in terms of the emulation theory which in such a way nicely complements the extended mind thesis: "If something counts as cognitive when it is performed in the head, it should also count as cognitive when it is performed in the world (mind leaks into the world). Also, if a process gives us an empirical discovery when it is performed in the world, it will also give us an empirical discovery when it is performed in the head (the world leaks into the mind)" [Dartnall, 2004, p. 402].

formulated in language" [Wheeler, 2004, p. 699] in these processes no language-specific computational device is used, but just "[...] the brain's basic modes of representation and computation" [Clark, 1997, p. 198]. A true situation of distributed cognition is occurring in the case of on-line thinking, like in our case of manipulative abduction and in other less expert and less creative cases, where the resources are not merely inner (neural) and embodied, but hybridly intertwined with the environment: in this case we come face to face with an abductive/adaptive process produced in the dynamical inner/outer coupling where internal elements are "directly *causally locked onto* the contributing external elements" [Wheeler, 2004, p. 705].

It is extremely important to note that both sides of the process contribute to the result: both intrinsic inner and outer constraints. For example in the case of language learning and rehearsed recapitulation, external signs and symbols are needed, but so too are the "productive internal capacities" to contribute to its aspects of infinite productivity and systematicity that perform the language-like combinatorial syntax.[67] Through re-embodiment I have illustrated above that the inner aspect (inner rehearsal, when for instance we make an arithmetical addition internally projecting the arithmetical written standard method on our "visual buffer", or when conversing with ourselves) can run alone and disentangle itself from the perception-action cycle typical of on-line thinking, by means of mere pattern-completion neural resources amenable to connectionist modelling (cf. above section 3.4). In this case inner representations, like for example perceptually-based visual, auditory, kinesthetic, motor, etc. imageries, fruit of a suitable past training of neural networks due to previous on-line processes, can furnish the necessary internal surrogate structures.

The origin of these inner representational structures can be accounted for by remembering how Brooks once fundamentally observed that at the root of the more basic forms of cognition it can be hypothesized that the "world serves as its own best model" [Brooks, 1991, p. 145]. [Thomas, 1999] says than rather than storing inner analogues from the external world, human brains generate them by running their perceptual abilities off-line. [Logan, 2006, p. 150] – following [Donald, 2001] – also speculates that "[...] as a starting point it was assumed that before the advent of speech hominid thought processes as inherited from our earliest ancestors were percept-based": mimetic culture, before the emergence of verbal language, would have had a perceptual basis. The human-like ancestors, to defend themselves from predators and to increase their food supply acquired tool making, control of fire, group foraging and coordinated hunting techniques, giving rise to a complex social organization which became too great to be handled merely through percept-based thought. Beyond mere perceptual-based culture,[68] verbal language emerged to deal with the new information overload [Logan, 2000] caused by the richness and

[67] On this problem cf. below subsection 3.6.7.

[68] The so called *preverbal* proto-languages [Logan, 2006, p. 157] of hominid mimetic culture – percept-based – are considered by [Donald, 1991] as related to 1) manual praxic articulation (tool-using); 2) socio-emotional organization and interaction; 3) preverbal mimetic communication (hand signals, mime, gestures and suitably related vocal tones to express various meanings as prosodic verbalizations).

challenges of the new social organization, furnishing at the same time the medium which made the mutual emergence of concept-based thought and new ways of managing things that are remote in both time and space possible.

Moreover, verbal language and its proper essential generativity would have emerged from a form of protolanguage [Bickerton, 1990] consisting of a limited verbal lexicon without syntax: [Logan, 2006, p. 158] usefully notes that this kind of language probably constituted a further step with respect to the proto-language (here with a hyphen, a language possibly endowed with a "through doing" protosemantic and protosyntax) – for example in tool-making and tool-using. In this way tools acquired the status of protosemantic elements – as intended by Donald (cf. the previous footnote) and furnished a kind of preadaptation for the generative grammar of spoken language, which probably arose 50-100 thousand years ago: "Many of the cognitive features usually identified exclusively with language were already present in mimesis: for instance, intentional communication, recursion, and differentiation of reference" [Donald, 1991, p. 200].[69] Of course the mind came into being thanks to verbal language and hence conceptual thought: "Syntactilized verbal language extended the effectiveness of human brain and created the mind. Language is a tool [...]. The human mind is the verbal extension of the brain, a bifurcation of the brain which vestigially retains the perceptual features of the hominid brain while at the same time becoming capable of abstract conceptual thought" [Logan, 2006, p. 162]: a kind of process made up, to use a metaphor, with the help of a software and not a hardware stratagem.[70]

I have noted that [Dartnall, 2005, p. 136] further contends that humans can perform operations in their heads that they would normally have performed in the world and consequently they can also make empirical discoveries in an internal way through the off-line deployment of their sensory abilities.[71] The inner operations are analogues of the inner/outer operations and there are no epistemological differences in the two cases; Dartnall usefully provides a further clarification which is epistemologically obvious but sometimes disregarded: "When we scan inner images we employ perceptual mechanisms that we normally employ in processing information about the world. This normal 'employment', however, takes place in our heads, not in the world. We perform operations in our heads on things in the world (frogs and foxes) and perform the same operations on things in our heads (images of frogs and foxes). This is weaker than internalism which says that we perform

[69] Unlike Donald, who sees speech as emerging from mimetic communication, [Deacon, 1995] contends that speech would have coevolved with it.

[70] It has to be recalled that, contrarily to the hypothesis above, Chomsky contended that humans possess a hardwired generative grammar which permits the quick and universal acquisition of speech by young children.

[71] [Weiskopf, 2008] usefully stresses that different mechanisms are at work when humans interact with the environment than when they use their natural, biological cognitive resources, so that we cannot speak of the same process being carried in the two different systems. A defense of the extended cognition hypothesis against several recent criticisms is provided by [Chemero and Silberstein, 2008]: they argue that extended cognition hypothesis does not derive from armchair theorizing and it neither disregards the results of neural sciences, nor minimizes the importance of the brain in the production of cognition.

operations in our heads that we normally perform in the world" [Dartnall, 2005, p. 140]. I would emphasize that this observation also suggests that "recapitulating" inner rehearsal capacities and endowments "in turn" can of course suitably enter further on-line processes – abductively *anticipating* and *emulating* [Grush, 2004a; Grush, 2007] various external aspects already internalized – to mirror new external representations and to favor subsequent cognitive manipulations and results.[72]

Finally, in the case of language Clark sees off-line thinking (like in the case of silently speaking to ourselves) as a – necessary – fundamental psychological ability that allows human beings to think "about" their own thoughts (second-order thinking), because he contends that only the capability to formulate thoughts in words renders a thought a stable object for evaluation and treatment. I agree with Wheeler who criticizes this conviction: language would not be the "only" route to this kind of second-order process, because many non-linguistic animals could also, in principle, present this capability through the use of other inner states (i.e. model-based representations).[73] Language would not be the "ultimate artifact", like Clark contends, because the same kind of intimate interlocking (almost invisible) with language that we see in human beings as users is also present with other cognitive tools: the example below of geometrical reasoning in subsection 3.6.3 is impressive, even if of course these kind of model-based skills seem less diffuse than the linguistic ones.[74]

It is worth quoting Clark's general summary of the ways in which body and world share the problem solving "load" with the biological brain, so that the mind can be properly considered as "extended". They are deeply illustrated in the first four chapters of the recent [Clark, 2008, p. 81], and all conform to the Principle of Ecological Assembly (PEA) I have quoted in section 1.6 of chapter one, devoted to the concept of manipulative abduction:

- The complex interplay between morphology and control and the value of "ecological control systems" in which goals are not achieved by micro managing every detail of the desired action or response but by making the most of robust, reliable sources of relevant order in the bodily or worldly environment of the controller.
- The use of "deictic pointers" and active sensing routines that retrieve information from worldly sources just in time for problem-solving use and the possible role of whole sensorimotor cycles in the construction of phenomenal experience.

---

[72] On the abductive role of anticipations in the so-called emulation and simulation theories cf. [Barsalou *et al.*, 2003; Decety, 1996; Frith and Dolan, 1996; Grush, 2004a; Hesslow, 2002; Jeannerod, 2001], cf. also chapter four of this book, section 4.7.4.

[73] On animal cognition and animal abductive capabilities cf. chapter five.

[74] Wheeler also presents the ultimateness of other tools/artifacts, different from language, taking advantage of some well-known Heideggerian ideas [Heidegger, 1926], even if it is highly questionable whether Heidegger considers language as a tool or just as a constitutive precondition of meaningful experience: "[...] Heidegger observes, not only are the hammer, nails, and work-bench in this way not part of the engaged carpenter's phenomenal world, neither, in a sense, is the carpenter! The carpenter becomes absorbed in his activity in such a way that he has no awareness of himself as a subject over and against a world of objects. So, in the domain of smooth and uninterrupted skilled tool-use there are, phenomenologically speaking, no subjects and no objects; there is only the experience of the ongoing task (e.g., hammering) [...] there's nothing special about language when compared with more familiar tools and artifacts" [Wheeler, 2004, p. 701].

- The use of open perceptual channels as a means of stabilizing an ongoing organism-environment relation rather than as transducers leading to internal recapitulations of the external scene.
- Our propensity to incorporate bodily and tool-based extensions and substitute sensory strategies deep into our problem-solving routines.
- The use of material symbols to augment our mental powers by adding problem-simplifying structure to our external and internal environments.
- The repeated and nested use of space, environmental structuring, and epistemic actions in online problem solving.
- The potential role of nonbiological media as support for an agents dispositional beliefs.

3.6.6 External Diagrammatization and Iconic Brain Coevolution

Following our previous considerations it would seem that diagrams can be fruitfully seen from a semiotic perspective as external representations expressed through icons and symbols, aimed at simply "mimicking" various humans' internal images. However, we have seen that they can also play the role of creative representations human beings externalize and manipulate not just to mirror the internal ways of thinking of human agents but to find room for concepts and new ways of inferring which cannot – at a certain time – be found internally "in the mind".

In summary, we can say that

- diagrams as external iconic (often enriched by symbols) representations are formed by external materials that either mimic (through reification) concepts and problems already internally present in the brain or creatively express concepts and problems that do not have a semiotic "natural home" in the brain;
- subsequent internalized diagrammatic representations are internal re-projections, a kind of recapitulations (learning), in terms of neural patterns of activation in the brain ("thoughts", in Peircean sense), of external diagrammatic representations. In some simple cases complex diagrammatic transformations – can be "internally" manipulated *like* external objects and can further originate new internal reconstructed representations through the neural activity of transformation and integration.

I have already stressed that this process explains – from a cognitive point of view – why human agents seem to perform both computations of a connectionist type such as the ones involving representations as

- (I Level) patterns of neural activation that arise as the result of the interaction (also presemiotic) between body and environment (and suitably shaped by the evolution and the individual history): pattern completion or image recognition,

and computations that use representations as

- (II Level) derived combinatorial syntax and semantics dynamically shaped by the various artificial external representations and reasoning devices found or constructed in the semiotic environment (for example iconic representations); they are – more or less completely – neurologically represented contingently as patterns of neural activations that "sometimes" tend to become stabilized meaning structures and to fix and so to permanently belong to the I Level above.

It is in this sense we can say the "System of Diagrammatization", in Peircean words, allows for a self-controlled process of thought in the fixation of originally vague beliefs: as a system of learning, it is a process that leads from "absolutely undefined and unlimited possibility" [Peirce, 1931-1958, 6.217] to a fixation of belief and "by means of which any course of thought can be represented with exactitude" [Peirce, 1931-1958, 4.530]. Moreover, it is a system which could also improve other areas of science, beyond mathematics, like logic, it "[...] greatly facilitates the solution of problems of Logic. [...] If logicians would only embrace this method, we should no longer see attempts to base their science on the fragile foundations of metaphysics or a psychology not based on logical theory" [Peirce, 1931-1958, 4.571].

As already stressed the I Level originates those sensations (they constitute a kind of "face" we think the world has), that provide room for the II Level to reflect the structure of the environment, and, most important, that can follow the computations suggested by the iconic external structures available. It is clear that in this case we can conclude that the growth of the brain and especially the synaptic and dendritic growth are profoundly determined by the environment. Consequently we can hypothesize a form of coevolution between what we can call the *iconic brain* and the development of the external diagrammatic systems. Brains build iconic signs as diagrams in the external environment learning from them new meanings through interpretation (both at the spatial and sentential level) after having manipulated them.

When the fixation is reached – imagine for instance the example above, that fixes the sum of the internal angles of the triangle, cf. above subsection 3.6.2 – the pattern of neural activation no longer needs a direct stimulus from the external spatial representation in the environment for its construction and can activate a "final logical interpretant", in Peircean terms. It can be neurologically viewed as a fixed internal record of an external structure (a fixed belief in Peircean terms) that can exist also in the absence of such external structure. The pattern of neural activation that constitutes the I Level Representation has kept record of the experience that generated it and, thus, carries the II Level Representation associated to it, even if in a different form, the form of *semiotic memory* and not the form of the vivid *sensorial experience* for example of the triangular construction drawn externally, over there, for instance in a blackboard. Now, the human agent, via neural mechanisms, can retrieve that II Level Representation and use it as an internal representation (and can use it to construct new internal representations less complicated than the ones previously available and stored in memory).

At this point we can easily understand the particular *mimetic* and *creative* role played by external diagrammatic representations in mathematics:

1. some concepts, meanings, and "ways of [geometrical] inferring" performed by the biological human agents appear hidden and more or less tacit and can be rendered explicit by building external diagrammatic mimetic models and structures; later on the agent will be able to pick up and use what was suggested

by the constraints and features intrinsic and immanent to their external semiotic materiality and the relative established conventionality: artificial languages, proofs, new figures, examples, etc.;

2. some concepts, meanings, and "new ways of inferring" can be discovered only through a problem solving process occurring in a distributed interplay between brains and external representations. I have called this process externalization (or disembodiment) of the mind: the representations are mediators of results obtained and allow human beings

   (a) to re-represent in their brains new concepts, meanings, and reasoning devices picked up outside, externally, previously absent at the internal level and thus impossible: first, a kind of alienation is performed, second, a recapitulation is accomplished at the neuronal level by re-representing internally that which has been "discovered" outside. We perform cognitive geometric operations on the structure of data that synaptic patterns have "picked up" in an analogical way from the explicit diagrammatic representations in the environment;
   (b) to re-represent in their brains portions of concepts, meanings, and reasoning devices which, insofar as explicit, can facilitate inferences that previously involved a very great effort because of human brain's limited capacity. In this case the thinking performance is not completely processed internally but in a hybrid interplay between internal (both tacit and explicit) and external iconic representations. In some cases this interaction is between the internal level and a computational tool which in turn can exploit iconic/geometrical representations to perform inferences (cf. above subsection 3.6.1).

An evolved mind is unlikely to have a natural home for complicated concepts like the ones geometry introduced, as such concepts do not exist in a definite way in the natural (not artificially manipulated) world: so whereas evolved minds could construct spatial frameworks and perform some simple spatial inferences in a more or less tacit way by exploiting modules shaped by natural selection, how could one think exploiting explicit complicated geometrical concepts without having picked them up outside, after having produced them?

Let me repeat that a mind consisting of different separated implicit templates of thinking and modes of inferences exemplified in various exemplars expressed through natural language cannot come up with certain mathematical and geometrical entities without the help of the external representations. The only way is to extend the mind into the material world, exploiting paper, blackboards, symbols, artificial languages, and other various semiotic tools, to provide *semiotic anchors*[75] for finding ways of inferring that have no natural home within the mind, that is for finding ways of inferring and concepts that take us beyond those that natural selection and previous cultural training could enable us to possess at a certain moment.

Hence, we can hypothesize – for example – that many valid spatial reasoning habits which in human agents are performed internally have a deep origin in the

[75] [Enfield, 2005; Callagher, 2005] point out the role of the body itself as and "anchoring" of cognitive processes, for instance in the case of human gestures linked to the expression of meanings.

past experience lived in the interplay with iconic systems at first represented in the environment. As I have just illustrated other recorded thinking habits only partially occur internally because they are hybridized with the exploitation of already available or suitably constructed external diagrammatic artifacts.

3.6.7 Delegated and Intrinsic Constraints in External Agents and the Role of Anchors in Conceptual Blending

We have said that through the cognitive interplay with external representations the human agent is able to pick up and use what suggested by the constraints and features intrinsic to their external materiality and to their relative established conventionality: artificial languages, proofs, examples, etc. Let us consider the example above (section 3.6.2) of the sum of the internal angles of a triangle. At the beginning the human agent – that is an interpretant in Peircean sense – embodies a sign in the external world that is in this case an icon endowed with "intentional" delegated cognitive conventional and public features – meanings – that resort to some already known properties of the Euclidean geometry: a certain language and a certain notation, the definition of a triangle, the properties of parallel lines that also hold in case of new elements and "auxiliary" constructions obtained through manipulation, etc. Then she looks, through diagram manipulations, for possible necessary consequences that occur over there, in the diagram/icon and that obey both

- the conventional *delegated* properties and
- the properties *intrinsic* to the materiality of the model.

This external model becomes a kind of autonomous cognitive *agent* offered to new interpretants of the problem/object in question. In its presence the competent reasoner is induced to trace series of interpretants in some directions and not in others, because the features of the external materiality at play dispose movement along certain paths and not others. They confront us both as a cluster of constraints and as a possibility. The model can be picked up later and acknowledged by the human agent through fixation of a new neural configuration – a new "thought" (in the case the new result concerning the sum of the internal angles).

In a recent article [Hutchins, 2005], taking advantage of various amazing and very interesting examples ranging from everyday to scientific cases, further analyzes the role of constraints through the association of conceptual structure and material structure in what he calls "conceptual blending" as a key cognitive strategy. First of all it is noted that in the external representations embodied in material artifacts – which form a "blended" space – some aspects can be manipulated and other parts remain stable.[76] Empirical results have shown the relevance of stability, portability,

[76] This perspective on conceptual blending and integration is further developed by [Fauconnier and Turner, 2003; Fauconnier, 2005]. It is especially expanded the analysis of "running the blend", seen as the cause of the formation of additional emergent structures, in some cases concerning mathematics and natural language, and of the requirement for simplicity in these processes.

and simplicity of representations in facilitating reasoning and their role as anchors for thoughts and at the same time as sources of new inferences and results. Hutchins describes many cases, for example the role of a "line" for people queuing at the theater creates a spatial memory for the order of arrival of clients: the blend – which originates the queue – consists of the mixture of the line and of the directional ordering. Like Brooks said: "the world is its own best model" [Brooks, 1991]. From the perspective of traditional representationalism we can consider the two inputs to the blend a mental (neural) conceptual structure on the one hand and a mental representation (neural) of the material structure on the other. Hutchins prefers to adopt a different view and avoids giving a separate mental representation of the material structure as an input space:

> Another alternative is to say that the physical objects themselves are input to the conceptual blending process. This is what I intend when I speak of "material anchors" for conceptual blends. What is at stake here is the boundary of the conceptual blending process. Shall the conceptual blending process be an entirely conceptual process that operates on ("real space" as delivered to the process in the form of) the output of perceptual processes, or shall the conceptual blending process include the perceptual processes and therefore include bodily interaction with the physical world. [...] First there is the selectivity of perception that produces a filtered conceptual representation of the physical world. Second, there is selective projection in the process by which the prior conceptualization of the world (the "real space" representation) is blended with the other conceptual input. Is there any evidence that these are two separate processes? It seems preferable to assume that the selective attention to, and projection of, structure from the material world to the blended space is the perceptual process. That is, that selective perception is a conceptual process [Hutchins, 2005, p. 1559–1561].

The main "emergent" property of the blend is the stabilization of representations of the conceptual relationships at stake (sequential relations among persons in the queue), thus enabling further inferential chances and providing full "cultural" models (habits, in Peircean terms). In summary, the material anchor does not have a cognitive value merely because of its intrinsic quality, but because of the way it is used: "If conceptual elements are mapped onto a material pattern in such a way that the perceived relationships among the material elements are taken as proxies (consciously or unconsciously) for relationships among conceptual elements, then the material pattern is acting as a material anchor. The conceptual relationships that are mapped into material structure range from the minimum case of individuation (this element is distinguishable from all others) to extremely complex systems of relationships (the scales on a sliderule, for example)" [Hutchins, 2005, p. 1562].[77]

The distinction above between delegated and intrinsic and immanent properties is also clear if we adopt the Peircean semiotic perspective. Peirce – speaking about

[77] [Fauconnier and Turner, 2003] illustrate the clock face, other gauges, and the method of loci taking advantage of Hutchins' idea of conceptual blending. They also extend the analysis to graves, headstones, dead people, money, and spoken and also written language. In the case of language, Hutchins says, the contribution of the material aspects, like a written word, just furnishes a minimal criterium of individuation unlike in the case of larger linguistic units such as grammatical forms.

the case of syllogistic logic, and not of geometry or algebra – deals with this problem by making an important distinction between what is going on in the brain of the logical human agent and the autonomous power of the chosen external system of representation or diagrammatization [Hoffmann, 2003]. The presence of this "autonomous power" explains why I attribute to the external system of representation a status of cognitive agency similar to the one of a human person, even if of course lacking aspects like direct intention and responsibility. Any diagram, Peirce says, makes use

> [...] of a particular system of symbols – a perfectly regular and very limited kind of language. It may be a part of a logician's duty to show how ordinary ways of speaking and of thinking are to be translated into that symbolism of formal logic; but it is no part of syllogistic itself. Logical principles of inference are merely rules for the illative transformation of the symbols of the particular system employed. If the system is essentially changed, they will be quite different [Peirce, 1931-1958, 2.599].

Of course the argumentation above also holds for our case of iconic geometric representation. This distinction integrates the one I have introduced above in the two levels of representations, and in some sense blurs it by showing how the *hybrid* character of the system composed by the two levels themselves, where the whole package of sensorial and kinesthetic/motor abilities are involved.

The construction of the diagram also depends on those delegated semiotic properties that are embedded in what Peirce calls "precept" as he says in the passage we have already quoted above and not only on the constraints expressed by the materiality of the model itself.[78] Semiotic delegation is made possible by humans' instinctual nature plus cultural inheritances and individual training as they can perform cognitive *inner* actions able to form various "precepts" that can interact with the material objects. These actions, that occur internally, are contrasted with the actions that instead are immediately related to the world external to the body: "Human instinct is no whit less miraculous than that of a bird, the beaver, or the ant. Only, instead of being directed to bodily motions, such as singing and flying, or to the construction of dwelling, or to the organization of communities, its theater is the plastic inner world, and its products are the marvelous conceptions of which the greatest are the ideas of number, time and space" [Peirce, 1966, 318]. In terms of traditional philosophical concepts, it is an activity that relates to the "imagining" of what might be in "fantasy". I must stress that, in this perspective, this inner activity is experience and action in itself and no less experience and action than that performed in the external world.[79]

Pickering depicts the role of some externalities (representations, artifacts, tools, etc.) in terms of a kind of non-human agency that interactively stabilizes with human agency in a dialectic of resistance and accommodation [Pickering, 1995, p. 17 and p.

---

[78] It is worth noting that this process is obviously completely related to the Peircean idea of pragmatism [Hoffmann, 2004], that he simply considers "the experimental method", which is the procedure of all science.

[79] On the Peircean emphasis on diagrammatic reasoning as a case of distributed cognition cf. [Skagestad, 1993].

22]. The two agencies, for example in scientific reasoning, originate a co-production of cognition the results of which cannot be presented and identified in advance: the outcome of the co-production is intrinsically "unpredictable". Latour's notions of the de-humanizing effect of technologies are based on the so-called "actor network theory",[80] which also stresses the semiotic role of externalities like the so-called non human agents. The actor network theory basically maintains that we should think of science, technology, and society as a field of human and non-human (material) agency. Human and non-human agents are associated with one another in networks, and they evolve together within these networks. Because the two aspects are equally important, neither can be reduced to the other: "An actor network is simultaneously an actor whose activity is networking heterogeneous elements and a network that is able to redefine and transform what is it made of [...]. The actor network is reducible neither to an actor alone nor to a network" [Callon, 1997, p. 93].

The whole process can be seen as a kind of experiment and, at the same time, an operation of thought. Peirce M. J. Pickering and S. Garrod is still of help: "There is not reason why 'thought' [...] should be taken in that narrow sense in which silence and darkness are favorable to thought. It should rather be understood as covering all rational life, so that an experiment shall be an operation of thought" [Peirce, 1931-1958, 5.420]. In this sense thought can be conceived of as a semiotic process occurring in the publicly observable domain of natural processes (including human actions) as in the publicly inaccessible realm of someone's individual consciousness [Colapietro, 2005, p. 416]. In this perspective the interplay between internal and external representation is a kind of experiment like "[...] conversation in which the topic being discussed is, by various contrivances, afforded by the opportunity to speak back, to object to the ways it is being spoken about" (*ibid.*). The object investigated becomes – through semiotic cognitive delegations – an interlocutor and the process transforms apparently "mute, objects, brute things" [Backhtin, 1982, p. 351] in a critical source.[81]

The operation on a diagram has reduced complexity enabling concentration on essential relations and has revealed new data. Moreover, through manipulations of the diagram new perspectives are offered to the observation, or interesting anomalies with respect the internal expectations are discovered. In the case of mathematicians, Peirce maintains, the diagram "[...] puts before him an icon by the observation of which he detects relations between parts of the diagram other than those which were used in its construction" [Peirce, 1976, III, p. 749]: "unnoticed and hidden relations among the parts" are discovered [Peirce, 1931-1958, 3.363]. This activity is

[80] This theory has been proposed by Callon, Latour himself, and Law [Callon, 1994; Callon, 1997; Latour, 1987; Latour, 1988; Callon and Latour, 1992; Law, 1993].

[81] Colapietro further observes that to make this dialogue possible at least three presuppositions are necessary: "We must suppose that reality can be other than our representation of it. We must also suppose that human experimenters are rational subjects whose unique individuality is largely of a privative character (individuals, so far as they are anything apart from others, and apart from what they and the others with whom their lives are so intimately and inescapably bound up, are mere negations [Peirce, 1931-1958, 5.317]. Finally, we must suppose that human beings are autonomous agents who can exercise an indeterminable measure of effective control over their future conduct" [Colapietro, 2005, p. 416].

a kind of "thinking through doing": "In geometry, subsidiary lines are drawn. In algebra permissible transformations are made. Thereupon, the faculty of observation is called into play. [...] Theorematic reasoning invariably depends upon experimentation with individual schemata" [Peirce, 1931-1958, 4.233].

We have said that firstly the human agent embodies structured signs in the external world that is in this geometrical case an icon endowed with "intentional" delegated cognitive conventional and public features – meanings – that resort to some already known properties of the Euclidean geometry: these features can be considered a kind of immanent rationality and regularity [Hoffmann, 2004] that establishes a disciplinary field to envisage conclusions. The system remains relative to the chosen conventional framework. These features are real as long as there is no serious doubt in their adequacy: "The 'real,' for Peirce, is part of an evolutionary process and while 'pragmatic belief' and unconscious habits might be doubted from a scientific point a view, such a science might also formulate serious doubts in its own representational systems" [Hoffmann, 2004, p. 295].

Paavola, Hakkarainen, and Sintonen [Paavola *et al.*, 2006] consider the interplay between internal and external aspects of abductive reasoning in the framework of the interrogative model of "explanation-seeking why-questions" and in the light of the perspective of distributed cognition. They emphasize interaction with the "environment" and show the importance of the heuristic strategies and of their trialogic nature (inquirer and fellow inquirers, object of inquiry, mediating artifacts and processes), also taking advantage of Davidson's ideas [Davidson, 2001] – as already stressed by Wirth [1999; 2005] – concerning triangulation.[82]

Let us imagine we choose a different representational system still exploiting material and external diagrams. Through the manipulation of the new symbols and diagrams we expect very different conclusions. An example is the one of the non-Euclidean discoveries. In Euclidean geometry, by adopting the postulate of parallels we necessarily arrive to the ineluctable conclusion that the sum of internal angles of a triangle is 180°, but this does not occur in the case of the non-Euclidean geometry where a different selected representational system – that still uses Euclidean icons – determines quite different possibilities of constructions, and thus different results from iconic experimenting.[83]

[82] Cf. also Arrighi and Ferrario's [2008] study on mutual understanding that emphasizes the role in abductive reasoning of the collaborative processes involved in interaction with other speakers and with the entities of the environment; they explicitly refer to my approach called "manipulative framework" and to the "strategic framework" described by Paavola and his collaborators. The fact that science is inherently seen as a social process is emphasized by [Addis and Gooding, 2008], who argue that abduction does not work in isolation from other inference mechanisms and use game theory to relate the abductive system to actions that produce new information. To suitably model this process an interesting computational model is proposed in order to display various aspects of collective belief-revision leading to consensus-formation.

[83] I have illustrated this problem in detail in [Magnani, 2002a]. Sections 2.9 and 2.11 of chapter two of this book illustrated an example of this process of cognitive delegation to external diagrams just taking advantage of the discovery of non-Euclidean geometry.

3.7 Mimetic Minds as Semiotic Minds

I contend that there are external representations that are representations of other external representations. In some cases they carry new scientific knowledge. To make an example, Hilbert's *Grundlagen der Geometrie* is a "formal" representation of the geometrical problem solving through diagrams: in Hilbertian systems solutions of problems become proofs of theorems in terms of an axiomatic model. In turn a calculator is able to re-represent (through an artifact) (and to perform) those geometrical proofs with diagrams already performed by human beings with pencil and paper. In this case we have representations that *mimic* particular cognitive performances that we usually attribute to our minds (cf. the first sections of this chapter).

We have seen that our brains delegate cognitive (and epistemic) roles to externalities and then tend to "adopt" and recapitulate what they have checked occurring outside, over there, after having manipulated – often with creative results – the external invented structured model. A simple example: it is relatively neurologically easy to perform an addition of numbers by depicting in our mind – thanks to that brain device that is called visual buffer – the images of that addition *thought* as it occurs concretely, with paper and pencil, taking advantage of external materials. We have said that mind representations are also over there, in the environment, where mind has objectified itself in various semiotic structures that mimic and enhance its internal representations.

Turing adds a new structure to this list of external objectified devices: an abstract tool, the (Universal) Logical Computing Machine (LCM), endowed with powerful mimetic properties. We have concluded the subsection 3.4.1 remarking that the creative "mind" is in itself extended and, so to say, both internal and external: the mind is *semiotic* because transcends the boundary of the individual and includes parts of that individual's environment, and thus constitutively artificial. Turing's LCM, which is an externalized device, is able to mimic human cognitive operations that occur in that interplay between the internal mind and the external one. Indeed Turing already in 1950 maintains that, taking advantage of the existence of the LCM, "Digital computers [...] can be constructed, and indeed have been constructed, and [...] they can in fact mimic the actions of a human computer very closely" [Turing, 1950, p. 435].

In the light of my perspective both (Universal) Logical Computing Machine (LCM) (the theoretical artifact) and (Universal) Practical Computing Machine (PCM) (the practical artifact) are *mimetic minds* because they are able to mimic the mind in a kind of universal way (wonderfully continuing the activity of disembodiment of minds and of semiotic delegations to the external materiality our ancestors rudimentary started). LCM and PCM are able to re-represent and perform in a very powerful way plenty of cognitive skills of human beings. Universal Turing Machines are discrete-state machines, DSM, "with a Laplacian behavior" [Longo, 2002; Lassègue, 1998; Lassègue, 1999]: "[...] it is always possible to predict all future states") and they are equivalent to all formalisms for computability (what is thinkable is calculable and mechanizable), and because universal they are able to simulate – that is to *mimic* – any human cognitive function, that is what is usually called mind. A natural consequence of this perspective is that Universal Turing machines do not represent (against classical AI and modern cognitivist computationalism) a "knowledge" of the mind and of human intelligence. Turing is perfectly aware of the fact that brain is not a DSM, but as he says, a "continuous" system, where instead a mathematical modeling can guarantee a satisfactory scientific intelligibility (cf. his studies on non-Laplacian mathematical models of morphogenesis).

We have seen that our brains delegate meaningful semiotic (and of course cognitive and epistemic) roles to externalities and then tend to "adopt" what they have checked occurring outside, over there, in the external invented structured and model. And a large part of meaning formation takes advantage of the exploitation of external representations and mediators. Our view about the disembodiment of mind certainly involves that the Mind/Body dualist view is less credible as well as Cartesian computationalism. Also the view that mind is computational independently of the physical (functionalism) is jeopardized. In my perspective on human cognition in terms of mimetic minds we no longer need Descartes dualism: we only have *semiotic brains* that make up large, integrated, material cognitive systems like for example LCMs and PCMs. These are new independent semiotic agencies that constitute real artificial minds aiming at "universally" imitating human cognition. In this perspective what we usually call mind simply consists in the union of both the changing neural configurations of brains together with those large, integrated, and material cognitive systems the brains themselves are continuously building in an infinite semiotic process.

Minds are material like brains, in so far as they take advantage of intertwined internal and external semiotic processes. It seems to me at this point we can better and more deeply understand Peirce's semiotic motto "man is an external sign" in the passage we have completely quoted above in subsection 3.5.1: "[...] as the fact that every thought is a sign, taken in conjunction with the fact that life is a train of thoughts, proves that man is a sign; so, that every thought is an *external sign*, proves that man is an external sign" [Peirce, 1931-1958, 5.324]. The only problem seems "how meat knows": we can reverse the Cartesian motto and say "sum ergo cogito".

We have seen that our brains delegate meaningful cognitive (and epistemic) roles to externalities and then tend to "adopt" what they have checked occurring outside, over there, in the external invented structures and models. And a large part of meaning formation takes advantage of the exploitation of external representations and mediators. We have said that PCMs can be considered mimetic minds (they are ideal "practical" – in Turing's sense – agents): what is in turn the cognitive status of "logical agents" from the point of view of their demonstrative aspect? I will treat this important problem in chapter seven, where various aspects concerning logical systems as ideal externalizations in demonstrative frameworks will be illustrated.

3.8 "Symbols" as Memory Mediators. Maximizing Abducibility through Psychic Energy Mediators

We have seen in the previous sections that recent research in the area of distributed cognition acknowledges the distinction between internal and external representations. To illustrate this process I have taken advantage of some paleoanthropological results on the birth of material culture, that provide an evolutionary perspective on the origin of intelligent behaviors, and of Turing's ideas about the passage from what he called unorganized brains to organized, mature, ones. Unorganized brains (in Turing's sense) organize themselves through semiotic activity that is reified in the external environment and then re-projected and reinterpreted through new configurations of neural networks and chemical processes. In the following sections, also considering some psychoanalytic insights, I will describe the centrality to semiotic cognitive information processes of *disembodiment of mind* from the point of view of cognitive interplay between internal and external representations. Humans continuously delegate and distribute cognitive functions to the environment to lessen their limits. They build models, representations, and other various mediating structures, that are thought to be good for thinking. In building various mediating structures and designing activities, such as models or representations, humans alter the environment and thus create those *cognitive niches* I will describe in chapter six. What is the role of this environment modification in psychoanalysis? In the following sections I will illustrate some fundamental aspects of the interplay above in the light of Jungian observations about the cognitive and therapeutic role of what we can call external *energy mediators*.

3.8.1 Mythologization of External "Observations"

Let us consider this interplay between internal and external representations taking advantage of some stimulating speculations made by Jung. Jung says "Psychic existence can be recognized only by the presence of contents that are *capable of consciousness*. We can therefore speak of an unconscious only insofar as we are able to demonstrate its contents. The contents of the personal unconscious are chiefly feeling-toned complexes: they constitute the personal and private side of psychic life. The contents of the collective unconscious, on the other hand, are known as *archetypes*" [Jung, 1968a, pp. 3-5]. Usually the contents of the collective unconscious have not yet been submitted to conscious elaboration and so they are an "immediate datum of psychic experience". They are "altered" by becoming conscious and by being perceived: the archetype, "a hypothetical and irrepresentable model", "takes its color from the individual consciousness in which it happens to appear" (*ibid.*, p. 5).

In the case of the primitive human beings unconscious psyche "[...] has an imperative need – or rather [...] an irresistible urge – to assimilate all outer sense experience to inner, psychic events. It is not enough for the primitive to see the sun rise and set; this external observation must at the same time be a psychic happening: the sun in its course must represent the fate of a god or hero who, in the last analysis, dwell nowhere except in the soul of man" (*ibid*, p. 6).[84] This process leads to a consciousness that in those primitive creatures is largely undifferentiated so we can guess that what we call "external" representations may not be seen by those individuals as external in the same way that we see them: it is only our perspective that sees that kind of consciousness as formed through the intervention of "externalities" (arrived at through "outer sense experience", Jung says).

In the light of the interplay I have illustrated in the previous sections of this chapter we can say that rough external representations (Jung says "observations"), merely arrived at through the senses, are rapidly "internalized" and become "mythologized". "Summer and winter, the phases of the moon, the rainy season, and so forth, are in no sense allegories" of those primary observations, because they become "symbolic expressions of the inner" and "become accessible to man's consciousness by way of projection, that is mirrored in the events of nature" (*ibid.*, p. 6). Let us reiterate that of course the "subjectivity" – so to speak – of the primitives is so large and extended that those external seen events, the – "primary observations" – which are suitably mythologized, are not "thought" of as "external", in the way we modern humans think. Of course in the case of the primitives those events that "we" clearly see as "external" are on the whole events of their large undifferentiated psyche ("a suffocating atmosphere of egocentric subjectivity", in Junghian terms), which of course does not allow a net distinction between inner and outer. It has to be noted that also in healthy modern humans who trust astrology some hesitation in discriminating between inner and outer events is present: the effects of that "partécipation mystique" already described by Lévi-Bruhl are still significantly in operation [Lévi-Bruhl, 1923].

This Junghian remark about the hesitation in discriminating between inner and outer events is confirmed in the framework of the catastrophe theory. Thom sees astrology as an intermediate level between magic, where narrative cognition that makes the world intelligible is controlled by the will of man (the "magician"), and science, where "[...] control is determined by the internal generativity of formal language describing external situations, a generativity over which man has no hold, once the initial conditions are laid down" [Thom, 1988, p. 33] (cf. also this book, chapter eight, subsection 8.5.2). Astrology depicts human situations as supposedly

[84] It is interesting to note that also René Thom recognizes this fact in the framework of his mathematical catastrophe theory: "Men (as well as prehominids) were early incited by group living to build up some representation of the behavior of their kind, a representation, in particular, of the paths of their affective regulation. As a result, any external entity thought of as being individuated tends, by empathy, to be imagined after the manner of a living being" [Thom, 1988, p. 16].

ruled by the mathematically determined course of the heavenly bodies, "a belief for which geocentrism was no doubts largely responsible" [Thom, 1988, p. 33].[85]

For the primitive "[...] knowledge of nature is essentially the language and outer dress of an unconscious psychic process. But the very fact that this process is unconscious gives us the reason why man has thought of everything except the psyche in his attempts to explain myths" (*ibid*, pp. 6-7).[86] Indeed "[...] anyone who descends into the unconscious gets into a suffocating atmosphere of egocentric subjectivity, and in this blind alley is exposed to the attack of all the ferocious beasts which the caverns of the psychic underworld are supposed to harbor" (*ibid.*, p. 20).

Reflecting upon the processes of mind externalization and disembodiment and subsequent re-internalization Jung of course speaks of projection: "The projection is so fundamental that it has taken several thousand years of civilization to detach it in some measure from the outer object" (*ibid.*, p. 6). Once formed – and in presence of an established, significant and more or less stable separation between the self and the external world – the internalized representations (mythologized) can be re-externalized ("projected") so giving rise to new "external representations" that go

[85] On the basis of these considerations Thom also further details that Lévi-Bruhl's classic notion of participation, (the possibility that two spatially separate beings can constitute the same being), and the notions associated with "magic", cannot actually be thought of as "pre-logic":

> For example a sorcerer may be at one and the same time a man sleeping in a hut and a tiger hunting in the jungle some distance away [...] if the tiger is wounded by hunters in the jungle, then the man-sorcerer in his hut, will reveal a wound in the homologous place on his body. A belief of this kind justifies the statement that the man-sorcerer, and the tiger have their "local somatic maps" identified, in spite of the fact that these maps relate to beings separated by several kilometres. From this viewpoint, it can be said that the act of magic is characterized essentially by an "action at a distance" which can be interpreted as a modification of the usual topology of space-time. In other words, the linking up between local maps which define usual space will not be fixed, but could be modified at the pleasure or the will of certain men (magicians or sorcerers) thanks to the use of specific procedures (magical rites, sacrifices, etc.). Further, the topology of the space will cease to be the same for all, given that the perceptual experiences of an observer can themselves be affected by magical action [Thom, 1980, p. 132].

The somatic local maps, predominant in primitives and animals, emanate from the individual as forms of control of the external world and identification of one's own body relating to the need to acknowledge the presence of local morphological accidents of which many have a "pregnant" character (cf. also chapter eight of this book, section 8.3): "This conception of a flexible and individual space-time, which will cease to be a universal frame valid for all men, clearly conflicts head on with the basic postulate of all modern science that there exists a universal space-time valid and isomorphic for all. It is without doubt this essential difference that Lévi-Bruhl had wished to signify in speaking of "pre-logical mentality" – unhappy words, for logic has in principle, nothing to do with the representation of space" [Thom, 1980, p. 133]. Further details on the cognitive problem of spatial frameworks are illustrated in the following chapter, subsection 4.7.

[86] "He simply didn't know that the psyche contains all the images that have ever given rise to myths, and that our unconscious is an acting and suffering subject with an inner drama which primitive man rediscovers, by means of analogy, in the processes of nature both great and small" (*ibid.*, p. 7).

beyond the previous ones, and that can mimic what is already internal. Through this process the externalities at play acquire new mental/psychic contents.

Recent research on "identity" processes in individual human beings have stressed the role of the non-human environment, for example in the case of plants, animals, wind, and water. Let us consider the case of trees. Jung sees the tree as participating in the processes of the formation of the psyche in so far as it constitutes an archetype in the collective unconscious. Darwinians sometimes speak of an innate emotional affiliation of humans with other living organisms so that their preferences would have been shaped over millennia through interactions with features of the environment helpful to the survival of the species in its early development (for example they offer prospect in predation and refuge). The so-called phenomenological approaches rely on various metaphors: roots, trunk, and canopy mirror the infernal, earthy, and heavenly domains, flowers, fruits, and colors supply subsidiary arguments for human identity – in this respect trees certainly offer more than grass, the most universal and successful of plants. Gibsonian ecological psychologists emphasize that trees are the source for humans of multiple innate affordances[87] that provide various action possibilities (climbing, hide, and seek), the possibility of making artifacts (rope swing, tree forts), and of satisfying human needs (shelter, food, fuel, and medicine) [Sommer, 2003].[88] Finally, ecopsychologists maintain that beyond the individual self there is an ecological self that is "nurtured through the contact with the natural environment" [Sommer, 2003, p. 191] (from this perspective trees are important for city residents because they provide contact with natural rhythms, life forms, seasonal markers, and the gentle motion and sound of rustling leaves).

Moreover, through anthropomorphic interpretation (for example in children) an external object such as a tree or a squirrel is perceived as being similar to oneself and humanlike in certain respects, where the identity of both the object and the observer is reached [Gebhard *et al.*, 2003]. This interpretation also includes moral aspects, such as freedom and an affectionate and caring nature, which are attributed in an isomorphic way to organisms or natural objects making it possible to grant greater independence and environmental moral value to them [Kahn, 2003]. In turn, in the case of physiomorphism, human experience can be interpreted in terms of nonhuman nature or natural objects. Both anthropomorphism and physiomorphism can start a never ending, cyclic process of mirrors:

> Thus we may draw upon experiences with natural objects to understand ourselves (physiomorphysm), but in turn our representations of these natural objects will have arisen by interpreting them in terms of ourselves and our personal experience (anthropomorphism). [...] But attaching subjective meaning to an object, the object and the self become mentally intertwined and a unique relationship between the two is

[87] The concept of affordance is illustrated below in chapter six of this book.

[88] In this last case, in modern humans, the satisfaction of needs occurs through the mediation of sophisticated cultural schemes (such as in the exploitation of trees to counter pollution, to protect privacy and limit noise and to shape the person-home-neighbourhood interplay on which survival of the individual depends).

established. It is perhaps in this manner that external objects contribute to the formation of personal identity (cit., pp. 104-105).[89]

I have contended above that there are two basic kinds of external representations active in this process of externalization of the mind: *creative* and *mimetic*. Mimetic external representations mirror concepts, emotional tones, and other structures that are already *present* or *re-presented* in the brain and need to be enhanced, solved or further complicated, etc., in order for them to sometimes give rise to new concepts, meanings, etc. I quoted above the Junghian passage stating that "Psychic existence can be recognized only by the presence of contents that are *capable of consciousness*": now it is clear that consciousness is basically made of suitably internalized sensory data.

In sum, a "differentiated" psyche is a hybrid product of the interplay between internal and external sensory representations. Conversely, external representations are continuously re-built through the delegation of psychic contents, like, for example, in the case of the ancient religious images such as the Trinity or the mystery of the Virgin birth. These representations are placed and "produced" over there, outside, hybridized with and within material supports[90] from the external environment, autonomized with respect to their human origin but more or less available,[91] they can suitably be "picked up" by human beings and re-represented in their brains. They are picked up to favor extension and "consolidation of consciousness" and "to keep back the dangers of unconscious", the "peril of the souls'" given the fact that "mankind always stands on the brink of actions it performs itself but does not control. The whole world wants peace and the whole world prepares for war" (*ibid.* p. 23). They are also picked up through imitation and sometimes to contrast individuation processes. "[...] the more beautiful, the more sublime, the more comprehensive the image that has evolved and been handed down by tradition, the further removed it is from individual experience. We can just feel our way into it and sense something of it, but the original experience is lost, [the images] have stiffened into mere objects of belief" (*ibid.*, p. 7). In these externalized images that became symbols of dogmatic archetypical ideas – "collective unconscious has been channelled [...] and flows along like a well-controlled stream in the symbolism of creed and ritual" (*ibid.*, p. 12).

I have said that for Jung consciousness is basically made of suitably internalized external "observations": in sum, psyche is a hybrid product of the interplay between internal and external representations. Conversely, once representations are

[89] In this perspective perceiving an object as humanlike can be related to the problem of its possible *moral worth* (thus activating a kind of micro morality at the personal level) and, consequently, for example in the case of animals or trees, it can lead to the more extended awareness of the need for their protection and preservation (a "biocentric" perspective that expresses a kind of transpersonal macromorality): killing plants can be seen as analogous to killing humans [Rest *et al.*, 1999]. Human bodily and mental characters play, through anthropomorphism, the role of *moral mediators* [Magnani, 2007d], that can permit nature to be moralized, at least at the level of micromorality.

[90] On the re-externalization in *artifacts* cf. the following section.

[91] Jung says that "[...] archetypal images are so packed with meaning in themselves that people never think of asking what they really do mean" (*ibid.*, p. 13).

internalized, we have them at our private disposal for re-building external representations through delegation (projections) of their psychic contents. First, this process presents the growth in complexity of the psyche where archetypes manifest themselves through their capacity to – unconsciously – organize images and ideas; later on archetypes, thanks to the above external/internal interplay, "by assimilating ideational material whose provenance in the phenomenal world is not to be contested, they become visible and *psychic*" [Jung, 1972b, p. 128].

From the perspective of the individual psyche Jung's remark is central and clearly stated. Acknowledging the importance of the above hybrid interplay he says that the word "projection" is not really appropriate, and the role of the senses is fundamental "[...] for nothing has been cast out of the psyche; rather, the psyche has attained its present complexity by a series of acts of introjections. Its complexity has increased in proportion to the despiritualization of nature" [Jung, 1968a, p. 25]. Even more clearly, Jung points out the relevance of the hybridization processes: "The organization of these particles [of light] produces a picture of the phenomenal world which depends essentially upon the constitution of the apperceiving psyche on the one hand, and upon that of the light medium on the other" [Jung, 1972b, p. 125].

I have already said that we can hypothesize a form of coevolution between the structural complexity of the psyche and that of the cognitively delegated cognitive external systems. Brains build external representations in the environment learning new meanings from them through interpretation (both at the model-based and sentential level), after having manipulated them through motor actions.[92] When the internal fixation of a new meaning is reached through internalization – like in the above example, concerning the formation of a new religious icon, that for example fixes the character of a suitable rite – the pattern of neural activation no longer needs a direct stimulus from the external representation in the environment for its construction. In this last case the new meaning can be neurologically viewed as a fixed internal record of an external structure (a fixed belief in Peircean terms) that can also exist in the absence of such external structure. In the examples I will illustrate in the following sections it will be clear how, for instance, a mimetic representation (for example embedded in an artifact) can become creative by giving rise to new meanings and ideas in the hybrid interplay between psyche and suitable cognitively delegated environments.

3.8.2 Cognitive/Affective Delegations to Artifacts

In chapter one I have introduced the concept of sentential, model-based, and manipulative abduction mainly at the level of logical and scientific reasoning. Manipulative abduction is also important in explaining how Jung conceives of "hypothesis" generation and creativity in human beings. We now have a good conceptual tool at

[92] Representations and inferences can be sentential (based on natural language), model-based (that is sensory-related, formed for example by visualizations, analogies, thought experiments, etc.) or hybrid (a mixture of the two aspects above together with various manipulations of the world) [Magnani, 2001b].

our disposal which can shed light on Jung's considerations illustrated in "On Psychic Energy". The manipulation of external mediators (and of the related external representations) is active at the level of scientific discovery but it can be viewed as a general more or less creative way of producing/extending cognition, in the widest sense of the term, also involving emotional, affective, attentional, doxastic, and other aspects. Jung acknowledges this fact: "The apperceiving consciousness has proved capable of a high degree of development, and constructs instruments with the help of which our range of seeing and hearing has been extended by many octaves. Consequently the postulated reality of the phenomenal world as well as the subjective world of consciousness has undergone an unparalleled expansion" [Jung, 1972b, p. 228].

Furthermore, the creative construction of artifacts as cognitive/affective mediators can be seen in the framework of the "psychic energy" flow: the secret of cultural development is the *mobility and disposability of psychic energy* [Jung, 1967], that appears – as a true "life-process" – in phenomena like "instincts, wishing, willing, affect, attention capacity to work", sexuality, morality, etc.[93] This "psychic energy flow" is at the core of the various possible processes of progression and regression as adaptations to the environment and to the inner world, as key tools able to satisfy the demand of individuation.[94] Psychic energy and its capacity to be extended to the external world is appropriately compared by Jung to the fundamental primitive idea of *mana*, in its capacity to externalize and delegate meanings, and potentially consume everything.

In this perspective "values" can be considered quantitative estimates of energy that people can attribute to external things in various ways. For example, in the case of ethics, the recent tradition of moral philosophy classifies things that are endowed with values attributed by humans as endowed with "intrinsic values", a mechanism which I have illustrated in detail in a recent book on morality in our technological world [Magnani, 2007d]. These evaluations are of course attributed by consciousness, but also by the unconscious [Jung, 1972c, p. 10].

Once externalized and stabilized, values are available to be picked up. From this perspective the constellations of psychic elements grouped around feeling toned contents (complexes) are related to both inner experience (intertwined with innate aspects of an individual's character and dispositions) and suitably sensory *external environment* representations picked up when available (*ibid*, p. 11). These constellations centrally relate to both conscious and unconscious value quantities (affective intensities). Moral and other values (sexual, esthetic, rational, etc.) externalized in artifacts, icons, etc. normally belong to an established collective framework, shared by a relatively stable group of human beings: of course "repression" or "displacement of affect" can give rise to false estimates so that "subjective evaluation is

[93] It has to be noted that this mobility and disposability of energy is granted by the fact that Jung contends that man, more than other animals, "possesses a relative surplus of energy that is capable of application apart from the natural flow" [Jung, 1972a, p. 47].

[94] In "civilized man" psychic energy is strongly canalized in that rationalism of consciousness, "otherwise so useful to him", but which in turn can become a possible obstacle to the "frictionless transformations of energy" [Jung, 1972a, p. 25].

therefore completely out of the question in estimating unconscious value intensities" (*ibid.* p. 10), in turn giving rise to various disequilibria.

3.8.3 Artifacts as Memory Mediators

An example of psychic energy delegation presented by Jung, that clearly expresses human culture as a "transformer of energy"[95] is the one related to an artifact build by the Wachandi of Australia:

> They dig a hole in the ground, oval in shape and set about with bushes so that it looks like a woman's genitals. Then they dance round this hole, holding their spears in front of them in imitation of an erect penis. As they dance round, they thrust their spears into the hole, shouting "Pulli nira, pulli nira, wataka!" (non fossa, non fossa, sed cunnus!). During the ceremony none of the participants is allowed to look at a woman [Jung, 1972a, p. 43].

It is a process of semiotic delegation of meanings to an external natural object – the ground, which applies energy for special purposes through the building of a mimetic artifact: an "analogue of the object of instinct", Jung says [1972a, p. 42]. The artifact is an analogue of female genitals, an "object of natural instinct", that through the reiterated dance, in turn mimicking the sexual act, suggests that the hole is in reality a vulva. This artifact makes possible and promotes the related inferential cognitive processes of the rite. Once the representations at play are externalized (representations which are psychic values, from the Junghian psychoanalytic perspective), they can be picked up in a sensory way (and so learnt) by other individuals not previously involved in its construction. They can in turn manipulate and reinternalize the meanings semiotically embedded in the artifact:

> The mind then busies itself with the earth, and in turn is affected by it, so that there is the possibility and even a probability that man will give it his attention, which is the psychological prerequisite for cultivation. Agriculture did in fact arise, though not exclusively, from the formation of sexual analogies [Jung, 1972a, p. 43].

Artifacts are produced by individuals and/or small groups and left over there, in the environment, perceivable, sharable, and more or less available. It is in this sense that we can classify an artifactual mediator of this psychoanalytic type as a *memory mediator* (as a kind of "memory store", in Leyton's sense),[96] which mediates and

[95] Animals also make artifacts, and Jung is aware of this fact when he acknowledges the role of what he calls "natural culture": "When the beaver fells trees and dams up a river, this is a performance conditioned by its differentiation. Its differentiation is a product of what one might call 'natural culture', which functions as a transformer of energy" [Jung, 1972a, p. 42].

[96] [Leyton, 1999; Leyton, 2001] introduces this concept in a very interesting new geometry where forms are no longer memoryless as in classical approaches such as the Euclidean and the Kleinian one, in terms of groups of transformations. From this mathematical perspective artifacts, in so far as they are expressed through icons, visual and other non linguistic configurations, are "memory stores" in themselves [Leyton, 2006]. Of course in our case memory has to be intended in an extended Junghian sense, going beyond the explicit, linguistic, or model-based aspects which are the main focus of the recent tradition of cognitive science: specific implicit structures are also at play.

makes available the *story* of its origin and the *actions* related to it, which can be learnt and/or re-activated when needed.[97]

Let us come back to the artifact above. Primitive minds are not a "natural home" (cf. above, sections 3.3.3 and 3.6.6) for thinking on agriculture: together with the cognitive externalization and the artifact – and the subsequent recapitulations – certain actions can be triggered, actions that otherwise would have been impossible with only the help of the simple available "internal" resources. The whole process actualizes an example of that manipulative abduction I have described above. When created for the first time it is a creative social process, however, when meanings are subsequently picked up through the process involving the symbolic genital artifact and suitably reproduced, it is no longer creative, at least from the collective point of view, but it can still be creative from the perspective of individuals' new cognitive achievements and learning. It is possible to infer (abduce) from the artifacts the events and meanings that generated them, and thus the clear and reliable cognitive hypotheses which can in turn trigger related motor responses. They yield information about the past, being equivalent to the story they have undergone. In terms of Gibsonian [Gibson, 1979] affordances we can say that artifacts as memory mediators – as reliable "external anchors" – afford the subject in terms of energy stimuli transduced by sensorial systems, so *maximing abducibility* (they maximize "recoverability" in Leyton's sense – cf. the following subsection) and actively providing humans with new, often unexpected, opportunities for both "psychic" and "motor" actions.

3.8.3.1 Speech as an Artifact and Hybrid Mediators

Speech can also be seen as an external artifactual tool in so far as it consists in the outward flow of thoughts formulated for communication [Jung, 1967]. I have illustrated in sections 3.4.2 and 3.6.5 that speech/language is able to "scaffold" thoughts that are externalized and open to the interplay of communication. Through this artifactual interplay new meanings can arise usually shaped by (and embedded in) suitable narratives. Often speech (and related narrative) is hybridized with suitable manipulations of other external iconic aspects, such as drawings and various representations contained in suitable material artifacts. For example, in the interplay of the communicative environment which characterizes the psychoanalytic therapeutic setting – speech (together with the help of the other "tools" described above) grants "explanations" of patient's events (internal, external – and "externalized" –, hybrid), also thanks to the mediating empathic endowments of the analyst.

[97] It is interesting to note that [Turner, 2005] identifies a range of "affordances" offered by a variety of mediating artifacts including the life stories of recovering alcoholics in AA meeting (affording rehabilitation), patients' charts in a hospital setting (affording access to a patient's medical history), poker chips (affording gambling) and "sexy" clothing (affording gender stereotyping) [Cole, 1996]. In this perspective mediating artifacts embody their own "developmental histories" which is a reflection of their use. I will illustrate in details the relationships between abduction and affordances in chapter six.

In "The Transcendent Function" Jung says that "visual types" will be more inclined to look for inner images and further analyze their externalization in material supports. "Audio-verbal types" hear inner words and will note these down in writing. Again, others will directly use "hands" (or, more rarely, bodily movements) to spontaneously build artifacts (for instance "plastic materials") of various kinds to try to give "expression to the contents of the unconscious" [Jung, 1972c, pp. 83–84]:

> Often it is necessary to clarify a vague content by giving it a visible form. This can be done by drawing, painting, or modeling. Often the hands know how to solve the riddle with which the intellect has wrestled in vain [Jung, 1972c, pp. 83–86].

3.8.4 Artifacts as Symbols That Maximize Abducibility

3.8.4.1 Symbols as Memory Mediators

I have contended that a primitive mind is unlikely to have a natural home for complicated concepts like those of agriculture, because such concepts do not exist in a definite way in the natural (not artificially manipulated) world. Jung says that in these cases we aim at doing something that "exceeds our powers" [Jung, 1972a, p. 45]: humans always resorted to "external" magical formalities and religious ceremonies, which can release deep emotion and cognitive forces. In other words, whereas primitive minds could construct knowledge about human and animal genitals and reproduction, and perform some trivial inferences about them in a more or less tacit way by exploiting modules shaped by natural selection, how could they think of exploiting more explicit, sophisticated concepts involving agriculture? It is necessary to "disembody" the mind, and after having built an artifact through the hybrid internal/external interplay, to pick the new meanings up, once they are available over there, like the Wachandi did with the help of the hole in the ground.

A primitive mind consisting in different separated implicit templates of thinking and modes of inferences exemplified in various exemplars – for example expressed through natural language – and merely shaped by natural selection, cannot come up with certain agricultural entities without the help of external representations. The only way is to extend the mind into the material/artifactual world, exploiting the ground, tools and bodily movements which are suitably enriched through cognitive delegations, to provide *semiotic anchors* for finding ways of inferring that have no natural home within the mind, that is for finding ways of thinking that take humans beyond those that natural selection and cultural training could enable us to possess at a certain moment.

The activity of delegation to external objects of cognitive values through the construction of artifacts is certainly semiotic in itself (and of course a diversion of libido/psychic energy, in Junghian terms), the result is the *emergence* of new intrinsic meanings, expressed by what Jung calls a *symbol.* It is to be recalled that artifacts are the fruit of the hybridization of both internal and external constraints. First of all this result expresses the "quality" of the cognitive aspects – in our example above – delegated by Whacandi's "minds" to the external materiality, which gives birth to the hybrid interplay. Second, it expresses the particular cognitive "reactions" triggered in other individuals by the materiality at hand (that specific ground, the tools, the shapes made possible by the specific bodies of the dancers, etc.).

Jung also nicely stresses the protoepistemic rôle that can be played by magical artifactual externalizations in creative reasoning, and he is aware that these magical externalizations constitute the ancestors of the scientific artifacts, like those – mainly explicit – I have described in my book [Magnani, 2001c] concerning the discovery of new geometrical properties through external diagrams: Jung says "Through a sustained playful interest in the object, a man may make all sorts of discoveries about it which would otherwise have escaped him. [...] Not for nothing is magic called the 'mother of science'" [Jung, 1972a, p. 46]. Alchemy, which always provided external symbolism related to "flows of energy", furnishes plenty of examples, which support this conviction.[98]

Progressively, what possible meaning that can be seen and learnt through the Whacandi artifact and the related rite can become completely internalized and fixed so that referral to this externality – and learning from it – is no longer needed. Once internalized, the knowledge and the templates of action are already available at the brain level of suitably trained neural networks with their electrical and chemical pathways. When fixed and internalized they provide an immediate and ready "disposable energy": for example "We no longer need magical dances to make us 'strong' for whatever we want to do, at least not in ordinary cases" [Jung, 1972a, p. 45].

The semiotic process of externalization leads to the formation of a new meaning, which, as I have already said, Jung calls a symbol:[99] "The Wachandi's hole in the earth is not a sign for the genitals of a woman, but a symbol that stands for the idea of the earth woman who is to be made fruitful" [Jung, 1972a, p. 45]. The artifact is a symbol, formed in the hybrid interplay between internal and external semiotic representations, and furnishes a "working potential in relation to the psyche" [Jung, 1972a, p. 46]. Another example is given by "[...] those South American rock-drawings which consist of furrows deeply engraved in the hard stone. They were made by the Indians playfully retracting the furrow again and again with stones, over hundreds of years. The content of the drawings is difficult to interpret, but the activity bound up with them is incomparably more significant" (*ibid.*).

3.8.4.2 Abducibility Maximization through Symbols

Through artifacts, the natural niche is transformed in a "cognitive niche"[100] by human consciousness and unconscious: symbols are provided as "libido analogues"

[98] Scientific historians have clearly illustrated the importance of Newton's early research on alchemy in the origin of classical Newtonian physics and of the concept of action-at-the-distance [Dobbs, 1983], cf. chapter two of this book, section 2.3.

[99] The Junghian concept of a symbol is a little different from the one used in semiotics: it is of course richer in psychoanalytic value. However, it has to be noted that also for Peirce's semiotics a symbol is just a kind of iconic sign, which is conventional and fixed, like the ones used in mathematics and logic, unlike generic signs, that are generally arbitrary and flexible.

[100] Cognitive niches are fully described in chapter six, subsection 6.1.3.

that convert energy "ad infinitum" [Jung, 1972a, pp. 49–50], and thus cognition and culture. It is clear that symbols are artifacts that exhibit a *maximization of abducibility* ([Leyton, 1999; Leyton, 2001] speaks of the"maximization of recoverability") regarding past history, that is, of all the events that originate them: they are collective, stable, more or less available and sharable, related to both unconscious and conscious dimensions, firm anchors for thinking and for triggering action, which escape the fleeting nature of internal subjective representations.

The maximization of recoverability of symbols clearly explains their *abductive* force in the analytic treatment. Symbolic artifacts made by individuals under treatment are obviously connected to the objective (already) externalized ones – which are certainly more or less conscious and explicit in individuals – but are clearly provided (created) through the long history of humanity. Individuals' symbols share their archetypal primordial origin with these. Through suitable manipulations of their external iconic character, together with the help of the externalization of speech and affective signs in the communicative mediating environment of the therapeutic setting, the constructed symbols can grant "explanations" of various psychic events (based on internal, external, and hybrid representations). These explanations can emerge thanks to the fact that symbols are memory mediators and, moreover, maximize abducibility (recoverability) of their past history, that is of all the psychic events that originate them. Of course the explanations found convert psychic energy flows (information, both cognitive and affective) by themselves and so activate and further reshape the processes of progression and regression, which the fundamental therapeutical exigency of Junghian "individuation" requires. Symbols maximize abducibility because they maximize recoverability, they mean "the best possible expression" of something not yet grasped by consciousness.

Re-stating Jung's views in the light of abduction and distributed cognition and taking advantage of Leyton's ideas has offered the new perspective concerning external artifactual symbols as abducibility maximizers: a new notion I consider both provocative and promising. Some external symbols of the right sort are optimal for the storage of semiotic meanings and their participation in thinking, like in the case of psychoanalytic setting. However, an artifact (or any machine) is optimal only against a set of assumptions about what alternatives are possible. I think this issue is worth to be further studied, both from cognitive and ethical point of view.

These "symbols" go beyond the already considerable abducibility/recoverability force of various semiotic (for instance iconic) externalizations related to more basic survival needs – also present in many animals – (like caches of food, hunting landmarks, etc.) that are more constrained, if compared to symbols, in their capacity to trigger actions by promptly recovering various information and skills. These last externalizations, often plastically shaped by learning and not a simple fruit of instinctual endowments, are more or less widespread in the human and non-human animals' collectives, but they are normally merely behavior oriented to the direct satisfaction of basic instincts, and thus they do not involve the broad cognitive role of symbols (in a Junghian sense). These artifacts basically play the role of strong cognitive remodeling of the human and animal niches and at the same time present a regulative function of higher cognitive skills.

To make a simple example, the huge success of some symbols of this kind, for instance the religious ones, can be certainly explained in terms of their capacity to maximize abducibility in a very extended domain of human "minds", even if endowed with various degrees of cognitive skills. They trigger good hypotheses/thoughts, but the "user" of those symbols available over there in the close environment is in general very passive. I guess the concept of "moral mediator" I have introduced in my book [Magnani, 2007d] can be exploited to the aim of clarifying the use of kinds of maximizing symbols especially in general social and collective situation where moral issues are at stake. Of course, a similar role is played by the epistemic mediators I have illustrated in chapters one and three, that maximize abducibility in smaller collectives, like the scientific ones: successful manipulative abduction reaches relevant hypotheses the agent – here active and interactive – has exploited suitable artifacts, external representations and cognitive delegations and a smart manipulations of them.

It may not be superfluous, at this point, to say a few words about the frequently heard objection that the constructive method is simply "suggestion". The method is based, rather, on evaluating the symbol (i.e. dream-image or fantasy) not merely *semiotically*, as a sign for elementary instinctual processes, but *symbolically* in the true sense, the word "symbol" being taken to mean the best possible expression for a complex fact not yet clearly apprehended by consciousness [Jung, 1972c, p. 75].

We can clearly acknowledge the fact that the inferential process described in the passage above is in itself an explanatory abduction, because the problem, in front of the available data which constitute the "unconscious material" [Jung, 1972c, p. 77], is that of creating or selecting the suitable symbol, among the several available, and then appropriately managing it, with the help of the construction of explanatory narratives, with the aim of correctly reverting the psychic energy of both the analysed and the analyst.

The Junghian focus on symbolic artifacts called mandala, made by patients, not merely based on the tradition of religious ones, but free creations determined by certain archetypical ideas unknown to their creators, is related to the problem of "seeing" for example stages of individuation, where step by step patients "give a mind to that part of the personality which has remained behind" [Jung, 1968c, p. 350]. They are presented as "ideograms" of unconscious contents. In this case the hybrid character of the process is at stake, where the (relative) *autonomous* role of the external materiality is patent: "[...] one can paint very complicated pictures without having the least idea of their real meaning. While painting them, the picture seems to develop out of itself and often in opposition to one's conscious intentions" [Jung, 1968c, p. 352]. They function as "magical" meanings related to the collective unconscious, like "icons, whose possible efficacy was never consciously felt by the patient" [Jung, 1968b, p. 361].

For Jung "the making of a religion" is a primary interest of the primitive mind and strongly relates to the production of symbols, which "enable man to set up a spiritual counterpole to its primitive instinctual nature" [Jung, 1972a, p. 59], like already Vico clearly stated. For Vico, the most "savage, wild, and monstrous men" did not lack a "notion of God," for a man of that sort, who has "fallen into despair of all the succours of nature, desires something superior to save him" [Vico, 1968, 339, p. 100]. This desire led those "monstrous men" to invent the idea of God as a protective and salvific agent outside themselves; this shift engendered the first rough concept of an external world, one with distinctions and choices and thus established conditions for the possibility of free will. In the mythical story, the idea of God supplies the first instance of "elbow room" for free will illustrated by [Dennett, 1984]: through God, men can "hold in check the motions impressed on the mind[101] by the body" and become "wise" and "civil."

According to Vico, it is God that gives men the "conatus" of consciousness and free will:

> [...] these first men, who later became the princes of the gentile nations, must have done their thinking under the strong impulsion of violent passions, as beasts do. We must therefore proceed from a vulgar metaphysics, such as we shall find the theology of the poets to have been, and seek by its aid that frightful thought of some divinity which imposed form and measure on the bestial passions of these lost men and thus transformed them into human passions. From this thought must have sprung the conatus proper to the human will, to hold in check the motions impressed on the mind by the body, so as either to quiet them altogether, as becomes the wise man, or at least to direct them to better use, as becomes the civil man. This control over the motion of their bodies is certainly an effect of the freedom of human choice, and thus of free will, which is the home and seat of all the virtues, and among the others of justice. When informed by justice, the will is the fount of all that is just and of all the laws dictated by justice. But to impute conatus to bodies is as much as to impute to them freedom to regulate their motions, whereas all bodies are by nature necessary agents (*ibid.*, 340, p. 101.)

Free will, then, leads to "family": "Moral virtue began, as it must, from conatus. For the giants, enchanted under the mountains by the frightful religion of the thunderbolts, learned to check their bestial habits of wandering wild through the great forests of the earth, and acquired the contrary custom of remaining hidden and settled in their fields. [...] And hence came Jove's title of stayer or establisher. With this conatus, the virtue of the spirit began likewise to show itself to them, restraining their bestial lust from finding satisfaction in the sight of heaven, of which they had a mortal terror" (*ibid.*, 504, p. 171.) "The new direction took the form of forcibly seizing their women, who were naturally shy and unruly, dragging them into their caves, and, in order to have intercourse with them, keeping them there as perpetual lifelong companions" (*ibid.*, 1098, p. 420.).

[101] Indeed "That is, the human mind does not understand anything of which it has had no previous impression [...] from the senses" (*ibid.*, 363, p. 110). "And human nature, so far as it is like that of animals, carries with it this property, that senses are its sole way of knowing things" (*ibid.*, 374, p. 116). Again, humans "in their robust ignorance" know things "by virtue of a wholly corporeal imagination" (*ibid.*, 376, p. 117). Aristotle had already contended that "nihil est in intellectu quod prius non fuerit in sensu."

Summary

The main thesis of this chapter is that the externalization/disembodiment of mind is a significant cognitive perspective able to unveil some basic features of creative abductive thinking and its cognitive and computational problems. Its fruitfulness in explaining the semiotic interplay between internal and external levels of cognition is evident and was traced back to some seminal thoughts Turing provided about what he called the transition from "unorganized" to "organized" brains. I maintained that various aspects of creative meaning formation could take advantage of the research on this interplay: for instance study of external mediators can provide a better understanding of the processes of explanation and discovery in science and in some areas of artificial intelligence related to mechanizing discovery processes.[102]

From the paleoanthropological perspective we have learnt that an evolved mind is unlikely to have a *natural home* for new concepts and meanings, as such concepts and meanings do not exist in the artificial and natural world as it is already known. Analogously, from this perspective, we have seen how the cognitive referral to the central role of the relation between meaningful behavior and dynamical interactions with the environment becomes critical to the problem of modeling up-to-date artificial systems devoted to performing creative and explanatory tasks: I contend that the epistemological role of those artifacts, such as computers, which I called "mimetic minds", can be further studied, taking advantage of research on hypercomputation. The imminent construction of new types of universal "abstract" and "practical" machines will constitute important and interesting new "mimetic minds" externalized and available over there, in the environment, as sources of the mechanisms underlying the emergence of new meaning processes. They will provide new tools for creating meaning in classical areas like analogical, visual, and spatial inferences, both in science and everyday situations, thereby extending the epistemological and psychological theory.

Finally, the externalization/disembodiment of mind is a significant cognitive perspective able to unveil some aspects of creative meaning formation central to psychoanalytic research and therapy. I have highlighted some Junghian analysis regarding the role of certain external artifacts where the mobility and disposability of psychic energy are seen as the secret of cultural development both at the collective and individual level. I have contended that symbols, in a psychoanalytic sense, are artifacts/tools that maximize abducibility, because they maximize the recoverability of something hidden, not yet grasped by consciousness. The new concept of maximization of abducibility promises to stimulate further research about the relationships between abduction and external mediators.

[102] On recent achievements in the area of machine discovery simulations of model-based creative tasks cf. [Magnani *et al.*, 2002a]. Cf. also section 2.7 of chapter two.

Chapter 4
Neuro-multimodal Abduction
Pre-wired Brains, Embodiment, Neurospaces

In chapter three I have illustrated the main features of the so-called disembodiment of mind from the point of view of the cognitive interplay between internal and external representations, where the problem of the continuous interaction between on-line and off-line intelligence can be properly addressed. I consider this interplay critical in analyzing the relation between meaningful semiotic internal resources and devices and their dynamical contact with the externalized semiotic materiality already embedded in the artificialized environment. Hence, minds are "extended" and artificial in themselves. It is from this distributed perspective that I will further stress how abduction is essentially *multimodal*, in that both data and hypotheses can have a full range of verbal and sensory representations, involving words, sights, images, smells, etc., but also kinesthetic experiences and other feelings such as pain, and thus all sensory modalities. The presence of kinesthetic aspects plainly demonstrates that abductive reasoning is basically manipulative. Again, both linguistic and non linguistic signs have an intrinsic semiotic life, as particular configurations of neural networks and chemical distributions (and in terms of their transformations) at the level of human brains, and as somatic expressions. However they can also be delegated to many external objects and devices, for example written texts, diagrams, artifacts, etc. We can also see, in this regard, how unconscious factors take part in the abductive procedure, which consequently acquires the character of a kind of "thinking through doing".

Hence, abductive cognition is occurring in a "distributed" framework and in a hybrid way, that is, in the interplay between internal and external signs. First of all we can say that all representations are brain structures and that abduction certainly is a neural process in terms of transformations of neural representations. We can reconceptualize abduction neurologically as a process in which one neural structure representing the explanatory (or non-explanatory and instrumental) target generates another neural structure that constitutes a hypothesis. Some fMRI studies on problem solving where the insight involved is more or less creative and accompanied by Aha! experience and on non-insight based problem solving interestingly show the different brain areas involved and the role played by emotional aspects. The neuro-multimodal perspective also clearly demonstrates how the classical view of

L. Magnani: Abductive Cognition, COSMOS 3, pp. 219–264.
springerlink.com

abduction based on logic (from the classical Peircean syllogistic model to the more recent non-standard deductive models) only captures limited properties of this cognitive process (fallacious vs. non-classical truth-value preservation, stabilization in axiomatic theories of standard modalities of inferring, etc.). Indeed, from the neuro-multimodal perspective even propositions "are" internal neural structures consisting of neural connections and spiking behaviors. Of course, I repeat, propositions and other semiotic aggregates, according to this externalist view, can also be represented in external supports, where they have to obey the proper constraints of the materiality at play.

Furthermore, cognitive abductive performances are to a large extent neurally hardwired: "mind" results from the rigid execution of the DNA program. When based on merely pre-wired genetic endowments, however, this execution interacts with the world and the phenotype is basically a creation of both genotype and environment. Thus, organisms are equipped with various ontogenetic mechanisms that permit them to acquire information and better adapt to the environment. For instance, the immune system in vertebrates and brain-based learning in animals and humans, which are mechanisms characterized by plasticity. I also describe how some well-known research on neurons, which in primates and humans react to observation of actions, raises interesting questions concerning the so-called "embodied cognition" and is able to shed light on the role of abduction in guessing intentions, in mentation and metamentation, in affective attunement and in other emotional and empathic appraisals.

The main body of the chapter illustrates that "abduction" is central to understanding some features of action and decision making. Abduction prompts action and plays a key role in decision making. Peirce teaches that the neurological perspective, depicted in this chapter, also increases knowledge about the distinction between thought and motor action, seeing both aspects as fruit of brain activity. We can say that thought possesses an essential "motoric" component reflected in brain action but not in actual movement. On the basis of this analysis I can further illustrate some problems related to the role of abduction in decision-making, both in deliberate and unconscious cases, and its relationship with both hardwired and trained emotions. I also contend that ethical deliberation, as a form of practical reasoning, shares many aspects with hypothetical explanatory reasoning (selection and creation of hypothesis, inference to the best explanation) as it is described by abductive reasoning in science. Of course in the moral case we have reasons that support conclusions instead of explanations that account for data, like in epistemological settings. To support this perspective, I propose a novel analysis of the "logical structure of reasons", which supports the thesis that we can look to scientific thinking and problem solving for models of practical reasoning. The distinction between "internal" and "external" reasons, originally proposed by Searle, is fundamental: internal reasons are based on a desire or on an intention, whereas external reasons are, for instance, based on external obligations and duties which we can possibly recognize as such. Some of these external reasons can be grounded in epistemic mediators of various types. Finally, it is important to illustrate why it is difficult to "deductively" grasp practical reasoning, at least when we are aided only by classical logic;

complications arise from the intrinsic multiplicity of possible reasons and from the fact that in practical reasoning we can often hold two or more inconsistent reasons at the same time.

The third and last part of the chapter, on "Spatial Frameworks, Anticipation, and Geometry", fulfils the exigence of presenting how abduction is at the basis of mammalian cognitive systems representing the location of objects within the spatial framework. Various aspects of this fundamental abductive activity are illustrated:

1. the concepts of spatial egocentric and allocentric mapping and its plasticity,
2. the role of the hippocampus in performing internal subsequent representations such as abductive hypotheses of space,
3. the discovery of a neuronal spatial map in the medial dorsocaudal entorhinal cortex, anchored by external landmarks, which constitutes the basic spatial input exploited by the hippocampus to construct more specific and context-dependent spatial firing in its place cells.

Related to the cognitive problem of spatiality and of the genesis of space is the description of the abductive role of the *Abschattungen* (adumbrations), as they are described in the framework of the philosophical tradition of phenomenology. The adumbrations naturally lead to the analysis of the so-called "anticipations", which share various features with visual and manipulative abduction, such as the way that they are conjectural and nonmonotonic, which means that incorrect anticipations have to be replaced by more plausible ones. The problem of anticipation is further complicated in the framework of the so-called emulation theory, which contends that emulation circuits are able "to hypothesize" any forward mapping from control signals to the anticipated – and so abduced – consequences of executing the control command. They "mimic" the body and its interaction with the environment enhancing motor control through sensorimotor abductive hypotheticals, and nicely explain the emergence of implicit and explicit agency. The abductive character of the concept of anticipation will be further reworked in chapter eight (subsection 8.2.1) linking it to the concept of attractor of the dynamical system theory.

4.1 Multimodal Abduction

As I have already stressed, Peirce considers inferential any cognitive activity whatever, not only conscious abstract thought; he also includes perceptual knowledge and subconscious cognitive activity. For instance in subconscious mental activities visual representations play an immediate role.[1] Many commentators criticized this Peircean ambiguity in treating abduction at the same time as inference and perception. It is important to clarify this problem, because perception and imagery are kinds of that model-based cognition which we are exploiting to explain abduction: I contend that we can render consistent the two views [Magnani, 2006c], beyond Peirce, but perhaps also within the Peircean texts, partially taking advantage of the

[1] Cf. [Queiroz and Merrell, 2005].

concept of *multimodal abduction*, which depicts hybrid aspects of abductive reasoning. [Thagard, 2005; Thagard, 2007] observes, that abductive inference can be visual as well as verbal, and consequently acknowledges the sentential, model-based, and manipulative nature of abduction I have illustrated above. Moreover, both data and hypotheses can be visually represented:

> For example, when I see a scratch along the side of my car, I can generate the mental image of a grocery cart sliding into the car and producing the scratch. In this case both the target (the scratch) and the hypothesis (the collision) are visually represented. [...] It is an interesting question whether hypotheses can be represented using all sensory modalities. For vision the answer is obvious, as images and diagrams can clearly be used to represent events and structures that have causal effects.

Indeed hypotheses can be also represented using other sensory modalities:

> I may recoil because something I touch feels slimy, or jump because of a loud noise, or frown because of a rotten smell, or gag because something tastes too salty. Hence in explaining my own behavior my mental image of the full range of examples of sensory experiences may have causal significance. Applying such explanations of the behavior of others requires projecting onto them the possession of sensory experiences that I think are like the ones that I have in similar situations. [...] Empathy works the same way, when I explain people's behavior in a particular situation by inferring that they are having the same kind of emotional experience that I have in similar situations [Thagard, 2007].

Thagard illustrates the case in which a colleague with a recently rejected manuscript is frowning: other colleagues can empathize by remembering how annoyed they felt in the same circumstances, projecting a mental image onto the colleague that is a non-verbal representation able to explain the frown. Of course a verbal explanation can be added, but this just complements the empathetic one. It is in this sense that Thagard concludes that abduction can be fully multimodal, in that both data and hypotheses can have a full range of verbal and sensory representations.

Thagard also insists on the fact that Peirce noticed that abduction often begins with puzzlement, but philosophers rarely acknowledged the emotional character of this moment: so not only is emotion an abduction itself, like Peirce maintains, but it is at the starting point of most abductive processes, no less than at the end, when a kind of positive emotional satisfaction is experienced by humans.

4.2 Neuroabduction: Internal and External Semiotic Carriers

Some basic aspects of this constitutive *hybrid* nature of multimodal abduction – involving words, sights, images, smells, etc. but also kinesthetic experiences and other feelings such as pain, and thus all sensorimotor modalities, clearly show the usefulness of accounting for all this semiotic activity from a neurological perspective. A neural structure can be seen as a set of neurons, connections, and spiking behaviors, and their interplay, and the behavior of neurons as patterns of activation, like maintained by the connectionist tradition, also endowed with an important exchange of chemical information. In this perspective it is clear we can say all representations are brain structures and abduction is a neural process in terms of transformations of neural representations. Thagard concludes: "Hence we can reconceptualize abduction neurologically as a process in which one neural structure representing the explanatory target generates another neural structure that constitutes a hypothesis" [Thagard, 2007].[2]

As I have already said, it is important to note that, in both humans and other vertebrates, neurons fire and provide electrical inputs to other neurons suitably connected by excitatory and inhibitory links, as illustrated in the connectionist neural-network models of brain. However, the direct effects of real neurons on each other are chemical rather than electrical, in that various molecules ("neurotransmitters", with excitatory and inhibitory effects) are emitted from one neuron and then passed to another neuron, where they trigger chemical reactions that generate the electrical functions of the stimulated neuron. Neural network are Turing-complete, that is they can compute any function that a Turing machine can, but of course they are also directly – physically – responsible for the organisms' cognitive capabilities.

The chemical messengers also include hormones and other molecules, and this neurochemistry is certainly central: it involves the action of the so-called "neuromodulators", which can also travel to parts of the body such as the adrenal glands, which in turn release other hormones that travel back to the brain and influence

[2] Recent research depict the localization in human brain of different abductive performances that emphasize some aspects of this multimodality at least from the neurological point of view. [Jung-Beeman *et al.*, 2004; Bowden and Jung-Beeman, 2003] have provided a neurological study on insight accompanied by Aha! experience in various problem solving abductive settings in their relationships with unconscious processing. The research depicts the different cognitive and neural processes that lead to insight vs. non insight solutions. There is different hemispheric involvement in subjects that solve verbal problems (recognizing distant or novel semantic – or associative – relations, extracting themes, comprehending indirect language such as jokes, metaphors and unconnected discourse, forming coherent memories for stories). fMRI data reveal increased activity in the right hemisphere (RH) anterior superior gyrus (aSTG) for insight relative to non insight solutions. The same area is involved during the initial effort; usually people cannot report the processing that enabled them to restructure the problem and to overcome the impasse. Scalp EEG recordings revealed a sudden burst of high-frequency (gamma-band) neural activity in the same area beginning 0,3s prior to insight solutions. This right anterior temporal area is associated with making connections across distantly related information during comprehension. Of course all problem solving relies on a largely shared cortical network, but the sudden flash occurs when the agents exploit distinct neural and cognitive processes that allow them to abductively envisage connections that previously eluded them. The researchers also detected an insight effect in small clusters in or near bilateral amygdala or parahippocampal gyrus, that is plausible if memory interacts with insight solutions differently from how it interacts with noninsight solutions. The authors are not arguing that the left hemisphere (LH) solves noninsight problems, or that the LH is conscious and the RH unconscious: it actually seems that people make conscious decisions and reach explicit results influenced by partially independent activation in each hemisphere of a population of many thousands of neurons, without the need of an executive "homunculus or grandmother cell" [Bowden and Jung-Beeman, 2003, p. 736]. This population is divided among the two hemispheres which work through distant mechanisms (LH → semantic coding, RH → coarse novel semantic coding). Other cortical areas such as the prefrontal cortes and the anterior cingulate (AC) may also be involved in this process of providing insight and noninsight solutions.

the firing of the appropriate neurons. Of course we can agree with Thagard on the fact that these neurochemical mechanisms can further clarify deep aspects of cognition, like emotions, problem solving, and decision-making, which can only be completely understood with difficulty through the traditional connectionist neural-network models. Hence, most synapses are chemical, neurotransmitter pathways that "[...] provide the brain with a kind of organization that is useful to accomplish different functions. [...] Neurotransmitters provide a course kind of wiring diagram, organizing general connections between areas of the brain that need to work together to produce appropriate reactions to different situations" [Thagard, 2002a] in a more flexible way.

Emotions can be easily thought of within this framework, they are neural structures that involve complicated interactions among sensory processes linking bodily and various other cognitive processes distributed in many brain areas. Let us see an example that comes form the area of scientific reasoning. In Thagard's terms introduced above, first of all a target meaningful assembly of possibly anomalous data (made available through the sensory processes), is internally emotionally transformed and marked as disturbing or puzzling. Through further synaptic connections it is coordinated to the spiking behavior of the emotional structures: in this way emotion furnishes immediate abductive appraisals of the bodily states, and provides a kind of "explanation" of them. Of course a marked emotional state can be coordinated to neural structures which in turn express other abductive verbal or sensory reactions, to which a further emotional experience of satisfaction or pleasure can in turn be associated. Also every unconscious activity of cognition can be reinterpreted in the meaningful terms of neural transformations, the study of which seems ultimately most approachable through new techniques and experimentation.

The neuro-multimodal perspective shows, resorting to basic biological levels, how the classical perspective of abduction based on logic (from the classical Peircean syllogistic model to the more recent non-standard deductive models – cf. [Magnani, 2007a]) only captures limited properties of this cognitive process (fallacious or truth-value preservation, stabilization in axiomatic theories of standard modalities of inferring, etc.). Indeed, in the neuro-multimodal perspective propositions too "are" internal neural structures consisting of neural connections and spiking behaviors. Consequently, beliefs and desires, that traditional philosophy interpreted as propositional attitudes, can be usefully seen as brain structures, and moreover, in this extended framework the concept of inference can be reinterpreted to encompass non-verbal representations from all sensory modalities and their hybrid combination, going beyond its merely logical meaning in terms of arguments on sentences.

Of course propositions, and other semiotic aggregates, can also be represented in *external* supports, where they have to obey the proper constraints of the materiality at play. In the previous chapter I have already illustrated that in the case of humans and other organisms, it is only through the interplay between internal and external representations that the many kinds of cognitive processes can work. Furthermore, an interplay between sentential/symbolic and model-based aspects is often at work in important cognitive processes as in the creative abductive ones. For example, when mathematicians mentally represent suitably composed or discovered sets of equations or symbolic structures to solve a problem, taking advantage of the mental recapitulations of visual images already eventually represented externally. Donald says

> [...] they cannot evaluate what they have done until they break the circle of symbols and relate them to other mental structures, usually visual images, that reside outside the equations themselves. Driven by a deeper intuition and a need for closure or differentiation, they evaluate the new symbolic expression and modify it out of necessity. The success of a truly new symbolic expression can therefore be judged only by part of the mind that intuits the successful clarification of its own inner state [Donald, 2001, p. 278].

Signs are everywhere in our human artificialized world, on blackboards, in books, on videos, as propositions, icons, symbols, configurations, diagrams, etc. But many kinds of signs are also expressed at the neural level. This means that semiotic minds/brains are not self-sufficient neural devices, like eyes, but the hybrid product of the brain-culture symbiosis [Donald, 2001, p. 202]. The most efficient move for individual human brains has been to become part of this external semiotic materiality – thus they can far exceed their capacity of coping with or remembering it – and enable themselves to evolve with it, tracking what is over there and fulfilling various degrees of need.

In this endeavor human beings always contribute to the modification of that external semiotic materiality, and in some cases favor its enhancement and creative change. The baby is already encapsulated in a world of model-based signs that operate at a pre-linguistic level, but very soon she can take advantage of subsequent encapsulation into a whole semiotic culture (also including language), mediated by parents, family, tribal customs or institutions, etc., that takes further control of his/her cognitive development.[3]

Starting from the "virtual" reality established at the evolutionary birth of human language, through the construction of an oral-mythic culture and scientific theories, right up until modern technological artifacts, which ideally continue the material culture of the hominids, human beings have always endowed themselves with powerful external devices able to overcome their brain biological limitations. These prostheses have enabled them to also build narratives that in turn have allowed the undertaking of big projects. In particular technological advances have provided external devices and artifacts, like computers, "specifically" designed to help us reason, remember, mime various internal representations and internally re-represent the previously externally stored representations. These are truly formidable devices,

[3] It is well-known that Vygotsky already focused the attention to this interplay between external and internal representations (and the process of re-internalization) in his pioneering studies on children's language acquisition (Outside/Inside principle) [Vygotsky, 1978; Vygotsky, 1986]. The developmental rule is that brain first represents external action and only later reconstructs it so that it will occur internally. On the problem of the so-called affective attunement in human infants cf. chapter five, subsection 5.7.3, on some suggestions coming from Thom's catastrophe theory on how natural syntactical language is seen in attunement of infants to mother's language cf. chapter eight, subsections 8.4.1 and 8.5.1.

which I have called in the previous chapter "mimetic minds" (section 3.7, cf. also [Magnani, 2006c]). At the same time these tools have produced the possibility to aggregate externalized thoughts in various ways, taking advantage of the external constraints of the electronic materiality at hand, of course different from the neural one available through brains.

Donald contends that this process is also manifest when we analyze the role played by memory media in the functioning of the human conscious mind: "The biological memory records, known as engrams, differ from the external symbols, or exograms, in most of their computational properties. [...] The conscious mind is thus sandwiched between two systems of representation, one stored inside the head and the other outside. [...] In this case, the conscious mind receives simultaneous displays from both working memory and the external memory field. Both displays remain distinct in the nervous system" [Donald, 2001, pp. 309–311].

4.3 Pre-wired Brains and Embodiment

4.3.1 The Pre-wired Brain

Cognitive performances are to a large extent neurally pre-specified. Research in genetics shows substantial evidence for genetic pre-wiring of a great deal of brain structure in both human and non-human animals. Important gene expression patterns of the brain occur before experience-dependent input. Hence, the brain is *pre-wired* even if not with total specificity, and with a great degree of chance involved.[4] Ramus says:

> Indeed whether a neurone will grow dendrites or produce a particular molecular cue at a given time depends on the expression of particular genes, which itself depends on many internal factors like which other genes the neurone currently expresses, as well as external factors like which molecules surround it and in what concentration (which can in particular specify the neurones' position within the brain and within its neural structure) [Ramus, 2006, p. 255].

[Baum, 2006, p. 336] observes, taking advantage of a computational metaphor, that "[...] mind results from the execution of the DNA program. But [...] this execution proceeds in interaction with the world. Culture is in a sense part of the working memory, the data space, used by this execution. [...] Evolution naturally discovers programs that interact with the environment to learn". [Marcus, 2004, p. 84] adds that "[...] the same genes that are used to adjust synapses based on internal instruction can be reused by external instruction". It is in this perspective that we can easily

[4] The term "wired" can be easily misunderstood. Generally speaking, I accept the distinction between cognitive aspects that are "hardwired" and those which are simply "pre-wired". By the former term I refer to those aspects of cognition which are fixed in advance and not modifiable. Conversely, the latter term refers to those abilities that are built-in prior the experience, but that are modifiable in later individual development and through the process of attunement to relevant environmental cues: the importance of development, and its relation with plasticity, is clearly captured thanks to the above distinction. Not all aspects of cognition are pre-determined by genes and hardwired components. Cf. also [Barrett and Kurzban, 2006].

understand why some authors consider a revolution in evolutionary biology "[...] to actually admit what geneticists and evolutionists have claimed all along without seeing: the phenotype is a creature of both genotype and environment. It is the mediator of all genetic and environmental influence, in both development and, as the object of selection, evolution" [West-Eberhard, 2003, p. 525].

Of course many neuroanatomical modules need considerable input from the external world (already at the level of utero biochemical factors), fine tuning, and experience-driven maintenance to get their – restructured – functional adult abilities, and also to preserve the structure itself. In this sense pre-wiring does not imply hardwiring, but the so-called "brain plasticity" endowments have limits and genome-directed constraints.[5] All environmental experience has to pass through the rigid features of sensory receptors imposed by the neural structure (especially in sub-cortical and sensory areas), and it is only in this way that it can be seen as the "execution" of a genetic program that codes for proteins: not everything is learnable.[6] In the framework of genetic influences on brain processes concepts like innateness, domain-specificity, and evolutionary selection, which are rather vague, could be reshaped in a more neurogenetical satisfactory scientific way, for instance taking advantage of research between genes and brain development, and its anomalies and cognitive dysfunctions.

Organisms are equipped with various ontogenetic mechanisms that permit them to acquire information and thus better adapt to the environment: for instance, the immune system in vertebrates and brain-based learning in animals and humans. Plasticity characterizes these mechanisms, which, West-Eberhard says, "mimic" selection: "Learning itself can be fine-tuned under selection and then can mimic selection to create, and rapidly spread, novel adaptive traits" [West-Eberhard, 2003, p. 337].[7] The role of these mechanisms is to provide organisms with supplementary devices to acquire information and thus afford various environmental contingencies that are not – and cannot be – specified at the genetic level [Odling-Smee *et al.*, 2003, p. 255]. A genetically specified initial set of behaviors is elaborated through experience of a relevant environment. These ontogenetic mechanisms are therefore a sort of *on-board* system allowing flexibility and plasticity of response to an ever-changing environment, which are at the core of the notion of cognition that is at the basis of our treatment [Godfrey-Smith, 2002].[8]

[5] On the so-called brain "cross-modal" plasticity, which deals with changes and reorganization of cortical functions in an individual brain in front of sensory deprivation cf. [Bavelier and Neville, 2002].

[6] Genetic control also operates throughout life and has on-line effects (for instance on coding of neurotransmitters, their receptors and other molecules involved in neurotransmission pathways) [Ramus, 2006, p. 258].

[7] On Edelman's neural Darwinism, which sees neurons as populations submitted to "loosely" Darwinian effects, cf. subsection 6.1.4, chapter six.

[8] Godfrey-Smith defines cognition as the capacity of coping with a range of possible behavioral options with different consequences for the organism's chance to survive. This definition allows him to embrace a broader notion of cognition which extends it to many animal and plant behaviors. I will discuss this thesis in chapter six, subsection 6.4.4 and in chapter eight 8.5.1, devoted to the interplay between abduction and affordances.

In the case of human beings and other mammals, bigger brains allow the storage of information which could not be pre-defined by genes [Aunger, 2002, pp. 182–193]. Flexibility and plasticity of response to an ever-changing environment are connected to the necessity of having other means for acquiring information, more readily and quickly than the genetic one. I posit that niche construction plays a fundamental role to in meeting this requirement. Plasticity and flexibility depend on niche construction as far as various organisms may alter local selective pressure via niche construction itself, and thus increase their chances of survival. More specifically, cognitive niches are crucial in developing ever more sophisticated forms of flexibility, because they constitute an additional source of information favoring behavior and development control. In this case, epigenesis is therefore augmented, and, at a genetic level, it is favored by genes regulating *epigenetic openness* [Sinha, 2006]. Epigenetic openness is closely related *phenotypic plasticity* [Godfrey-Smith, 2002]; the flexible response of living organisms (humans in particular) leans on sensitivity to environmental clues, and this process of attunement to relevant aspects of the environment cannot be separated from niche construction (cf. subsection 6.1.3, chapter six, on the interesting concept of *cognitive niche*).[9]

What are the consequences of the scientific results I have previously illustrated on the concept of abduction? We have to observe that of course all the abductive performances I have illustrated have neural correlates. Recent research in neurology has especially increased knowledge about cognitive skills in various organisms, mainly dealing with perception and motor control in non-human animals, and some important new scientific perspectives could be useful to further detail certain aspects – especially non-verbal – of abductive thinking which have already been emphasized in philosophical, logical and psychological investigations. In the following subsection I will describe how some well known research on neurons, which in primates and humans react to observation of actions, raise interesting questions concerning the so-called "embodied cognition".

4.3.2 Embodiment and Intentionality

Neurons in area F5 in the monkey ventral premotor cortex do not code elementary motions but goal-related actions [Rizzolatti *et al.*, 1988]. A subclass of them – the very famous mirror neurons – fire both when a monkey observes an action performed by another individual and when it executes the same or a similar action, like grasping, tearing, holding or manipulating objects. A subset of mirror neurons

[9] The relationships between developmental plasticity and evolution are described in [West-Eberhard, 2003]. For an illustration of the recent controversies and conceptual confusions about phenotypic plasticity and genetic assimilation cf. [Pigliucci *et al.*, 2006, p. 2366]: the authors contend that phenotypic plasticity is not a threat to Modern Synthesis but an expansion of it, it is (in part) a developmental process, not an evolutionary one. "As such, it can be the target of natural selection (an evolutionary mechanism, though of course not the only one), and yields – under certain conditions – the evolutionary outcome of genetic assimilation or phenotypic accommodation. Once one recognizes the clear hierarchical distinctions among these concepts, most fears about an imminent overthrow of the Modern Synthesis should dissipate."

respond not only when the monkey executes or observes an action but also when it hears that same action is performed by another agent. Neither the sight of the other agent alone nor the sight of the object alone are sufficient to fire the mirror neurons' response, as in the case of mimicking actions without a target object. Another subset of mirror neurons can generalize and react in the presence of various, similar actions and not only in the presence of a specific one, so they present a kind of inductive hypothesizing pre-wired character. There is also evidence in favor of their existence in humans.

The discovery of mirror neurons has given rise to speculation on various aspects of social cognition such as intentions, action, empathy, mind-reading, emergence of language, and the discussion has also included the problem of affective attunement that I have cited in the previous subsection. What it is important to note is that in the experiments on mirror neurons the interaction between agents, and the object of the action is independent of the self-other distinction. [Gallese, 2006] contends that the interaction would instead be "we-centric", and would also underpin the understanding of others as rudimentary and implicit "intentional" agents.

I am inclined to partially follow [Tummolini *et al.*, 2006], in saying that mirror neurons do not underpin a "representation" of the other agent, neither taking advantage of a self-representation nor of a representation in terms of a "we-agent". Indeed, frankly, I do not think the interaction can be considered "we-centric": how is it possible to conjecture a "we" if not in terms of an already hypostatized "me" and "you"? The problem is that there is neither the we-centric situation nor the self/other one, at least, not as they are understood in our ordinary folk psychology. The observed interaction in primates cannot say anything about the self/non-self distinction, it seems to be basically agentless. It can simply affect changes and transformations in the individual's behavior that we (as researchers) see. Moreover, if representations are at play, they are representations in a Picwickian sense, like the ones I will describe in the following chapter of this book, acting in many cases of animal behavior, that is, they are simply perceptual-, motor-, and behavior-directed and not mentalistic.

I have said that many researchers on mirror neurons hypothesize that the interaction at play would also underpin, in the case of children, the understanding of others as rudimentary and implicit "intentional" agents. Of course in the absence of the standard agentive and objective units it is difficult to hypothesize a kind of (implicit) intention attribution in the perspective of the standard psychological sense of the concept. Rather, in infants, only something similar to the case of some animals can be hypothesized where visual and motor "representations" carry perceptual representations among external events, so that actions are understood as motor processes that lead to certain outcomes, that is, like "self-propelled beings that make things happen" [Tummolini *et al.*, 2006, p. 106]. Of course this framework involves a kind of intentionality that cannot be described as fully psychological. In such a way, through the equivalence between actions perceived and actions performed, animals and infants can abductively form preferential hypotheses in terms of "intentions", which explain some perceptual events linking different motor schemata.

These hypotheses enable the organisms to predict the consequences of the actions perceived.[10]

It has also been contended that the process leading to the abduction of subsequent self-correcting "hypotheses" that in turn lead to affective attunement,[11] which is formed through the interplay of facial imitation (cf. [Meltzoff and Brooks, 2001]) is also at play and underpinned by mirror neurons. In affective attunement, the coordinated process of exchange of a full range of somatosensory perceptual information (visual, auditory, and tactile) can be seen in this light. A process in which the slow formation of a kind of "selfness", can be hypothesized, suitably accompanied by the first social feeling of an identity of "being-like-you" [Gallese, 2006].

This embodied and subpersonal way of "understanding" and feeling events is present also in adult humans. In this case it is either implicit, isolated, and basically unconscious, or accompanied by that effect of mentalization that leads to the interpretation of beings like full intentional agents, showing a movement from the attribution of a kind of external – "in the world" – intentionality to the agent-oriented intentionality of folk psychology. The shift is from "goals as relational structures in the world to goals as intentional relations between agents and the world" (*ibid.*), in terms of belief/desire and intentions mental/brain states. In this last case it is obvious to note that full representational mechanisms are needed, for example propositional ones, which go far beyond the basic visual and motor framework, and involve other regions of the brain, embracing the sensorimotor ones. I agree with [Gallese, 2006] that this distinction between "behavior readers", like non-human primates and other animals on the one hand and "mind readers", like humans on the other, does not have to be taken as a sharp distinction depicting a strong discontinuity, where the role of propositional attitudes in the formation of the self-other distinction is overstated.[12]

In the experiments on primates it can be clearly seen how information (model-based, visual or auditory), conveyed through perception from the outside, is stored through mirror neurons and coupled with them, and made available to enable suitable actions and new behaviors (i. e., appropriate to the ecological situation at stake). Mirror neurons are simulators of the planned action, used to predict its consequences and thus achieve a better control of the action itself: information is picked up outside and through their neural visual and motor processing, it is so to speak "socially" re-externalized in the action performed, which of course is endowed

[10] On the distinction between desires, beliefs, and intentions and on the various skillful abductive metarepresentational processes that are at play in human intention attribution – where intentions are intended as private or joint (in groups of people, the so-called "we-intentions") mental states of an agent – cf. the issues described in Malle, Moses, and Baldwin [2001]. The inferential process – based on both the observed behavior and external information (multi-sensory: visual-kinesthetic, auditory, etc.) and including other cues in the immediate context – involves an expert "explanatory" selective abduction that has to discriminate among a very large range of possible intentions that are consistent with a given action.

[11] Cf. chapter five of this book, subsection 5.7.3 and chapter eight, subsection 8.5.1.

[12] An analysis of the problem of embodiment in neuroscience, in traditional cognitive science and in some aspects of phenomenology is given in [Gallese, 2005]. Recent research suggests that both the understanding of intentionality in others – engagement in intentional exchanges, and understanding how the mind of others has been affected by the communicative act, is also present in people that lack language, such as deaf non signers [Donald, 2001, p. 144].

with a visual dynamic structure made available over there, in the environment, for other watchers and listeners, in a cycle of further possible repetitions and modifications/enhancements.[13] The representations carried out are of course not propositional and so they cannot carry "understanding" in the current meaning of the term. They can activate cognition exactly in the sense of the term we will encounter, in the following chapter, in the case of nonlinguistic animals. Of course it is perfectly plausible to guess that these representations ground (the neuroscientists say "scaffold") further development of more explicit abductive intentional attribution and agentive understanding, even at the conscious level.

The entire process of non-declarative and non-conceptual *embodied* simulation implied by the same brain areas of mirror neurons seen in non-human primates and infants can be seen, as many authors do, like a grounding embodied process – both from the ontogenetic and the phylogenetic perspective – of full maturation in adult human beings. [Iacoboni, 2003] demonstrates how intention can be understood through imitation, a perspective that sheds light on the formation of complex social cognitive skills like intentions themselves, planned actions, empathy, mind-reading, and the emergence of language. The embodied and prereflexive effect carried out by the sensorimotor system also seems to be particularly important in bodily recognition and emotion sharing (for instance, the experience of painful sensations). It is also supposed to be the foundation of empathy because the embodied process of pretension is both a-centered and, at the same time, the basis of self-other distinction.

The objectual "self" that is being formed through mirror neurons becomes the "pseudoself" that is being circumscribed.[14] Both emotions and empathy can acquire various levels of awareness and meta-awareness through the more complicated intervention of larger brain areas in human adults, which enact possible new ways of abductive emotional and empathic appraisals. Finally, it has to be remembered that also spatial mapping plays an important role in the formation of the self-subject-as-object and consequently of its distinction from other objects (and possible agents) (cf. below section 4.7).

Further insight on the problem of mentation, metamentation and the reflexive mind is illustrated by [Bogdan, 2003], who also takes into account an evolutionary perspective. Bogdan contends that, given the fact that the pressures on (and opportunities for) planning and problem solving are much greater in the precultural social brain (genetically fixed) than in the mere manipulative "mechanical domain", at the origin of mentation there would be the re-use of patterns tracked by interpretation in the social domain (thus not the ones merely derived from physical skills operating in the mechanical domains) already seen in primates: they slowly become objects of mental rehearsal. I sum, social life (rather than physical work) and asocial agency directed to conspecifics (rather than mechanical agency directed at the physical world) would be the main forces behind the evolution of primate mentation.

[13] On mirror neurons as a clear example of sensorimotor brain structure cf. chapter three, section 3.6.4.

[14] Research on imitation deficit in autism seems to confirm the presence of dysfunctions in mirror motor-sensory areas of children affected by that disease [Oberman *et al.*, 2005].

Interpretation and mental rehearsal – thinking about thoughts – with the obvious function of imagining nonactual situations, would be in turn crucial to the evolution of propositional attitudes and of explicit metathoughts, that are at the heart of metamentation, thinking about representations of agency and intentionality.

4.4 Actions vs. Thoughts?

Adult and normal human brain circuitry compute the operations of human intelligence and cognition through hierarchical structures, embedded sequences and hash coding. Naturalistically considered, these are physical and biological events in-the-world that embody the semiotic life of human and many non-human animals. From this neurological perspective an interesting conclusion on our traditional dichotomous view of organisms' actions and thoughts can be derived.

Let us show a fragment of brain neurocomputational mechanisms, as illustrated by Granger in a recent article [2006], which illustrates brain circuitry at work at the level of striatal cortex. "It results that two separated pathways from cortex through matrisomes involve different subpopulations of cells: (1) MSN_1 neurons project to GPi [pallidum, *pars interna*] $\rightarrow$ thalamus $\rightarrow$ cortex; (2) MSN_2 neurons insert an extra step: GPe [pallidum, *pars externa*] $\rightarrow$ GPi $\rightarrow$ thalamus $\rightarrow$ cortex. MSN and GP projections are inhibitory (GABAergic), such that cortical excitatory activation of MSN_1s causes inhibition in GPi cells, which otherwise inhibit thalamus and brainstem regions. Hence MSN_1 cells disinhibit, or enhance, cortical and brainstem activity. In contrast, the extra inhibitory link intercalated in the MSN_2 pathway causes MSN_2s to decrease the activity of cortex and brainstem neurons. These two pathways through MSN_1 and MSN_2 cells are thus called *go* and *stop* paths, for their opposing effect on their ultimate cortical and motor targets" [Granger, 2006, pp. 16-17].

Coordinated operation over time can yield a sophisticated combination of activated "go" and withheld "stop" *motor* responses (for example to stand or walk), or correspondingly complex *thought* (cortical responses). The cortex $\rightarrow$ striosome path triggers a kind of evaluation signal, provisionally settled in default values, corresponding to an expected experiential (sensory) reward for the given action, which, if the reward is not present, can lead to modifications which in turn trigger through other complicated paths other new possible actions and thoughts. Of course thoughts can in turn "mediate" further actions. At the level of the thalamocortical system, "core" circuit cells that respond to a particular input pattern, in turn elicit activation patterns which involve an effect of "clustering", and so of abductive generalization by preventing fine distinction between the members of the cluster. Through subsequent "samples" across time, refinements that activate subclustering are originated. The process is then directed to a learning process that generates the synaptic weights that can be assimilated to those "representations" we know through cognitive science.[15]

I think this example can imply two important theoretical considerations. First, both brain events, motor and thought responses, are material/biological processes

[15] For details on the role of other brain circuits cf. [Granger, 2006, pp. 19-24].

of the organism, the first originating action at the phenomenal level of the individual, the second originating thought, which we hypothesize in its physico/chemical aspects. In both cases, so to speak, the organisms "act". Second, both "processes" are subsequently modified by calculation based on the experiential outcomes, so that the action or thought chosen can be revised like in the epistemological schema of the abductive process I have illustrated in chapter one.[16] In this neural naturalistic perspective the *quasi*-ontological dichotomy between actions and thoughts, even if justified at a different, not neural, epistemological level – typical of the received philosophical and cognitive tradition – vanishes: obviously, both actions and thoughts are, so to speak, "actions" of the organisms, both phenomenal and not phenomenal levels are "performed" in the organism's body. This conclusion is also acknowledged by Edelman who, in the perspective of its neural Darwinism, quotes Peirce:

> Peirce pointed out that sensations are immediately present to us as long as they last. He noted that other elements of of conscious life, for example, thoughts, are actions having a beginning, middle, and end covering some portion of past or future. This fits our proposal that thought has an essential motoric component reflected in brain action but not in actual movement.[17] [...] This view of thought as being essentially motoric is consistent with the known interactions of the frontal and parietal cortex with basal ganglia, the subcortical regions involved in motor programs [Edelman, 2006, p. 123 and p. 168, footnote 3].

4.4.1 Decision Making and Action

Rational and deliberate decisions made by an agent are basically mediated by a conscious processing of high-level cognitive functions like the sentential ones (for example natural language) and various model-based ways of reasoning, but can also be intertwined with "thinking through doing" and action-based cognition, all able to carry an adequate amount of suitable knowledge. This kind of decision making has to be distinguished from that in which thoughts or cognitive actions enter the cognitive process without self-awareness, even if it has to be said that many decision making processes are the fruit of a hybrid blending between conscious and unconscious cognitive aspects [Piller, 2000; Thagard, 2001].[18] In rational and deliberate

[16] Cf. also [Magnani, 2001b].

[17] Thom further stresses that spoken a word is at the stage of emission an "action" in the literal sense of a "muscular motor field (a chreod in Waddington's sense) affecting the muscles of the thorax, the glottis, the vocal chords and the mouth" [Thom, 1980, p. 236]. On Thom's catastrophe theory and abduction cf. chapter eight of this book.

[18] The analysis of the concept of affordance [Gibson, 1979] also provides an alternative account of the role of the environment and of external – also artifactual – objects and devices, as the source of action possibilities (constraints for allowable actions). Artifactual cognitive objects and devices extend, modify, or substitute "natural" affordances actively providing humans and many animals with new *opportunities for action* [Norman, 1988]. Neuropsychological and physiological evidence on how simple visual information affords and potentiates action – and the highly integrated nature of visual and motor representations – is described in [Tucker and Ellis, 2006; Knoblich and Flach, 2001; Derbyshire *et al.*, 2006]. For further details on the interplay between affordances, actions, and decisions [Magnani and Bardone, 2008] and this book, chapter six.

decision making strategies (strategic principles)[19] are at work [Brogaard, 1999], as already stressed by the supporters of the general theory of games. These strategic principles always mediate between a given situation and a continuous range of future possibilities or choices that are described and provided by suitable pieces of hypothetical and generalized knowledge about events and processes (and "possible" events and processes).

Of course decision processes in humans constitutively occur in the presence of incomplete information and knowledge, like Peirce pointed out: "[...] the sum of all that will be known up at any time, however advanced, into the future, has a ratio less than any assignable ratio to all that may be known at a time still more advanced" [Peirce, 1931-1958, 5.330]. As I have already illustrated in chapter two (subsection 2.1) – where the abduction as an ignorance preserving kind of cognition is stressed – abduction plays a key role in decision making processes: one of the central aims of abduction is to recommend a course of action, in fact abduction usually provides previously unavailable more or less reliable hypothetical knowledge able to explain data which in themselves are considered inadequate to trigger a decision for action (indeed the abductive hypotheses describe what will happen, if an action is carried out). Nevertheless, even after having extracted new knowledge through abduction, the best rational and deliberate decision is never reached like it would be in the ideal situation where all the possible required knowledge is available: the best performance resorts to a simple activity of "maximization", where the presence of the available information and the smart use of strategic principles are the tools that make humans able to pursue this target.

I have said in the previous subsection that, both brain events, – motor and thought [inner] responses – are material/biological processes of the organism, the first originating action at the phenomenal level of the individual, the second originating thought, which we can only hypothesize. In both cases, so to speak, the organisms "act". Both "processes" are subsequently modified by calculation based on the experiential outcomes, so that the action or thought chosen can be revised as in my epistemological schema of the abductive process. From this neurological perspective the traditional dichotomous view between organism's actions and organism's thoughts can be weakened.[20]

From this perspective abduction recommends decisions

1. which leads to a course of *immediate motor actions*,

   - that, for instance by acquiring new information with respect to future interrogation and control, further trigger internal *thoughts* "while" modifying the environment (thinking through doing). Peirce provides the example of a cook making an apple pie for her master: "Throughout her whole proceedings she pursues an idea or dream without any particular thisness or thatness

[19] I will describe in more detail the status of strategies in abductive reasoning in chapter seven, subsection 7.3.2.2.

[20] Cf. Edelman's remarkable words about this issue, I have already quoted above, where : "essential motoric" aspect of thought is emphasized [Edelman, 2006, p. 123 and p. 168, footnote 3].

– or, as we say, *hecceity* – to it, but this dream she wishes to realize in connection with an object of experience, which as such does possess hecceity; and since she has to act, and action only related to this and that, she has to be perpetually making random selection, that is, taking whatever comes handiest" [Peirce, 1931-1958, 1.341] (I think this case is related to the role of manipulative abduction in my epistemological schema),

or

- that are more defined and not immediate, less random, and more pragmatical and less intertwined with a continuous interplay with subsequent "thoughts"; in this case action derives from a classical planning determined by inner established thoughts/hypotheses.

but abduction also recommends decisions/actions

2. (*as mere thoughts*), which – internally – lead to a further course of thinking detached from any immediate motor action to reach new hypothetical knowledge (of course, later on, open to make recommendations on further motor action).

4.4.2 Decision and Emotion

In the previous section I have said that rational and deliberate decision is basically mediated by a conscious processing of high-level cognitive functions like the sentential ones (for example natural language) and various model-based ways of reasoning, but it can also be intertwined with "thinking through doing". Emotion, an important model-based aspect of cognition, plays a pivotal role in decision making: emotions speed up the process and lead directly to actions. However using them to make choices is usually considered irrational because of their disadvantages: in the throes of strong feeling, we may be blind to some options, overlook critical information, or, when participating in a group charged with making a collective decision, fail to engage or connect with others who do not share our emotional state [Thagard, 2001, 356–357].

I think it is important to understand, however, that emotions are not inherently irrational. For example, they can be useful tools in moral decision making if they are successfully intertwined with learned cultural behaviors so that they become "intelligent emotions." Emotions can be developed, and Picard points out, "Adult emotional intelligence consists of the abilities to recognize, express, and have emotions, coupled with the ability to regulate these emotions, harness them for constructive purposes, and skillfully handle the emotions of others."[21] There is ongoing debate about the use of the expression "emotional intelligence": while the word *intelligence* implies something innate, many aspects typical of emotional intelligence are actually skills that can be learned.[22]

[21] [Picard, 1994, p. 49], cf. also [Nussbaum, 2001].

[22] On the neurological and cognitive role of emotions, cf. [Damasio, 1994; Damasio, 1999]. A cognitive theory of moral emotions in terms of "coherence" is illustrated in [Thagard, 2000; Thagard, 2001]. For recent work on emotional intelligence, see [Ben-Ze'ev, 2000], [Matthews *et al.*, 2002] and [Moore and Oaksford, 2002].

Antonio Damasio differentiates conscious "feeling" from unconscious "emotion" (of course, only conscious individuals "feel") [Damasio, 1999]: the genesis of emotions also relates to an individual animal's need to respond with its whole body – to run away from danger, for example, or to care for offspring. Emotions can communicate information about a situation and trigger a response even in absence of consciousness; in turn, this holistic response seems to have influenced the evolutionary formation of self and consciousness.[23]

Happiness, sadness, fear, anger, disgust, and surprise all can be viewed as *judgments* about a person's general state; a man who unexpectedly comes across a tiger on the loose, for example, would be understandably afraid because the large carnivore threatens his instinct to stay alive. In fact, all emotions are connected to goal accomplishment:[24] people become angry when they are thwarted, for instance, and feel pleased when they are successful. In this sense, emotion is a *abductive summary appraisal* of a problem-solving situation. Moreover, it provides cognitive "focalization" of the situation and readies us for action. On the contrary, we can consider emotions just as *physiological reactions* rather than cognitive judgments. Damasio refers to the signals that the body sends to the brain as *somatic markers*. Neuroscience taught us that emotions depend on interaction between bodily signals and cognitive appraisal. That is, they involve "both" judgments about how the current situation is affecting our goals and neurological assessment of our body's reaction to that situation.

Emotions are represented in the brain, but they cannot be represented like concepts because this conceals their links to a judgment that involves abductive processes, physiology, and feeling. We can imagine emotions as patterns of activation across many neurons in many brain areas, including those sites involved in cognitive judgments, like the prefrontal cortex, as well as those that receive input from bodily states, like the amygdala. Put another way, emotion activates neurons in different areas of the brain, areas that may have either inferential or sensory functions (cf. [Wagar and Thagard, 2004].).

We have already pointed out that emotions play an important role in decision making (a striking example is given by the case of moral deliberations). Damasio hypothesizes that in some cases of brain damage to the ventromedial, bottom-middle, or prefrontal cortex areas, people loose the ability to make effective decisions because they cannot discern the future consequences of their actions, especially in social contexts. That part of the brain provides connections between areas of the cortex involved in judgment and areas involved in emotion and memory, the amygdala and hippocampus.

---

[23] [Modell, 2003]. On consciousness, conscious will, and free will, cf. my book [Magnani, 2007d], chapter three, section "Critical Links: Consciousness, Free Will, and Knowledge in the Hybrid Human."

[24] On the role and nature of goals cf. [Thagard and Millgram, 2001]. The paper illustrates a theory (in terms of deliberative coherence) and a computational model of decision making that sees it not only as a process of choosing actions but also of evaluating goals.

Computer simulations of decision making have made it clear that we need more neurologically "realistic" models involving the role of emotions.[25] The GAGE neurocomputational program [Wagar and Thagard, 2004] aims at filling this gap. It models the above cognitive situation, which is due to a brain lesion, using groups of computational spiking neurons corresponding to each of the crucial brain areas involved: 1) vetromedial frontal cortex, 2) amygdala, 3) nucleus accumbens (a region strongly associated with rewards). So GAGE is capable of taking into account both cognitive aspects of judgment and appraisal performed by the ventromedial prefrontal cortex and physiological input mediated by the amygdala.[26]

Feelings serve as decisional inferences only if they are intertwined with learned cultural behaviors and therefore become "intelligent emotions" or, as some ethicists say, "appropriate" emotions.[27] Hume was wrong to view emotions as separate from intellect, says Johnson [1956], who maintains that the two dimensions are actually blended. I do not think we need to blend rational and emotional aspects; emotions are clearly distinguishable, even though we may be tempted to lump them together with purely cognitive functions because they can function in decision-making under certain conditions – that is, when they are not just raw products of evolution but are, instead, shaped further by knowledge and information. In this sense a decision lead by certain trained emotions, even if not fully deliberate and conscious, must not be considered as instinct-based. This decision derives from feelings built in a past history of conscious cognitive choices which have been able to reshape emotions giving rise to certain sophisticated feelings. Consider the example of a husband and his positive emotional attitude toward his marriage – it can be hypothesized that his strong commitment to his wife relates to his understanding of marriage, which in his case is stable and well formed, even if tacit and unreflective.

I have already pointed out that I agree wholeheartedly with Peirce that all thinking is in signs and that these signs take many forms – icons, indices, or symbols and so on, as we mentioned before. If all *inference* is, in fact, a form of sign activity – as Peirce contends – and we use the word *sign* to include feelings, images, conceptions, and other representations, then we must include unconscious thought among the model-based ways of moral thinking. Indeed, it is not only conscious abstract thought that we can consider inferential: we can characterize many cognitive activities that way.

Martha Nussbaum has emphasized the cognitive value of emotions and further clarified their moral role; in her work she improves and updates the Greek Stoic view, which holds that emotions are evaluative judgments that ascribe great importance to certain things and persons outside a person's own control. In this perspective, such things then have the power to determine human flourishing

[25] They of course will still lack the possibility of "feeling" emotions: indeed they will not have bodily inputs.

[26] Some meta-ethicists call moral intuitionism the view of emotions as central in justifying moral beliefs [Sinnott-Armstrong, 1996]. [Ben-Ze'ev, 2000] maintains that optimal moral behavior is that which combines emotions and intellectual reasoning, a complex integration that requires the so-called "emotional intelligence".

[27] Cf. [Thomson, 1999, pp. 148-149]. In this case [Oatley, 1992] speaks of "learned spontaneity."

(*eudaimonia*).[28] Put another way, through emotions, people acknowledge that external things/persons they do not fully control are very important for their own flourishing. Emotions are always seen as involving thought of an "intentional" object combined with thought of that object's salience, value, and importance in the framework of what Nussbaum calls the "cognitive-evaluative" view. This perspective contrasts with the "adversary view" that considers emotions mere "non-reasoning movements" that would derive "bodily" from an animal part of our nature rather than "mentally" from a specifically human part. Unlike the Stoic approach, it is difficult to see emotions as judgments in this view, and it would seem hard "to account for their urgency and heat given the facts that thoughts/judgments are usually imagined as detached and calm" [Nussbaum, 2001, p. 27]: emotions are fundamentally seen as irrational and a bad guide to action in general and to moral action in particular.

In the "cognitive-evaluative" perspective, it is very easy to think of emotions more broadly in a way that goes beyond the Stoic starting point; consider, for example, the cognitive role of emotions in animals and the evaluative appraisal they perform in moral life. Of particular interest is the role that elements of culture – social norms, for example – play in shaping certain feelings like compassion that I refer to above as "trained emotions": "Human deliberative sociability also affects the range of emotions of which humans are capable, [...]" [Nussbaum, 2001, p. 148] just as individual history influences the perception of that effect and cognitively embeds emotions in a complex of personal narratives. In this last sense, to acknowledge the influence that social constructions have on emotions is to see that emotions consist of elements we have not ourselves constructed.

Finally, I must note that Michael Gazzaniga, citing James Q. Wilson's prediction, provides a brain-based account of moral reasoning centered in the areas of emotion. It has been found that brain regions normally involved in emotional processing may be activated by one type of moral judgment but not another. In this sense, when someone is willing to act on a moral belief, it is because the emotional side of her brain was activated as she checked the moral question at hand. If, on the other hand, she decides not to act, the emotional part of the brain does not become active.[29]

---

[28] Greek eudaimonistic ethical theories are concerned with human flourishing: *eudaimonia* is taken to include everything to which a person attributes intrinsic value. Nussbaum retains this spelling, rather than using the English word "eudaemonistic," because she wants to refer to the ancient Greek concept and avoid more recent connotations associated with the idea, "namely, the view that the supreme good is happiness or pleasure" [Nussbaum, 2001, p. 31].

[29] [Gazzaniga, 2005]. Other studies exploiting neuroimaging have dealt with the neuroanatomy and neuroorganization of emotion, social cognition, and other neural processes related to moral judgment in normal adults and in adults who exhibit aberrant moral behavior [Greene and Haidt, 2002]. [Moll *et al.*, 2002; Moll *et al.*, 2005] have established that moral emotions differ from basic emotions in that they are interpersonal: the neural correlates that are more interested in moral appraisal appear to be the orbital and medial sectors of the prefrontal cortex and the superior sulcus region, which are also critical regions for social behavior and perception.

4.5 The Agent-Based and Abductive Structure of Reasons in Moral Deliberation

This section illustrates in detail that "abduction", that is the reasoning to hypotheses, is central to the problem of "inferring reasons" in decision making, as a fundamental kind of practical reasoning. Moral deliberation will be our example. We have seen that in abduction we usually base our guessing of hypotheses on incomplete information, and so we are facing nonmonotonic inferences: we reach defeasible conclusions from limited information, and these conclusions are always withdrawable.[30] It is in this sense that both explanatory and instrumental abductive reasoning constitutes a possible useful model of practical reasoning: ethical deliberations are always adopted on the basis of incomplete information and on the basis of the selection of particular abduced hypotheses which play the role of *reasons*. Hence, ethical deliberation shares some aspects with hypothetical explanatory reasoning as it is typically illustrated by abductive reasoning in scientific settings. To support this perspective on the "logical structure of reasons" we will provide an analysis based on the distinction between "internal" and "external" reasons and on the difficulties in "deductively" grasping practical reasoning, at least with the only help of classical logic.

In a previous work [Magnani, 2007d, chapters six and seven] devoted to introduce the methodological problems of ethical deliberation, I contend, following Rachels, that morality is the effort to guide one's conduct by reasons, that is, to do what there are the best reasons for doing while giving equal weight to the interests of each individual who will be affected by one's conduct. "The logical structure of reasons" is the title of chapter four in Searle's book *Rationality in Action* [2001]. I plan to use Searle's conceptual framework to better understand what exactly are "reasons" in the specific case of ethics. Whereas Searle deals with rational decision making, many of his conclusions appear to be appropriate for ethical cases, too.

By criticizing the classical model of rational decision making (which always requires the presence of a desire as the condition for triggering a decision), Searle establishes the fundamental distinction between *internal* and *external* reasons for action: those that are internal might be based on a desire or on an intention, for instance, while external reasons might be grounded in external obligations and duties. When I pay my bill at the restaurant, I am not doing so to satisfy an internal desire, so this action does not arise from internal motivations; instead, it is the result of my recognition of an external obligation to pay the restaurant for the meal it has provided. Analogously, if an agent cites a reason for a past action, it must have been the reason that the agent "acted on". Finally, reasons can be for future action, and this is particularly true in ethics where they do not always trigger an action – in this case, however, they must still be able to motivate an action: they are reasons an agent can "act on".

Searle's anti-classical emphasis on "external reasons" does not have to appear strange: in [2007d] I have often stressed the fact that human beings not only

[30] For further specifications cf. chapter one, section 1.3 and chapter two, section 2.1, where the fallacious character of abduction is described.

delegate cognitive roles but also and moral worth to external objects that consequently acquire the status of deontic moral structures. This also occurs when we articulate ideas in verbal statements – promises, commitments, duties, and obligations, for example – that then exist "over there", in the external world. Imagine the deontic role that concrete buildings (like for instance the ones whose shapes restrict routes people can follow) or abstract institutions (for example, some modern constitutions usually compel us to consider equality of citizens as important) can play in depicting duties and commitments we can (or have to) respect. Human beings are bound to behave in certain ways as spouses, tax payers, teachers, workers, drivers, and so on. All these external factors can become – Searle says – reasons/motivators for prior intentions and intentions-in-action of human beings.

Many things around us are human made, artificial – not only concrete objects like a hammer or a PC, but also human organizations, institutions, and societies. Economic life, laws, corporations, states, and school structures, for example, can also fall into that category. We have also projected many intrinsic values on things like flags, justice rituals, or ecological systems, and as a result, these external objects have acquired a kind of autonomous automatism "over there" that conditions us and distributes roles, duties, moral engagements – that is, it supplies potential "external reasons". Non-human things (as well as so-to-say "non-things" like future human beings and animals, etc.) become moral clients as well as human beings, so that current ethics must pay attention not only to relationships between human beings, but also to those between human and non-human entities.

Moreover, we can observe how external things we usually consider to be morally inert can be transformed into those *moral mediators* which express the idea of a distributed morality. For example, we can use animals to highlight new, previously unseen moral features of other living objects, as we can do with the earth or with (non natural) cultural objects; we can also use external "tools" like writing, narratives, others persons' information, rituals, and various kinds of institutions to morally reconfigure social orders. Hence, not all moral tools are inside the head along with the emotions we experience or the abstract principles we refer to; many are over there, even if they have not yet been identified and represented internally, distributed in external objects and structures which function as ethical devices available for acknowledgment by every human agent. These delegations to external structures – thus transformed in moral mediators – encourage or direct ethical commitments, and, they favor the predictability in human behavior that is the foundation for conscious will, free will, freedom, and of the ownership of our own destinies: if we cannot anticipate other human beings' intentions and values, we cannot ascertain which actions will lead us to our goals, and authoring our own lives becomes impossible.

Let us return to the role played by reasons in ethical reasoning. Intentional states with a propositional content have typical *conditions of satisfaction* and *directions of fit*.

1. First, mental and linguistic entities have directions of fit: for example, a belief has a *mind-to-world* direction of fit. For example, if I believe it is raining, my belief is satisfied if and only if it is raining "[...] because it is the responsibility

of the belief to match an independently existing reality, and it will succeed or fail depending or whether or not the content of the belief in the mind actually does fit the reality of the world" [Searle, 2001, p. 37]. On the other hand, a desire (or an order, promise, or intention) has a *world-to-mind* direction of fit: "[...] if my belief is false, I can fix it up by changing the belief, but I do not in that way make things right if my desire is not satisfied by changing the desire. To fix things up, the world has to change to match the content of the desire" [Searle, 2001, p. 38].

2. Second, other objects (not mental and not linguistic) also have a direction of fit similar to the ones of beliefs. A map, for example, which may be accurate or not, has a *map-to-world* direction of fit, whereas the blueprints for a house have a *world-to-blueprint* direction of fit because they can be followed or not followed [Searle, 2001, p. 39]. Needs, obligations, requirements, and duties are not in a strict sense linguistic entities, but they have propositional contents and directions of fit similar to the ones of desires, intentions, orders, commitments, and promises that have a *world-to-mind, world-to-language* direction of fit. Indeed, an obligation is satisfied if and only if the world changes to match the content of the obligation: if I owe money to a friend, the obligation will be discharged only when the world changes in the sense that I have repaid the money.

When for example we apply the moral principle of the *wrongness of discriminating against the handicapped* to the a specific moral case (for example the recent famous case of Baby Jane Doe [Rachels, 1999] where the parents had to decide (and so they had to choose/infer the right "reason") in favor or against a fundamental surgical operation), we resort to a kind of "external" reason that we have to "internalize" – that is, recognize as a reason worth considering for a possible deliberation. If we instead exploit strong personal feelings like pity and compassion to guide our reasoning, we would decide for or against the operation based on a completely "internal" reason. We have to note, of course, that external reasons are always observer-relative. It is only human intentionality that furnishes meaning to a particular configuration of things in the external moral or non-moral world. The objective fact that, say, I have an increased white blood cell level acquires a direction of fit that is a direction for action only if related to a human being's interpretation (for example only "in the light" of a diagnosed disease, that same fact can trigger the decision for a therapy).

Searle also discusses the so-called collective intentionality that enables people to create common institutions such as those involving money, property, marriage, government, and language itself, an intentionality that gives rise to new sets of "conditions of satisfaction", duties, and commitments. In our perspective we say these external structures have acquired a kind of delegated intentionality because they have become moral mediators, they have acquired a kind of moral "direction", as I have illustrated in [Magnani, 2007d, chapter six]. In those cases, when we have to deal with a moral problem through moral mediators, evaluating reasons of any kind immediately involves manipulating non-human externalities in natural or artificial environments by applying old and new behavior *templates* that exhibit some uniformities. This moral process is still hypothetical (abductive): these templates are embodied hypotheses of moral behavior (either pre-stored or newly created in the mind-body system) that, when appropriately employed, make possible what can be called a moral "doing" (cf. also [Magnani, 2006e]).

I contend that external moral mediators are a powerful source of information and knowledge; they redistribute moral effort by managing objects and information in new ways that transcend the limits and the poverty of the moral options immediately represented or found internally (for example exploiting resources in terms of merely internal/mental moral principles, utilitarian envisaging, and model-based moral reasoning – emotions, for example).

It follows from the previous discussion that many entities can play the role of deontic moral structures. This fact can lead to a re-examination of the concept of duty. In this perspective duties can be also grounded on trained emotional habits, visual imagery, embodied ways of manipulating the world, exploitation of moral mediators – as we have just seen, endowed with a sufficient ethical worth in a collective.

4.5.1 The Ontology of Reasons

What are these "reasons" that, following Searle, are the basis of rational actions and the basis of moral action? A reason answers the question "Why?" with a "Because"; it can be a statement, like a moral principle, as in the answer to "Why should we perform surgery on a handicapped baby?": "Because of the wrongness of discriminating against the handicapped". In reality, reasons are "expressed" by the statements-*explanations* in so far as they are *facts* in the world (the fact that it is raining is the reason I am carrying an umbrella). They are also represented by *propositional intentional states* such as desires (my desire to stay dry is the reason I am carrying the umbrella), and, finally, by *propositionally structured entities* such as obligations, commitments, needs, and requirements, like in the case of our moral "principle" of "the wrongness of discriminating against the handicapped". Usually good reasons explain and usually explanations give reasons. Searle also distinguishes between reasons that justify my action and thus explain why it was the right action to perform, and the reasons that explain why in fact I did it.

1. First of all, in rational decision making, when we must provide a reason for an intentional state, we have to make an intelligent selection from a range of reasons that exist either internally or externally – in the latter case, we must take the external reason, recognize it as good, and internalize it. With respect to our ideal of an ethical deliberation sustained by "reasons", we can affirm that it is not unusual for the "deliberator" to have limited knowledge and inferential expertise at his or her disposal. For instance, she may simply not have important pieces of information about the moral problem she has to manage, or she may possess only a rudimentary ability to compare reasons and ascertain data. Ethical reasoning is so abductive and defeasible: because it is impossible to obtain all information about any given ethical situation, every instance of moral reasoning occurs without benefit of full knowledge, so we must remember that any reason can be rendered irrelevant or inappropriate by new information. Generally

speaking, as illustrated above, these reasons can take three different forms: external *facts* in the world, such as empirical data; internal *intentional states* such as beliefs, desires, or emotions; and *entities* in the external world like duties, obligations, and commitments with the direction of fit upward (world-to-mind). External facts must be internalized and "believed", while external entities must be internalized and adopted ("recognized") as good and worth of consideration. The same happens in the case of rational moral deliberation.

2. Second, we must remember that maintaining a flexible, open mind is particularly important when we lack the ethical knowledge necessary to confront new or extreme concrete situations.

When evaluating an ethical case, we have at hand all the elements of rational moral decision making: the problem we face, the "reasons", and the agents involved. Every reason, Searle says, contributes to a "total reason" that is ultimately a composite of every good reason that has been considered – beliefs, desires, obligations, or facts, for example. As already observed, first, "rationality" requires the agent to recognize the facts at hand (I have to believe that it is raining) and the obligations undertaken (I have to adopt the principle of the sanctity of human life) without denying them (which would be obviously irrational) [Searle, 2001, p. 115]. Second, reasons can be more than one, indeed I need at least one motivator, but in some cases there are many, and these reasons often conflict with one another; it then becomes necessary to appraise their relative weights in order to arrive at the prior intention and the intention-in-action.

In abductive reasoning, this kind of appraisal is linked to evaluating various inferred explanatory, non-explanatory, and instrumental hypotheses/reasons, and, of course, it varies depending on the concrete cognitive and/or epistemological situation. In the section 1.3 of chapter one, we have illustrated that epistemologically using abduction as an inference to the best explanation simply requires evaluating competing hypotheses (that express competing "reasons" in the ethical case). The best [total] reason would be the one that creates prior intention and intention-in-action.

What criteria can we adopt to choose the reason(s) that will become the motivator(s)? Thagard [2000] has proposed a framework in terms of coherence, in which ethical deliberation is seen as involving conflicting reasons (deductive, explanatory, deliberative, analogical) that can be appraised by testing their relative "coherence". This "coherence view" is terrifically interesting because it reveals multidimensional character of ethical deliberations. The criteria for choosing the most coherent "reason/motivator" represent a possible abstract cognitive reconstruction of an ideal of "rationality" in moral decision making, but they can also describe the behavior of real human beings. Human beings usually take into account just a fraction of the possible knowledge when performing ethical judgments. For example, when making judgments, it is common for utilitarians to employ only what Thagard calls "deliberative" coherence or for Kantians to privilege principles over consequences. Psychological resources are limited for any agent, so it is difficult to mentally process all levels of ethical knowledge simultaneously in an attempt to calculate and maximize the overall coherence of the competing moral options. The "coherence" model accounts for these "real" cases of human moral reasoning by showing they fit only "local areas" of the coherence framework: in general, real human beings come to immediate conclusions through one moral aspect (for instance, the "consequentialist" one) and disregard the possible change in coherence weight that could result from considering other levels (for instance, the "Kantian" one).

Searle interprets rationality in decisions naturalistically: "Rationality is a biological phenomenon. Rationality in action is that feature which enables organisms, with brains big and complex enough to have conscious selves, to coordinate their intentional contents, so as to produce better actions than would be produced by random behavior, instinct, tropism, or acting on impulse" [Searle, 2001, p. 142]. I agree, but I would add that rationality is a product of a hybrid organism. This notion obviously derives from the fact that even the external tools and models we use in decision making – an externalized obligation, a computational aid, and even Thagard's "coherence" model described above – are products of biological human beings, but at the same time these tools constitutively affect human beings, who are, as we already know, highly "hybridized".[31]

4.5.2 Abduction in Practical Agent-Based Reasoning

Searle considers "bizarre", and I strongly agree with him – that feature of our intellectual tradition, according to which true statements that describe how things are in the world can never imply a statement about how they ought to be: in reality, to make a simple example, to say something is true is already to say you ought to believe it, that is other things being equal, you ought not to deny it. Also, logical consequence can be easily mapped to the commitments of belief. Given the fact that logical inferences preserve truth, "The notion of a valid inference is such that, if p can be validly inferred from q, then anyone who asserts p ought not deny q, that anyone who is committed to p ought to recognize its commitment to q" (*ibid.*, p. 148.). This means that normativity is more widespread than expected.

Certainly, theoretical reasoning can be seen as a kind of practical reasoning where deciding what beliefs to accept or reject is a special case of deciding what to do. The reason it is difficult to "deductively" grasp practical reasoning is related to the intrinsic multiplicity of possible reasons and to the fact that we can hold two or more inconsistent reasons at the same time.[32] The following example illustrates how practical contexts are refractory to logical modeling. Given the fact that we

[31] Further details on coherence, truth, and the development of scientific knowledge are illustrated in [Thagard, 2006]. Searle [2001] calls the means and ways of performing an action (for instance to fulfill an obligation) "effectors" and "constitutors". An obligation to another person is an example. I know I own you some money: "I can drive to your house" and "give you the money" – effector and constitutor.

[32] Searle "reluctantly" declares that it is impossible to construct a formal logic of practical reasoning "adequate to the facts of the philosophical psychology" [Searle, 2001, p. 250]. I think that many types of non-standard logic (deontic, nonmonotonic, dynamic, ampliative, adaptive, etc.) reveal interesting aspects of practical reasoning by addressing the problem of defeasibility of reasons and of their selection and evaluation.

consider it a duty to do p and that I also feel committed not to do p, we cannot infer that I am committed to do (p and not p). I am a physician committed to not killing a patient in a coma, but at the same time my compassion for the patient commits me to the opposite duty. This does not mean that I want to preserve the life of the patient and, at the same time, I want to kill him – that would lead to an inconsistent moral duty. All this represents an unwelcome consequence of the fact that commitment to a duty is not closed under conjunction [Searle, 2001, p. 250].

In practical reasoning, we are always faced with desires, obligations, duties, commitments, needs, and requirements, etc., that are at odds with one another. Moreover, even if I consider it a duty to do p and I believe that (if p then q), I am not committed to do q as a duty: I can be committed to killing a patient in a coma and at the same time believe this act will cause pain for his friends, but I am not committed to causing this pain. *Modus ponens* does not work for duty/belief mixture [Searle, 2001, pp. 254–255].

The examples above illustrate the difficulties that arise when classical logic meets practical reasoning. They further stress the importance we attribute to abductive explanatory inferences in agent-based practical settings, where creating, selecting, and appraising hypotheses are central functions.

4.6 Picking Up Information

We have seen that it is difficult to hypothesize internal cognitive states in nonlinguistic organisms.[33] The approach in terms of the so-called active perception, that minimizes the role of representations, can help us overcome this impasse. From this perspective we can first of all say, at least in the case of human beings, that abduction is a complex process that works through *imagination*: for example, it suggests a new direction in reasoning by shaping new ways for explaining (cf. the templates mentioned above). Imagination should not, however, be confused with intuition. Peirce describes abduction as a dynamic modeling process that fluctuates between states of doubt and states of belief. To assuage doubt and account for anomalies, the agent gathers information that relates to the "problem," to the agent's evolving understanding of the situation, and to its changing requirements. When I use the word "imagination" here, I am referring to this process of knowledge gathering and shaping, a process that Kant considered invisible to us and yet which leads us to see things we would not otherwise have seen: it is "a blind but indispensable function of the soul, without which we should have no knowledge whatsoever" [Kant, 1929, A78-B103,p. 112]. For example scientific creativity, it is pretty obvious, involves seeing the world in a particular new way: scientific understanding permits us to see some aspects of reality in a particular way and creativity relates to this capacity to shed new light.

We can further analyze this process using the *active perception* approach [Aloimonos *et al.*, 1988; Ballard, 1991], a theory developed in the area of

[33] The whole problem of animal abduction and of the so-called mindless cognition will be treated in the following chapter.

computer vision [Thomas, 1999]. This approach seeks to understand cognitive systems in terms of their environmental *situatedness*: instead of being used to build a comprehensive inner model of its surroundings, the agent's perceptual capacities are seen as simply used to obtain "whatever" specific pieces of information are necessary for its behavior in the world. The agent constantly "adjusts" its vantage point, updating and refining its procedures, in order to uncover a piece of information. This resorts to the need to specify how to efficiently examine and explore and to "interpret" an object of a certain type. It is a process of attentive and controlled perceptual exploration through which the agent is able to collect the necessary information: a purposeful examination is carried out, actively picking up information rather than passively transducing (cf. [Gibson, 1979]).

As suggested for instance by Lederman and Klatzky, this view of perception may be applied to all sense modes: for example, it can be easily extended to the haptic mode [1990]. Mere passive touch, in fact, tells us little, but by actively exploring an object with our hands we can find out a great deal. Our hands incorporate not only sensory transducers, but also specific groups of muscles and musculature which, under central control, move in appropriate ways: lifting something tells us about its weight, running fingers around the contours provides shape information, rubbing it reveals for instance texture, as already stressed by Peirce in the quotation I reported above, when dealing with the hypothesizing activity of what I call manipulative abduction [Peirce, 1931-1958, 5.221].[34]

Nigel Thomas suggests we think of the fingers together with the neural structures that control them so that we can consider the afferent signals they generate as a sort of knowledge-gathering (perceptual) *instrument*: a complex of physiological structures capable of active testing for some environmental property [Thomas, 1999]. The study of manipulative abduction that I outlined above can benefit from this approach. For example, the analysis of particular epistemic mediators (optical diagrams) in non-standard analysis, I have illustrated in chapter one (section 1.7), and their function in grasping and teaching abstract and difficult mathematical concepts stresses the activity on picking up described above. In this case the external models (mathematical diagrams) do not give all available knowledge about a mathematical object, but compel the agent to engage in a continual epistemic dialogue between the diagrams themselves and her internal knowledge either to enhance personal understanding of existing information or to facilitate the creation of new knowledge.

As I have already noted in chapter two (section 2.8) human beings and other animals make frequent use of perceptual reasoning and kinesthetic abilities. Usually the "computations" these tasks require are not fully conscious. Mathematical cognition uses verbal explanations, but it also involves nonlinguistic notational devices

[34] A kind of inner touch can be very significant in human brains and so furnish a kind of "imagery" that drives thoughts: the famous case is Helen Keller, who became deaf and blind at the age of eighteen months, and so, lacking normal auditory and visual channels, did not meet the standard requirement for acquiring language. Later on, with the help of Annie Sullivan, she was able to learn language and an adult personality mainly exploiting haptic modes. This extraordinary story tells us that her brain plastically employed an anatomical path usually not at work in normal linguistic communication, which uses very different neural patterns (cf. [Donald, 2001, pp. 232-250]).

and models that require our perceptive and kinesthetic capacities. In chapter two I have illustrated how geometrical constructions are a prototypical case of this kind of extra-linguistic system that functions in a model-based, manipulative, and abductive way.

4.7 Spatial Frameworks, Anticipation, and Geometry

4.7.1 Abduction and Neurospaces

Many parts of the mammalian cognitive system represent the location of objects within the spatial framework. The egocentric ones are spaces where the objects are located in a framework that is referenced to a sensory receptor or body surface, e.g. retinal axes of the visual cortex, head-centered axes of the partial cortex. In many vertebrates these spaces are formed by taking advantage of the available cues in natural environment and/or of the artificially made cues and landmarks suitably externalized in the material environment by the individual or group in question. The cues become part of the internal representation through the perceptual system so furnishing the basic data which make the abductive formation of the spatial mapping possible. Of course these sources, even if reliable, often are not easily accessible.[35] Another spatial framework, called allocentric, is instead referenced to the environment and so not centered on the organism.[36]

Research in neuroscience has stressed the pre-wired role of the hippocampus, a paradigmatic example of archicortex, in building a prereflexive allocentric spatial framework in mammals. Many empirical investigations and mathematical and computational models of this mechanism have been provided in the last decades [O'Keefe, 1999]. Hippocampus receives input from the "[. . .] entorhinal cortex and the septum. The entorhinal cortex in turn receives inputs directly or indirectly from many neocortical areas and is believed to be the major conduit of sensory information into the hippocampus. In contrast, the septum gets its input primarily from the hypothalamus and brainstem, and is thought to convey information about actual or intended movements and about the animal's bodily states and needs" (*ibid.*, pp. 53-54). Models of hippocampus functions have been recently studied to detect the effects, for example of lesions, on the dentate gyrus and CA3. It has resulted

[35] More information on the exploitation of spatial external representations is detailed in [Freska, 2000, pp. 1126-1127]).

[36] More details on the high plasticity in animal spatial egocentric and allocentric mapping, also focusing the attention on switching between different mappings depending on the local situation, is provided by [Roberts, 2001]. The role of external representations, as "spatial products", in human cognition, is illustrated in [Liben, 2001; Tversky, 2001; Tversky *et al.*, 2006]. On the role in chicks' spatial orientation of non-geometrical information given by landmarks and of geometrical information given by the shape of the enclosure cf. [Vallortigara *et al.*, 2005; Chiesa *et al.*, 2006], who also stress the role in these tasks of the spatial logics of a dual brain. The research also aims at suggesting that the reliance on the use of geometric information of the spatial scale of the environment is not an exclusive endowment of humans: animals do (implicit) geometry (also in the case of fish [Sovrano *et al.*, 2005]).

that both lesions strongly reduce the efficacy of the exploration associated with displayed objects and affect detection of spatial novelty (cf. [Lee *et al.*, 2005], and [Hasselmo, 2005]). Finally, the role of the hippocampus seems involved not only in the representation of the environment but also in the building of the primitive cognitive pre-conditions for the representation of the subject-as-object and the further related various levels of consciousness.

In the process of hippocampal abductive formation of spatial mapping it is important to stress the complicated interplay between internal and external sensory representations (visual, acoustic, and proprioceptive), together with the fact that dynamical motor aspects, which also involve manipulations of objects, are central and basically independent of human language [Freska, 2000]. In the case of rats, as the animal moves around a known environment the hippocampal mapping system continuously updates various parameters through a comparison between the (suitable, decoupled) internal navigation representations (a kind of spatial imagination), and actual representations derived from the sensory input. Internal subsequent representations are abductive hypotheses of space (Husserl would say, anticipations, cf. subsection 4.7.4). Given the fact I have often remarked that various cognitive biological agents must reason on the basis of incomplete or uncertain information, an appraisal of the necessary implications of spatial relations and actions is only slowly and progressively reached through a cycle of continuous updating of spatial mappings.

The process occurs in such a way that deviations from the expected sensory data provide an internal signal for exploration to the aim of abductively re-constructing and modifying the spatial representation of the environment. The internal signals for exploration are also driven by a motivational system activated for instance by hunger and by food in a particular location. Of course the neural, settled "representation" obtained can be reliably re-used when exploring the same niche: spatial relations can be replicated by superimposing them upon the sensory data at hand and thus implicitly filling in missing relations not explicitly available. Even the well-known three-dimensionality of space can be accounted for by the pre-wired brain limitations on the representational properties.

[O'Keefe, 1999] contends that the results on hippocampal spatial mapping in vertebrates would furnish a kind of naturalistic confirmation of the Kantian philosophical concept of spatial "pure intuition". Furthermore, it can be hypothesized that the historical perplexities of philosophers and geometricians over the centuries regarding the famous Euclidean fifth postulate,[37] can be finally explained considering the intrinsic neural features of the mammalian hippocampal mapping system. The perplexities were mainly due to the fact that the postulate appeals to infinity, and so to properties that cannot be verified by the experience accumulated through human perception. Studies on the hippocampus show that the abductive formation of a directional system – which projects a set of parallel lines, which never meet, upon the given environment – is intrinsic and allows the organism to compute a direction

[37] Cf. my book [Magnani, 2001c] and chapter two of this book, sections 2.9 and 2.11.

based on cue distribution, and this direction is independent of location within the environment [O'Keefe, 1999, pp. 59–62].[38]

A further consequence of this neural perspective concerns the hypothesis that the strong development of temporal and prefrontal neocortices has been driven in part by the existence of the hippocampal map. Indeed, the formation of the self-other distinction (and thus the idea of "agent") and the related formation of the image of a self-referenced body are probably related to the development of the frontal cortex which is in turn based on the spatial schemes furnished by the hippocampal map. The progressive acknowledgement of the pregnant fact that there is nothing in the mapped spatial framework corresponding to an entity in the location of the place where the organism actually is, possibly leads to "filling" the empty site with an object marker. Later on other complicated cognitive abilities would intervene to support a self-referenced image, also based on the recognition that agents are certainly objects that can change location in spatial mapping without external influence. It would be thanks to this nonlinguistic background that, finally, in humans, the abductive formation of the higher awareness of the self can be implemented, typical of the mentalized idea of a free and intentional agent. In conclusion, it is plausible to affirm that a spatio-temporal packet of sensory stimuli, which maintains its integrity as the organism moves or as it itself moves through a succession of sites within the map, can be the precondition for more articulated and more or less conscious ideas of "subject" and "object".[39]

Moser's research group has recently provided some of the most groundbreaking insights so far concerning the computation of spatial location and spatial memory in rat brains [Fyhn *et al.*, 2004], by discovering a neuronal spatial map in the medial dorsocaudal entorhinal cortex. The key unit of the map is the "grid cell", which is activated whenever the position of the animal coincides with any of the vertex of a regular grid of equilateral triangles spanning the surface of the environment. The "geometrical" map is anchored by external landmarks, but persists in their absence, suggesting that grid cells may be part of a generalized, path-integration-based map of the spatial environment. The grid cells constantly update the mammal's sense of its location, even in absence of the external sensory input, so providing a constant and very general abductive appraisal of the situation at hand: indeed "The firing pattern – which is similar regardless of whether the rat is in a familiar, well-lit room or in a strange location that is pitch-dark – must be a pure cognitive construct. Although grid cell firing patterns are updated and calibrated by sensory input from the vestibular, visual and other sensory systems, they do not depend on external sensory cues" [Knierim, 2007, p. 48]. Grid cells seem to constitute the basic spatial input that is exploited by the hippocampus to construct more specific and context-dependent spatial firing in its place cells. Place cells merely fire when a mammal occupies a single, particular location but each grid cell – which projects a latticework of perfect

[38] It seems the components of the mapping system require a large number of hippocampal identical harmonic oscillators, which are in variable ways intrinsic endowments of many living organisms.

[39] On the role of the auditory systems in spatial cognition of vertebrates cf. [de Cheveigné, 2006] and this book, chapter five, section 5.5.2.

equilateral triangles across the environment, the corners of which are sensible to the rat's presence – will fire when it is any one of the many locations that are arranged in a stunningly uniform hexagonal grid (as if the cell were related to a number of alarm tiles spaced at specific, regular distances). Because the grids projected by the brain's grid cells overlap, the grid cell system fires whenever the rat moves, thus updating the animal's location [Knierim, 2007, pp. 44–45].

4.7.2 Adumbrations: Perceptions and Kinesthetic Sensations Intertwined

Also the philosophical tradition of phenomenology fully recognizes the protogeometrical role of kinesthetic data in the generation of the so-called "idealities" (and of geometrical idealities). The objective space we usually subjectively experience has to be put in brackets by means of the transcendental reduction, so that pure lived experiences can be examined without the compromising intervention of any psychological perspective, any "doxa". By means of this transcendental reduction, we will be able to recognize perception as a structured "intentional constitution" of the external objects,[40] established by the rule-governed activity of consciousness (similarly, we will see that space and geometrical idealities, like the Euclidean ones, are "constituted" objective properties of these transcendental objects).

The modality of appearing in perception is already markedly structured: it is not that of concrete material things immediately given, but it is mediated by sensible schemata constituted in the temporal continual mutation of adumbrations. So at the level of "presentational perception" of pure lived experiences, only partial aspects (*adumbrations* [*Abschattungen*]) of the objects are provided. Therefore, an activity of unification of the different adumbrations to establish they belong to a particular and single object (noema) it is further needed.[41]

The analysis of the generation of idealities (and geometrical idealities) is constructed in a very thoughtful philosophical scenario. The noematic appearances are the objects as they are intuitively and immediately given (by direct acquaintance) in the constituting multiplicity of the so-called adumbrations, endowed with a morphological character. The noematic meaning consists of a syntactically structured categorical content associated with judgment. Its ideality is logical. The noema consists of the object as deriving from a constitutive rule or synthetic unity of the appearances, in the transcendental sense [Petitot, 1999]. To further use the complex Husserlian philosophical terminology, we can say: hyletic data (that is immediate given data) are vivified by an intentional synthesis (a noetic apprehension) that transforms them into noematic appearances that adumbrate objects, etc.

[40] A recent article [Overgaard and Grünbaum, 2007] deals with the relationship between perceptual intentionality, agency, and bodily movement and acknowledges the abductive role of adumbrations. In the remaining part of this section I will try to clarify their meaning.

[41] On the role of adumbrations in the genesis of ideal space and on their abductive and nonmonotonic character cf. below section 4.7.4.

4.7.3 The Genesis of Space

As illustrated by Husserl in *Ding und Raum* [1907] [Husserl, 1973] the geometrical concepts of point, line, surface, plane, figure, size, etc., used in eidetic descriptions are not spatial "in the thing-like sense": rather, in this case, we deal with the problem of the generation of the objective space itself. Husserl observes: it is "senseless" to believe that "the visual field is [...] in any way a surface on objective space" (§48, p. 166), that is, to act "as if the oculomotor field were located, as a surface, in the space of things" (§67, p. 236).[42] What about the phenomenological genesis of geometrical global three-dimensional space?

We have to start dealing with the problem of the treatment of adumbrations. The adumbrative aspects of things are part of the visual field. To manage them a first requirement is related to the need of gluing different fillings-in of the visual field to construct the temporal continuum of perceptive adumbrations in a global space: the visual field is considered not translation-invariant, because the images situated at its periphery are less differentiated than those situated at its center (and so resolution is weaker at the periphery than at the center), as subsequently proved by the pyramidal algorithms in neurophysiology of vision research.

Perceptual intentionality basically depends on the ability to realize kinesthetic situations and sequences. In order for the subject to have visual sensations of the world, he/she must be able not only to possess kinesthetic sensations but also to freely initiate kinesthetic sequences: this involves a bodily sense of agency and awareness on the part of the doer [Overgaard and Grünbaum, 2007, p. 20]. The kinesthetic control of perception is related to the problem of generating the objective notion of three-dimensional space, that is, to the phenomenological constitution of a "thing",[43] as a single body unified through the multiplicity of its appearances. The "meaning identity" of a thing is of course related to the continuous flow of adumbrations: given the fact that the incompleteness of adumbrations implies their synthetic consideration in a temporal way, the synthesis in this case, *kinetic*, involves eyes, body, and objects.

Visual sensations are not sufficient to constitute objective spatiality. Kinesthetic sensations[44] (relative to the movements of the perceiver's own body)[45] are required. Petitot continues:

> Besides their "objectivizing" function, kinesthetic sensations share a "subjectivizing" function that lets the lived body appear as a proprioceptive embodiment of pure experiences, and the adumbrations as subjective events. [...] There exists an obvious

[42] Moreover, Husserl thinks that space is endowed with a double function: it is able to constitute a phenomenal extension at the level of sensible data and also it furnishes an intentional moment. Petitot says: "Space possesses, therefore, a noetic face (format of passive synthesis) and a noematic one (pure intuition in Kant's sense)" [Petitot, 1999, p. 336].

[43] Cf. also [Husserl, 1931, §40, p. 129] [originally published in 1913].

[44] Husserl uses the terms "kinestetic sensations" and "kinesthetic sequences" to denote the subjective awareness of position and movement in order to distinguish it from the position and movement of perceived objects in space. On some results of neuroscience that corroborate and improve several phenomenological intuitions cf. [Pachoud, 1999, pp. 211–216] and [Barbaras, 1999; Petit, 1999].

[45] The ego itself is only constituted thanks to the capabilities of movement and action.

> equivalence between a situation where the eyes move and the objects in the visual field remain at rest, and the reciprocal situation where the eyes remain at rest and the objects move. But this trivial aspect of the relativity principle is by no means phenomenologically trivial, at least if one does not confuse what is constituting and what is constituted. Relativity presupposes an *already* constituted space. At the preempirical constituting level, one must be able to discriminate the two equivalent situations. The kinesthetic control paths are essential for achieving such a task [Petitot, 1999, pp. 354–355].

Multidimensional and hierarchically organized, the space of kinesthetic controls includes several degrees of freedom for movements of eyes, head, and body. Kinesthetic controls are kinds of *spatial* gluing operators. They are able to compose, in the case of visual field, different partial aspects – identifying them as belonging to the same object, (cf. Figure 4.1), that is constituting an ideal and transcendent "object". They are realized in the pure consciousness and are characterized by an intentionality that demands a temporal lapse of time.

With the help of very complex eidetic descriptions,[46] that further develop the operations we sketched, Husserl is able to explain the constitution of the objective parametrized time and of space. Discussing the intertwined role played by the kinesthetic systems composed by eyes, head, and body (that is not restricting himself to the oculomotor kinesthetic level), and binocular vision and stereopsis, Husserl is able to account for the formation of objective three-dimensional space and of the three-dimensional things inside it. He stresses that stereopsis derives from the fact that the two binocular images are not identical, so their differences are "intentionally" deciphered by the visual system as depth values[47] thereby the third dimension is constituted.

Of course, when the three-dimensional space (still inexact) is generated (by means of two-dimensional gluing and stereopsis) it is possible to invert the phenomenological order: the visual field is so viewed as a portion of surface in $\mathbf{R}^3$, and the objective constituted space comes first, instead of the objects as they are intuitively and immediately given by direct acquaintance. So the space is in this case an objective datum informing the cognitive agent about the external world where she can find objects from the point of view of their referentiality and denotation. The kinesthetic system "makes the oculomotor field (eventually enlarged to infinity) the mere projection of a three spatial thingness" [Husserl, 1973, section 63, p. 227]. Adumbrations now also appear to be consequences of the objective

[46] Husserl considered it impossible to give a scientific account (for instance mathematical, neurobiological, or in terms of dynamical systems) of operations and events at the level of descriptive eidetics and of morphological essences in perception. The so-called naturalized phenomenology tries to fill this gap [Petitot *et al.*, 1999]. For instance Petitot demonstrates that there exists "[...] a geometrical descriptive eidetics able to assume for perception the constitutive tasks of transcendental phenomenology and to mathematize the correlations between the kinetic noetic synthesis and the noematic morphological *Abschattungen*" [Petitot, 1999, p. 371]; the mathematical apparatus he exploits is partially derived from Thom's theory of catastrophes [Thom, 1975] I will reconsider in chapter eight of this book.

[47] This point of view is confirmed by recent research on stereopsis [Ninio, 1989].

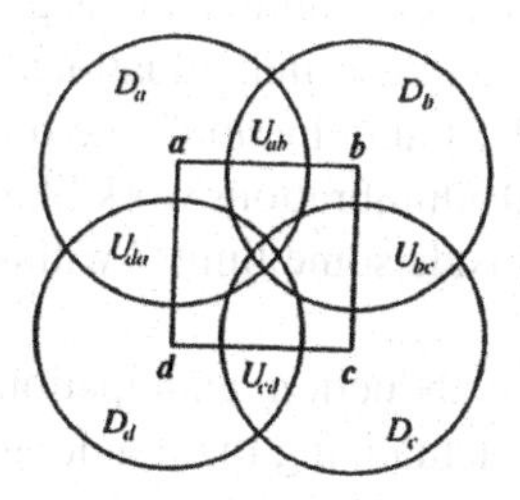

Fig. 4.1 Scanning square S with corners a, b, c, d. To each position p corresponds a token Dp of the visual field D centered on p (focalization on p). The neighboring Dp overlap. (From [Petitot *et al.*, 1999], ©1999 by the Board of Trustees of the Leland Stanford Junior University, Stanford University Press, Stanford, reprinted by permission).

three-dimensional space, as continuous transformations of two-dimensional images as if the body were embedded in the space $\mathbf{R}^3$.[48]

4.7.4 Anticipations as Abductions

Of course adumbrations, the substrate of gluing operations that give rise to the two-dimensional space, are multiple and infinite, and there is a potential co-givenness of some of them (those potentially related to single objects). They are incomplete and partial so for the complete givenness of an object a temporal process is necessary. *Anticipations* are the operations necessary to manage adunbrations that have to be performed by objective transcendence. Adumbrations, not only intuitively presented, can be also represented at the level of *imagination* (on the role of imagination cf. the following section).

Just because incomplete, anticipations correspond to a kind of non-intuitive intentional expectation. When we see a spherical form from one perspective (as an adumbration), we will assume that it is effectively a sphere, but it could be also a hemisphere (an example already employed by Locke).

Anticipations share with visual and manipulative abduction (cf. chapter one, this book) various features: they are highly conjectural and nonmonotonic, so wrong anticipations have to be replaced by other plausible ones. Moreover, they constitute an activity of "generate and test" as a kind of action-based cognition: the finding of adumbrations involves kinesthetic controls, sometimes in turn involving manipulations of objects; but the activity of testing anticipations also implies kinesthetic controls

[48] The role of adumbrations in objectifying entities can be hypothesized in many cases of nonlinguistic animal cognition dealing with the problem of reification and the formation of a kind of "concept", cf. chapter five of this book, section 5.5. In human adults objects are further individuated and reidentified by using both spatial aspects, such as place and trajectory information and static-property information (in this last case exploiting what was gained through previous adumbration activity); adults use this property information to explain and predict appearances and disappareances: "If the same large, distinctive white rabbit appears in the box and later on in the hat, I assume it's the same rabbit" [Gopnik and Meltzoff, 1997].

and manipulations. Finally, not all the anticipations are informationally equivalent and work like attractors for privileged individuations of objects. In this sense the whole activity is toward "the best anticipation", the one that can display the object in an optimal way. Prototypical adumbrations work like structural-stable systems, in the sense that they can "vary inside some limits" without altering the apprehension of the object.

Like in the case of selective abduction, anticipations are able to select possible paths for constituting objects, actualizing them among the many that remain completely tacit. Like in the case of creative abduction, they can construct new ways of aggregating adumbrations, by delineating the constitution of new objects/things. In this case they originate interesting "attractors" that give rise to *new* "conceptual" generalizations.

Some of the wonderful, philosophical Husserlian speculations are being further developed scientifically from the neurological and cognitive perspective in current cognitive science research. [Grush, 2004a; Grush, 2007] has built an emulation theory based on control theory where forward models as emulators (shared by humans and many other animals) are used to illustrate, in the case of humans, various cognitive processes like perception, imagery, reasoning, and language.[49] He contends that simulation circuits are able to hypothesize forward mapping from control signals to the anticipated – and so abduced – consequences of executing the control command. In other words, they mimic the body and its interaction with the environment, enhancing motor control through sensorimotor abductive hypotheticals: "For example, in goal-directed hand movements the brain has to plan parts of the movement before it starts. To achieve a smooth and accurate movement proprioceptive/kinesthetic (and sometimes visual) feedback is necessary, but sensory feedback per se is too slow to affect control appropriately [Desmurget and Grafton, 2002]. The "solution" is an emulator/forward model that can predict the sensory feedback resulting from executing a particular motor command" [Svensson and Ziemke, 2004, p. 1310].

The control theory framework is also useful to describe the emergence of implicit and explicit agency [Grush, 2007]. The humans' understanding of themselves as explicit agents is accomplished through an interplay between the standard egocentric point of view and the so-called "simulated alter-egocentric" point of view, which represents the agent itself as an entity in the environment. In sum two emulators work in tandem when an agent has to conceive its own actions as objective: the first is egocentric and it maintains the egocentric representation of the environment, the second is alter-egocentric and it recalls knowledge of the agent's own action as an entity in the environment. The organism coordinates the two fundamental emulators simultaneously, representing the egocentric environment and at the same time representing the first representation of the scene from the alter-egocentric point of view. This perspective leads to a new and interesting reinterpretation of mirror neurons, I agree with:

> If this is correct, then mirror neurons are, in a sense, not mirror neurons at all. They have one precise function: to fire when there is a representation of another agent

[49] This approach is related to the so-called "simulation theory" (cf. chapter three, subsection 3.6.3.1) which does not posit anything corresponding to an emulator.

> performing some action. Such a representation occurs when the monkey observes another animal perform the action. In that case the representation is a straight-forward perceptual representation. However, when the monkey is itself performing an action a representation of "another" agent performing an action is also maintained in its cognitive system. In this case, this representation is a mock perceptual representation, the situation as perceived by an alter-ego. In fact this alter-ego perceives the monkey itself as another agent. The mirror neurons are the crucial link between the agent's implicit representation of itself and its capacity to represent itself explicitly and objectively as an agent [Grush, 2007, pp. 65–66].

Given the fact that motor imagery can be seen as the off-line driving force of the emulator via efference copies, it is noteworthy that the emulation theory can be usefully extended to account for visual imagery as the off-line operator behind an emulator of the motor-visual loop. In these systems a kind of *amodal* spatial imagery can be hypothesized: "Modal imagery [...] is imagery based on the operation of an emulator of the sensory system itself, whereas amodal imagery is based on the operation of an emulator of the organism and its environment: something like arrangements of solid objects and surfaces in egocentric space. I show how the two forms of emulation can work in tandem" [Grush, 2004a, p. 386]. It is important to note that amodal imagery is neither sentential nor pictorial because the amodal environment space/objects emulators are closely tied to the organism's sensorimotor engagement with the environment.

A discussion of amodal emulators, considered puzzling because they seem entirely independent of any sensory "tags", is provided by [Sathian, 2004], who proposes clearer interpretation of them as multi-modal emulators, systems which receive inputs from more than one sensory modality and are able to coordinate their transformation and integration. Amodal systems, following Sathian, would be better illustrated as conceptual and linguistic rather than as perceptual or as the substrate for either imagery or sensorimotor emulation. However, amodal emulators also refer to cases, like those mentioned by Grush "where an object cannot currently be sensed by any sensory modality (because it is behind an occluder, is silent and odorless, etc.) yet it is represented as being at a location. I think it is safe to say that our representation of our own behavioral (egocentric) space allows for this, and it is not clear how a multisensory system, in which tags for specific modalities were always present, could accomplish this" [Grush, 2004b, p. 434]. Coherent with the Kantian idea of "pure intuition", I would say![50]

4.7.5 *The Genesis of Geometrical Idealities*

What about the genesis of Euclidean geometry? Husserl declares:

> Geometry and the sciences most closely related to it [the prescientific world] have to do with space-time and the shapes, figures, also shapes of motion, alterations of

[50] On Grush's approach cf. the detailed discussion illustrated in [Clark, 2008, chapter seven] in the framework of the theory of the extended mind; a treatment of current cognitive theories, such as the sensorimotor theory of perception, which implicitly furnish a scientific account of the phenomenological concept of anticipation, is given in chapter eight of the same book.

deformation, etc., that are possible within space-time, particularly as measurable magnitudes. It is now clear that even if we know almost nothing about the historical surrounding world of the first geometers, this much is certain as an invariant, essential structure: that it was a world of "things" (including the human beings themselves as subjects of this world); that all things necessarily had to have a bodily character. [...] What is also clear [...] is that these pure bodies had spatiotemporal shapes and "material" qualities (color, warmth, weight, hardness, etc.) related to them. Further it is clear that in the life of practical needs certain particularizations of shape stood out and that a technical praxis always aimed at the production of particular preferred shapes and the improvement of them according to certain directions of gradualness [Husserl, 1978, pp. 177–178, originally published in 1939].

4.7.5.1 Prescientific World

At the origins of geometry, as the science of what is absolutely objective, there is a *prescientific world*,[51] where first of all geometrical protoidealities are produced: it is a world of things/bodies endowed with spatial shapes, shapes of motion and deformations, material qualities, and "disposed of according to an inexact space and time" [Derrida, 1978, p. 122]; by practical needs some of these shapes are perceived, restored and perfected (rigid lines and surfaces). These pregeometrical things of the prescientific world present a merely morphological character (for instance they are more or less smooth surfaces). This pregeometrical, sensible world, is fundamentally a world of interactions with external things/bodies and of manipulations of them.

4.7.5.2 Imagination

An act of *imagination* leads to pure morphological types (like roundness, under which the geometrical ideality of the "circle" will be constructed), that still are of a sensible order (that is they are not yet geometrical idealities, emancipated from any sensible/imaginative intuitiveness): they are a kind of pure but sensible ideality. Husserlian imagination is different from Kant's[52] in that it realizes a kind of "method of variation"[53] of shapes [Husserl, 1978, p. 178] that produces the

[51] Of course we do not have to confuse the prescientific world with the more elementary prepredicative world of appearances and primordial and immediately given experiences. The prescientific world is already characterized by predications, values, empirical manipulations and techniques of measurement.

[52] "According to Kant, geometry is not imaginary because it is grounded on the universal form of pure sensibility, on the ideality of sensible space. But according to Husserl, on the contrary, geometrical ideality is not imaginary because it is uprooted from *all* sensible ground in general. [...] Husserl remains then nearer to Descartes than to Kant. It is true for the latter, as has been sufficiently emphasized, that the concept of sensibility is no longer derived from a 'sensualist' definition. We could not say this is always the case for Descartes or Husserl" [Derrida, 1978, pp. 124-125, footnote 140].

[53] Cf. below subsection 4.7.5.5, "Diagram Constructions as Epistemic Mediators".

essential form. As clearly stated in the *Crisis*: "fantasy can transform sensible shapes only into other sensible shapes" [Husserl, 1970, §9a, p. 25] [1954] uprooting morphological pregeometrical idealities from pure sensible reality. Imagination operates on empirical ways of measurement too (for instance in surveying, design for buildings – for instance altars, pathways) that can be considered a further step in the direction of pure geometrical idealities: "the rough estimate of magnitudes is transformed into the measurement of magnitudes by counting the equal. [...] Measuring belongs to every culture" [Husserl, 1978, p. 178]:

> The art of measuring discovers *practically* the possibility of picking out as standard measures certain empirical basic shapes, concretely fixed on empirical rigid bodies which are in fact generally available; and by means of the relations which obtain (or can be discovered) between these and other body-shapes it determines the latter intersubjectively and in practice univocally – at first within narrow spheres (as in the art of surveying land), then in new spheres where shape is involved [Husserl, 1970, §9a, p. 28].

The philosopher (candidate geometer), "proceeding from the practical, finite, surrounding world (of the room, the city, the landscape, etc.), and temporally the world of periodical occurences: day, month, etc." (*ibid.*) can create geometry by means of idealizations ("limit-shapes emerge toward which the particular series of perfecting tend", [Husserl, 1970, §9a, p. 26]) and the abductive anticipatory structure of intentionality, beyond every sensible and factual level. A horizon of "open infinity" is disclosed. Geometry definitely presents itself like a completely created eidetic science. The whole activity is attributed to the "ancients", governed by the Platonic doctrine of ideas; moreover, Husserl notes exact concepts of geometry have the character of Ideas in the Kantian sense.

4.7.5.3 Orality, Writing, Historicity

First orality, then language and writing are the ways for exposing and making public the generated ideal objectivities: this means those objectivities are constitutively *historical*, so that historicity coincides with the openness of the infinite task of developing geometrical idealities. Ideal objects are *traditional* objects, and so they possess historicity as one of their multiple eidetic components: they are "sedimentations of a truth meaning".

Sedimentation describes the cumulative character of human experience: not every "abiding possession" of mine is traceable to a self-evidence of my own. Those derived from social context are the sedimentations of someone else's experience that, of course, either I can repeat, given the suitable circumstances, or I can taken for granted; every sedimentation produces a traditionalization. So geometrical idealities: "If each geometer tried seriously to repeat all the mental processes on which the work of its predecessors was based, he would have no time of energy left for advancing the discipline [...] In a vast and cumulative enterprise like geometry, especially in its more advanced stages, it would be counterproductive for such an ideal to be realized" [Carr, 1981, p. 252]. In case the tradition were interrupted, the entire process of traditionalization could be started up again.

It is clear that the characterization of history is in this case totally philosophical, and related to the phenomenological explanation of the birth of geometry. Every effort to find the "origins" of geometry is unavoidably noetic, a kind of "reduction", endowed with noetic characters: "We can also say that history is from the start nothing other than the vital movement of the coexistence and the interweaving of original formations and sedimentations of meaning" [Husserl, 1978, p. 174]. Of course this does not mean that an historical approach founded on facts is forbidden. Nevertheless "historicism is mistaken in principle":

> In any case, we can now recognize from all this that historicism, which wishes to clarify the historical and epistemological essence of mathematics from the standpoint of the magical circumstances or other manners of apperception of a time-bound civilization, is mistaken in principle. For romantic spirits the mythical-magical elements of the historical and prehistorical aspects of mathematics may be particularly attractive; but to cling to this merely historically factual aspect of mathematics is precisely to lose oneself to a sort of romanticism and to overlook the genuine problem. The internal-historical problem, the epistemological problem. Also, one's gaze obviously cannot then become free to recognize that facticities of every type, including those involved in the historicist objection, have a root in the essential structure of what is generally human, through which a teleological reason running throughout all historicity announces itself [Husserl, 1978, p. 180].

Husserl contends that "facticities of every type [...] have a root in the essential structure of what is generally human", and that "human surrounding world is the same today and always" (*ibid.*). Of course this does not hold when we consider the possible evolutionary character of this surrounding world. A similar kind of possibility was advanced by Helmholtz and Poincaré, when they hypothesized the famous "fantastic worlds" in which there are beings educated in an environment quite different from ours [Poincaré, 1902, pp. 64–68]. Their different "experience" will lead these beings to classify phenomena in a different way than we would, that is a non-Euclidean way, because it is more convenient, even though the same phenomena could described in a Euclidean way. In fact, Poincaré says that these worlds can be described "without forsaking the use of ordinary geometrical language" [Poincaré, 1902, p. 71].

4.7.5.4 Axiomatics

Finally, definitional expressions and self-evident *axioms* are created, [Husserl, 1978, p. 170] and so the classical axiomatic structure of Euclidean geometry, with the realization of "the highly impressive idea of a systematically deductive theory" [Husserl, 1970, §8, p. 21], where deduction is able to derive, potentially, and from a finite number of concepts and propositions, all the forms that exist in space.

4.7.5.5 Diagram Constructions as Epistemic Mediators

With the institution of geometrical idealities the road to the mathematization of nature, where "nature itself is idealized under the guidance of the new mathematics" [Husserl, 1970, §9, p. 23] is opened: geometry is constantly and practically applied to the world of sense experience, also as a means for technology.[54] The limit-shapes are acquired as tools that can be used and applied habitually. Geometry "becomes in a certain respect a general method of knowing the real" [Husserl, 1970, §9b, p. 33]:

Like all cultural acquisitions which arise out of human accomplishment, they remain objectively knowable and available without requiring that the formulation of their meaning be repeatedly and explicitly renewed. On the basis of sensible embodiment, e.g., in speech and writing, they are simply apperceptively grasped and dealt with in our operations. Sensible "models" function in a similar way, including especially the drawings on paper which are constantly used during work, printed drawings in textbooks for those who learn by reading, and the like. [...] Serving in the methodical praxis of mathematicians, in this form of long-understood acquisitions, are significations which are. So to speak, sedimented in their embodiments. And thus they make mental manipulation possible in the geometrical world of the ideal objects [Husserl, 1970, §9a, pp. 26–27].

Geometrical constructions (that Husserl calls, in the passage above, "models"), used in education, are sedimented in their embodiments. In chapter one I have called these kinds of external cognitive objects *epistemic mediators*. In this perspective geometrical constructions are sedimented epistemic mediators.

We have to notice that these mediators are also important for abductively discovering new geometrical theorems and properties. Following the Husserlian phenomenological point of view, this is due to the fact that they are analogous to the sensible intuitable shapes that were at the origins of geometry, as I have already illustrated. To increase geometrical knowledge it is possible to use these models, that are "sensible"

> [...] according to universal operations which can be carried out with them, to *construct* not only more and more shapes which, because of the method which produces them, are intersubjectively and univocally determined. For in the end the possibility emerges of producing constructively and univocally, through an a priori, all-encompassing systematic method, *all* possible *conceivable* ideal shapes [Husserl, 1970, §9a, p. 27].

The horizon of constructions offers a collection of epistemic mediators that are able to perform creative geometrical developments according to abductive manipulations guided by "universal operations". It is by means of constructions, again, like in Kant, that the "exactness" of new geometrical idealities is attained and, moreover:

> This occurs not only in particular cases, according to an everywhere similar method which, operating on sensible intuitable shapes chosen at random, could carry out idealizations everywhere and originally create, in objective and univocal determinateness, the pure idealities which correspond to them (*ibid.*).

[54] A very simple example is given by the fact it is used to improve methods of measurement increasing the approximation of geometrical tools to the geometrical ideals.

The operations performed by constructions are analogous to the ones Husserl illustrates in the case of the so-called eidetic variation (or method of variation) and consist in trying out various objects that instantiate some essences to see whether they also instantiate others. For instance, using an "empirical example" of an ideal geometrical object, we can find properties of the objects not yet discovered. [Føllesdal, 1999, p. 190] describes another case: Bolzano constructed an example of a continuous function which is not differentiable; Weierstrass thirty years later constructed another more complex non differentiable function – two ways of adding new properties to the ideal concept of function by means of empirical, tangible examples. Consequently, eidetic variation re-echo that "passing beyond" we already explained when Kant examined geometrical constructions and manipulations of a concrete triangle:

> For I must not restrict my attention to what I am actually thinking in my concept of a triangle (this is nothing more than the mere definition); I must pass beyond it to properties which are not contained in this concept, but yet belong to it [Kant, 1929, A718-B746, p. 580].

Like in Kant, in Husserl the role of imagination is still central, together with the acknowledgment of importance of the constraints imposed by the external materiality cognitively exploited (cf. previous chapter, subsection 3.6.7):

> [...] in actual drawing and modelling he [the geometer] is restricted, in fancy he has a perfect freedom in the arbitrary recasting of the figures he has imagined [...]. The drawings therefore follow normally *after* the constructions of fancy and the pure eidetic thought built upon these as a basis, and serve chiefly to fix stages in the process already previously gone through, thereby making it easier to bring it back to consciousness again [Husserl, 1931, §70, pp. 199-200].

The imagination has to be devoted to look for a "perfect clearness" but also has to be enriched by the systematic exploitation of the "best observations" in primordial intuition.

4.7.5.6 Geometry Model of Phenomenology

Like in the case of Kant, in Husserl too geometry can be considered a model of philosophy. This is explicitly stated in *Ideas I*: "[...] the position for the phenomenologist [...] is essentially the same [as the geometer]" (*ibid.*). There are infinite essential forms of the phenomenological kind and the phenomenologist can make a limited use of the primordial order of givenness. Only the main types of primordial data are at his disposal. Consequently he has to operate with the help of "fancy": "But naturally not for all possible special forms, just as little as the geometer depends on drawings and models for the infinite variety of his corporeal types" (*ibid.*).

This relationship between the two eidetic activities and the role of imagination leads us to say that the "element which makes up the life of phenomenology as of all eidetical science is 'fiction' " (Husserl says in the footnote "A sentence which should be particularly appropriate as a quotation for bringing ridicule from the naturalistic side on the eidetic way of knowledge!" – cit., §70, p. 201).

4.8 Non-conceptual and Spatial Abilities

The importance of abductive manipulative skills can also be understood by considering the recent tradition of cognitive research in *dynamical systems*. The importance and the validity of this tradition is controversial in the cognitive science community, yet – as I will better illustrate in chapter eight of this book – I think it can provide many interesting suggestions to further develop the abductive cognitive aspects of external and bodily epistemic mediators. It is well-known that cognitive science has been dominated by the approach that considers cognition as an operation of a special mental computer, situated in the brain.[55] Sensory organs discharge representations of the state of the environment to the mental computer. The system processes a specification of appropriate thought, reasoning or actions. In the last case it is the body which carries out the action. Representations are considered like static structures of symbols, and cognitive procedures are sequential transformations from one structure to the next.

Following the cognitive approach suggested by research in dynamical systems, we immediately learn that cognitive processes and their context have to be considered as unfolding "continuously and simultaneously in real time" [van Gelder and Port, 1995; van Gelder, 1999]. This approach [Clark, 1997] tries to explain how a cognitive system can generate its own change, displaying its self-organizing character. Hence, a cognitive system is not a computer, but a dynamical system, a whole system including the nervous system, body (with its movements, feelings, and emotions), and environment: every cognitive process takes place in real biological hardware and this *embeddedness*, and the various "interactions" involved, become its central aspect.

Many studies in the area are related to the way infants acquire simple body skills and actions, showing how thought is embodied: thought grows from actions and from control of the body, as already guessed and partially described by [Piaget, 1952] and [Piaget and Inhelder, 1956], who illustrated the role of sensorimotor period of here-and-now bodily existence. For example we have a felt embodied understanding of bilateral geometrical symmetry that can be considered one of the basic experiential levels from which the highest levels of human art and language are developed. Experiments show that exploration of the environment and control of body in children is the fundamental step that favors the cognitive constructions of hidden embodied templates and patterns of kinematic skills but also of speech abilities: "[...] what infants sense and what they feel in their ordinary looking and moving are teaching their brain about their bodies and about their worlds" [Thelen, 1995].

When explained in dynamic terms [Saltzman, 1995], the activity of perceptual motor category formation can be easily seen as highly abductive and foundational for all cognitive development. Of course, interactions between the body and the environment are not the only form of cognitive mediation with the world. It is obvious that social communication too (and of course manipulations of the conditions of possibility of this communication) provides rich information to many of our senses.

---

[55] It is the so-called CRUM (Computational-Representational Understanding of Mind) illustrated and criticized by [Thagard, 1996].

Other research that involves the problem of embodiment comes from the field of robotics [Chrisley, 1995], where the study of human cognition emphasizes the role of acts in real-time and real-space environments, going beyond the computer/brain model to expand the analogy to the robot/body. Robotic computation can help to learn how embeddedness and embodiment of non-conceptual abilities in an intentional system can be delineated and understood (cf. also [Dorigo and Colombetti, 1998]).

A very interesting kind of formation of non-conceptual hidden templates, patterns, and skills, is illustrated by the generation of *spatial abilities* in children, adults, non-human mammals and birds. This investigation also challenges some certainties of the propositional view of the mind. Non-conceptual spatial patterns are abductively formed in cognitive situations that are highly integrated with external objects, environment, and body movements [Foreman and Gillett, 1997]. Many cognitive cases are studied using traditional experimentation in psychology and ethology, but also taking advantage of a neurobehavioral approach: 1. egocentric ways of encoding space in comparison to the allocentric ones, capable of guaranteeing navigational spatial skills, 2. spatial mapping in people, 3. homing behavior of various mammals and birds [Bovet, 1997; Papi, 1992], 4. use of landmarks and/or gestures in orientation and perceptual localization (to reach external objects) [Bloch and Morange, 1997], 5. several ways – landmark, route, and survey based – of children's way finding [Blades, 1997], 6. people navigating in large scale environments [Gärling *et al.*, 1997], 7. spatial abilities in mechanical reasoning [Hegarty and Kozhevnikov, 2000].

Many of these cognitive behaviors and attitudes are related in various ways to the exploitation of formation of geometrical shapes and frameworks. For example in the orienting and reaching behavior, distant objects are generally viewed as parts of a spatial situation and are treated by reference to a spatial frame often endowed with geometrical characters. Different spatial frameworks with various geometrical features are present at the level of exploration, as a sensorimotor activity, organized along a body-centred referent, entailing the position of the sensory receptors, the direction of the displacement, and gravitational forces (the so-called "personal" spaces), and at the level of more abstract representations, such as cognitive maps, that are independent of the subject's actual position.

Finally, it is interesting to note that in mammalian species the sensorimotor activity of exploration is displayed in the presence of novel and/or unexpected events and situations. This kind of investigatory behavior seems to be part of a general process of knowledge. Also environment and interactive "punishments" in cases of non-human animals and robots spatial explorations can be cognitively seen as playing the role of contradicting some expected outcomes and as an important step that encourages and supports the emergence of more or less stable spatial templates.[56]

We can guess that some abduced patterns and templates acquired and formed during exploration constitute intermediate steps between body-centred non-conceptual abilities and further representational and more general and abstract frameworks in

[56] On recent research concerning the neural correlated of this process cf. above section 4.7.

humans as well, like for example cognitive maps and diagrams and geometrical knowledge and generalizations. It can be reasonably hypothesized that during exploration, feedback arising from the trajectory itself (vestibular, muscular, and other information) is matched with the ever-changing visual scenes of the environment. This matching would inform the subject that it is moving in a stable environment whose properties are invariant whatever their perceptual appearance [Thinus-Blanc *et al.*, 1997].

Summary

I have contended in this chapter that abduction is essentially *multimodal*, in that both data and hypotheses can have a full range of verbal and sensory representations, involving words, sights, images, smells, etc., but also kinesthetic and motor experiences and other feelings such as pain. Using a methodological approach to rethink abduction in mere neurological terms is a particularly useful exercise. It shows that both linguistic and non linguistic signs have an internal semiotic life as particular configurations of neural networks and chemical distributions (and in terms of their transformations) at the level of human brains, and as somatic expressions, not disregarding the fact they can also be delegated to many external objects and devices. The revaluation of neural and multimodal aspects of abduction is compelling, given the fact that abductive cognition is occurring in a "distributed" framework and in a hybrid way, that is, in the interplay between internal and external signs I have described in the previous chapter. It also shows how the classical perspective on abduction based on logic captures important but limited properties of this cognitive process, basically ignoring the instinctual, model-based and embodied aspects. These last features can only be accounted for if we acknowledge that abductive performances are to a considerable extent neurally hardwired and embodied, resulting from the rigid execution of the DNA program. Of course organisms are also equipped with various ontogenetic mechanisms that – made possible by pre-wired genetic endowments – permit them to plastically acquire and produce information and knowledge and thus better adapt to the environment (for example by forming logical templates of reasoning). If we aim at shedding light on the role of abduction in guessing intentions, in mentation and metamentation, in affective attunement and in other emotional and empathic appraisals, going beyond logical models is mandatory and cognitive processes allowed by both hardwired and pre-wired aspects have to be considered.

Finally, I would like to reiterate the importance of the sections devoted to the distinction between thought and motor action, seeing both aspects as fruit of brain activity. This approach highlights the role of internal and external "reasons" in ethical deliberation, as a form of practical reasoning, which shares many aspects with hypothetical reasoning as it is described by abductive reasoning. In this case abduction nicely accounts for the choice of internal and external "reasons" in decision making processes. I also believe it can help us to overcome some sterile consequences of the old-fashioned assumptions that prohibit arrival at an "ought" from an "is" in the tradition of moral philosophy.

I have concluded the chapter showing how abduction also deserves full attention when performed in mammalian cognitive systems representing the location of objects within the spatial framework. I believe spatial cognition does not have to be merely considered in neurological, biological and psychological terms. We need not fear the synthesis of scientific results with philosophical intuitions, such as those provided by the phenomenological tradition in terms of adumbrations and anticipations. The consideration of these philosophical results in turn acquires further clarification in the framework of the description of agency due to the so-called emulation theory. This framework presents emulation circuits as abductive devices able to hypothesize a forward mapping from control signals to the anticipated consequences of executing the control command.

Chapter 5
Animal Abduction
From Mindless Organisms to Artifactual Mediators

The first two sections of this chapter are strictly related to some seminal Peircean philosophical considerations concerning abduction, perception, inference, and instinct which I consider are still important to current cognitive research. Peircean analysis helps us to better grasp how model-based, sentential and manipulative aspects of abduction have to be seen as intertwined. Moreover, Peircean emphasis on the role of instincts in abduction provides a perfect philosophical introduction to the problem of animal hypothetical cognition in the remaining part of the chapter.

1. First, Peirce explains to us that *perceptions* are abductions, and thus that they are hypothetical and withdrawable. Moreover, given the fact that judgments in perception are fallible but indubitable abductions, we are not in any psychological condition to conceive that they are false, as they are unconscious habits of inference. *Unconscious* cognition legitimately enters the abductive processes (and not only in the case of some aspects of perception, as we will see). The same happens in the case of emotions, which provide a quick – even if often highly unreliable – abductive appraisal/explanation of given data, which is usually anomalous or inconsistent.
2. Second, Peirce contends that perception is the fruit of an abductive "semiotic" activity that is *inferential* in itself. The philosophical reason is simple, Peirce stated that all thinking is in signs, and signs can be icons, indices, or symbols. The concept of sign includes "feeling, image, conception, and other representation": *inference* is in turn a form of sign activity, that is, the word inference is not exhausted by its logical aspects and refers to the effect of various sensorial activities.
3. Third, *iconicity hybridates logicality*: the sentential aspects of symbolic disciplines like logic or algebra coexist with model-based features – iconic. Sentential features like symbols and conventional rules[1] are intertwined with the spatial configuration, like in the case of "compound conventional signs". What I have called sentential abduction is in reality far from being strongly separated

[1] Written natural languages are intertwined with iconic aspects too.

L. Magnani: Abductive Cognition, COSMOS 3, pp. 265–316.
springerlink.com

by model-based aspects: iconicity is always present in human reasoning, even if often hidden and implicit.

It is from this perspective that [sentential] syllogism and [model-based] perception are seen as rigorously intertwined. Consequently, there is no sharp contrast between the idea of abduction (both creative and selective) as perception and the idea of abduction as something that pertains to logic. Both aspects are inferential in themselves and fruit of sign activity. Taking the Peircean philosophical path we return to observations made in the previous chapter: abduction is basically *multimodal* in that both data and hypotheses can have a full range of verbal and sensory representations, involving words, sights, images, smells, etc. but also kinesthetic and motor experiences and feelings such as pain, and thus all sensory modalities.

As I argue in the second section of the chapter, the Peircean architectural view of abduction is completed and further enhanced by his consideration of the role of instincts seen as "inherited habits, or in a more accurate language, inherited dispositions" [Peirce, 1931-1958, 2.170]:

1. Peirce says abduction even takes place when a new born chick picks up the right sort of corn. This is another example of what could be called spontaneous abduction – analogous to the case of some unconscious/embodied abductive processes in humans;
2. not only hypothesis generation has to be considered as a largely instinctual endowment of human beings, a fact which can be justified thanks either to:

   a. the idealistic notion of *synechism*, according to which everything is continuous and the future is in some measure continuous with the past, and where mind and matter are not entirely distinct. Just as it is – analogously – for instinct and inference. The human mind would have developed under the same metaphysical laws that govern the universe leading us to hypothesize that the mind has a tendency to find true hypotheses concerning the universe. Peirce thinks that thought is not necessarily connected with the brain and in fact it appears in the work of bees, crystals and throughout the purely physical world;
   or to
   b. the naturalistic view, which involves at least two aspects: *evolutionary/adaptive* and *perceptual*. The capacity to guess correct hypotheses can be seen as instinctive and enrooted in our evolution and from this perspective abduction is undoubtedly a property of naturally evolving organisms;

3. nature "fecundates" – in Peircean words – the mind because it is through its disembodiment and extension in nature that in turn nature affects the mind. If we argue a conception of mind as extended (cf. chapter three of this book), it is simple to consider its instinctual part as shaped by evolution through the constraints found in nature itself. It is in this sense that the mind's guesses – both instinctual and reasoned (in this last case shaped through the coevolution between nature and cognitive niches) – can be classified as plausible hypotheses

about nature/the external world. The mind grows *together with* the representational delegations to the external world that mind itself has made throughout the history of culture by constructing so-called cognitive niches.

At this point an obvious conclusion can be derived. The idea that there would be another sharp conflict (present in Peirce's texts too) between views of abduction (and of practical reasoning) in terms of *heuristic strategies* and in terms of instinct (insight, perception), appears old-fashioned. The two aspects simply coexist at the level of the real organic agent, it all depends on the cognitive/semiotic perspective we adopt: either

1. we can see an organic agent as a practical agent that mainly takes advantage of its implicit endowments in terms of guessing right, hardwired by evolution, where of course instinct or hardwired programs are central,
   or
2. we can see it as the user of explicit and more or less abstract semiotic devices (language, logic, visualizations, etc.) internally stored and/or externally available – hybrid – which realize heuristic plastic strategies that in some organisms are "conscious". These strategies, used for guessing right, exploit various and contextual relevance and plausibility criteria built up during cultural evolution and made available in cognitive niches, where they can potentially be taught and learnt.

I can conclude that instinct vs. inference represents a conflict we can overcome simply by observing that the work of abduction is partly explicable as a biological phenomenon and partly as a more or less "logical" operation related to "plastic" cognitive endowments of all organisms. I entirely agree with Peirce: a guess in science and the appearance of a new hypothesis is also a biological phenomenon and as such it is related to instinct, in the sense that we can compare it to a chance variation in biological evolution (even if of course the evolution of scientific guesses does not conform to the pattern of biological evolution). An abduced hypothesis introduces a change (and an opportunity) in the semiotic processes to advance new perspectives in the coevolution of the organism and the environment.

The resulting idea that abduction is partly explicable as a biological instinctual phenomenon and partly as a more or less "logical" operation related to "plastic" cognitive endowments of all organisms naturally leads to the remaining sections (starting from section 5.3) of this key chapter. Many animals – traditionally considered "mindless" organisms – make up a series of signs and are engaged in making, manifesting or reacting to a series of signs.[2] Through this semiotic activity – which is considerably model-based – they are at the same time engaged in "being cognitive agents" and therefore in thinking intelligently. An important effect of this semiotic activity is a continuous process of "hypothesis generation" that can be seen at the level of both instinctual behavior, as a kind of "hardwired" cognition,

[2] For a recent survey on the current research on the origin, evolution, dynamics, and learning of signaling systems, in humans and other animals, both from an individual and a "network" perspective, cf. [Skyrms, 2008].

and representation-oriented behavior, where nonlinguistic pseudothoughts drive a plastic model-based cognitive role.

This activity is at the root of a variety of human and non-human abductive performances, which I analyze in the light of the concept of affordance. Another important character of the model-based cognitive activity above is the externalization of artifacts that play the role of mediators in animal languageless reflexive thinking. That is, the interplay between internal and external representations exhibits a new cognitive perspective on the mechanisms underlying the semiotic emergence of abductive processes in important areas of model-based thinking of mindless organisms. To illustrate this process I will take advantage of the case of affect attunement which exhibits an impressive case of model-based communication. Again, a considerable part of abductive cognition occurs through an activity consisting in a kind of reification in the external environment followed by re-projection and reinterpretation through new configurations of neural networks and their chemical processes. Analysis of the central problems of abduction and hypothesis generation helps to address the problems of other related topics in model-based animal cognition, like pseudological and reflexive thinking, the role of pseudoexplanatory guesses in plastic cognition, the role of reification and beliefs and the problem of the relationship between abduction and perception and rationality and instincts.

5.1 Iconicity and Logicality in Reasoning

5.1.1 Perception vs. Inference

We should remember, as Peirce noted and I already stressed in chapter one (cf. section 1.5), that abduction plays a role even in relatively simple visual phenomena. What I have called *visual abduction* [Magnani *et al.*, 1994; Magnani, 1996], a special form of non verbal abduction – a kind of model-based cognition – occurs when hypotheses are instantly derived from a stored series of previous similar experiences. It covers a mental procedure that falls into the category called "perception". Peirce considers *perception* a fast and uncontrolled knowledge-production process. Perception is a kind of vehicle for the instantaneous retrieval of knowledge that was previously assembled in our mind through inferential processes: "[...] a fully accepted, simple, and interesting inference tends to obliterate all recognition of the uninteresting and complex premises from which it was derived" [Peirce, 1931-1958, 7.37]. Perception is abductive in itself: "Abductive inference shades into perceptual judgment without any sharp line of demarcation between them" [Peirce, 1992-1998, p. 224]. In chapter one I have called this kind of image-based hypothesis formation *visual* (or *iconic*) *abduction* (cf. also [Magnani *et al.*, 1994; Magnani, 1996]). It plays an important cognitive role in both everyday reasoning and science, where it is well known it can provide epistemically substantial shortcuts to dramatic new discoveries.

If perceptions are abductions they are basically withdrawable, just like the scientific hypotheses abductively found. They are "hypotheses" about data we can accept (sometimes this happens spontaneously) or carefully evaluate. Moreover, the fact they are, as we will see in the following subsection, *inferences*, in the Peircean sense, and of course withdrawable, does not mean they are controlled (deliberate), like in the case of explicit inferences, for example in logic and other types of rational or fully conscious human reasoning. Perception involves semiosis and is abductive, and it is able to correct itself when it falls into error, and consequently it can be censured. However, we have to carefully analyze the proper character of this kind of controllability, following Peirce's considerations on the so-called "perceptual judgment" ("The seven systems of metaphysics", 1903):

> Where then in the process of cognition does the possibility of controlling it begin? Certainly not before the *percept* is formed. Even after the percept is formed there is an operation, which seems to me to be quite uncontrollable. It is that of judging what it is that the person perceives. A judgment is an act of formation of a mental proposition combined with an adoption of it or act of assent to it. A percept on the other hand is an image or moving picture or other exhibition. The perceptual judgment, that is, the first judgment of a person as to what is before his senses, bears no more resemblance to the percept than the figure [cf. figure 5.1] I am going to draw is like a man.
>
> I do not see that it is possible to exercise any control over that operation or to subject it to criticism. If we can criticize it at all, as far as I can see, that criticism would be limited to performing it again and seeing whether with closer attention we get the same result. But when we so perform it again, paying now closer attention, the percept is presumably not such it was before. I do not see what other means we have of knowing whether it is the same as it was before or not, except by comparing the former perceptual judgment to the later one. I would utterly distrust any other method of ascertaining what the character of the percept was. Consequently, until I am better advised, I shall consider the *perceptual judgment* to be utterly beyond control [Peirce, 1992-1998, II, p. 191].

Fig. 5.1 In [Peirce, 1992-1998, II, p. 191].©1998 Indiana University Press, reprinted by permission.

In summary, judgments in perception are fallible but indubitable abductions – we are not in any condition to psychologically conceive that they are false, as they are unconscious habits of inference.

Nevertheless, percept and perceptual judgment are not unrelated to abduction because they are not entirely free

> [...] from any character that is proper to interpretations [...]. The fact is that it is not necessary to go beyond ordinary observations of common life to find a variety of widely different ways in which perception is interpretative. The whole series of hypnotic phenomena, of which so many fall within the realm of ordinary everyday observation, – such as waking up at the hour we wish to wake much nearer than our waking selves could guess it, – involve the fact that we perceive what we are adjusted for interpreting though it be far less perceptible that any express effort could enable us to perceive [...]. It is a marvel to me that the clock in my study strikes every half an hour in the most audible manner, and yet I never hear it [...]. Some politicians think it is a clever thing to convey an idea which they carefully abstain from stating in words. The result is that a reporter is ready to swear quite sincerely that a politician said something to him which the politician was most careful not to say. It is plainly nothing but the extremest case of Abductive Judgment [Peirce, 1992-1998, II, p. 229].

The fact that perception functions as a kind of "abstractive observation" [Peirce, 1992-1998, II, p. 206], so that "perceptual judgments contain general elements" [Peirce, 1992-1998, II, p. 227] relates it to the expressive power of icons. It is analogous to what is occurring in mathematics when the reasoner "sees" – through the manipulations and constructions on an external single diagram (icon) – that some properties are not merely single but of a general nature: perception functions as "an abstractive observation".[3] Peirce eloquently says that it is "[...] a very extraordinary nature of Diagrams that they show – as literally as Percept shows the Perceptual Judgment to be true, – that a consequence does follow, and more marvelously yet, that it would follow under all varieties of circumstances accompanying the premises" [Peirce, 1976, pp. 317-318].[4]

One more example that supports this interpretative nature is given by the fact that the perception of tone arises from the activity of the mind only after having noted the rapidity of the vibrations of the sound waves, but the possibility of individuating a tone happens only after having heard several of the sound impulses and after having judged their frequency. Consequently the sensation of pitch is made possible by previous experiences and cognitions stored in memory, so that one oscillation of the air would not produce a tone.

For Peirce all knowing is *inferring* and inferring is not instantaneous, it happens in a process that needs an activity of comparisons involving many kinds of patterns in a more or less considerable lapse of time. We have already seen in chapter one that all sensations or perceptions participate in the nature of a unifying hypothesis, that is, in abduction, in the case of emotions too: for example the various sounds made by the musical instruments of the orchestra strike upon the ear, and cause a peculiar musical emotion, completely distinct from the sounds themselves. Emotion is in this case considered by Peirce the same thing as a hypothetic inference, an abduction.

The analogy between abduction and emotion is strong in Peircean writings and Peirce was impressed by the elicitation of an emotion by a complex cognition.

[3] I have illustrated this important cognitive and epistemological issue in chapter two, sections 2.9 and 2.11, and in chapter three, subsection 3.6.3.

[4] Cf. [Turrisi, 1990]. Other considerations on abduction and perception are given in [Tiercelin, 2005].

Emotion detects a kind of inconsistency among a set of beliefs, so what beliefs to abandon has to be determined in producing an explanation that resolves the inconsistency. Peirce said emotions are simple predicates that are elicited by complex predicates like in the case of the anxiety that is triggered by the thought that someone has died. In this sense emotions resort to a wonderful example of model-based reasoning, where a signal is sent to the rest of the brain if a certain event occurs. Peirce would have agreed with the current view that in humans, emotions – that can be hardwired or learned through long experience with other human beings and certain situations – are typically non-intentional abductive signals to themselves that "[...] allow humans, who have slender computational resources, to choose among multiple goals, and to act – despite their limited and often incorrect knowledge, and despite their limited physical powers" [Oatley and Johnson-Laird, 2002, p. 171].

They can sometimes provide the first indication of an inconsistency, like in the case of anxiety when a loved-one's lateness for an appointment exceeds a certain time but also in high-level cognitive settings such as scientific reasoning. They are also useful to enter mental models of other individuals and tell you about their suitability for possible agreements in the future:

> One woman, for instance, waited for a new colleague in one restaurant, while he sat for over an hour in a different located restaurant in the same chain waiting for her. The fact that he had "stood her up", she said, would be at the back of her mind the next time she had dealings with him. Indeed, it was, even though she stated in her diary that his explanation was convincing. He waited longer than she had in the restaurant, and he was the one who had to phone to find out what had gone wrong. She knew he had been no more in fault than her. Nonetheless the emotion of distrust provided a new kind of forward consistency for her in her relations with this man. This evaluation was compelling even though she held explicit beliefs that were inconsistent with it. The emotion overruled the propositional inconsistency [Oatley and Johnson-Laird, 2002, p. 176].

Of course abductive results can also cause emotions, for example depression can be the sign of the invalidity of certain previously abduced hypotheses coupled with some basic aspects of a certain individual's life.

5.1.2 Iconicity Hybridates Logicality

5.1.2.1 Is Perception an Inference?

Let us consider some basic philosophical aspects related to this problem of perception. In the following passage, which Peirce decided to skip in his last of the seven Harvard Lectures (14 May 1903), perception is clearly considered a kind of abduction: "A mass of facts is before us. We go through them. We examine them. We find them a confused snarl, an impenetrable jungle. We are unable to hold them in our minds. [...] But suddenly, while we are poring over our digest of the facts and are endeavoring to set them into order, it occurs to us that if we were to assume something to be true that we do not know to be true, these facts would arrange themselves luminously. That is *abduction* [...]".[5] This passage seems to classify abduction as emerging in "perceiving" facts and experiences, and not only in the conclusions of an "inference" [Hoffmann, 1999, pp. 279-280]. In this sense perceptual and inferential views would be contrasted and a kind of inconsistency would arise, like many of the researchers on Peirce contend. It is well-known that in Peirce the inferential side is first of all expressed and denoted by the syllogistic framework.[6] Following this point of view the genesis of an – abductive – perceptual judgment would have to be located at the level of the premises of the famous Peircean syllogistic schema that depicts abduction as the fallacy of the affirming the consequent. Moreover, it would be at the level of this perceptual side, and *not* at the level of the logico-sentential one that the proper creative virtues of abduction are disclosed. The explaining solution would emerge in perceiving facts and experience and not in the conclusion of the logical inference ("the initial conceiving of a novel hypothesis is not the product of an inferential transition" [Kapitan, 1997, p. 2]).[7]

I further think that the two – often considered contrasting – views more simply and coherently can coexist, beyond Peirce,[8] but also in the perspective of the orthodoxy of Peircean texts: the prevailing Peircean *semiotic* conception of inference as a form of sign activity, where the word sign includes "feeling, image, conception, and other representation" offers the solution to this potential conflict.[9] In this perspective the meaning of the word inference is not exhausted by its "logical" aspects but is referred to the effect of various sensorial activities. One more reason that supports my contention is that for Peirce the sentential aspects of symbolic disciplines like logic or algebra coexist with model-based features – iconic. Sentential features like symbols and conventional rules are intertwined with the spatial configuration; in Peirce's terms:

> The truth, however, appears to be that all deductive reasoning, even simple syllogism, involves an element of observation; namely, deduction consists in constructing an icon or diagram the relations of whose parts shall present a complete analogy with those of the parts of the object of reasoning, of experimenting upon this image in the imagination and of observing the result so as to discover unnoticed and hidden relations among the parts [Peirce, 1931-1958, 3.363] .

[5] Cf. "Pragmatism as the logic of abduction", in [Peirce, 1992-1998, pp. 227-241], the quotation is from footnote 12, pp. 531-532.

[6] Cf. section 1.4, chapter one, this book.

[7] It has to be said that some authors (for example [Hoffmann, 1999, p. 280]) contend that, in order to explain abduction as the process of forming an explanatory hypothesis within Peirce's concept of "logic", it is necessary to see both sides as coming together.

[8] It is well-known that in later writings Peirce seems more inclined to see abduction as both insight and inference.

[9] [Anderson, 1987, p. 45] maintains that "Peirce quite explicitly states that abduction is both an insight and an inference. This is a fact to be explained, not to be explained away". Anderson nicely solves this problem by referring to Peirce's complicated theory of the three fundamental categories, Firstness, Secondness, and Thirdness: abduction, as a form of reasoning is essentially a third, but it also occurs at the level of Firstness "as a sensuous form of reasoning" (p. 56 ff.).

In another passage, which refers to the "conventional" character of algebraic formulas as icons, the idea is even clearer and takes advantage of the introduction of the idea of the "compound conventional sign":[10]

> Particularly deserving of notice are icons in which the likeness is aided by conventional rules. Thus, an algebraic formula is an icon, rendered such by the rules of commutation, association, and distribution of the symbols; that it might as well, or better, be regarded as a compound conventional sign [Peirce, 1966, pp. 787 and pp. frm-e6-28 CSP].

It seems for Peirce that iconicity of reasoning, and consequently of abduction are fundamental, like it is clearly stressed in the following passage: "I said, Abduction, or the suggestion of an explanatory theory, is inference through an Icon" [Peirce, 1986, p. 276]. Moreover, induction and deduction are inferences "through an Index" and "through a Symbol" (*ibid.*).

To conclude, it would seem that there is not an inferential aspect of abduction, characterized by the syllogistic model, separated from (or contrasted with) the perceptual one, which would be "creative" instead, like many authors contend.[11] I consider the two aspects consistent, and both are perfectly understandable in the framework of Peircean philosophy and semiotics.[12]

A further evidence of the fact that the two aspects of abduction are intertwined derives from the study of children's early word acquisition [Roberts, 2004]. Children form knowledge and expectations about the symbolic functioning of a particular word in routine events where model-based perceptual and manipulative aspects of reasoning are predominant and furnish suitable constraints: they generate abductions that help to acquire the content-related symbolic functioning, going beyond what was already experienced. These abduced hypotheses are "practical", about knowing how to use a word to direct attention in a certain way. These hypotheses need not be verbalized by the children, who only later on acquire a more theoretical status through a systematization of their knowledge. It is at this level that they are expressed verbally and concern causal frameworks rather than specific causal mechanisms – for instance of natural kind terms.

To deepen the particular "inferential" status of abduction we have illustrated above further problems regarding the relationship between sentential and model-based aspects of abduction have to be analyzed.

---

[10] Cf. subsection 3.6.4, chapter one of this book. [Stjernfelt, 2007] provides a full analysis of the role of icons and diagrams in Peircean philosophical and semiotic approach, also taking into account the Husserlian tradition of phenomenology.

[11] For example [Kapitan, 1997; Hoffmann, 1999].

[12] On the contrary, some authors (for example [Hoffmann, 1999; Hoffmann, 2004; Paavola, 2004]), as [Frankfurt, 1958, p. 594] synthesized, find a central paradox in "[...] that Peirce holds both that hypotheses are the products of a wonderful imaginative faculty in man and that they are product of a certain sort of logical inference". Furthermore, some commentators seem to maintain that "creative" aspects of abduction would exclusively belong to the perceptual side.

5.1.2.2 Syllogism vs. Perception

The following is a frequently quoted passage by Peirce on perception and abduction related to the passage on "perceptual judgment" that I reported above at the beginning of this section:

> Looking out of my window this lovely spring morning I see an azalea in full bloom. No no! I do not see that; though that is the only way I can describe what I see. *That* is a proposition, a sentence, a fact; but what I perceive is not proposition, sentence, fact, but only an image, which I make intelligible in part by means of a statement of fact. This statement is abstract; but what I see is concrete. I perform and abduction when I so much as express in a sentence anything I see. The truth is that the whole fabric of our knowledge is one matted felt of pure hypothesis confirmed and refined by induction. Not the smallest advance can be made in knowledge beyond the stage of vacant staring, without making an abduction in every step.[13]

The classical interpretation of this passage stresses the existence of a vicious circle [Hoffmann, 1999, p. 283] on the one hand, we learn that the creativity of abduction is based on the genesis of perceptual judgments. On the other hand, it is now said that any perceptual judgment is in itself the result of an abduction. Or, as Peirce says, "[...] our first premises, the perceptual judgments, are to be regarded as an extreme case of abductive inference, from which they differ in being absolutely beyond criticism" [Peirce, 1931-1958, 5.181].

Surely it can be maintained that for Peirce perception on the whole is more precisely the act of subsuming sense data or "percepts" under concepts or ideas to give rise to perceptual judgments: we have just said that he in turn analyzed this act of subsuming as an abductive inference depicted in syllogistic terms

> (P1) A well-recognized kind of object, *M*, has for its ordinary predicates P[1], P[2], P[3], etc.
>
> (P2) The suggesting object, *S*, has these predicates P[1], P[2], P[3], etc.
>
> (C) Hence, *S* is of the kind *M* [Peirce, 1931-1958, 8.64].

In this abductive inference – which actually is merely selective – the creative act "would" take place in the second premise: if we distinguish in abduction an inferential part and a perceptual one – cf. the previous subsection – (that is the genesis of a perceptual judgment), and if we understand according to Peirce the arising of a perceptual judgment for itself as an abductive inference, then in explaining the possibility of abduction we get an infinite regress:

> On its side, the perceptive judgment is the result of a process, although of a process not sufficiently conscious to be controlled, or, to state it more truly, not controllable and therefore not fully conscious. If we were to subject this subconscious process to logical analysis, we should find that it terminated in what that analysis would represent as an abductive inference, resting on the result of a similar process which a similar logical

[13] Cf. the article "The proper treatment of hypotheses: a preliminary chapter, toward and examination of Hume's argument against miracles, in its logic and in its history" [1901] (in [Peirce, 1966, p. 692]).

> analysis would represent to be terminated by a similar abductive inference and so on *ad infinitum*. This analysis would be precisely analogous to which the sophism of Achilles and the Tortoise applied to the chase of the Tortoise by Achilles, and it would fail to represent the real process for the same reason. Namely, just as Achilles does not have to make the series of distinct endeavors which he is represented as making, so this process of forming the perceptual judgment, because it is sub-conscious and so not amenable to logical criticism, does not have to make separate acts of inference, but performs its act in one continuous process [Peirce, 1931-1958, 5.181].

This recursiveness, and the related vicious circle, even if stressed by many commentators, does not seem to me really important. I think we can give a simpler explanation of this conflict between the inferential and perceptual side of abduction by recalling once again the Peircean *semiotic* conception of inference as a form of sign activity, where the word sign includes "feeling, image, conception, and other representation". It is interesting to note that recent research on non-deductive reasoning stresses the presence of *Gestalt* phenomena not only in the realm of perception but also in the realm of logical inference [Pizzi, 2006].[14]

5.1.2.3 Explicit, Uncontrolled, and Unconscious Inferences in Multimodal Abduction

As I have maintained in the previous subsections, I think that two contrasting views of abduction such as inference and perception can coherently coexist: I have already contended that the prevailing Peircean *semiotic* conception of inference as a form of sign activity offers the solution to the conflict. We also said that for Peirce the sentential aspects of logic, even if central, coexist with model-based features – iconic. Abduction can be performed by words, symbols, and logical inferences, but also by internal processes that treat external sensuous input/signs through a merely unconscious mechanisms which give rise to abductive actions and reactions, like in the case of the humble Peircean chicken (cf. below subsection 5.2.1) or of human emotions and other various implicit ways of thinking. In these last cases sentential aspects do not play any role.

We can say, following Thagard [2005; 2007] that abduction is fundamentally performed in a *multimodal* way:[15] for example, we consciously perform a perceptual judgment about the azalea, and in this case also concepts, ideas and statements certainly play a central abductive role, but – Peirce says, they are only *part* of the whole process: "what I perceived is not proposition, sentence, fact, but only image,

[14] The exploitation of this analogy is fruitful: from the perspective of logic a *Gestalt* effect can be seen as the derivability of incompatible indifferent conclusions on the basis of the same background information, such as it occurs in counterfactual and ampliative – inductive and abductive – reasoning. It can be concluded that beyond the structural analogies between counterfactual, inductive, and abductive reasoning there is a net of unequivocal logical relations between the various forms of conditionals expressing contextually different applications of the considered schema of argument. A kind of general theory of rational inference in which counterfactual, inductive and abductive conditionals can be treated as special cases of what we could call a rational conditional.

[15] On the concept of multimodal abduction see chapter four, section 4.1, in this book.

which I made intelligible in part by means of a statement of fact".[16] It is in this way that perceptions acquire "meanings": they nevertheless remain "hypotheses" about data we can accept (sometimes this happens spontaneously) or carefully submit to criticism. It is in this sense that the visual model of perception does not work in isolation from other modes of perception or from other persons or sources of experience [Gooding, 1996]. As I have already illustrated in the first subsection of this chapter perceptions are withdrawable "inferences", even if not controlled (deliberate), like we control explicit inferences for example in logic and other types of rational human "reasoning" and argumentation.

Being creative is not a peculiarity of perceptual/visual abduction, like – as I have already said – some commentators seem to maintain [Kapitan, 1997]. Moreover, perception and cognition alike are inherently inferential. If awareness, whether propositional or perceptual, is semiotic, then all awareness involves the interpretation of signs, and all such interpretation is inferential: semiosis not only involves the interpretation *of linguistic* signs, but also the interpretation of *non-linguistic* signs. Abduction of course embraces much of these semiotic performances.

Both linguistic and non linguistic signs

1. have an internal semiotic life, as particular configurations of neural networks and chemical distributions (and in terms of their transformations) at the level of human brains, and as somatic expressions,
2. but can also be delegated to many external objects and devices, for example written texts, diagrams, artifacts, etc.

In this "distributed" framework those central forms of abductive cognition that occur in a hybrid way, that is in the interplay between internal and external signs, are of special interest, as I have illustrated in chapter three.

5.2 Instinct vs. Heuristic Strategies

5.2.1 The Peircean Abductive Chicken and Animal Hypothetical Cognition

An example of instinctual (and putatively "unconscious") abduction is given by the case of animal embodied kinesthetic/motor abilities, capable of leading to some appropriate cognitive behavior; Peirce says abduction even takes place when a new born chick picks up the right sort of corn. This is another example, so to say, of spontaneous abduction – analogous to the case of some unconscious/embodied abductive processes in humans:

> When a chicken first emerges from the shell, it does not try fifty random ways of appeasing its hunger, but within five minutes is picking up food, choosing as it picks, and

[16] Cf. "The proper treatment of hypotheses: a preliminary chapter, toward an examination of Hume's argument against miracles, in its logic and in its history" (1901) (in [Peirce, 1966, p. 692]).

> picking what it aims to pick. That is not reasoning, because it is not done deliberately; but in every respect but that, it is just like abductive inference.[17]

What happens when the abductive reasoning in science is strongly related to extra-theoretical actions and manipulations of "external" objects? When abduction is "action-based" on *external models*? When thinking is "through doing", as illustrated in the simple case above[18] of distinguishing the simple textures of cloth by feeling? To answer these questions I have delineated the features of what I have called *manipulative abduction* (cf. above chapter one and [Magnani, 2001b, chapter three]) by showing how we can find in scientific and everyday reasoning methods of constructivity based on external models and actions, where external things, usually inert from the semiotic point of view, acquire central cognitive status.

5.2.2 *Instinct-Based Abduction*

Stressing the role of iconic dimensions of semiosis in the meantime celebrates the virtues of analogy, as a kind of "association by resemblance", as opposed to "association by contiguity":

> That combination almost instantly flashed out into vividness. Now it cannot be contiguity; for the combination is altogether a new idea. It never occurred to me before; and consequently cannot be subject to any *acquired habit.* It must be, as it appears to be, its analogy, or resemblance in form, to the nodus of my problem which brings it into vividness. Now what can that be but pure fundamental association by resemblance? [Peirce, 1931-1958, 7.498].

Hypothesis selection is a largely instinctual endowment[19] of human beings which Peirce thinks is given by God or related to a kind of Galilean "*lume naturale*": "It is a primary *hypothesis* underlying all abduction that the human mind is akin to the truth in the sense that in a finite number of guesses it will light upon the correct hypothesis" [Peirce, 1931-1958, 7.220]. Again, the example of the innate ideas of "every little chicken" is of help to describe this human instinctual endowment:

> How was it that man was ever led to entertain that true theory? You cannot say that it happened by chance, because the possible theories, if not strictly innumerable, at any rate exceed a trillion – or the third power of a million; and therefore the chances are too overwhelmingly against the single true theory in the twenty or thirty thousand years during which man has been a thinking animal, ever having come into any man's head. Besides, you cannot seriously think that every little chicken, that is hatched, has

[17] Cf. the article "The proper treatment of hypotheses: a preliminary chapter, toward and examination of Hume's argument against miracles, in its logic and in its history" [1901] (in [Peirce, 1966, p. 692]).

[18] Cf. section 1.6.2, chapter one of this book.

[19] Instinct is of course in part conscious: it is "always partially controlled by the deliberate exercise of imagination and reflection" [Peirce, 1931-1958, 7.381].

> to rummage through all possible theories until it lights upon the good idea of picking up something and eating it. On the contrary, you think the chicken has an innate idea of doing this; that is to say, that it can think of this, but has no faculty of thinking anything else. The chicken you say pecks by instinct. But if you are going to think every poor chicken endowed with an innate tendency toward a positive truth, why should you think that to man alone this gift is denied? [Peirce, 1931-1958, 5.591].

Paavola [2005] illustrates various Peircean ways of understanding the nature of instinct: *naturalistic* (for getting food and for reproducing), *theistic* (also related to the concept of *agapastic* evolution, the idea that the law of love is operative in the cosmos [Peirce, 1992-1998, I, pp. 352-371]), and *idealistic*. The last case is related to the so-called synechism, according to which everything is continuous and the future will be in some measure continuous with the past, mind and matter are not entirely distinct, and so it is – analogously – for instinct and inference [Santaella, 2005, p. 191].[20] The human mind would have been developed under those metaphysical laws that govern the universe so that we can consequently hypothesize that the mind has a tendency to find true hypotheses concerning the universe.

The naturalistic view of instinct involves at least two aspects: *evolutionary/ adaptive* and *perceptual* – as a "certain insight" [Peirce, 1931-1958, 5.173]: the instinctual insight that leads to a hypothesis is considered by Peirce to be of "the same general class of operations to which Perceptive Judgments belong (*ibid.*) (cf. above section 5.1). Hence, Peirce considers the capacity to guess correct hypotheses as instinctive and enrooted in our evolution and from this perspective abduction is surely a property of naturally evolving organisms:

> If you carefully consider with an unbiased mind all the circumstances of the early history of science and all the other facts bearing on the question [...] I am quite sure that you must be brought to acknowledge that man's mind has a natural adaptation to imagining correct theories of some kind, and in particular to correct theories about forces, without some glimmer of which he could not form social ties and consequently could not reproduce his kind [Peirce, 1931-1958, 5.591].

5.2.3 Mind and Matter Intertwined

Peirce says "Thought is not necessarily connected with brain. It appears in the work of bees, of crystals, and throughout the purely physical world; and one can no more deny that it is really there, than that the colours, the shapes, etc., of objects are really there" [Peirce, 1931-1958, 4.551]. It is vital to explain the meaning of this important statement.

First of all it has to be noted that instincts themselves can undergo modifications through evolution: they are "inherited habits, or in a more accurate language, inherited dispositions" [Peirce, 1931-1958, 2.170]. Elsewhere Peirce seems to maintain that instinct is not really relevant in scientific reasoning but that it is typical of

[20] I think some of the ideas of the traditional synechism can be usefully deepened in the framework of current research on the so-called multiple realizability thesis, which admits that mind can be "realized" in several material supports, cf. [Shapiro, 2004; Baum, 2006].

just "the reasoning of practical men about every day affairs". So as to say, we can perform instinctive abduction (that is not controlled, not "reasoned") in practical reasoning, but this is not typical of scientific thinking:

> These two [practical and scientific reasoning] would be shown to be governed by somewhat different principles, inasmuch as the practical reasoning is forced to reach some definite conclusion promptly, while science can wait a century or five centuries, if need be, before coming to any conclusion at all. Another cause which acts still more strongly to differentiate the methodeutic of theoretical and practical reasoning is that the latter can be regulated by instinct acting in its natural way, while theory of how one should reason depends upon one's ultimate purpose and is modified with every modification of ethics. Theory is thus at a special disadvantage here; but instinct within its proper domain is generally far keener, and surer, and above all swifter, than any deduction from theory can be. Besides, logical instinct has, at all events, to be employed in applying the theory. On the other hand, the ultimate purpose of pure science, as such, is perfectly definite and simple; the theory of purely scientific reasoning can be worked out with mathematical certainty; and the application of the theory does not require the logical instinct to be strained beyond its natural function. On the other hand, if we attempt to apply natural logical instinct to purely scientific questions of any difficulty, it not only becomes uncertain, but if it is heeded, the voice of instinct itself is that objective considerations should be the decisive ones.[21]

I think that the considerations above do not mean, like some commentators seem to maintain [Rescher, 1995; Hoffmann, 1999; Paavola, 2005], that instinct – as a kind of mysterious, not analyzed, guessing power – "does not" operate at the level of conscious inferences like for example in the case of scientific reasoning. I think a better interpretation is the following that I am proposing here: certainly instinct, which I consider a simple and not a mysterious endowment of human beings, is at the basis of both "practical" and scientific reasoning, in turn instinct shows the obvious origin of both in natural evolution. If every kind of cognitive activity is rooted in a hybrid interplay with external sources and representations, which exhibit their specific constraints and features, as I have illustrated in the previous chapters and as would Peirce certainly agree, it does not appear surprising that "[...] the instincts conducive to assimilation of food, and the instincts conducive to reproduction, must have involved from the beginning certain tendencies to think truly about physics, on the one hand, and about psychics, on the other. It is somehow more than a *mere* figure of speech to say that nature fecundates the mind of man with ideas which, when those ideas grow up, will resemble their father, Nature" [Peirce, 1931-1958, 5.591]. Hence, from an evolutionary perspective instincts are rooted in humans in this interplay between internal and external aspects and so it is obvious to see that externalities ("Nature") "fecundate" the mind.

Beyond the multifarious and sometimes contrasting Peircean intellectual strategies and steps in illustrating concepts like inference, abduction, perception and

[21] Cf. Arisbe Website, http://members.door.net/arisbe/. The passage comes from MS L75 Logic, regarded as semeiotic (The Carnegie application of 1902).

instinct, which of course are of great interest for the historians of philosophy,[22] the perspective I have illustrated seems to me to be able to clearly focus on some central recent cognitive issues which I contend also implicitly underlie Peircean thoughts: nature fecundates the mind because it is through a disembodiment and extension of the mind in nature that in turn nature affects the mind. If we contend a conception of mind as "extended", it is simple to grasp its instinctual part as shaped by evolution through the constraints found in nature itself. It is in this sense that the mind's guesses – both instinctual and reasoned – can be classified as plausible hypotheses about nature/the external world because the mind grows up together with the representational delegations to the external world that has made itself throughout the history of culture by constructing the so-called cognitive niches.[23] In this strict perspective hypotheses are not merely made by pure unnatural chance.[24]

Peirce says, in the framework of his synechism that "[...] the reaction between mind and matter would be of not essential different kind from the action between parts of mind that are in continuous union" [Peirce, 1931-1958, 6.277]. This is clearly seen if we notice that "[...] habit is by no means a mental fact. Empirically, we find that some plants take habits. The stream of water that wears a bed for itself is forming an habit" [Peirce, 1931-1958, 5.492]. Finally, here the passage we already quoted at the beginning of this subsection, clearly establishing Peirce's concerns about the mind: "Thought is not necessarily connected with brain. It appears in the work of bees, of crystals, and throughout the purely physical world; and one can no more deny that it is really there, than that the colours, the shapes, etc., of objects are really there" [Peirce, 1931-1958, 4.551].

To conclude, instinct vs. inference represents a conflict we can overcome simply by observing that the work of abduction is partly explicable as an instinctual biological phenomenon and partly as a "logical" operation related to "plastic" cognitive endowments of all organisms. I entirely agree with Peirce: a guess in science, the appearance of a new hypothesis, is also[25] a biological phenomenon and so it is related to instinct: in the sense that we can compare it to a chance variation in biological evolution [Peirce, 1931-1958, 7.38], even if of course the evolution of scientific guesses does not conform to the pattern of biological evolution [Colapietro, 2005, p. 427]. An abduced hypothesis introduces a change (and a chance) in the semiotic processes to advance new perspectives in the co-evolution of the organism and the environment: it is in this way that they find a continuous mutual variation. The organism modifies its character in order to reach better fitness; however, the environment (already artificially – culturally – modified, i.e. a cognitive niche), is equally continuously changing and very sensitive to every modification.

[22] For example, in the latest writings at the beginning of XIX century Peirce more clearly stresses the instinctual nature of abduction and at the same time its inferential nature [Paavola, 2005, p. 150]. On the various approaches regarding perception in Peircean text cf. [Tiercelin, 2005].

[23] The concept of cognitive niche is illustrated in detail in chapter six of this book.

[24] This is not a view that conflicts with the idea of God's creation of human instinct: it is instead meant on this basis, that we can add the theistic hypothesis, if desired.

[25] Of course this conclusion does not mean that artifacts like computers do not or cannot perform abductions.

5.2.4 Peircean Chickens, Human Agents, Logical Agents

It is certainly true that Peirce is also convinced that there is a gap between logic and scientific reasoning on one side and practical reasoning on the other (he also rejects the possibility of a practical logic and consequently of a logic of abductive reasoning in practical contexts): "In everyday business reasoning is tolerably successful but I am inclined to think that it is done as well without the aid of theory as with it" [Peirce, 2005, p. 109]. "My proposition is that logic, in the strict sense of the term, has nothing to do with how you think [...]. Logic in the narrower sense is that science which concerns itself primarily with distinguishing reasonings into good and bad reasonings, and with distinguishing probable reasonings into strong and weak reasonings. Secondarily, logic concerns itself with all that it must study in order to draw those distinctions about reasoning, and with nothing else" (*ibid.*, p. 143). We have illustrated that the role of instinct at the level of human unconscious reasoning is obvious, this kind of cognition has been wired by the evolution (like also happens in the case of some animals, for example the Peircean chicken above), and in some organisms cannot be even partially accessed by consciousness.

However, today we have at our disposal many logics of abduction: [Gabbay and Woods, 2005] contend that these logics are just formal and somewhat idealized descriptions of an abductive agent. A real human agent (the every day "business reasoner") can be considered a kind of biological realization of a nonmonotonic paraconsistent base logic and surely the strategies provided by classical logic and some strictly related non standard logics form a very small part of individual cognitive skills, given the fact that human agents are not in general dedicated to error avoidance like "classical" logical agents [Magnani, 2007a].[26] The fact that human beings are error prone does not have to be considered as something bad from the evolutionary point of view, like Quine contends [1969]: for example, hasty generalizations (and many other fallacies) are bad ways of reasoning but can be the best means for survival (or at least for reaching good aims) in particular contexts [Woods, 2004; Magnani and Belli, 2006].

Questions of relevance and plausibility regarding the activity of guessing hypotheses (in an inductive or abductive way) are embedded at the level of the implicit reasoning performances of the practical agent. It will be at the level of the formal model, as an idealized description of what an abductive/inductive agent does, that for example questions of economy, relevance and plausibility, in so far as they are heuristics strategies, will be explicitly described.[27]

In summary, from a semiotic point of view, the idea that there is a conflict (or a potential conflict) (present in Peirce's texts too) between views of abduction – and of practical reasoning – in terms of heuristic strategies or in terms of instinct (insight, perception) [Hoffmann, 1999; Paavola, 2004; Paavola, 2005], appears

[26] Cf. also chapter seven of this book.

[27] On the role of strategies, plausibility, and economy of research and their relationships with Peircean Grammar, Critic, and Methodeutic cf. [Paavola, 2004]. A detailed and in-depth description of these difficult aspects of philosophical and semiotic issues of Peirce's approach is given in [Kruijff, 2005].

old-fashioned. The two aspects simply coexist at the level of the real organic agent, it depends on the cognitive/semiotic perspective we adopt:

1. we can see it as a practical agent that mainly takes advantage of its implicit endowments in terms of guessing right, wired by evolution, where of course instinct or hardwired programs are central, or
2. we can see it as the user of explicit and more or less abstract semiotic devices internally stored or externally available – or hybrid – where heuristic plastic strategies (in some organism they are "conscious") exploiting relevance and plausibility criteria – various and contextual – for guessing hypotheses are exploited.

What is still important to note is that these heuristic strategies and reasoning devices are determined and created at the level of individuals and through the interplay of both the internal and the external agency already endowed with those cognitive delegated representations and tools occurring in the continuous semiotic activity of "disembodiment of mind" that I have illustrated in chapter three. This interplay of course occurs both at the contextual level of learning in the individual history and at the level of the evolutionary effects. The efficiency of these strategies in terms of "naturalistic" – and so instinctual – characters is just guaranteed by this interplay.

In the first case the role of instinct is clear in the sense that the cognitive skills have been wired thanks to the evolutionary interplay between organic agents and their environments. In the second case the role of instinct is still at stake, but in so far as it is at the origins of the historical process of formation of heuristic strategies, and thus of reasoning devices, there is the same interplay between organism and environment, internal and external representations, but in terms of explicit tools sedimented in historical practices and learnt – and possibly improved – by the individuals. Sedimented heuristics are *context-dependent*, to make an example, scientists, in their various fields, make use of many heuristic strategies that are explicitly stated, learnt, and enhanced. These reasoning processes, even if far from being considered merely instinctual, still serve what Peirce calls "the probable perpetuation" [Peirce, 1992-1998, II, pp. 464 f.] of the race.

For example, when we model abduction through a computational logic-based system, the fundamental operation is to search, which expresses the heuristic strategies [Thagard, 1996]. As I have already illustrated in chapter one (section 1.4), when there is a problem to solve, we usually face several possibilities (hypotheses) and we have to select the suitable one (selective abduction). Accomplishing the assigned task requires that we have to search through the whole space of potential solutions to find the desired one. In this situation we have to rely on heuristics, that are rules of thumb expressed in sentential terms. The well-known concept of *heuristic search*, which is at the basis of many computational systems based on propositional rules, can perform this kind of sentential abduction (selective). We have pointed out that other computational tools can be used to this aim, like neural and probabilistic networks, and frames-like representations also able to imitate both sentential, model-based, and hybrid ways of reasoning of real human agents, but less appropriate to model the traditional concept of heuristic strategy.[28]

5.3 Mindless Organisms and Cognition

I have illustrated in the first two sections of this chapter some seminal Peircean philosophical considerations concerning abduction, perception, inference, and instinct. I think they are of strong philosophical importance in current cognitive research. Peircean analysis helps us to better grasp how model-based, sentential and manipulative aspects of abduction have to be seen as intertwined. Moreover, they show how we can exploit the concept of abduction, as a basic kind of human cognition, not only as a conceptual tool helpful in delineating the first principles of a new theory of science, but also in the unification of interdisciplinary perspectives, which would otherwise remain fragmented and dispersed, and thus devoid of the necessary philosophical analysis.

The present chapter aims at illustrating how Peircean emphasis on the role of instincts in abduction provides a deep philosophical framework which in turn supplies an anticipatory and integrated introduction to the problem of animal hypothetical cognition. In a sense, the results of recent research on animal cognition present a confirmation of the epistemological utility of the theory of abductive cognition I am presenting in this book, a theory which also delineates an "ideal" framework very useful for the interdisciplinary dialogue, where the role of abstract and general concepts is central.

Philosophy itself has for a long time disregarded the ways of thinking and knowing of animals, traditionally considered "mindless" organisms. Peircean insight regarding the role of abduction in animals was a good starting point, but only more recent results in the fields of cognitive science and ethology about animals, and of developmental psychology and cognitive archeology about humans and infants, have provided the actual intellectual awareness of the importance of the comparative studies.

Sometimes philosophy has anthropocentrically condemned itself to partial results when reflecting upon human cognition because it lacked in appreciation of the more "animal-like" aspects of thinking and feeling, which are certainly in operation and are greatly important in human behavior. Also in ethical inquiry a better understanding of animal cognition could in turn increase knowledge about some hidden aspects of human behavior, which I think still tend to evade any ethical account and awareness. Scientists too often disregard the ethical concern.

In the recent [Magnani, 2007d] I maintain that people have to learn to be "respected as things", sometimes, are. Various kinds of "things", and among them work of arts, institutions, symbols, and of course animals, are now endowed with intrinsic moral worth. Animals are certainly morally respected in many ways in our technological societies, but certain knowledge about them has been disregarded. It

[28] A more detailed description of the role of heuristic and strategies in abductive (and deductive) reasoning is illustrated in subsection 7.3.2.2 of chapter seven of this book.

is still difficult to acknowledge respect for their cognitive skills and endowments. Would our having more knowledge about animals happen to coincide with having more knowledge about humans and infants, and be linked to the suppression of constitutive "anthropomorphism" in treating and studying them that we have inherited through tradition? Consequently, would not novel and unexpected achievements in this field be a fresh chance to grant new "values" to humans and discover new knowledge regarding their cognitive features? (cf. also [Gruen, 2002]). Darwin already noted that studying cognitive capacities in humans but also in non-human animals "[...] possesses, also, some independent interest, as an attempt to see how far the study of the lower animals throws light on one of the highest psychical faculties of man" – the moral sense [Darwin, 1981].

Among scientists it is of course Darwin [1985] who first clearly captured the idea of an "inner life" (the "world of perception" included) in some humble earthworms [Crist, 2002]. A kind of mental life can be hypothesized in many organisms: Darwin wanted "to learn how far the worms acted consciously and how much mental power they displayed" [Darwin, 1985, p. 3]. He found levels of "mind " where it was not presumed to exist. It can be said that this new idea, which bridges the gap between humans and other animals, in some sense furnishes a partial scientific support to that metaphysical synechism claimed by Peirce contending that matter and mind are intertwined and in some sense indistinguishable.[29]

5.3.1 Worm Intelligence, Abductive Chickens, Instincts

Let us consider the behavior of very simple creatures. Earthworms plug the opening of their burrow with leaves and petioles: Darwin recognized that behavior as being too regular to be random and at the same time too variable to be merely instinctive. He concluded that, even if the worms were innately inclined to construct protective basket structures, they also had a capacity to "judge" based on their tactile sense and showed "some degree of intelligence" [Darwin, 1985, p. 91]. Instinct alone would not explain how worms actually handle leaves to be put into the burrow. This behavior seemed more similar to their "having acquired the habit" [Darwin, 1985, p. 68]. Crist says: "Darwin realized that 'worm intelligence' would be an oxymoron for skeptics and even from a commonsense viewpoint 'This will strike everyone as very improbable' he wrote [Darwin, 1985, p. 98]. [...] He noted that little is known about the nervous system of 'lower animals', implying they might possess more cognitive potential than generally assumed" [Crist, 2002, p. 5].

It is important to note that Darwin also paid great attention to those external structures built by worms and engineered for utility, comfort, and security. I will describe later on in this chapter the cognitive role of artifacts in both human and non-human animals. Artifacts can be illustrated as *cognitive mediators* [Magnani, 2001b] which are the building blocks that bring into existence what it is now called

[29] The recent discovery of the cognitive roles (basically in the case of learning and memory) played by spinal cord further supports this conviction that mind is extended and distributed and that it can also be – so to say – "brainless" [Grau, 2002].

a "cognitive niche":[30] Darwin maintains that "We thus see that burrows are not mere excavations, but may rather be compared with tunnels lined with cement" [Darwin, 1985, p. 112]. Like humans, worms build external artifacts endowed with precise roles and functions, which strongly affect their lives in various ways, and of course their opportunity to "know" the environment.

I have said their behavior cannot be accounted for in merely instinctual terms. Indeed, the "variability" of their behavior is for example illustrated by the precautionary capacity of worms to exploit pine needles by bending over pointed ends: "Had this not effectually been done, the sharp points could have prevented the retreat of the worms into their burrows; and these structures would have resembled traps armed with converging points of wire rendering the ingress of an animal easy and its egress difficult or impossible" [Darwin, 1985, p. 112]. Cognitive *plasticity* is clearly demonstrated by the fact that Darwin detected that pine was not a native tree! If we cannot say that worms are aware like we are (consciousness is unlikely even among non-human vertebrates), certainly we can acknowledge in this case a form of material, interactive, and embodied, manifestation of awareness in the world.

Recent research has also demonstrated the existence of developmental plasticity in plants [Novoplansky, 2002]. For example developing tissues and organs "inform" the plant about their states and respond according to the signals and substrates they receive. The plant adjusts structurally and physiologically to its own development and to the habitat it happens to be in (for example a plasticity of organs in the relations between neighboring plants can be developed) Mackey, J. M. L. [Sachs, 2002; Grime and Mackey, 2002].

In the following sections I am interested in further improving knowledge on abduction and model-based thinking. By way of introduction let me quote again the interesting Peircean passage again about hypothesis selection and chickens, which touches on both ideas, showing a kind of completely language-free, model-based abduction:

> How was it that man was ever led to entertain that true theory? You cannot say that it happened by chance, because the possible theories, if not strictly innumerable, at any rate exceed a trillion – or the third power of a million; and therefore the chances are too overwhelmingly against the single true theory in the twenty or thirty thousand years during which man has been a thinking animal, ever having come into any man's head. Besides, you cannot seriously think that every little chicken, that is hatched, has to rummage through all possible theories until it lights upon the good idea of picking up something and eating it. On the contrary, you think the chicken has an innate idea of doing this; that is to say, that it can think of this, but has no faculty of thinking anything else. The chicken you say pecks by instinct. But if you are going to think every poor chicken endowed with an innate tendency toward a positive truth, why should you think that to man alone this gift is denied? [Peirce, 1931-1958, 5.591].

and again, even more clearly, in another related passage we have already quoted above

[30] A concept introduced by Tooby and DeVore [1987] and later on reused by Pinker [1997; 2003]. I will illustrate it in the following chapter.

> When a chicken first emerges from the shell, it does not try fifty random ways of appeasing its hunger, but within five minutes is picking up food, choosing as it picks, and picking what it aims to pick. That is not reasoning, because it is not done deliberately; but in every respect but that, it is just like abductive inference.[31]

From this Peircean perspective hypothesis generation is a largely instinctual and *nonlinguistic* endowment of human beings and, of course, also of animals. It is clear that for Peirce abduction is rooted in the instinct and that many basically instinctual-rooted cognitive performances, like emotions, provide examples of abduction available to both human and non-human animals. Also cognitive archeology [Mithen, 1996; Donald, 2001] acknowledges that it was not language that made cognition possible: rather it rendered possible the integration in social environments of preexistent, separated, domain-specific modules in prelinguistic hominids, like complex motor skills learnt by imitation or created independently for the first time [Bermúdez, 2003]. This integration made the emergence of tool making possible through the process of "disembodiment of mind" that I illustrated in in chapter three of this book. Integration also seeks out established policies, rituals, and complicated forms of social cognition, which are related to the other forms of prevalently nonlinguistic cognitive behaviors.

5.3.2 *Nonlinguistic Representational States*

It can be hypothesized that some language-free, more or less stable, *representational* states that are merely model-based[32] are present in animals, early hominids, and human infants. Of course tropistic and classically conditioned schemes can be accounted for without reference to these kind of model-based "representations", because in these cases the response is invariant once the creature in question has registered the relevant stimuli.

The problem of attributing to those beings strictly nonlinguistic model-based inner "thoughts", beliefs, and desires, and thus suitable ways of representing the world, and of comparing them to language-oriented mixed (both model-based and sentential) representations, typical of modern adult humans, appears to be fundamental to comprehending the status of animal presumptive abductive performances.

Of course this issue recalls the traditional epistemological Kuhnian question of the incommensurability of meaning [Kuhn, 1962]. In this case it refers to the possibility of comparing cognitive attitudes in different biological species, which express potentially incomparable meanings. Such problems already arose when dealing with the interpretation of primitive culture. If we admit, together with some ethologists,

---

[31] Cf. the article "The proper treatment of hypotheses: a preliminary chapter, toward and examination of Hume's argument against miracles, in its logic and in its history" [1901] [Peirce, 1966, p. 692].

[32] They do not have to be taken like for example visual and spatial imagery or other internal model-based states typical of modern adult humans, but more like action-related representations and thus intrinsically intertwined with perception and kinesthetic/motor abilities. Saidel [2002] interestingly studies the role of these kinds of representations in rats.

animal behaviorists, and developmental psychologists, that in nonlinguistic organisms there are some intermediate representations, it is still difficult to make an analogy with those found in adult humans. The anthropologists who carried out the first structured research on human primitive cultures and languages already stressed this point, because it is difficult to circumstantiate thoughts that can hold in beings but only manifest themselves in superficial and external conducts (cf. Quine [1960]).

A similar puzzling incommensurability already arises when we deal with the different sensorial modalities of certain species and their ways of being and of feeling to be in the world. We cannot put ourselves in the living situation of a dolphin, which lives and feels by using echolocations, or of our cat, which "sees" differently, and it is difficult to put forward scientific hypotheses on these features using human-biased language, perceptive capacities, and cognitive representations. The problem of the existence of "representation states" is deeply epistemological: the analogous situation in science concerns for example the status of the so-called theoretical terms, like quarks or electrons, which are not directly observable but still "real", reliable, and consistent when meaningfully legitimated/justified by their epistemological unavoidability in suitable scientific research programs [Lakatos, 1970].

I have already said that commitment to research on animal cognition is rare in human beings. Unfortunately, even when interested in animal cognition, human adult researchers, victims of an uncontrolled, "biocentric" anthropomorphic attitude, always risk attributing to animals (and of course infants) their own concepts and thus misunderstanding their specific cognitive skills [Rivas and Burghardt, 2002].

5.4 Animal Abduction

5.4.1 "Wired Cognition" and Pseudothoughts

Nature writes programs for cognitive behavior in many ways. In certain cases these programs draw on cognitive functions and sometimes they do not. In the latter case the fact that we describe the behavioral effect as "cognitive" is just a metaphor. This is a case of *instinctual* behavior, which we should more properly name "hardwired cognition".

Peirce spoke – already over a century ago – of a wide semiotic perspective, which taught us that a human internal representational medium is not necessarily structured like a language. I plan to develop and broaden this perspective. Of course this conviction strongly diverges from that maintained by the intellectual traditions which resort to the insight provided by the modern Fregean logical perspective, in which thoughts are just considered the "senses of sentences". Recent views on cognition are still influenced by this logical perspective, and further stress the importance of an isomorphism between thoughts and language sentences (cf. for example Fodor's theory [1987]).

Bermúdez clearly explains how this perspective also affected the so-called *minimalist view* on animal cognition (also called *deflationary view*) [Bermúdez, 2003, p. 27]. We can describe nonlinguistic creatures as thinkers and capable of goal-directed actions, but we need to avoid assigning to them the type of thinking common to linguistic creatures, for example in terms of belief-desire psychology: "Nonlinguistic thinking does not involve propositional attitudes – and, a fortiori, psychological explanation at the nonlinguistic level is not a variant of belief-desire psychology" (*ibid.*). Belief-desire framework should only be related to linguistic creatures. Instead, the problem for the researcher on animal cognition would be to detect how a kind of what we can call "general belief" is formed, rather than concentrating on its content, as we would in the light of human linguistic tools.

Many forms of thinking, such as imagistic, empathetic, trial and error, and analogical reasoning, and cognitive activities performed through complex bodily skills, appear to be basically model-based and manipulative. They are usually described in terms of living beings that adjust themselves to the environment rather than in terms of beings that acquire information from the environment. In this sense these kinds of thinking would produce responses that do not seem to involve sentential aspects but rather merely "non-inferential" ways of cognition. If we adopt the semiotic perspective above, which does not reduce the term "inference" to its sentential level, but which includes the whole arena of sign activity – in the light of Peircean tradition – these kinds of thinking promptly appear full, inferential forms of thought. Let me recall that Peirce stated that all thinking is in signs, and signs can be icons, indices, or symbols, and, moreover, all *inference* is a form of sign activity, where the word sign includes "feeling, image, conception, and other representation" [Peirce, 1931-1958, 5.283].

From this perspective human and the most part of non-human animals possess what I have called *semiotic brains* (cf. chapter three of this book and [Magnani, 2007c]), which make up a series of signs and which are engaged in making or manifesting or reacting to a series of signs: through this semiotic activity they are at the same time occasionally engaged in "being cognitive agents" (like in the case of human beings) or at least in thinking intelligently.[33]

From a perspective similar to the one I have followed, where a broad concept of cognition is adopted, [Godfrey-Smith, 2002] contends that all the mechanisms that enable organisms to coordinate their behavior with complex environmental conditions by tracking and dealing with them[34] involve some degree of intelligence "so there is no opposition between intelligence and what is often referred to as 'instinct'". From this perspective abductive endowments – that are also complex – range across a multiple set of adaptive and maladaptive skills, from instincts to high-level cognitive functions. They are "sets of capacities" that regard perception, internal representations, memory, learning, decision-making and of course actions. They shade into each other and shade off into non-cognitive parts of the biological organisms. From this angle an idea of "continuity" is adopted with regards to cognition, even if cognition is expressed in various ways, often presenting opposite and

[33] Research in biosemiotics has provided new knowledge about the semiotic aspect of cognition in various organisms; cf. the first issue of the journal *Biosemiotics*, 2008, and the recent collection [Barbieri, 2008].

[34] The environment is complex is so far as it is heterogeneous and problematic but also in so far as it provides suitable opportunities – that is "affordances", cf. chapter six, this book.

incommensurable characteristics. Examples of cases where broad animal capacities are shared with humans typically express this idea of continuity of cognition. These shared capacities can be seen in the evolution of associative learning, in the development – beyond the immediate effect of conditioning – of very complicated and variegated forms of spatial memory and of cognitive spatial maps (which can be inner and/or taking advantage of suitable outer natural or artifactual landmarks),[35] and in social intelligence.

From this point of view a low-level variety of cognition and intelligence exists and it helps in meeting the challenges of complex environments and if so, it is selectively favored.[36] It would be better to use only the term cognition, given the fact the term intelligence has strong connotations in common usage, where it is strictly related to humans and expresses complicated endowments like consciousness and planning skills. Cognition in a broader sense is "discovered and rediscovered by evolution many times", and exists at various levels starting from those which can also be classified as proto-cognitive and which are still embedded in higher-level organisms, such as in many vertebrates and in humans themselves. This is evident in the case of endocrine and immune systems. Godfrey-Smith's main thesis is that the function of cognition is to enable the agent to deal with environmental complexity, even if, "In some cases it is hard to distinguish cognition from other control systems in the body, and hard to distinguish behavior from such things as growth, development and the regulation of metabolism" (*ibid.*). Organisms exploit or consume information, that is semiotic systems consume semiotic resources in the complex environment, including those resources that they and other organisms have themselves previously embedded in it (one of the reasons why the environment is complex is because it is suitably complicated by – and dependant on – the various cognitive niches which all organisms build in it.)[37]

For example, spatial imaging and analogies based on perceiving similarities – fundamentally context-dependent and circumstantiated – are ways of thinking in which the "sign activity" is of a nonlinguistic sort, and it is founded on various kinds of implicit naïve physical, biological, psychological, social, etc., forms of intelligibility. In scientific experimentation on prelinguistic infants a common result is the detection of completely language-free working ontologies, which only later on, during cognitive development, will become intertwined with the effect of language and other "symbolic" ways of thinking.[38]

With the aim of describing the kinds of representations which would be at work in these nonlinguistic cognitive processes Dummett [1993] proposes the term *protothought*. I would prefer to use the term *pseudothought*, to minimize the

---

[35] Cf. chapter four, section 4.7.1.

[36] The "lowest" level would be represented by cases of flexibility of behavior or control of development through the fixed response to an environmental cue – in the absence of proper inner processing, but where the response is not directly driven by the physical properties of the cue but is instead characterized by a kind of arbitrariness.

[37] On the role of organisms in building "cognitive niches" see chapter six in this book.

[38] [Godfrey-Smith, 2002] also stresses the fact that internal representations do not have to be intended as necessarily requiring language.

hierarchical effect that – ethnocentrically – already affected some aspects of the seminal work on primitives of an author like Lévi-Bruhl, who introduced the concept of "pre-logical" primitive mentality [1923]. An example of the function of model-based pseudothoughts can be hypothesized in the perception of space in the case of both human and non-human animals. The perceived space is not necessarily three-dimensional and merely involves the apprehension of movement changes, and the rough properties of material objects. Dummett illustrates the case of the car driver and of the canoeist:

> A car driver or canoeist may have to estimate the speed and direction of oncoming cars and boats and their probable trajectory, consider what avoiding action to take, and so on: it is natural to say that he is highly concentrated in thought. But the vehicle of such thoughts is certainly not language: it would be said, I think, to consist in visual imagination superimposed on the visual perceived scene. It is not just that these thoughts are not in fact framed in words: it is that they do not have the structure of verbally expressed thoughts. But they deserve the name of "protothoughts" because while it would be ponderous to speak of truth or falsity in application to them, they are intrinsically connected with the possibility of their being mistaken: judgment, in a non-technical sense, is just what the driver and the canoeist need to exercise [Dummett, 1993, p. 122].

5.4.2 Plastic Cognition in Organisms' Pseudoexplanatory Guesses

To better understand what the study of nonlinguistic creatures teaches us about model-based and manipulative abduction (and go beyond Peirce's insights on chickens' "hardwired" abductive abilities), it is necessary to acknowledge the fact that it is difficult to attribute many of their thinking performances to innate releasing processes, trial and error or to a mere reinforcement learning, which do not involve complicated and more stable internal representations.

Fleeting and evanescent (not merely reflex-based) pseudorepresentations are needed to account for many animal "communication" performances even at the level of the calls of "the humble and much-maligned chicken", like Evans says:

> We conclude that chicken calls produce effects by evoking representations of a class of eliciting events [food, predators, and presence of the appropriate receiver]. This finding should contribute to resolution of the debate about the meaning of referential signals. We can now confidently reject reflexive models, those that postulate only behavioral referents, and those that view referential signals as imperative. The humble and much maligned chicken thus has a remarkably sophisticated system. Its calls denote at least three classes of external objects. They are not involuntary exclamations, but are produced under particular social circumstances [Evans, 2002a, p. 321].

In sum, in nonlinguistics animals, a higher degree of *abductive* abilities has to be acknowledged: chicken form separate representations faced with different events and they are affected by prior experience (of food, for example). They are mainly due to internally developed plastic capacities to react to the environment, and can be thought of as the fruit of learning. In general this plasticity is often accompanied by the suitable reification of external artificial "pseudorepresentations" (for example landmarks, alarm calls, urine-marks and roars, etc.) which artificially modify the environment, and/or by the referral to externalities already endowed with delegated cognitive values, made by the animals themselves or provided by humans.

The following is an example of not merely reflex-based cognition and it is fruit of plasticity: a mouse in a research lab perceives not simply the lever but the fact that the action on it affords the chance of having food; the mouse "desires" the goal (food) and consequently acts in the appropriate way. Some authors contend this is not the fruit of innate and instinctual mechanisms, merely a trial and error routine, or brute reinforcement learning able to provide the correct (and direct) abductive appraisal of the given environmental situation. Instead it can be better described as the fruit of learnt and flexible thinking devices, which are not merely fixed and stimulus driven but also involve "thought".[39] "Pseudothought" – I have already said – is a better term to use, resorting to the formation of internal structured representations and various – possibly new – links between them. The mouse also takes advantage in its environment of an external device, the lever, which the humans have endowed with a fundamental predominant cognitive value, which can afford the animal: the mouse is able to cognitively pick up this externality, and to embody it in internal, useful representations.

Another example of plastic cognition comes from the animal activity of reshaping the environment through its mapping by means of seed caches:

> Consider, for example, a bird returning to a stored cache of seeds. It is known from both ethological studies and laboratory experiments that species such as chickadees and marsh tits are capable of hiding extraordinary number of seeds in a range of different hiding places and then retrieving them after considerable periods of time have elapsed ([Sherry, 1988], quoted in [Bermúdez, 2003, p. 48]).

It is also likely to hypothesize that this behavior is governed by the combination of a motivational state (a general desire for food) and a memory of the particular location, and how to get to it.[40] The possibility of performing such behavior is based on structured internal pseudorepresentations originating from the previous interplay between internal and external signs suitably picked up from the environment in a step-by-step procedure.

To summarize, in these cases we are no longer observing the simple situation of the Peircean, picking chicken, which "[...] has an innate idea of doing this; that is to say, that it can think of this, but has no faculty of thinking anything else".

[39] On this problem cf. also the following subsection about the distinction between classical and instrumental conditioning. In chapter eight, section 8.4, an analogous case is analyzed in the framework of Pavlovian conditioning: in this perspective the emergence of the relationship between the food and its index, the lever, coincides with the emergence of what we can call a "semiotic" activity.

[40] Of course the use of concepts like "desire", deriving from the "folk-psychology" lexicon, has to be considered merely metaphorical.

This "cognitive" behavior is the one already described by the minimalist contention that there is no need to specify any kind of internal content. It is minimally – here and now and immediately related to action – goal-directed, mechanistic, and not "psychological" in any sense, even in a metaphorical one, as we use the term in the case of animals [Bermúdez, 2003, p. 49].

On the contrary, the birds in the example above have at their disposal flexible ways of reacting to events and evidence, which are explainable only in terms of a kind of *thinking* "something else", to use the Peircean words, beyond mere – so say – "mechanistic" (hardwired) responses. They can choose between alternative behaviors founding their choice on the basis of evidence available to be picked up. The activity is "abductive" in itself: it can be *selective*, when the pseudoexplanatory guess, on which the subsequent action is based, is selected among those already internally available, but it can also be *creative*, because the animal can form and excogitate for the first time a particular pseudoexplanation of the situation at hand and then creatively act on the basis of it. The tamarins quickly learn to select the best hypothesis about the tool – taking into account the different tools on offer – that has to be used to obtain the most food in "varied" situations. To avoid "psychological" descriptions, animal abductive cognitive reaction at this level can be seen as an emergent property of the whole organism, and not, in an anthropocentric way, as a small set of specialized skills like we usually see them in the case of humans. By the way, if we adopt this perspective it is also easier to think that some organisms can learn and memorize even without the brain.[41]

As I will illustrate in subsection 5.4.4, animals occupy different environmental niches that "directly" afford their possibility to act, like Gibson's original theory teaches, but this is only one of the ways the organism exploits its surroundings to be suitably attuned to the environment. When behaviors are more complicated other factors are at stake. For example, animals can act on a goal that they cannot perceive – the predator that waits for the prey for example – so the organism's appraisal of the situation includes factors that cannot be immediately perceived,

Well-known dishabituation experiments have shown how infants use model-based high-level physical principles to relate to the environment. They look longer at the facts that they find surprising, showing what expectations they have; animals like dolphins respond to structured complex gestural signs in ways that can hardly be accounted for in terms of the Gibsonian original notion of immediate affordance. A similar situation can be seen in the case of monkeys that perform complicated technical manipulations of objects, and in birds that build artifacts to house beings that have not yet been born. The problem here is that organisms can dynamically abductively "extract" or "create" – and further stabilize – affordances not previously available, taking advantage not only of their instinctual capacities but also of the plastic cognitive ones (cf. below subsection 5.4.4).

[41] It is interesting to note that recent neurobiological research has shown that neural systems within the spinal cord in rats are quite a bit smarter than most researchers have assumed, they can, for example, learn from experience [Grau, 2002]. Cf. also footnote 29, p. 284.

5.4.3 Artifacts and Classical and Instrumental Conditioning

Other evidence supports the assumption about the relevance of nonlinguistic model-based thinking beyond the mere reflex-based level. The birth of what is called material culture in hominids, I will quote in the following subsection, and the use of artifacts as external cognitive mediators in animals, reflect a kind of *instrumental* thought that cannot be expressed in terms of the minimalist conception. The instrumental properties are framed by exploiting artificially made material cognitive tools that *mediate* and so enhance perception, body kinesthetic skills, and a full-range of new cognitive opportunities. Through artifacts more courses of action can be selected, where – so to say – "sensitivity" to the consequences is higher. In this case actions cannot be accounted for solely in terms of the mere perceptual level.[42]

The difference has to be acknowledged between sensitivity to consequences, which is merely due to innate mechanisms and/or classical conditioning (where behavior is simply modified in an adaptive way on the basis of failures and successes), and the more sophisticated sensitivity performed through some doxastic/representational intermediate states:

> In classical conditioning, a neutral stimulus (e. g., the sound of a bell) is followed by an unconditioned stimulus (e. g., the presentation of food) that elicits a reaction (e. g., salivation). The outcome of classical conditioning is that the conditioned response (the salivation) comes to be given to the conditioned stimulus (the sound of the bell) in the absence of the unconditioned stimulus. In instrumental operant conditioning the presentation of the reinforcing stimulus is contingent on the animal making a particular behavioral response (such as a pecking lever). If the behavioral response does not occur, the reinforcing stimulus is withheld. Classical conditioning behavior is not outcome-sensitive in any interesting sense, since it is not the behavior that is reinforced [Bermúdez, 2003, p. 167].

It is evident that instrumental conditioning is also important in (and intertwined with) tool and artifact construction where for example the ability to *plan* ahead (modifying plans and reacting to contingencies, such as unexpected flaws in the material and miss-hits) is central.

5.4.4 Affordances and Abduction

Gibson's eco-cognitive concept of "affordance" [Gibson, 1979] and Brunswik's interplay between proximal and distal environment also deal with the problem of the so-called model-based pseudothoughts, which concern any kind of thinking far from the cognitive features granted by human language.[43] These kinds of cognitive tools typical of infants and of many animals (and still operating in human adults in various forms of more or less unexpressed thinking) are hypothesized to express the

[42] This sensitivity is already present in birds like ravens [Heinrich, 2002].

[43] A detailed illustration of the relationships between affordances and abduction is given in [Magnani and Bardone, 2008] and in the following chapter of this book.

organic beings' implicit skills to act in a perceived environment as a distal environment ordered in terms of the possibilities to *afford* the action in response to local changes.

Different actions will be suitable to different ways of apprehending aspects of the external world. The objectification of the world made possible by language and other highly abstract organizing cognitive techniques (like mathematics) is not needed. An affordance is a resource or chance that the environment presents to the "specific" organism, such as the availability of water or of finding recovery and concealment. Of course the same part of the environment offers different affordances to different organisms. The concept can be also extended to artificial environments built by humans, my cat affords her actions in the kitchen of my house differently than me, for example I do not find affordable to easily jump through the window or on the table! I simply cannot imagine the number of things that my cat Sheena is possibly "aware" of (and her way of being aware) in a precise moment, such as the taste of the last mouse she caught and the type of memory she has of her last encounter with a lizard:[44] "Only a small part of the network within which mouseness is nested for us extends into the cat's world" [Beers, 1997, p. 203]. It is possible to imagine – but this is just science fiction – that we can "afford" the world like a dolphin only artificializing us by using prosthetic instruments and tools like sonar, fruit of modern scientific knowledge.

It can be hypothesized that in many organisms the perceptual world is the only possible model of itself and in this case they can be accounted for in terms of a merely reflex-based notions: no other internal more or less stable representations are available. In the case of affordance in the sensitive organisms described above the coupling with the environment is more flexible because it is important in coupling with the niche to determine what environmental dynamics are currently the most relevant, among the several ones that afford and that are available. An individual that is looking for its prey and at the same time for a mate (which both immediately afford it without any ambiguity) is contemporarily in front of two different affordances and has to *abductively* select the most suitable one weighting them. Both affordances and the more or less plastic processes of their selection in specific situations can be stabilized, but both can also be modified, increased, and substituted with new ones. In many animals, still at the higher level on not-merely reflex-based cognitive abilities, no representational internal states need be hypothesized [Tirassa *et al.*, 1998].

The etheromorphism of affordances is also important: bats use echolocation, and have a kind of sensory capacity that exceeds that of any man-made systems; dolphins can for example detect, dig out, and feed on fish and small eels buried up to

---

[44] The point of view of Gibson has been taken into account by several people in the computational community, for example by Brooks in robotics [1991]. "Vision is not delivering a high level representation of the world, instead it cooperates with motor controls enabling survival behavior in the environment. [...] While it is very sensible that the main goal of vision in humans is to contribute to moving and acting with objects in the word, it is highly improbable that a set of actions can be identified as the output of vision. Otherwise, vision must include all sort of computations contributing to the acting behavior in that set: it is like saying that vision should cover more or less the whole brain activity" [Domenella and Plebe, 2005, pp. 369–370].

45 cm beneath the sandy seabed and are able to detect the size, structure, shape, and material composition of distant objects.[45] They can also discriminate among aluminum, copper, and brass circular targets, and among circles, squares, and triangular targets covered with neoprene [Roitblat, 2002]. These amazing cognitive performances in dolphins are processed through complex computations that transform one dimensional waves (and multiple echoes), arriving at each of their two ears, into representations of objects and their features in the organism's niche. The process is "multimodal" because dolphins also interface with their world using visual and other auditory signals, vocal and behavioral mimicry, and representational capabilities. It even seems that significant degrees of self-awareness are at work, unique to nonhuman animals [Herman, 2002]. I have already said that it is easy to imagine that we could afford the world in a similar way only by hybridizing ourselves using artificial instruments and tools like sonar: the fruit of modern scientific knowledge.

It is important to note recent research based on Schrödinger's focusing on energy, matter and thermodynamic imbalances provided by the environment,[46] draws the attention to the fact that all organisms, including bacteria, are able to perform elementary *cognitive functions* because they "sense" the environment and process internal information for "thriving on latent information embedded in the complexity of their environment" (Ben Jacob, Shapira, and Tauber [2006, p. 496]). Indeed Schrödinger maintained that life requires the consumption of negative entropy, i.e. the use of thermodynamic imbalances in the environment. As a member of a complex superorganism – the colony, a multi-cellular community – each bacterium possesses the ability to sense and communicate with the other units comprising the collective and performs its work within a distribution task. Hence, bacterial communication entails collective sensing and cooperativity through interpretation of chemical messages, distinction between internal and external information, and a sort of self vs. non-self distinction (peers and cheaters are both active).

In this perspective "biotic machines" are *meaning*-based forms of intelligence to be contrasted with the *information*-based forms of artificial intelligence: biotic machines generate new information, assigning contextual meaning to gathered information. Self-organizing organisms like bacteria are afforded – through a real cognitive act – by "relevant" information that they subsequently couple with the regulating, restructuring, and *plastic* activity of the contextual information (intrinsic meaning) already internally stored, which reflects the intra-cellular state of the cells. Of course the "meaning production" involved in the processes above refers to structural aspects of communication that cannot be related to the specific sentential and model-based cognitive skills of humans, primates, and other simpler animals,

[45] In the case of echolocation a kind of acoustic power is reached through *production* processes. It has been hypothesized that a similar productive system was exploited by some hominids who used their voices to navigate within caves, using particular resonances [Reznikoff, 1987]: indeed the voice has a controllable power, directivity, and spectral properties, which are appropriate to excite resonances modes in specific environments like caves [de Cheveigné, 2006, p. 258].

[46] On the strict relationship between the concept of the cognitive niche and thermodynamics cf. chapter six, section 6.1.3.

but still shares basic functions with these like sensing, information processing, and collective abductive contextual production of meaning. As stressed by Ben Jacob, Shapira, and Tauber

> In short, bacteria continuously sense their milieu and store the relevant information and thus exhibit "cognition" by their ability to process information and responding accordingly. From those fundamental sensing faculties, bacterial information processing has evolved communication capabilities that allow the creation of cooperative structures among individuals to form super-organisms [Ben Jacob *et al.*, 2006, p. 504].

In these cases, following [Godfrey-Smith, 2002], we can say that cognition "shades off" into other kinds of biological processes. It is not usually considered as genuine cognition, for example, when some bacteria adjust themselves to changing circumstances around them by using little internal magnets to distinguish north and south and thus move towards water or when they use external cues, through tactile exploration, to adjust their metabolic processes.[47]

Furthermore, as I will stress below in section 5.7.1, plants plastically take on rapid and reversible "habits" exploiting and building suitable cues in the environment, phenomena that enable us to see that "smart" control of plant behavior and "smart" control of plant development have much in common (like in the case of those fish that develop as male or female via perception of their relative size within a population – cf. [Francis and Barlow, 1993], quoted in [Godfrey-Smith, 2002]). This also means that in many animals cognition controls things other than behavior, such as growth, development, and metabolism.[48]

Organisms need to become *attuned* to the relevant features offered in their environment and many of the cognitive tools built to reach this target are the result of evolution. The wired and embodied perceptual capacities and imagistic, empathetic, trial and error, and analogical devices I have described above already fulfill this task. These capabilities can be seen as devices adopted by organisms that provide them with potential "abductive" powers: they can provide an overall appraisal of the situation at hand and thus orient action. They can be seen as providing abductive "pseudoexplanations" of what is occurring "over there", as it emerges through that material contact with the environment grounded in perceptual interplay. It is through this embodied process that affordances can arise both in wild and artificially modified niches. Peirce had already contended more than one hundred years ago that abduction even takes place when a new born chick picks up the right sort of corn. This is an example, so to say, of spontaneous abduction – analogous to the case of some unconscious/embodied abductive processes in humans.

[47] The case of a bacterium called *Haemophius influenzae*, which causes meningitis is impressive: it displays a range of plastic adaptive behavior that is caused by contingency genes [Moxon *et al.*, 1994; Moxon and Wills, 1999]. These mutable genes permit *Haemophius influenzae* to increase the rate of discriminate mutation [Jablonka and Lamb, 2005], and thus to survive in different microhabitats. *Haemophius influenzae* takes advantage of what can be considered a kind of genetically driven abductive guessing that clearly provides an adaptive response to an ever-changing environment.

[48] On the limitations of application regarding evolutionary standard schemas – such as in the case of the biological species concept – to bacteria, and the need for more "pragmatical" and less evolutionary accounts, cf. [Franklin, 2007].

The original Gibsonian notion of affordance deals with those situations in which the signs and clues the organisms can detect, prompt, or suggest a certain action rather than others. They are immediate, already available, and belong to the normality of the adaptation of an organism to a given ecological niche. Nevertheless, if we acknowledge that environments and organisms evolve and change, and so both their instinctual and cognitive plastic endowments, we may argue that affordances can be related to the variable (degree of) "abducibility" of a configuration of signs: a chair affords sitting in the sense that the action of sitting is a result of a sign activity in which we perceive some physical properties (flatness, rigidity, etc.), and therefore we can ordinarily "infer" (in Peircean sense) that a possible way to cope with a chair is sitting on it. So to say, in most cases it is a spontaneous abduction to find affordances because this chance is already present in the perceptual and cognitive endowments of human and non-human animals.

I maintain that describing affordances that way may clarify some puzzling themes proposed by Gibson, especially the claim concerning the fact that organisms directly perceive affordances and that the value and meaning of a thing is clear on first glance. As I have just said, organisms have at their disposal a standard endowment of affordances (for instance through their hardwired sensory systems), but at the same time they can plastically extend and modify the range of what can afford them through the appropriate cognitive abductive skills (more or less sophisticated). As maintained by several authors [Rock, 1982; Thagard, 1988; Hoffman, 1998; Magnani, 2001b], what we see is the result of an embodied cognitive abductive process. For example, people are adept at imposing order on various, even ambiguous, stimuli [Magnani, 2001b, p. 107]. Roughly speaking, we may say that what organisms *see* (or *feel* with other senses) is what their visual (or other senses') apparatus can, so to say, "explain". It is worth noting that this process happens almost simultaneously without any further mediation. Perceiving affordances has something in common with it. Visual perception is indeed a more automatic and "instinctual" activity, that Peirce claimed to be essentially abductive. Indeed – I repeat – he considers inferential any cognitive activity whatever, not only conscious abstract thought: he also includes perceptual knowledge and subconscious cognitive activity. For instance he says that in subconscious mental activities visual representations play an immediate role [Magnani, 2006b].

We also have to remember that environments evolve and change and so the perceptive capacities especially when enriched through new or higher-level cognitive skills, which go beyond the ones granted by the merely instinctual levels. This dynamics explains the fact that if affordances are usually stabilized this does not mean they cannot be modified and changed and that new ones can be formed.

It is worth noting that the history of the construction of artifacts and various tools can be viewed as a continuous process of building and crafting new affordances upon pre-existing ones or even from scratch. From cave art to modern computers, there has been a co-evolution between humans and the environment they live in. Indeed, what a computer can afford embraces an amazing variety of opportunities and chances comparing with the ones exhibited by other tools and devices. More precisely, a computer as a Practical Universal Turing Machine [Turing, 1969] can mimetically reproduce even some of the most complex operations that the human brain-mind systems carry out (cf. [Magnani, 2006c] and chapter three of this book).

The hypothetical status of affordances reminds us that it is not necessarily the case that just any organisms can detect it. Affordances are a mere potentiality for organisms. First of all perceiving affordances results from an abductive activity in which we infer possible ways to cope with an object from the signs and cues available to us. Some of them are stable and in some cases they are neurally hardwired in the perceptual system. This is especially true when dealing with affordances that have a high cognitive valence. Perceiving the affordances of a chair is indeed not neurally hardwired but strongly rooted and stabilized in our cultural evolution. The differences that we can appreciate are mostly *inter-species* – so to speak. A chair affords a child as well as an adult. But this is not the case of a cat. The body of a cat – actually, the cat can sit down on a chair, but also it can sleep on it – has been shaped by evolution quite differently from us.

In higher-level cognitive performances there is something different, since *intra-species* differences seem to be strongly involved. For instance, only a person that has been taught about geometry can infer the affordances "inside" the new manipulated construction built on a geometrical depicted diagram in front of him/her. He/she has to be an "expert". First of all, artificial affordances are intimately connected to culture and the social settings in which they arise and the suitable availability of knowledge of the individual(s) in question. Secondly, affordances deal with learning. There are some affordances like those of an Euclidean triangle that cannot be perceived without a learning process (for instance a course of geometry): people must be somehow *trained* in order to perceive them. Of course acknowledging this last fact places much more emphasis upon the dynamic and also evolutionary character of affordances. The abductive process at play in these cases is very complicated and requires higher level education in cognitive information and skills.

I have already noted that an artificially modified niche (at both levels of biotic and abiotic sources) can be also called "cognitive niche". Recently it has been contended that cognitive niche construction is a kind of evolutionary process in its own right rather than a mere product of natural selection. Through cognitive niche construction organisms not only influence the nature of their world, but also in part determine the selection pressure to which they and their descendants are exposed (and of course the selection pressures to which other species are subjected).

This form of feedback in evolution has been rarely considered in the traditional evolutionary studies [Day *et al.*, 2003]. On this basis a co-evolution between niche construction and brain development and its cognitive capabilities can be clearly hypothesized, a perspective further supported by some speculative hypotheses given by cognitive scientists and paleoanthropologists (for example [Donald, 1998; Donald, 2001; Mithen, 1996].[49] These authors first of all maintain that the birth of material culture itself was not just the product of a massive "cognitive" chance but also cause of it. In the same light the "social brain hypothesis" (also called

[49] I have treated this problem connecting it to some of Turing's insights on the passage from "unorganized" to "organized" brains in a recent article on the role of mimetic and creative representations in human cognition [Magnani, 2006c]. Cf. chapter three of this book.

"Machiavellian intelligence hypothesis" [Whiten and Byrne, 1988; Whiten and Byrne, 1997; Byrne and Whiten, 1988]), holds that the relatively large brains of human beings and other primates reflect the computational demands of complex social systems and not only the need of processing information of ecological relevance.

5.5 Perception as Abduction

5.5.1 Reifications and Beliefs

Some examples testify how animals are able to form a kind of "concept". These activities are surely at the basis of many possibilities to reify the world. Honey bees are able to learn/form something equivalent to the human concepts of "same" and "different"; pigeons, learn/form such concepts as tree, fish, or human [Gould, 2002; Cook, 2002].[50] Sea lions abduce among already formed equivalence classes: a pup's recognition of its mother "[...] depends on the association of many sensory cues with the common reinforcing elements of warmth, contact, and nourishment, while a female recognition of her sisters may depend on their mutual association with the mother" [Schusterman *et al.*, 2002].

Something more complicated than classical conditioning is at play when some animals are able to *reify* various aspects of the world using a kind of analogical reasoning. In this way they are able to detect similarities in a certain circumstance, which will be properly applied in a second following situation. Of course this capacity promotes the possibility to form a more contextual independent view about the objects perceived, for example it happens when recognizing similarities in objects that afford food. The mechanism is analogous to the one hypothesized by philosophers and cognitive scientists when explaining concept formation in humans, a process that of course in this case greatly takes advantage of the resources provided by language. This way of thinking also provides the chance of grasping important regularities and the related power to re-identify objects and to predict what has to be expected in certain out-coming situations [Schusterman *et al.*, 2002, p. 58]. It is a form of abduction by analogy, which forms something like general hypotheses from specific past event features that can be further applied to new ones.

Bermúdez [2003, chapter four] maintains that the process of ascribing thoughts to animals is a form of what Ramsey called "success semantic" [Ramsey, 1927]. When for example we are confronted by the evidence that a chicken abstains "[...] from eating such caterpillars on account of unpleasant experience" a pseudobelief[51] that something is poisonous can be hypothesized and equated to this event. "Thus any set of actions for whose utility P is a necessary and sufficient condition might be called a belief that P, and so would be true if P, i.e., if they are useful" [Bermúdez, 2003, p. 65]. Success semantics adopts a "thought/truth" condition for belief, respecting the idea that thoughts can be true or false because they represent states of

[50] On "naïve physics" and "naïve biology" in pigeons cf. [Watanabe, 2006].

[51] I have illustrated above at p. 289 the reasons that explain why I prefer to adopt the prefix "pseudo" instead of "proto" in terms such as pseudothought, pseudobelief and pseudological inference.

affairs as holding: thought is truth-evaluable. Utility condition of a belief is a state of affairs that when holding leads to the satisfaction of desires with which that belief is combined. The satisfaction condition is equally that state of affairs that "[...] extinguishes in the right sort of way the behavior to which the desire has given risen. [...] The utility condition of a belief in a particular situation is completely open to the third-person perspective of the ethologist or developmental psychologist [...] and provides a clear way of capturing how an adaptive creature is in tune with its environment without making implausibile claims at the level of the vehicle of representation" (pp. 65 and 68).

Hence, in success semantics the role of reinforcement through satisfaction is still relevant but it does not impede the fact that also internal pseudorepresentations can be hypothesized, especially when we are dealing with non-basic appetites. Indeed, following Bermúdez, we can say that in some cases representational states are at stake and are directly related to evolutionary pressures: "[...] the attunement of a creature to its environment niche is a direct function of the fact that various elements of the subpersonal representational system have evolved to track certain features of the distal environment" [Bermúdez, 2003, p. 69], like in the case of so-called "teleosemantics" [Dretske, 1988]. In other cases intelligent skills arise where it is difficult to hypostatize representational contents in situations where evolutionary notions do not play any role: here "Attunement to the environment arises at the level of organism, rather than at the level of subpersonal representational vehicles. That is to say, an organism can be attuned to the environment in a way that will allow it to operate efficiently and successfully, even if there has not been selective pressure for sensitivity at the subpersonal level to the relevant features of the distal environment" [Bermúdez, 2003, p. 69].

5.5.2 Perception as Abduction

The analysis of animal cognition I have illustrated in the previous sections naturally rejoins the problem of the link between abduction and perception described in the first section – 5.1 – of this chapter (Bermúdez says: "A body is a bundle of properties. But a body is a thing that has certain properties. The simple clustering of collocated features can be immediately perceived, but to get genuine reification there needs to be an understanding (which may or may not be purely perceptual) of a form of coinstantiation stronger that mere spatio-temporal coinstantiation" [Bermúdez, 2003, p. 73]. *Reification* that is behind coinstantiation is not necessarily a matter of the effect generated by the poietic activity of linguistic devices (names for example). Objects over there in the environment, grasped through perception, obey certain principles and behave in certain standard ways that can be reflected and ordered in creatures' brains. To perceive a body is to perceive a cluster of semiotic features that are graspable through different sensory modalities, "but" this process is far beyond the mere activity of parsing the perceptual array. This array has to be put in resonance – to be matched – with already formed suitable configurations of neural networks (endowed with their electrical and chemical processes), which combine the various semiotic aspects arrived at through senses.

These configurations are able for instance to maintain constant some aspects of the environment, like the edges of some standard forms, that also have to be kept constant with respect to kinesthetic aspects related to the motor capabilities of the organism in question. For example these neural configurations compensate variation of size and shape of a distal object with respect to an organism's movements. It is in this sense that we can say, by using a Kantian lexicon, that these neural configurations "construct" the world of the chaotic multiplicity gathered at the level of phenomena. The process is of course very different in different organisms – for example some creatures are not able to retain the size of an object through rotation – but still create a permanent cluster of other appropriate intertwined features.[52]

Perception is strongly tied up with reification. Through an interdisciplinary approach and suitable experimentation some cognitive scientists (cf. for example Raftopoulos [2001b; 2001a]) have recently acknowledged the fact that in humans perception (at least in the visual case) is not strictly modular, like Fodor [1984] argued, that is, it is not encapsulated, hardwired, domain-specific, and mandatory.[53] Neither is it wholly abductively "penetrable" by higher cognitive states (like desires, beliefs, expectations, etc.), by means of top-down pathways in the brain and by changes in its wiring through perceptual learning, as stressed by Churchland [1988]. It is important to consider the three following levels: visual sensation (bodily processes that lead to the formation of retinal image which are still useless – so to say – from the high-level cognitive perspective), perception (sensation transformed along the visual neural pathways in a structured representation), and observation, which consists in all subsequent visual processes that fall within model-based/propositional cognition. These processes "[...] include both post-sensory/semantic interface at which the object recognition units intervene as well as purely semantic processes that lead to the identification of the array – high level vision" [Raftopoulos, 2001b, p. 189].

On the basis of this distinction it seems plausible – like Fodor contends – to think there is a substantial amount of information in perception which is theory-neutral. However, also a certain degree of theory-ladenness is justifiable, which can be seen at work for instance in the case of so-called "perceptual learning". However, this fact does not jeopardize the assumption concerning the basic cognitive impenetrability of perception: in sum, perception is informationally "semi-encaspulated", and also semi-hardwired, but, despite its bottom-down character, it is not insulated from "knowledge". For example, it results from experimentation that illusion is a product of learning from experience, but this does not regard penetrability of perception because these experience-driven changes do not affect a basic core of perception.[54]

Higher cognitive states affect the product of visual modules only after the visual modules "[...] have produced their product, by selecting, acting like filters, which output will be accepted for further processing" [Raftopoulos, 2001a, p. 434], for

[52] On neural correlates of allocentric space in mammals cf. [Roberts, 2001; Freska, 2000; O'Keefe, 1999] and chapter three, section 4.7, this book.

[53] Challenges to the modularity hypothesis are illustrated in [Marcus, 2006].

[54] Evidence on the theory-ladenness of visual perception derived from case-studies in the history of science is illustrated in Brewer and Lambert [2001].

instance by selecting through attention, imagery, and semantic processing, which aspects of the retinal input are relevant, activating the appropriate neurons. I contend these processes are essentially abductive, as is also clearly stressed by Shanahan [2005], who provides an account of robotic perception from the perspective of a sensory fusion in a unified framework: he describes problems and processes like the incompleteness and uncertainty of basic sensations, top-down information flow and top-down expectation, active perception and attention.[55]

It is in this sense that a certain amount of *plasticity* in vision does not imply the full penetrability of perception. As I have already noted, this result does not have to be considered equivalent to the claim that perception is, so to say, not theory-laden. It has to be acknowledged that even basic perceptual computations obey high-level constraints acting at the brain level, which incorporate implicit and more or less model-based assumptions about the world, coordinated with motor systems. At this level, they lack a semantic content, so as they are not learnt, because they are shared by all, and fundamentally hardwired.

Human auditory perception should also be considered semi-encapsulated [de Cheveigné, 2006]. The human auditory system resembles that of other vertebrates, such as mammals, birds, reptiles, amphibians or fish, and it can be thought to derive from simple systems that were originally strictly intertwined with motor systems and thus linked to the sense of space.[56]

Hearing, which works in "dark and cluttered" [de Cheveigné, 2006, p. 253] environments, is complementary to other senses, and has both neural bottom-up and top-down characters. The top-down process takes advantage of descending pathways that send *active* information out from a central point and play a part in selectively "listening" to the environment, involving relevant motor aspects (indeed action is fundamental to calibrating perception). The role of hearing in the perception of space is central, complementing multichannel visual information with samples of the acoustic field picked up by the ears: cues to location of source by means of interaural intensity, difference and distance according to cues like loudness are two clear examples of the abductive *inferential* processes performed by hearing that provide substantial models of the scene facing the agent. The whole process is

[55] [Cohn *et al.*, 2002] propose a cognitive vision system based on abduction and qualitative spatio-temporal representations capable of interpreting the high level semantics of dynamic scenes. [Banerjee, 2006] present a computational system able to manage events that are characterized by a large number of individual moving elements, either in pursuit of a goal in groups (as in military operations), or subject to underlying physical forces that group elements with similar motion (as in weather phenomena). Visualizing and reasoning about happenings in such domains are treated through a multilayered abductive inference framework where hypotheses largely flow upwards from raw data to a diagram, but there is also a top-down control that asks lower levels to supply alternatives if the higher level hypotheses are not deemed sufficiently coherent.

[56] The example of a simple hypothetical organism equipped with two fins and two eyes [Szentagothai and Arbib, 1975] can explain this link between perception and action in the case of vision: "The right eye was connected to the left fin by a neuron, and the left eye to the right fin. When a prey appears within the field the right eye, a command is sent to the left fin to instruct it to move. The organism then turns towards the prey, and this orientation is maintained by bilateral activation until the prey is reached. *Perception* in this primitive organism is not distinct from action" [de Cheveigné, 2006, pp. 253–254].

abductive in so far as it provides selections of cues, aggregation of acoustic fragments according to source and an overall hypothetical meaningful explanation of acoustic scenes, that are normally very complex from the point of view of the plurality of acoustic sources. The auditory system of vertebrates which decouples perception from action (motor systems) – still at work together in acoustically rudimentary organisms – enhances economy, speed, and efficacy of the cognitive system by exploiting abstract models of the environment and motor plans.

High order physical principles are also important in reification: I have already cited the experiments on dishabituation in nonlinguistic infants and animals, which have shown that sensitivity to some physical principles starts at birth, and so before the acquisition of language both in phylogenetic and ontogenetic terms [Bermúdez, 2003, pp. 78–79]. In these results it is particularly interesting to see how nonlinguistic beings are able to detect that objects continue to exist even if not perceived, thus clearly showing a kind of reification at work in the perception of an organized world.

In the various nonlinguistic organisms different sets of spatial and physical principles give rise to different ontologies (normally shared with the conspecifics at a suitable stage of development). The problem is to recognize how they are structured, but also how they "evolve". Of course different properties – constant and regular in an appropriate lapse of time – will be salient for an individual at different times, or for different individuals at a given time. This way of apprehending is implicitly explanatory and thus still abductive (selective or creative) in itself and of course related to the doxastic states I introduced above. Consequently, the "intelligent" organism exhibits a suitable level of flexibility in responding. To make an example, when a mouse is in a maze where the spatial location of food is constant, it is in a condition to choose different paths (through a combination of heuristics and of suitable representations), which can permit it to reach and take the food.[57] This means that in mouse spatial cognition, various forms of prediction/anticipation are at play.

5.6 Is Instinct Rational? Are Animals Intelligent?

5.6.1 Rationality of Instincts

Instincts are usually considered irrational or at least a-rational. Nevertheless, there is a way of considering the behavior performances based on them as *rational*. Based on this conclusion, while all animal behavior is certainly described as rational, at the same time it is still rudimentarily considered instinctual. The consequence is that every detailed hypothesis on animal intelligence and cognitive capacities is given up: it is just sufficient to acknowledge the general "rationality" of animal behavior. Let us illustrate in which sense we have to interpret this apparent paradox. I think

[57] An illustration of the different spatial coordinate systems and their kinesthetic features in rat navigation skills (egocentric, allocentric, in terms of route in a maze space, etc.) is given in the classical Tolman, Ritchie, and Kalish [1946], O'Keefe and Nadel [1978], Gallistel [1990].

the analysis of this puzzling problem can further improve knowledge about model-based and manipulative ways of thinking in humans, offering at the same time an integrated view regarding some central aspects of organisms' cognitive behavior. Furthermore, I contend that the argument for reduction of the distinction between abductive inference and perception/instinct, I have provided in the first two sections of this chapter, offers the suitable philosophical tools to make up this integrated view.

Explanations in terms of psychological states obviously attribute to human beings propositional attitudes, which are a precondition for giving a *rational* picture of the explained behavior. These attitudes are a combination of beliefs and desires. Rational internal – doxastic – states characterize human behavior and are related to the fact that they explain why a certain behavior is appropriate on the basis of a specific relationship between beliefs, desires, and actions (cf. [2007d, chapter seven]). How can this idea of rationality be extended to nonlinguistic creatures such as human infants and several types of animals, where the role of instinct is conspicuous? How can the inferential transformations of their possible internal thoughts be recognized when, even if conceivable as acting in their nervous systems, these thoughts do not possess linguistic/propositional features?

The whole idea of rationality in human beings is basically related to the fact we are able to apply *deductive* formal-syntactic rules to linguistic units in a truth preserving way, an image that directly comes from the tradition of classical logic: a kind of rationality robustly related to "logico-epistemological" ideals. The computational revolution of the last decades has stressed the fact that rationality can also be viewed as linked to ways of thinking such as *abduction* and *induction*, which can in turn be expressed through more or less simple *heuristics*. These heuristics are usually well-assessed and shared among a wide community from the point of view of the criteria of applicability, but almost always they prove to be strongly connected in their instantiation to the centrality of language. Indeed cognitive science and epistemology have recently acknowledged the importance of model-based and manipulative ways of rational thinking in human cognition, but their efficacy is basically considered to be strictly related to their hybridization with the linguistic/propositional level. Consequently, for the reasons I have just illustrated, it is still difficult to acknowledge the rationality of cognitive activities that are merely model-based and manipulative, like those of animals.

At the beginning of this subsection I said that, when dealing with rationality in nonlinguistic creatures, tradition initially leads us to a straightforward acknowledgment of the presumptive and intrinsic "rationality" of their instincts. The background assumption is the seeming impossibility that something ineluctable like instinct cannot be at the same time intrinsically rational. Of course the concept of rationality is in this case paradoxical and the expression "rationality" has to be taken in a Pickwickian sense: indeed, in this case the organisms at stake "cannot" be irrational. A strange idea of rationality! Given the fact that many performances of nonlinguistic organisms are explainable in terms of sensory preconditioning (and so are most probably instinct-based – hardwired – and without learnt and possibly conscious capacities which enable them to choose and decide), the rationality of costs and benefits in these behaviors is expressed in the "non-formal" terms of Darwinian "fitness". For example, in the optimal foraging theory, "rationality" is related to the animal's capacity – hardwired thanks to evolution – to optimize the net amount of energy in a given interval of time. Contrarily to the use of some consciously exploited heuristics in humans, in animals many heuristics of the same kind are simply hardwired and so related to the instinctual adaptation to their niches.

The following example provided by Bermúdez can further clarify the problem. "Redshanks are shorebirds that dig for worms in estuaries at low tide. It has been noticed that they sometimes feed exclusively on large worms and at other times feed on both large and small worms. [...] In essence, although a large worm is worth more to the red shank in terms of quantity of energy gained per unit of foraging time than a small worm, the costs of searching exclusively for large worms can have deleterious consequences, except when the large worms are relatively plentiful" [Bermúdez, 2003, p. 117]. The conclusion is simple: even if the optimal behavior can be described in terms of a "rational" complicated version of expected utility theory, "[...] the behaviors in which it manifests itself do not result from the application of such a theory" (*ibid.*). We can account for this situation in our abductive terms: the alternatives which are "abductively" chosen by the redshanks are already wired, so that they follow hardwired algorithms developed through evolution, and simply instantiate the idea of abduction related to instincts present in Peircean insights.

The situation does not change in the case that we consider short-term and long-term rationality in evolutionary behaviors. In the case of the redshank we deal with "short-term" instinct–based rationality related to fitness, but in the case of animals that sacrifice their lives in a way that increases the lifetime fitness of other individuals we deal with "long-term" fitness. It has to be said that sometimes animals are also "hardwired" to use external landmarks and territory signs, and communicate with each other using these threat-display signals that consent them to avoid direct conflict over food. These artifacts are just a kind of instinct-based *mediators*, which are "instinctually" externalized and already evolutionarily stabilized. These mediators are similar to the cognitive, epistemic, and moral mediators that humans externalize thanks to their plastic high-level cognitive capacities, but less complex and merely instinct-based. I have fully described the role of epistemic mediators in scientific reasoning in chapter one of this book, and of moral mediators in ethics in [Magnani, 2007d].[58]

5.6.2 *Levels of Rationality in Animals*

Beyond the above idea of "rationality" in animals and infants as being related to tropistic behaviors connected to reflexes and inborn skills such as imprinting or classical conditioning, the role of intermediary internal representations has to be clearly acknowledged. In this last case we can guess that a "rational" intelligence

[58] See also below section 5.7.

closer to the one expressed in human cognition, and so related to higher levels of abductive behavior, is operating. We fundamentally deal with behaviors that show the capacity to choose among different outcomes, and which can only be accounted for by hypothesizing learnt intermediate representations and processes. In some cases a kind of decision-making strategy can also be hypothesized: in front of a predator an animal can fight or flee and in some sense one choice can be more rational than the other. In front of the data, to be intended here as the "affordances" in a Gibsonian sense, provided through mere perception and which present various possibilities for action, a high-level process of decision-making is not needed, but choice is still possible. With respect to mere pre-wired capacities the abductive behavior above seems based on reactions that are more flexible.

Bermúdez [2003, p. 121] labels Level 1 this kind of rationality. It differs from "rationality" intended as merely instinct-based, expressed in immutable rigid behaviors (called Level 0). Level 1 rationality (which can still be split in short-term and long-term) is for example widespread in the case of animals that entertain interanimal interactions. This kind of rationality would hold when we clearly see ir-rational animals, which fail to signal to the predator and instead flee, thus creating a bad outcome for group fitness (and for their own lifetime fitness: other individuals will cooperate with them less in the future and it will be less probable for them to find a mate).

To have an even higher level rationality (Level 2) we need to involve the possibility of abductively selecting among different "hypotheses" which make the organisms able to perform certain behaviors: a kind of capacity to select among different "hypotheses" about the data at hand, and to behave correspondingly. This different kind of "rational" behavior, is neither merely related to instincts nor simply and rudimentarily flexible, like in the two previous cases.

To make the hypothesis regarding the existence of this last form of rationality plausible, two epistemological pre-conditions have to be fulfilled. The first is related to the acknowledgment that model-based and manipulative cognitions are endowed with an "inferential" status, as I explained above when dealing with the concept of abduction, taking advantage of the semiotic perspective opened up by Peirce. The second relates to the rejection of the restricted logical perspective on inference and rationality I have described in the previous subsection, which identifies inferences at the syntactic level of natural and artificial/symbolic languages (in this last case, also endowed with the truth-preserving property, which produces the well-known isomorphism between syntactic and semantic/content level). I thhink the perspective in terms of multimodal abductive cognition, which avoids old-fashioned models of rationality, furnishes the suitable background for a unified analysis of the issue.

At this high-rationality level we can hypothesize in nonlinguistic organisms more than the simple selection of actions, seen as merely hardwired and operating at the level of perceptions like the theory of immediate affordances teaches, where a simple instrumental conditioning has attached to some actions a positive worth. Instead, in Level 2 rationality, complicated, relatively stable, internal representations that account for consequences are at work. In this case selecting is selecting – so to speak – for some "reasons": a bird that learns to press a lever in a suitable way to obtain food, which will then be delivered in a given site, acts by considering an association between that behavior and the consequences. A kind of instrumental pseudobelief about the future and about certain probable regularities is established, and contingencies at stake are represented and generalized in a merely model-based way. Then the organism internally holds representations with some stability and attaches utility scores to them: based on their choice a consequent action is triggered, which will likely satisfy the organism's desire. The action will be stopped, in a nonmonotonic way, only in the presence of out-coming obstacles, such as the presence of a predator.

Of course the description above suffers the typical anthropomorphism of the observer's "psychological" explanations. However, beliefs do not have to be considered explicit; nevertheless, some actions cannot be explained only on the basis of sensory input and from knowledge of the environmental parameters. Psychological explanations can be highly plausible when the goal of the action is immediately perceptible or when the distal environment contains immediately perceptible instrumental properties. This is obvious and evident in the case of human beings' abilities, but something similar occurs in some chimpanzees' behavior too. When chimpanzees clearly see some bananas they want to reach and eat, and some boxes available on the scene, they have to form an internal instrumental belief/representation on how to exploit the boxes. This "pseudobelief" is internal because it is not immediately graspable through mere perceptual content:

> Any psychological explanation will always have an instrumental content, but the component needs not take the form of an instrumental belief. [...] instrumental beliefs really only enter the picture when two conditions are met. The first is that the goal of the action should not be immediately perceptible and the second is that there should be no immediately perceptible instrumental properties (that is to say, the creature should be capable of seeing that a certain course of action will lead to a desired result). The fact, however, that one or both of these conditions is not met does not entail that we are dealing with an action that is explicable in non-psychological terms [Bermúdez, 2003, p. 129].

The outcomes are represented, but these "pseudorepresentations" lack in lower kinds of rationality. The following example is striking. A food source was taken away from chicken at twice the rate they walked toward it but advanced toward them at twice the rate they walked away from it: after 100 trials, this did not affect the creatures' behavior which failed to represent the two contingencies ([Hershberger, 1986] quoted in [Bermúdez, 2003, p. 125]). Chicken, which do not retreat from a certain kind of action faced with the fact that a repeated contingency no longer holds, are not endowed with this high level "pseudorepresentational" kind of abductive rationality.

The widespread diffusion of abductive reasoning is also confirmed by research into the cognitive basis of science dealing with the potential "rational" behavior of certain early hominids. [Carruthers, 2002b, p. 78] contests the discontinuity view supporting the idea according to which the human mind needs to be radically reprogrammed by immersion in an appropriate language-community and culture to acquire cognitive processes (and "in order for anything resembling science to become possible"). The evolutionary successful, and "social", art of tracking in hunter-gatherer communities would have helped them to develop imagination – linguistic and model-based – and thus, hypothetical reasoning (abductive) – endowed with a kind of explanatory, causal, and instrumental/predictive power. It would have come about because of its capacity to detect the behavior of animals through the few signs available, to reach the best explanation. This contention offers Carruthers the chance to state that "[...] anyone having a capacity for sophisticated tracking will also have the basic cognitive wherewithal to engage in science" [Carruthers, 2002b, 83]. The only difference lies in aims and beliefs, and of course in the fact that the development of science needed suitable props and aids, such as instruments, the printing press and a collective exchange of ideas. The human beings that created science would not have needed major cognitive re-programming.

5.7 Artifactual Mediators and Languageless Reflexive Thinking

5.7.1 Animal Artifactual Mediators

Even if the animal construction of external *artifactual mediators* is sometimes related to instinct, as I have observed in the subsection 5.6.1, it can also be the fruit of plastic cognitive abilities strictly related to the need to improve actions and decisions.[59] In this case action occurs through the expert delegation of cognitive roles to external tools, like in the case of chimpanzees in the wild, that construct wands for dipping into ant swarms or termite nests. These wands are not innate but highly specialized tools. They are not merely the fruit of conditioning or trial and error processes as is clearly demonstrated by the fact they depend on hole size and they are often built in advance and away from the site where they will be used.

The construction of handaxes by the hominids had similar features. It involved paleocognitive model-based and manipulative endowments such as fleeting consciousness, private speech, imposition of symmetry, understanding fracture dynamics, ability to plan ahead, and a high degree of sensorimotor control. I have already said in subsection 5.3.1 and in chapter three they represent one of the main aspects of the birth of *material culture* and technical intelligence and are at the root of what it has been called the process of a "disembodiment of mind" [Magnani, 2006c; Mithen, 1996].

From this perspective the construction of artifacts is an "actualization" in the external environment of various types of objects and structures endowed with a

[59] I have already stressed that plants also exhibit interesting plastic changes. In resource-rich productive habitats where the activities of the plants "generate" various resources above and below ground that strongly modify the environment, plants themselves exhibit various kinds of, so-called, morphological plasticity – that is, the replacement of existing tissues [Grime and Mackey, 2002, p. 300]. It is important to note that plant plasticity is particularly advantageous when responses are reversible rather than irreversible [Alpert and Simms, 2002]. On plants phenotypic plasticity, like their reaction to appropriate environmental cues, see also [Godfrey-Smith, 2002]; on plants capacity to build complicated niches (plasticity in dispersal, flowering timing, and germination timing) cf. [Donohue, 2005]. On animals artifacts cf. also the articles by James L. Gould [2007] and Jean Mandler [2007] contained in the recent [Margolis and Laurence, 2007]; the book also illustrates other interesting psychological, neurological, evolutionary, and philosophical issues concerning artifacts in general.

cognitive/semiotic value for the individual of for the group. Nonlinguistic beings already externalize signs like alarm calls for indicating predators and multiple cues to identify the location of the food caches, which obey the need to simplify the environment and which of course need suitable spatial memory and representations [Shettleworth, 2002; Balda and Kamil, 2002]. However, animals also externalize complicated artifacts like in the case of Darwin's earthworms that I have illustrated in subsection 5.3.1.

These activities of cognitive delegation to external artifacts is the fruit of expert behaviors that conform to innate or learnt embodied templates of cognitive doing. In some sense they are analogous to the templates of epistemic doing I have described in chapter one, which explain how scientists, through appropriate actions and by building artifacts, elaborate for example a simplification of the reasoning task and a redistribution of effort across time. For example, Piaget says, they "[...] need to manipulate concrete things in order to understand structures which are otherwise too abstract" [1974] also to enhance the social communication of results. Some templates of action and manipulation, which are implicit and embodied, can be *selected* from the set of the ones available and pre-stored, others have to be *created* for the first time to perform the most interesting creative cognitive accomplishments of manipulative cognition.

Manipulative "thinking through doing" is creative in particularly skilled animals, exactly like in the case of human beings, when for example chimpanzees make a "new" kind of wand for the first time. Later on the new behavior can possibly be imitated by the group and so can become a shared "established" way of building artifacts. Indeed chimpanzees often learn about the dynamic of objects from observing them manipulated by other fellows: a process that enhances social formation and transmission of cognition.

5.7.2 Pseudological and Reflexive Thinking

Among the various ways of model-based thinking present in nonlinguistic organisms, some can be equated to well-known inferential functional schemes which logic has suitably framed inside abstract and ideal systems. There are forms of pseudological uses of negation (for example dealing with presence/absence, when mammals are able to discern that a thing cannot have simultaneously two contrary properties), of *modus ponens* and *modus tollens* (of course both related to the presence of a pseudonegation), and of conditionals (cf. [Bermúdez, 2003, chapter seven]). Of course, these ways of reasoning are not truth preserving operations on "propositions" and so they are not based on logical forms, but it can be hypothesized that they are very efficient at the nonlinguistic level, even if they obviously lack an explicit reference to logical concepts and schemes.[60] They are plausibly all connected with innate abilities to detect regularities in the external niche. In addition, forms

[60] On the formation of idealized logical schemes in the interplay between internal and external representations cf. [Magnani, 2007a] and chapter seven of this book. Further results about the role of proto-logical and illogical performances humans share (or do not share) with some animals cf. [Yamazaki *et al.*, 2006] and chapter seven of this book, footnote 22 at p. 381.

of causal thinking are observed, of course endowed with an obvious survival value, related to the capacity to discriminate causal links from mere non-causal generalizations or accidental conjunctions.[61]

It is interesting to note in prelinguistic organisms the use of both "logical" and fallacious types of reasoning. For example the widespread use of "hasty generalization" shows that poor generalizations must not only be considered – in the perspective of a Millian abstract universal standard – as a bad kind of induction. Even if hasty generalizations are considered bad and fallacious in the light of epistemological ideals, they are often strategic to the adaptation of the organism to a specific niche [Magnani and Belli, 2006].[62]

An open question is the problem of how nonlinguistic creatures could possess second-order thoughts on thoughts (and so the capacity to attribute thoughts to others) and first – and second-order – desires (that is desires when one should have a specific first-order desire).[63] In human beings, self-awareness and language are the natural home for these cognitive endowments. Indeed, it is simple to subsume propositions as objects of further propositions for ourselves and for others, and consequently to make "reflexive" thinking possible. This kind of thinking is also sensitive to the inferences between thoughts, which are suitably internally represented as icons of written texts or as representations of our own or others' external voices. In addition, the use of external propositional representations favors this achievement, because it is easy to work over there, in an external support, on propositions through other propositions and then internally recapitulate the results.

If it is difficult to hypothesize that some animals and early infants can attribute beliefs and desires to other individuals without the mediation of language and of what psychologists call the "theory of mind", but it is still plausible to think that they can attribute goal-desires to other individuals. In this sense they still attribute a kind of intentionality, and are consequently able to distinguish in other individuals between merely instinctive and purposeful conducts.[64]

In human beings, intentional attitudes are attributed by interpreters who abductively undertake what Dennett [1987] calls the "intentional stance": they abduce hypotheses about "intentions". These attributions are "[...] ways of keeping track of what the organism is doing, has done, and might do" [Jamesion, 2002, p. 73]. However, animals too have the problem of "keeping track" of the behavior of other individuals. For example, it is very likely they can guess model-based abductive hypotheses about what other organisms are perceiving, even if those perceptions are

---

[61] Human prelinguistic infants show surprise in front of scenes when "action-at-a-distance" is displayed (it seems they develop a pseudothought that objects can only interact causally through physical contact) [Spelke, 1990]. Some fMRI experiments on "perceptual" causality are described in [Fugelsang *et al.*, 2005]: specific brain structures result involved in extracting casual frameworks from the world. In both children and adults these data show how they can grasp causality without inferences in terms of universality, probability, or casual powers.

[62] Cf. also chapter seven of this book, section 7.4.

[63] Cf. also the considerations on second-order cognitive dynamics given in [Clark, 2008, chapter three, section six], in the framework of the theory of the extended mind.

[64] Recent research on mirror neurons in primates and human beings support the neurological foundation of this ability [Rizzolatti *et al.*, 1988; Gallese, 2006], cf. also chapter six, section 6.4.3.

not comprehended and made intelligible through the semantic effect produced by language, like in humans.[65] The importance of this capacity to monitor and predict the conduct of conspecifics and/or predators is evident, but other individuals are not seen as thinkers, instead they are certainly seen as doers.

Recent research has shown in animals various capacities to track and "intentionally" influence other individuals' behavior.[66] Tactical deception takes advantage of the use of various semiotic and motor signs in primates: for example, some females, by means of body displacements not seen by a dominant male, can cheat him when they are grooming another non-dominant male [Tomasello and Call, 1997]. Ants, through externalized released pheromone, deceive members of other colonies: these signs/signals play the role of indirect exchanges of chemicals as units of cheating communication.[67] These activities of deception can be seen in the light of the ability to alter other individuals' sensory perceptions. The case of some jumping hunting spiders illustrated by Wilcox and Jackson is striking. By stalking across the web of their prey, they cheat it, through highly specialized signals, also suitably exploiting aggressive mimicry. The interesting thing is that they plastically adapt their cheating and aggressive behavior to the particular prey species at stake, all this by using a kind of trial and error tactic of learning, also reverting to old strategies when they fail [Wilcox and Jackson, 2002].

To conclude, it can be conjectured that, at the very least, emotions in animals can play a kind of reflexive role because they furnish an appraisal of the other states of the body, which arise in the framework of a particular perceptual scenario. This fact clearly refers to another kind of reflexivity, distant from the one that works in beings able to produce thoughts of thoughts, attribute thoughts to others (so possessing a "theory of mind"), monitor thoughts and belief/desire generation and engage in self-evaluation and self-criticism.[68] Also in adult humans emotions play this reflexive role, but in this case usually emotions can be considerably trained and/or intertwined with the effects produced by culture and thus language.[69] It seems researchers agree in saying that propositions/sentences are the only suitable mediators of second order thoughts. It is plausible to conclude that nonlinguistic creatures are excluded from many typically human ways of thinking, and it is obvious to guess that this reciprocally happens for humans, who do not possess various perceptual and cognitive skills of animals.

A picture offering a simpler view of the cognitive capabilities of organisms, to be considered now, after illustration of various cases in the previous sections, is provided by [Tirassa *et al.*, 1998]. They base their view on the theory of autopoiesis

[65] On the encapsulation of perception in language in humans cf. subsection 5.5.2 above.

[66] Of course these capacities can be merely instinct-based and the fruit of a history of selection of certain genetic "programs", and consequently not learned in particular environmental contingencies, like in the cases I am illustrating here.

[67] Cf. Monekosso, Remagnino, and Ferri [2004], that also illustrate a computational learning program which makes use of an artificial pheromone to find the optimal path between two points in a regular grid.

[68] Nevertheless, we have seen that nonlinguistic organisms "can" revise and change their representations.

[69] Cf. Magnani [2007d], chapter six.

(self-organization) proposed by Maturana and Varela [1980; 1991] and on the adaptivity of organisms seen as an ability to create a subjectively relevant set of environmental dynamics to cope with. They basically distinguish between cognitive architectures where the internal dynamics are entirely coupled with the environment (there would be no internal models of the environment in this case), and where some decoupling capability is demonstrated, exhibiting representational states. The first type includes organisms in which innate, reflex-based endowments produced by evolution, determine the whole realm of possible interactions with their subjective environment. In this case the coupling can be more or less flexible, it is more flexible when organisms are able to weigh up different affordances (cf. above subsection 5.4.4) The second type of cognitive architecture, which is decoupled from the external world and where learning is possible, involves representations (internal models of the subjective environment). They range from simple "deictic" ones (where reifying activity allows the possibility of singling an object out and perhaps labelling it as an individual entity), to base-level and meta-representational ones (the first where object permanence and spatial mapping in the environment are possible, the second where organisms, like human beings and maybe primates, can represent their own thoughts thus performing higher and more abstract cognitive skills).

5.7.3 Affect Attunement and Model-Based Communication

An interesting extension of the model I have introduced chapter three of this book[70] concerning "mimetic and creative representations" in the interplay between internal/external is furnished by the merely model-based case of some nonlinguistic and prelinguistic living beings. Human infants entertain a coordinated communication with their caregivers, and it is well known that many psychoanalysts have stressed the importance of this interplay in the further development of the self and of its relationships with the unconscious states. Infants' emotional states, as "signs" in a Peircean sense, are displayed and put out into the external world through the semiotic externalization of facial expressions, gestures, and vocalizations. The important fact here is that this cognitive externalization is performed in front of a living external "mediator", the mother, "the caregiver", endowed with a perceptual system that can grasp the externalized signs and send a feedback: she cognitively and affectively mediates the initial facial expression and the interplay among the subsequent ones. The interplay above is also indicated as a case of human *affective attunement* [Stern, 1985].

In general an agent can expect a feedback also after having "displayed" suitable signs on a non living object, like a blackboard, but it is clear that in this last case a different performance is at play, which involves explicit manipulations of the external object, and not a mere exchange of – mainly facial-based – sensations, like in affective attunement. The external delegated representation to a non living object shows more or less complicated "active" responses, which are intertwined with the

[70] Cf. also chapter six of this book, section 6.4.4.

agent's manipulations. For example, a blackboard presents intrinsic properties that limit and direct the manipulation in a certain way, and so does a PC, which has – with respect to the blackboard – plenty of autonomous possibility to react: usually the interplay is hybrid, taking into account both propositional, iconic (in a Peircean sense), and of course motor aspects.[71]

In affect attunement, the interplay is mainly model-based and mostly iconic (also taking advantage of the iconic force of gestures[72] and voice), meaningful words are also present, but the semiotic "propositional" flow is fully understood only on the part of the adult, not on the part of the infant, where words and their meanings are simply being learnt. The infant performs an "expressive" behavior based on appearances and gestures that are spontaneously externalized to get a feedback. Initially the expressions externalized are directly *mimetic* of the inner state but – through the interplay – where subsequent recapitulations of the mother's facial expressions are performed and are gradually, suitably picked up "outside" the mom's body, novel "social" expressions are formed. These expressions are shared with the mom and thus they are no longer arbitrary. Once stabilized, they constitute the expected affective "attunement" to the mom/environment, which is the fruit of a whole abductive model-based activity of subsequent "facial hypotheses". In this process, the external manifestation of the nonlinguistic organism is established as the quality of feelings that testify a shared affect. A new way of sharing affect is abductively *created*, which is at the basis of the further social expression of emotions.

[Bjiorklunf, 2006] analyzes the role of early development (maternal effects and changes in maternal behavior) in large-brained animal evolution, stressing the role of behavioral plasticity. The hypothesis is that in mammals differences in maternal behavior may have contributed substantially to the transmission of non-genetic characteristics across generations (epigenetic inheritance).[73] For example, in our human ancestors, such differences influenced the acquisition of symbolic functioning but they also furthered the evolution of other symbolic activities, with regards to social cognition, in great apes, thanks to newly created selective pressures. Female-infant bonding also seems to be related to the development of larger brains: it can be hypothesized that during evolution from smaller to larger brains in "Old World" primates, such as baboons, (where mother and infant alone would not survive), an emancipation from merely hormonal-based parenting behavior occurred. There would be a progression from brains mainly invested in olfactory processing (sufficient for basic processes like mating, pregnancy, and parturition), to those committed to complex visual processing, also reflecting the habit of life in large groups, where social-glue and foraging information are central [Curley and Keverne, 2005]. Large brains would permit some degree of sophisticated abductive and

[71] It has to be noted that for Peirce iconic signs are generally arbitrary and flexible but there are some symbols, still iconic, which are conventional and fixed, like the ones used in mathematics and logic.

[72] [Mitchell, 2002] contends infants need a connection between kinesthesis and vision. That is, without this connection the organism would not be able to connect the kinesthetic image it has of its own body with any visual image.

[73] On this concept cf. chapter five, section 6.1.3.

decision-making activities able to free primate groups from merely depending on hormonal determinism: sexual bonding acquires more than a reproductive role and "female-infant bonding is neither restricted by the hormones of pregnancy nor limited to the post-partum period" [Curley and Keverne, 2005, p. 565].

In the case of externalization of signs in non-human animals, when the sharing of affect is not at play, we are, for example, faced with the mere communication of useful information. Many worker honeybees socially externalize dances that express the site where they have found food to inform the other individuals about the location:

> [...] the waggle dances communicate information about direction, distance, and desirability of the food source. Each of these three dimensions of variation is correlated with a dimension of variation in the dance. The angle of the dance relative to the position of the sun indicates the direction of food source. The duration of a complete figure-of-eight circuit indicates the distance to the food source (or rather the flying time to the food source, because it increases when the bees would have to fly into a headwind). And the vigor of the dance indicates the desirability of the food to be found [Bermúdez, 2003, p. 152].[74]

The externalized figures performed through movements are agglomerative[75] signs that grant a cognitive – communicative – *mediator* to the swarm. Through this interplay with other bees, the dancers can get a feedback from the other individuals, which will help them later on to refine and improve their exhibition. In the case of animals, which perform these kinds of externalizations on a not merely innate basis, the true "creation" of new ways of communicating can also be hypothesized, through the invention of new body movements, new sounds or external landmarks, which can be progressively provided, if successful, as a cognitive resource to the entire group.

Related to both the infant affect attunement and bee dances illustrated above an epistemological remark is fundamental. When we speak about internal and external representations in the abductive interplay we put ourselves in the perspective of the researcher, who "sees" two or more different agents in the sense of folk psychology. Nevertheless, in the two examples, the agents (with the exception of the caregiver) are not reified in the sense that "they" do not perceive "themselves" as agents, like we instead do. Rather, for instance in the case of affect attunement, it is the process itself that is responsible for the formation of the infant's agentive status. A clarification of this problem can be found in some cognitive results derived form

[74] Bees would certainly find human communication very poor because we do not inform our fellows on the location of the closest restaurant by dancing!

[75] The theoretical distinction between agglomerative diagrammatic signs and discursive signs in reasoning, together with many other interesting clarifications of Peircean insights, also concerning mathematical proofs, are given in [Stenning, 2000]. On the cognitive advantages (and also disadvantages) – in humans – of diagrammatic dynamic reasoning over sentential reasoning cf. [Jones and Scaife, 2000]: in a watcher/user/learner better cognitive offloading is allowed by external diagrammatic dynamic representations and their "hidden" dependencies.

neurological research, which I have described in a forthcoming paper [Magnani, 2009a].[76]

Summary

The cognitive importance of abduction means that we must take great care over its intellectual management. As we have seen, there are several very sophisticated, classical philosophical issues related to this challenge that have to be taken into account, as illustrated in the first two Peircean sections of this chapter. They teach us something interesting about well-known conflicting views on abduction. *i*) There is no sharp contrast between the idea of abduction (both creative and selective) as perception and the idea of abduction as something that pertains to logic, both aspects are inferential in themselves, and fruit of sign activity; indeed *ii*) a suitable meaning of the concept of inference does not have to be exhausted by its logical aspects, but instead referred to the effect of various sensorial activities; thus *iii*) the dichotomy instinct vs. inference represents a conflict we can overcome simply by observing that the work of abduction is partly explicable as a biological phenomenon and partly as a "logical" operation related to "plastic" cognitive endowments of all organisms. From the same perspective it has been possible to see that iconicity hybridates logicality so that the sentential aspects of symbolic disciplines like logic or algebra coexist with model-based features – iconic. Moreover, no strong conflict between views of abduction (and of practical reasoning) in terms of *heuristic strategies* or in terms of instinct (insight, perception), has to be hypothesized.

The result that abduction is partly explicable as a biological phenomenon and partly as a more or less "logical" operation related to the "plastic" cognitive endowments of all organisms naturally led to the remaining sections. The main concern of this second part is that model-based reasoning represents a significant cognitive perspective able to unveil some basic features of abductive cognition in non-human animals. Its fertility in explaining how animals make up a series of signs and are engaged in making or manifesting or reacting to a series of signs in instinctual or plastic ways is evident. I have illustrated that a considerable part of this semiotic activity is a continuous process of "hypothesis generation" that can be seen at the level of both instinctual behavior and representation-oriented behavior, where nonlinguistic pseudothoughts drive a "plastic" model-based cognitive role. I also maintain that the various aspects of these abductive performances can be better understood by taking some considerations on the concept of affordance into account. From this perspective referral to the central role of the externalization of artifacts

[76] [Sinha, 2006, p. 115] speculates about the possible incorporation of the dual ontology of the human body (individual-biological and socio-cultural) in the genotype and expressed in the early stages of post-natal epigenetic development, "in the responsiveness of the human infant to the communicative actions of caretakers in the primary intersubjective semiotic circuit". I think it is more likely to think that a mere epigenetic openness driven by suitable regulatory genes has been exploited by the effect of the specific cultural normative inheritance, effect that we see in the semiotic capacities of human beings.

that act as mediators in animal languageless cognition becomes critical to the problem of abduction. Moreover, I tried to illustrate how the interplay between internal and external "pseudorepresentations" exhibits a new cognitive perspective on the mechanisms underlying the emergence of abductive processes in important areas of model-based inferences in the so-called mindless organisms.

Finally, in light of the considerations I outline in this chapter, it can be said that a considerable part of abductive cognition occurs through model-based activity that takes advantage of pseudoexplanations, reifications in the external environment, and hybrid representations. An activity that is intrinsically multimodal and distributed. Given the importance I have attributed in this and in the previous chapters to the cognitive role played by the external environment, a further step has to be accomplished. This brings us to the focus of chapter five: the link between cognition and external environment and their relationships in abductive reasoning can be satisfactorily made rigorous thanks to both the concept of affordance and that of the cognitive niche.

Chapter 6
Abduction, Affordances, and Cognitive Niches
Sharing Representations and Creating Chances through Cognitive Niche Construction

As we have seen in the previous chapters, humans continuously delegate and distribute cognitive functions to the environment to lessen their limits. They create models, representations and other various mediating structures, that are thought to be aid for thinking. The aim of this chapter is to shed light on these *design* activities. In the first part of the chapter I will argue that these design activities are closely related to the process of niche construction. I will point out that in building various mediating structures, such as models or representations, humans alter the environment and thus create *cognitive niches*. In this sense, I argue that a cognitive niche emerges from a network of continuous interplays between individuals and the environment, in which people alter and modify the environment by mimetically externalizing fleeting thoughts, private ideas, etc., into external supports. Cognitive niche construction may also contribute to making available a great portion of knowledge that would otherwise remain unexpressed or unreachable. This can in turn be useful in all those situations that require the transmission and sharing of knowledge, information and, more generally, of cognitive resources.

The section on cognitive niches also addresses the following important problem: a fundamental role in evolution of "non" genetic information has to be hypothesized but what kind of evolution can we hypothesize in this case? This process of selection selects for *purposeful* organisms, that is, niche-constructing organisms. Consequently, the process of transmission and selection of the extragenetic information that is embedded in cognitive niche transformations has to be considered *loosely* Darwinian for various reasons. The issue is also related to the description of the so-called neural Darwinism.

In dealing with the exploitation of cognitive resources embedded in the environment, the notions of *affordance*, originally proposed by Gibson [1979] to illustrate the hybrid character of visual perception, together with the proximal/distal distinction described by Brunswik [1952], are extremely relevant. The analysis of the concept of affordance also provides an alternative account of the role of external – also artifactual – objects and devices. Artifactual cognitive objects and devices extend, modify, or substitute "natural" affordances actively providing humans and many animals with new opportunities for action.

L. Magnani: Abductive Cognition, COSMOS 3, pp. 317–359.
springerlink.com

In order to solve various controversies on the concept of affordance and on the status of the proximal/distal dichotomy, I will still take advantage of some further useful insights that come from the study on abduction. Abduction describes all those human and animal hypothetical inferences that are operated through actions made up of smart manipulations to both detect new affordances and to create manufactured external objects that offer new affordances/cues. After presenting some theoretical muddles concerning affordance and proximal/distal distinction, the last part of the chapter will refer to abduction as the process which humans lean on in order to detect and design affordances and thus modify or even create cognitive niches.[1]

6.1 Cognitive Niches: Humans as Chance Seekers

6.1.1 Incomplete Information and Human Cognition

Humans usually make decisions and solve problems relying on incomplete information [Simon, 1955]. Having incomplete information means that 1) our deliberations and decisions are never *the best* possible answer, but they are at least *satisficing*; 2) our conclusions are always *withdrawable* (i.e. questionable, or never final). That is, once we get more information about a certain situation we can always revise our previous decisions and think of alternative pathways that we could not "see" before; 3) a great part of our job is devoted to elaborating conjectures or hypotheses in order to obtain more adequate information. Making conjectures is essentially an act that in most cases consist in manipulating our problem, and the representation we have of it, so that we may eventually acquire/create more "valuable" knowledge resources. Conjectures can be either the fruit of an abductive selection in a set of pre-stored hypotheses or the creation of new ones, like in scientific discovery (see section 1.3 in chapter one and chapter two). To make conjectures humans often need more evidence/data: in many cases this further cognitive action is the only way to simply make possible (or at least enhance) a reasoning to "hypotheses" which are hard to successfully produce.

Consider, for instance, diagnostic settings: often the information available does not allow a physician to make a precise diagnosis. Therefore, he/she has to perform additional tests, or even try some different treatments to uncover symptoms otherwise hidden. In doing so he/she is simply aiming at increasing the *chances* of making the appropriate decision. There are plenty of situations of that kind. For example, scientists are continuously engaged in a process of manipulating their research

[1] [Cunningham, 1988] already pointed out the importance of connecting abduction and affordances, in the framework of a semiotic approach to cognition, contrasted with the computational-representation model of the mind. He says: "But the world is not infinitely malleable to our sign structures and abductive process will be again instigated". One of the aims of this chapter is to fill this cognitive gap.

settings in order to get more valuable information, as illustrated by Magnani [2001b]. Most of this work is completely tacit and embodied in practice. The role of various laboratory artifacts is a clear example, but also in everyday life people daily face complex situations which require knowledge and manipulative expertise of various kinds no matter who they are, whether teachers, policy makers, politicians, judges, workers, students, or simply wives, husbands, friends, sons, daughters, and so on. In this sense, humans can be considered *chance seekers*, because they are continuously engaged in a process of building up and then extracting latent possibilities to uncover new valuable information and knowledge.

The idea I will try to deepen in the course of this chapter is the following: as chance seekers, humans are *ecological engineers*.[2] That is: humans like other creatures do not simply live their environment, but they actively shape and change it looking for suitable chances. In doing so, they construct *cognitive niches* (cf. [Tooby and DeVore, 1987] and Pinker [1997; 2003]) through which the offerings provided by the environment in terms of cognitive possibilities are appropriately selected and/or manufactured to enhance their fitness as chance seekers. Hence, this ecological approach aims at understanding cognitive systems in terms of their *environmental situatedness* (cf. [Clancey, 1997] and [Magnani, 2005]). Within this framework, chances are that "information" which is not stored internally in memory or already available in an external reserve but that has to be "extracted" and then *picked up* upon occasion.

Related to this perspective is also the so-called *Perceptual Activity Theory* (PA) (cf. Ellis [1995] and Ramachandran [1997]). I have already illustrated in section 4.6 of chapter four that what these studies suggest is that an observer actively selects the perceptual information it needs to control its behavior in the world [Thomas, 1999]. In this sense, we do not store descriptions of pictures, objects or scenes we perceive in a static way: we continuously adjust and refine our perspective through further *perceptual exploration* that allows us to get a more detailed understanding. As Thomas [1999] put it, "PA theory, like *active vision* robotics, views it [perception] as a continual process of active interrogation of the environment". As I will show in the following sections "the active interrogation of the environment" is also at the root of the evolution of our organism and its cognitive system. I will also describe this ecological activity exploiting the notion of abduction and its semiotic dimension.

6.1.2 Cognitive Niche Construction and Human Cognition as a Chance-Seeker System

It is well-known that one of the main forces that shape the process of adaptation is natural selection. That is, the evolution of organisms can be viewed as the result of a selective pressure that renders them well-suited to their environments. Adaptation is

[2] Clark too adopts the expression "epistemic engineers", due to [Sterelny, 2003], in the framework of the analysis of the role of cognitive niches in the evolution and development of extended human cognition.

therefore considered as a sort of *top-down process* that goes from the environment to the living creature (cf. [Godfrey-Smith, 1998]). In contrast to that, a small fraction of evolutionary biologists have recently tried to provide an alternative theoretical framework by emphasizing the role of niche construction (cf. [Laland *et al.*, 2000; Laland *et al.*, 2001; Odling-Smee *et al.*, 2003]).

According to this view, the environment is a sort of "global market" that provides living creatures with unlimited possibilities. Indeed, not all the possibilities that the environment offers can be exploited by the human and non-human animals that act on it. For instance, the environment provides organisms with water to swim in, air to fly in, flat surfaces to walk on, and so on. However, no creatures are fully able to take advantage of all of them. Moreover, all organisms try to modify their surroundings in order to better exploit those elements that suit them and eliminate or mitigate the effect of the negative ones.

This process of *environmental selection* (cf. Odling-Smee [1988]) allows living creatures to build and shape the "ecological niches". An ecological niche can be defined, following Gibson, as a "setting of environmental features that are suitable for an animal" [Gibson, 1979]. It differs from the notion of habitat in the sense that the niche describes *how* an organism lives its environment, whereas habitat simply describes *where* an organism lives.

In any ecological niche, the selective pressure of the *local* environment is drastically modified by organisms in order to lessen the negative impacts of all those elements which they are not suited to. Indeed, this does not mean that natural selection is somehow halted. Rather, this means adaptation cannot be considered only by referring to the agency of the environment, but also to that of the organism acting on it. In this sense, animals are ecological engineers, because they do not simply live their environment, but they actively shape and change it (cf. [Day *et al.*, 2003]).

6.1.3 What Are the Cognitive Niches?

It is important to deeply clarify the concept of cognitive niche that is at the basis of many issues illustrated in this chapter. The recent book by [Odling-Smee *et al.*, 2003] offers a full analysis of the concept of the cognitive niche mainly from a biological and evolutionary perspective. "Niche construction should be regarded, after natural selection, as a second major participant in evolution. [...] Niche construction is a potent evolutionary agent because introduces feedback into the evolutionary dynamics" [Odling-Smee *et al.*, 2003, p. 2].[3] By modifying their environment and by their affecting, and partly controlling, some of the energy and matter flows in their ecosystems, organisms (not only humans) are able to modify some of the

[3] Attention was drawn for the first time to the idea of niche construction by important researchers like Schrödinger, Mayr, Lewontin, Dawkins, and Waddington. Firstly in the field of physics and subsequently in the field of evolution theory itself. Waddington particularly stressed the influence of organism development.

natural selection pressures present in their local selective environments, as well as in the selective environments of other organisms. This happens particularly when the same environmental changes are sufficiently recurrent throughout generations and selective change: "Even though spiders' webs are transitory objects [...] the spiders' genes 'instruct' the spider to make a new one" [Odling-Smee *et al.*, 2003, p. 9]. The fact that spiders on a web are exposed to avian predators suggests that webs can be a source of selection that produces further phenotype changes in some species, such as the marking of their webs to enhance crypsis or the creation of dummy spiders probably to divert the attention of the birds which prey on them.[4]

In summary, general inheritance (natural selection among organisms influences which individuals will survive to pass on their genes to the next generation) is usually regarded as the only inheritance system to play a fundamental role in biological evolution, nevertheless, where niche construction plays a role in various generations, this introduces a second general inheritance system (also called *ecological inheritance* by Odling-Smee). The first system occurs once, through the process of reproduction (sexual for example), during the life of organisms; on the contrary, the second can in principle be performed by any organism towards any other organism ("ecological" but not necessarily "genetic" relatives), at any stage during their lifetime. Organisms adapt to their environments but also adapt to environments reconstructed by them or other organisms.[5] From this perspective acquired characteristics can play a role in the evolutionary process, even if in a non-Lamarckian way, through their influence on selective environments via cognitive niche construction. Phenotypes construct niches, which then become new sources of natural selection, possibly responsible for modifying their own genes through ecological inheritance feedback (in this sense phenotypes are not merely the "vehicles" of their genes). Of course we have to remember that humans are not unique in their capacity to modify their environment: other species are informed by a kind of protocultural and learning process that is very often intrinsically social, even if we have to say that animals seem to lack the ability to accumulate information as seen in the human cultural/technological case.

It has to be noted that cultural niche construction alters selection not only at the genetic level, but also at the ontogenetic and cultural levels as well. For example the construction of various artifacts challenges the health of human beings:

> Humans may respond to this novel selection pressure either through cultural evolution, for instance, by constructing hospitals, medicine, and vaccines, or at the ontogenetic level, by developing antibodies that confer some immunity, or through biological

[4] More examples, from birds to earthworms, are illustrated in [Odling-Smee *et al.*, 2003, chapters one and two].

[5] This perspective has generated some controversies, since it is not clear the extent to which modifications count as niche-construction, and so enter the evolutionary scene. The main objection regards how far individual or even collective actions can really have ecological effects, whether they are integrated or merely aggregated changes. On this point, see [Sterelny, 2005] and the more critical view held by [Dawkins, 2004]. For a reply to these objections, see [Laland *et al.*, 2005].

> evolution, with the selection of resistant genotypes. As cultural niche construction typically offers a more immediate solution to new challenges, we anticipate that cultural niche construction will usually favor further counteractive cultural niche construction, rather than genetic change [Odling-Smee *et al.*, 2003, p. 261].

However, if some counteractive cultural reactions fail to reduce natural selection pressures, usually because of costs, ignorance, or ethical limitations, genotypes that are better suited to the unchanged cultural modified environment could increase in frequency.

More powerful than sociobiology and evolutionary psychology, the theory of niche construction explains at the same time the role of cultural aspects (transmitted ideas), behavior (niche construction itself), and ecological inheritance (artifacts). Of course niche construction may also depend on learning. It is interesting to note that some species, many vertebrates for example, have evolved a capacity to learn from other individuals and to transmit this knowledge, thereby activating a kind of protocultural process which also affects niche construction skills: it seems that in hominids this kind of cultural transmission of acquired niche-constructing traits was ubiquitous. "This demonstrates how cultural processes are not just a product of human genetic evolution, but also a cause of human genetic evolution" [Odling-Smee *et al.*, 2003, p. 27].

From this viewpoint the notion of *docility* acquires an explanatory role in describing the way human beings manage ecological and social resources to make their own decisions. According to Herbert Simon, humans are docile in the sense that their fitness is enhanced by "[...] the tendency to depend on suggestions, recommendations, persuasion, and information obtained through social channels as a major basis for choice" [Simon, 1993, p. 156].[6] In other words, humans support their limited decision-making capabilities, counting on external data obtained through the senses from the social environment. The social context gives them the main data filter, available to increase individual fitness [Secchi, 2007]. Therefore, docility is a kind of attitude or disposition underlying those activities of cognitive niche construction, which are related to the delegation and exploitation of

[6] [Woods, 2010] touches upon the same problem when, analyzing fallacious reasoning, he stresses the fact (doxastic irresistibility) that "Whether full or partial, belief states are not chosen. They befall us like measles", in other words, "say so" induces belief; the problem is related to the effect of what they call *ad ignorantiam rule*: "Human agents tend to accept without challenge the utterances and arguments of others except where they know or think they know or suspect that something is amiss" [Gabbay and Woods, 2005, p. 27]. The individual agent also economizes by unreflective acceptance of anything an interlocuter says or argues for, short of particular reasons to do otherwise, by applying the *ad verecundiam* fallacy. According to it, the reasoner accepts her sources' assurances because she is justified in thinking that the source has good reasons for them (the fallacy would be the failure to note that the source does not have good reasons for his assurances). I have illustrated in chapter three (section 3.5.1) that Peirce contended, in a similar way, that it is not true that thoughts are in us because we are in them; "beings like us have a *drive* to accept the say so of others" [Woods, 2010]. On the relationships between fallacies and abduction cf. this book, chapter seven.

ecological resources. That is, docility is an adaptive response to (or consequence of) the increasing cognitive demand (or selective pressure) on those information-gaining ontogenetic processes, resulting from an intensive activity of niche construction. In other words, docility permits the inheritance of a large amount of useful knowledge while lessening the costs related to (individual) learning.[7]

It is noteworthy that all these information resources do not only come from other human beings. This is clearly an oversimplification. Indeed, the information and resources that we continuously exploit are – so to speak – *human-readable*. Both information production and transfer are dependent on various *mediating structures*, which are the result of more or less powerful cognitive delegations, namely, niche construction activities. Of course, it is hard to develop and articulate a rich culture as humans did, and still do, without effective mediating systems. Hence, we can say that, first of all, docility is more generally concerned with the tendency to lean on various *ecological* resources, which are released through cognitive niches. Secondly, social learning cannot be seriously considered without referring to the agency of those mediating structures, whose efficiency in storing and transmitting information far exceeds, from many perspectives, that of the human beings.

It is well-known that from the point of view of physics organisms are far-from-equilibrium systems relative to their physical or abiotic surroundings.[8] Apparently they violate the second law of thermodynamics because they stay alive, the law stating that net entropy always increases and that complex and concentrated stores of energy necessarily break down. It is said that they are open, dissipative systems [Prigogine and Stengers, 1984], which maintain their status far from equilibrium by constantly exchanging energy and matter with their local environments. Odling-Smee, Laland and Feldman quote Schrödinger who says that an organism has to "feed upon negative entropy [...] continually sucking orderliness from its environment" [Schrödinger, 1992, p. 73]. Creating cognitive niches is a means that an organism (which is always smartly and plastically "active", looking for profitable resources, and aiming at enhancing fitness) has to stay alive without violating the second law: indeed it "cannot" violate it. In this sense cognitive niches can be considered *obligatory*: "To gain the resources they need and to dispose their detritus, organisms cannot just respond to their environments [...] to convert energy in dissipated energy" [Odling-Smee *et al.*, 2003, p. 168].

Evolution is strictly intertwined with this process and so it has consequences not only for organisms but also for environments. Sometimes the thermodynamic costs are negligible (like in the heat loss caused by photosynthesis that is returned to the universe, "which is in effect infinite"– p. 169), sometimes they are not, in this case abiota of the environment have no capacity to contrast the niche-constructing activities of organisms (like for example, the atmosphere, which is in a

[7] Recent research on the so-called neurobiology of trust has focused on the role of oxytocin, a hormone and neurochemical, in enhancing individuals' propensity to trust a stranger when that person exhibits non-threatening signals [Kosfeld *et al.*, 2005].

[8] On bacteria and the second law of thermodynamics cf. chapter four, section 6.2.

new physical state of extreme disequilibrium in relation to exploitation of the Earth's limited resources). The only no-costs exception is when organisms die – and lose their far-from-equilibrium status): in this case the dead bodies are returned to the local environment in the form of dead organic matter (DOM), still a kind of niche construction, so to say, also called "ghost niche construction" [Odling-Smee *et al.*, 2003, p. 170]. Of course biota can resist any thermodynamic costs imposed on them by other niche-constructing organisms, often performing counteractive niche-constructing activities.

At an intermediate level between biota and abiota, *artifacts* – parts of a cognitive niche – present negative entropy because they are highly organized but have no active ability to defend themselves and prevent their dissipation. Of course, and this is the case for various human artifacts, they can originate niches that seem maladaptive[9] rather than adaptive: "Contrary to common belief, environmentally induced novelties [induced by phenotypes] may have greater evolutionary potential than do mutationally induced ones. They can be immediately recurrent in a population; are more likely than are mutational novelties to correlate with particular environmental conditions and be subjected to consistent (directional selection); and, being relatively immune to selection, are more likely to persist even if initially disadvantageous" [West-Eberhard, 2003, p. 498].[10]

A large part of the niche construction process is intrinsic to the Darwinian framework. The information that basically drives niche construction is of course at the level of semantic information encoded in DNA and provided by the evolutionary process as the result of natural selection. However, niche construction is also *active* and not reactive, *profitable* and not goalless, like natural selection and, moreover, it is always an informed selective process (governed by memory and learning). In this last sense niche construction is related to cognitive processes which are abductive in themselves, because it formulates hypotheses about the *chances* offered by the environment and the possible subsequent active *changes* in terms of niche. Finally, through inherited selected genes organisms are also informed about past natural selection, that is, their niche construction activities are informed a priori by past natural selection. This is not enough however, the semantic information being further retested and updated because of current natural selection pressures and then passed on in the genes of future generations: the best adaptive niche constructors have more chance of being selected.

Indeed, it has to be recognized that first of all "[...] evolution depends on two selective processes, rather than one. A blind process based on the natural selection of diverse organisms in populations exposed to environmental selection pressures, and a second process based on the semantically informed selection of diverse actions, relative to diverse environmental factors, at diverse times and places, by

[9] Cf. [Laland and Brown, 2006].

[10] In the case of modern humans the problem of managing these maladaptive artifactual niches immediately relates to the relationships between morality and knowledge in our technological world. I have fully analyzed this topic in the recent [Magnani, 2007d].

individual niche-constructing organisms" [Odling-Smee *et al.*, 2003, p. 185]. Of course the second process was not described by Darwin. In this process selection selects for *purposive* organisms, that is, niche-constructing organisms. West-Eberhard nicely observes: "Incorporation of environmental modifications into the genetic theory of natural selection greatly increases the power of Darwinian argument by showing that it does not depend entirely upon 'random mutation' but can capitalize on preexisting adaptive plasticity and re-organizational novelty in response to recurrent environmental induction" [West-Eberhard, 2003, p. 498].[11] In this sense variation is *blind* but also *constructed* because "[...] which variants are inherited and what final form they assume depend on various 'filtering' and 'editing' processes that occur before and during the transmission" [Jablonka and Lamb, 2005, p. 319]. This does not have any Lamarckian or teleological flavor.[12]

Finally, as I have already said, organisms can gain from themselves a posteriori – at the individual level – more information through learning from their own experi-

[11] An interesting example of active new selection pressure generated by a cognitive niche is that menopause and longevity (i.e. devoting more energy to raising existing children – also through acting as grandmother – rather that producing new offspring), may be not a simple manifestation of normal mammalian aging but instead the result of a successful adaptation "via whatever mechanism" [Allen *et al.*, 2005, p.]. Changes in diet involving extracting and hunting food that accompanied the origin of Homo could have placed young children in a position of requiring more assistance from maternal adults of both generations. Both female longevity and brain size which correlate significantly across mammal species, would have coevolutionarily conferred increased fitness to organisms (offering greater chances to store intergenerational information about resources) to cope with the variable environmental cognitive niches characterized by the presence of extended families. Finally, the "cognitive reserve hypothesis" further states that, in a niche already characterized by intergenerational transfer of information about food, the development of technological and social intelligence – language is probably fundamental in this case – could have formed an increased cognitive reserve in aging. This is reflected in favoring the longevity of old, healthy human brains and their capacity to resist injury and diseases like dementia and Alzheimer (but also in favoring, as a secondary factor, the selection of brain size itself).

[12] On the introduction of "a sense of purposefulness" in evolution cf. [Turner, 2004, pp. 348–349], who says that purposefulness is embodied in the phenomenon of homeostasis: "Evolution then becomes less a province of one class of arbiters of future function – genes – and more the result of a nuanced interplay between the multifarious specifiers of future function". [Kaplan *et al.*, 2000] further develop the "grandmother hypothesis" showing and more clearly how intelligence and brain size would have been coevolved with the dietary shift toward high-quality calorie-dense, difficult-to-acquire food resources. The attainment of skills and abilities requires time and thus an extended learning phase, during which productivity is low, later compensated by higher productivity during the adult period (for example males further enhancing the hunting-extraction-feeding cognitive niche to support women's reproduction) and an intergenerational flow of food from old to young. Lowered mortality rates and greater longevity are thus selected. Of course the new feeding niche further promotes social aspects (and partnership between men and women) such as food sharing, provision for juveniles, and a lowered predation risk, which in turn promote longevity and lengthening of the juvenile period. As we will see in the section 8.6 of chapter eight, [Bingham, 1999, p. 140] proposes the higher level speculative "coalition enforcement hypothesis", which, taking into account moral and violent aspects sometimes disregarded in evolutionary anthropology, aims at providing a unifying explanation of the various phenomena I have just illustrated, from lowered mortality to longevity, from brain size to dietary shifts.

ence (only possible if the organisms possess the required gene-informed subsystems which allow them to do this and develop the primordial forms of social instinct). On the role of learning, social learning, and culture as adaptation and maladaptation in evolution it is interesting to quote the recent work by [Richerson and Boyd, 2005]. The authors provide an in depth description of culture as an unusual system of phenotypic flexibility – through now dramatically increased non parental and parental cultural transmission – that can accumulate adaptive information more rapidly than selection could change gene frequencies. They usefully discuss the "big-mistake hypothesis" in explaining maladaptations, and contrast it to the explicit cultural evolutionary explanations, considered as empirically more adequate. Following the big-mistake hypothesis it seems that most of the information necessary to construct what we call culture is latent in genes shaped in Pleistocene environments, when decision-making systems evolved. However, in the post-Pleistocene epoch a sudden acceleration of cultural change modified the environments – thanks to cognitive niches – so that they are now far outside the ranges of evolved decision-making systems. The favored counter-hypothesis is that once cultural traditions create novel environments, these can affect the fitness of alternative genetically transmitted variants both in animals and humans, so that genes and culture are joined in a coevolutionary dance, as indicated by the cognitive niche theory: in this case the maladaptation generated by the cultural established tendency towards global cooperation might be explained in the following way: "Humans are quite adept at cooperating in large groups with strangers and near strangers, while the theory of selection on genes suggests that cooperation should be restricted to relatives and well-known nonrelatives" [Richerson and Boyd, 2005, p. 190].

Beyond the effect created by learning and culture in evolution, a kind of "supergenetic" transmission (a phenotype which affects genotypes at higher levels of organization – cells, organism, group) can be hypothesized. This transmission would be brought about through the range of possibilities granted by the genetic systems, making many phenotypes possible. Even if there is still a reluctance to recognize that it plays a role in evolution some hypotheses now focus on the the so-called "epigenetic inheritance systems" [Jablonka and Lamb, 2005, pp. 245–246]. Exploiting a musical analogy, they say:

> I have suggested that the transmission of information through the genetic system is analogous to the transmission of music through a written score, whereas transmitting information through non-genetic systems, which transmit phenotypes, is analogous to recording and broadcasting, through which particular interpretations of the score are reproduced. A piece of music can evolve through changes being introduced into the score, but also independently through the various interpretations that are transmitted through the recording and broadcasting system. [...] A recorded and broadcasted interpretation of a piece of music could affect the copying and future fate of the score in two different ways. First, a recorded interpretation could directly bias the copying errors that made. [...] A second, more indirect effect would occur if a new and popular interpretation affects which versions of a score are copied and used as the basis for a new generation of interpretations. [...] Epigenetic systems could have either or

> both types of effect on the genetic system: they could directly bias the generation of variation in DNA, or they could affect the selection of variants, or they could do both.[13]

At the level of the influence of learning and culture in human and animal evolution (which coincides with a fundamental part of niche construction), the authors distinguish between merely "behavioral" effects (for example thought, observation, imitation, and the role of lifestyles, not necessarily "social"), and "symbolic" ones, which are of course related to the effect – in the case of humans – of language and other semiotic social communication systems and information exchange.

We will see in the following sections that abductive cognition is fundamental in niche construction. Abductive cognition generates all the innovative hypotheses which lead to the creation of new niches and to the modification of the previous ones.

6.1.4 Extragenetic Information, Loosely Darwinian Effects, Baldwin Effect

I have already said that from the point of view of niche-construction evolution depends on two selective processes, a blind process based on the natural selection of diverse organisms in populations exposed to environmental selection pressures and a second process based on the semantically informed selection of diverse actions, relative to diverse environmental factors, at diverse times and places, by individual niche-constructing organisms. The second process was not described by Darwin . In this process selection selects for *purposive* organisms, that is, niche-constructing organisms. Consequently, the process of transmission and selection of the extragenetic information that is embedded in cognitive niche transformations has to be considered *loosely* Darwinian for three main reasons [Odling-Smee *et al.*, 2003, pp. 256-257.]:

1. extragenetically informed behavior patterns are broadly adaptive and maladaptive;
2. variants occurring during genetic evolution are random, whereas those of extragenetic information are not. They are smart variants, because the response to

[13] On the epigenetic inheritance system and the role of language as a cognitive niche embedded in a wider niche, like a semiotic network which includes symbolic and non-symbolic artifacts cf. [Sinha, 2006]. It must be stressed that all human artifacts – material and symbolic – are situated and can be re-situated in semiotic fields that have a cognitive value. In this framework epigenetic developmental processes are those "[...] in which the developmental trajectory and the final form of the developing behavior are a consequence as much of the environment information as of the genetically encoded information. [...] regulatory genes augmenting epigenetic openness can therefore be expected to have been phenogenotypically selected for in the human genome, permitting further adaptive selection for domain-specific learning in the semiotic biocultural complex, in particular for language". But it seems no innateness for language can be hypothesized in humans, as Clark also contends (cf. chapter three of this book, section 7.5): "Note however, that in an epigenetic perspective, any developmental predisposition for learning language is unlikely either to involve direct coding of, or to be dedicated exclusively to linguistic structure" [Sinha, 2006, p. 113]. On the so-called "extended phenotypes" and "extended organisms" cf. [Turner, 2004].

both internal and environmental cues is targeted appropriately to their behavioral repertoires. As I have said above variation is in this case *blind* but also *constructed* because "[...] which variants are inherited and what final form they assume depends on various 'filtering' and 'editing' processes that occur before and during the transmission" [Jablonka and Lamb, 2005, p. 319]. In this sense extragenetic information produces variants, which originate suitable cognitive stabilities in artifactual niches, which in turn "afford" humans and other organisms in many ways (cf. below section 6.6.2.).

3. extragenetic information
   a. when neurally stored – both consciously and unconsciously, it presents a strong interaction of its elements, which can also be seen in selective terms, as suggested by neural Darwinism theory, which sees neurons as diverse populations submitted to loosely Darwinian effects at the level of both neural development and moment-to-moment functioning which interfaces experience [Edelman, 1989; Edelman, 1993; Seth and Baars, 2005];
   b. when stored in material devices it no longer presents the characters of a more or less Darwinian evolving creative population, like in the case of neural cells: instead it has an evolutionary impact insofar as it causes modifications upon the environment, that persist.

Edelman's theory – which is also supported by the evidence provided by some AI devices called "Darwin automata", expressly built to test it – is controversial, as Rose clearly explains, because of the way the theory considers how extragenetic information is transformed through the dynamics of synaptic modifications:

> During development there is thus a superabundance of synaptic productions, a veritable efflorescence – but if synapses cannot make their appropriate functional connections with the dendrites of the neurons they approach, they become pruned away and disappear. This overproduction of neurons and synapses might seem wasteful. It has led to the argument that just during evolution "natural selection" still eliminates less-fit organisms, so some similar process of selection occurs within the developing brain – a process that the immunologist and theorist of human consciousness Gerald Edelman has called "neural Darwinism". However, this transference of the "survival of the fittest" metaphor from organisms to cells is only partially correct. It seems probable that the whole process of cellular migration over large distances, the creation of long-range order, requires the working out of some internal programmes of both individual cells and the collectivity of cells acting in concert. [...] Overproduction and subsequent pruning of neurons and synapses may at one level of magnification look like competition and selection; viewed on the larger scale, they appear as co-operative processes [Rose, 1993, p. 76].

Anyway, according to Edelman, from a loosely Darwinian perspective it is clear how more or less plastic and flexible brain functions, some of those subjectively experienced as perception and cognition, result from the operation of managed, internal (loosely) Darwinian systems (see, for example, [Jerne, 1967; Edelman, 1987; Changeux and Dehaene, 1989] and [Plotkin, 1993] for a review).

More simply,

> [...] such systems function by generating multiple alternative information structures and by competitively choosing among them on the basis of fit, congruence or interaction with other information structures – all within an individual animal. [...] Moreover, such structures inevitably undergo new rounds of internal Darwinian processing in each new mind to which they are transmitted. In the course of routine internal Darwinian processing, each new mind will generate potentially novel variants of the transmitted information structure and subject these variants to selection against this mind's partially idiosyncratic informational repertoire [Bingham, 1999, p. 147].

According to Edelman the brain would be a Darwinian "selection system that operates within an individual lifetime" [Edelman, 2006, p. 27] so as some synapses are strengthened and some are weakened through the experiential selection. What biases the brain system to yield adaptive responses is a process called reentry "[...] a continual signaling from one brain region (or map) to another and back again across massively parallel fibers (axons) that are known to be omnipresent in higher brains. Reentrant signal paths constantly change with the speed of thought. [...] consciousness is entailed by reentrant activity among cortical areas and the thalamus and by the cortex interacting with itself and with subcortical structures" [Edelman, 2006, p. 28 and p. 36]. It is from this perspective that – through core reentrant neural integrative processes – many sensory and motor signals are linked together, thus providing various perceptual categorizations (also connected to memory) which originate a scene" in the remembered present of primary consciousness, a scene with which an animal could lay plans", and of course motor outputs [Edelman, 2006, p. 36]. Selectionist brains are of course the effect of historical contingency, irreversibility, and the operation of non linear processes.

The number of independent rounds of refinement and, of course, the effective size and temporal persistence of cooperative hominid coalitions will affect (and have a causal impact on) the quality of extragenetic information.[14] In this sense, it is clear that the presence of enforcing coalitions will provide the adaptation context in which cooperative information processing and exchange may take place. In turn, those refined and highly successful information structures will constitute the bases (or primitives) for future developments, which are likely to be exponential or quasi-exponential. In addition, "long-term memory apparently structural involves changes in neural connections (see [Gazzaniga, 1995]), and references therein)" [Bingham, 1999, p. 148].

Hence, the perspective of cognitive niches does not preclude the possibility of regarding the extragenetic information selection and transmission at the neural level in a (loosely) Darwinian manner, especially to account for highly organized cognitive processes such as perception and consciousness. A clear example of the loosely Darwinian effects is provided by the case of perception, for example "[...] members of sets of internal information structures representing possible organizations of

[14] Indeed, the problem of extragenetic information can be fruitfully seen in the perspective of the so called "coalition enforcement hypothesis", I will illustrate in detail in chapter eight, section 8.6.

salient features of the external environment are competitively tested for quality of congruence with incoming sensory data" [Bingham, 1999, p. 146].

I think Edelman's neural Darwinism is certainly more satisfactory than Dawkins's conviction that cultural processes constitute an additional and apparently gene-independent Darwinian machine in terms of *memes* [Dawkins, 1977]. Edelman further observes that "Truth, though heterogeneous in its origins, is itself normative and thus worth caring about. Establishing truth requires many different means and methodologies. These cannot be traced directly back to evolution or the physiology of the brain" [Edelman, 2006, p. 149]. From this viewpoint "the search for truth" is, as if maintained by [Gould and Vrba, 1982], an *exaptation*: Edelman agrees that the claim, even metaphorical, that there is a physiology of truth is ill-founded. He also thinks that the famous Popperian (and similar) evolutionist models of how knowledge evolves is unconvincing, "given the evidence of our irrational behavior" [Edelman, 2006, p. 170, footnote 170]. In other words, the evolution of consciousness can be seen in loosely selective terms as fruit of the reentrant thalamocortical system, like neural Darwinism suggests: but consciousness is no guarantee of truth. Epistemology has to be founded on a sound pragmatic basis, brains have not evolved directly to achieve knowledge of truth.

For my part, I can say that my approach is strictly materialist and coherent with the loosely Darwinian approach of the theory of cognitive niches. For example, in the case of humans, I am referring to people's brains and bodies that make up – through hypothetical reasoning – large, integrated, material cognitive, and cultural systems such as artificial environments consisting of food, dwellings and furniture. From this perspective, largely described in the previous chapters, what we usually call "mind" simply consists of the hybrid product of interplay between the transformation activity of neural configurations (whether considered in terms of loosely Darwinian processes or in terms of merely rule-governed cooperative structures) and those large, integrated, material cultural, and cognitive systems that the brains themselves are building continuously. I have described in the first sections of the previous chapter how perception (for instance visual) can be philosophically described in abductive terms. These terms do not involve Darwinian processes but still show high cross-modal and sensory integration involving visual, auditory, olfactory, kinesthetic and somatosensory aspects, as hypothesized by the Edelmanian concept of dynamic reentrant neural interactions [Seth and Baars, 2005, p. 145].

We can say that partially divergent copies of the best stabilized perceptual abductions at our disposal (the "best fit" in Darwinian terms) from one round are then reproduced and retested competitively against new sensory information as merely retractable "perceptual hypotheses" in a second round, and so on. The abductive cyclic process is unconscious (performed by the cooperation of various synaptic transformations) and ultimately produces what organisms experience subjectively as the final "recognition" of their immediate physical environment. This in turn can be stored in long-term memory for future retrieval and use.[15] Abduction as perception still plays the Peircean role of providing a reliable framework for prediction and

[15] On the abductive character of visual perception cf. section 5.5, chapter five.

action by reducing ambiguities and uncertainties. In Edelman and Tononi's words we can say that "At any given time we experience a particular conscious state selected out of billions of possible states, each of which can lead to a specific behavioral consequence. The occurrence of a particular conscious state is therefore highly informative in the specific sense that information is the reduction of uncertainty among a number of alternatives" [Edelman and Tononi, 2000, p. 125].

Finally, The Baldwin effect basically contends that plastic phenotypes which are first favored by selection are subsequently substituted by fitter non-plastic genotypes – that is phenotypic plasticity is followed by genetic assimilation of the trait learnt – can be usefully explained in terms of niche construction, without going outside of the Darwinian framework. In other words the Baldwin effect states that learning and cultural activities in both human and non-human animals affect the direction and rate of evolution by natural selection (cf. [Lachapelle *et al.*, 2006], also referring to works by [Dennett, 1991; Dennett, 1995; Pinker, 1994]). [Deacon, 1995] proposes an *extended* version of the Baldwin effect in terms of niche construction, which is preferable to the highly questionable[16] narrow version (which resorts to the idea of genetic assimilation). The extended version encompasses but does not endorse the narrow version. Faced with some environmental conditions organisms engineer their environments and thus the available social and cultural ecology – niche construction is a concept that expresses more than just Dawkins's idea [1982] of the *extended phenotype* – in a way that is likely to be good for their fitness and so *de facto* they build new selective pressures which redirect the evolutionary trajectories of all the organisms affected.

6.1.5 Niche Construction and Distributed Human Cognition

My contention is that the notion of niche construction is fruitfully applicable to human cognition. More precisely, I claim that cognitive niche construction can be considered as one of the most distinctive traits of human cognition.[17] It emerges from a network of continuous interplay between individuals and the environment,

[16] "[...] the idea that a trait used to be learned can become innate in such a way that there is a given set of genes (or an *organ* or a *module*, depending on ones' terminology that code *specifically* for this trait") contrasts with the evidence that no trait is entirely innate "for no trait can unfold without the input of many environmental resources [...] there is *no such thing as a one-to-one correspondence between genes and phenotypes*" [Lachapelle *et al.*, 2006, p. 318].

[17] If we also recognize in animals, like many ethologists do, a kind of nonlinguistic thinking activity basically model-based (i.e. devoid of the cognitive functions provided by human language – cf. above chapter five), their ecological niches can be called "cognitive", when for example complicate animal artifacts like landmarks of caches for food are fruit of "flexible" and learned thinking activities which indeed cannot be entirely connected with innate endowments (cf. [Magnani, 2007b]). The psychoanalyst Carl Gust Jung, who is aware that also animals make artifacts, nicely acknowledges their cognitive role proposing the speculative expression "natural culture": "When the beaver fells trees and dams up a river, this is a performance conditioned by its differentiation. Its differentiation is a product of what one might call 'natural culture', which functions as a transformer of energy" [Jung, 1972a, p. 42]. I have illustrated Jung's ideas about cognitive externalizations and artifacts in psychoanalysis in chapter three, section 3.8.

in which they more or less tacitly manipulate what is occurring outside at the level of the various structures of the environment in a way that is suited to them. Accordingly, we may argue that the creation of cognitive niches is *the* way cognition evolves, and humans can be considered as ecological cognitive engineers.

Recent studies on distributed cognition seem to support my claim.[18] I have already described in this book that, according to this approach, cognitive activities like, for instance, problem solving or decision-making, cannot only be regarded as internal processes that occur within the isolated brain. Through the process of niche creation humans extend their minds into the material world, exploiting various external resources. For "external resources" I mean everything that is not inside the human brain, and that could be of some help in the process of deciding, thinking about, or using something. Therefore, external resources can be artifacts, tools, objects, and so on. Problem solving, such as general decision-making activity (cf. [Bardone and Secchi, 2006]), for example, are unthinkable without the process of connection between internal and external resources.

In other words, the exploitation of external resources is the process which allows the human cognitive system to be shaped by environmental (or contingency) elements. According to this statement, we may argue that external resources play a pivotal role in any cognitive process. Something important must still be added, and it deals with the notion of representation: the traditional notion of representation as a kind of abstract mental structure is old-fashioned and misleading.[19] If some cognitive performances can be viewed as the result of a smart interplay between humans and the environment, the representation of a problem is partly internal but it also depends on the smart interplay between the individual and the environment.

As I have already said, an alternative definition of the "ecological" niche that I find appealing in treating our problem has been provided by Gibson [1979]: he pointed out that a niche can be seen as a set of *affordances*. My contention is that the notion of affordance may help provide sound answers to the various questions that come up with the problem of ecological niches. The notion of affordance is fundamental for two reasons. First of all, it defines the nature of the relationship between an agent and its environment, and the mutuality between them. Second, this notion may provide a general framework to illustrate humans as chance seekers.

In order to illustrate the notion of affordance and its implications for my approach, we have to keep in mind the main aspects of abduction, especially those basic ones illustrated in chapters one, two, and three. I have posited that humans continuously exploit the latent possibilities and chances offered by the environment. This is carried out by an ecological activity called cognitive niche construction. A cognitive niche describes *how* humans exploit external resources and incorporate them into their cognitive systems. As mentioned above, in constructing cognitive niches, humans do not hold a rich model of the environment in their memory, but

[18] Cf. [Zhang, 1997; Hutchins, 1995; Clark and Chalmers, 1998; Wilson, 2004; Magnani, 2006c; Magnani, 2007b].

[19] Cf. [Zhang, 1997; Gatti and Magnani, 2005; Knuuttila and Honkela, 2005]. An analysis of the concept of representation in the perspective of the dynamical systems is illustrated in chapter eight (on the substantial antirepresentationalism of this tradition cf. subsection 8.5.2).

they pick up what they find outside upon occasion. In dealing with these activities that can be considered "semiotic", the role of abduction is a key point in my discussion (cf. [Magnani, 2006c]).

6.2 Affordances and Cognition: The Received View

As I have illustrated in the first part of this chapter, humans and some animals manipulate and distribute cognitive meanings after having delegated them to suitable environmental supports. The activity of cognitive niche construction reveals something important about the human and animal cognitive system.

As already mentioned, human cognition and its evolutionary dimension can be better understood in terms of its environmental situatedness. This means humans do not retain in their memory rich representations of the environment and its variables, but they actively manipulate it by picking up information and resources upon occasion, already available, or extracted/created and made available: information and resources are not only given, but they are actively sought and even manufactured. In this sense, I consider human cognition as a chance-seeker system. Consequently, in my terminology, chances are not simply information, but they are also "affordances", namely, environmental anchors that allow us to better exploit external resources.

6.2.1 The Notion of Affordance and Its Inferential Nature

One of the most disturbing problems with the notion of affordance is that any examples provide different, and sometimes ambiguous insights on it. This fact makes very hard to give a conceptual account of it. That is to say, when making examples everybody grasps the meaning, but as soon as one tries to conceptualize it the clear idea one got from it immediately disappears. Therefore, I hope to go back to examples from abstraction without loosing the intuitive simplicity that such examples provide to the intuitive notion.

Gibson defines "affordance" as what the environment offers, provides, or furnishes. For instance, a chair affords an opportunity for sitting, air breathing, water swimming, stairs climbing, and so on. By cutting across the subjective/objective frontier, affordances refer to the idea of agent-environment mutuality. Gibson did not only provide clear examples, but also a list of definitions (cf. [Wells, 2002]) that may contribute to generating possible misunderstanding:

1. affordances are opportunities for action;
2. affordances are the values and meanings of things which can be directly perceived;
3. affordances are ecological facts;
4. affordances imply the mutuality of perceiver and environment.

I contend that the Gibsonian ecological perspective originally achieves two important results. First of all, human and animal agencies are somehow hybrid, in the sense that they strongly rely on the environment and on what it offers. Secondly,

Gibson provides a general framework about how organisms directly perceive objects and their affordances. His hypothesis is highly stimulating: "[...] the perceiving of an affordance is not a process of perceiving a value-free physical object [...] it is a process of perceiving a value-rich ecological object", and then, "physics may be value free, but ecology is not" [Gibson, 1979, p. 140]. These two issues are related, although some authors seem to have disregarded their complementary nature. It is important here to clearly show how these two issues can be considered two faces of the same medal. Let us start our discussion.

6.2.2 Affordances Are Opportunities for Action

Several authors have been extensively puzzled by the claim repeatedly made by Gibson that "an affordance of an object is directly perceived".[20] During the last few years an increasing number of contributions has extensively debated the nature of affordance as opportunity for action. Consider for instance the example "stairs afford climbing". In this example, stairs provide us with the opportunity of climbing; we climb stairs because we perceive the property of "climbability", and that affordance emerges in the interaction between the perceiver and stairs (cf. [Chemero, 2003; Stoffregen, 2003]). In order to prevent from any possible misunderstanding, it is worth distinguishing between "affordance property" and "what" and object affords [Natsoulas, 2004]. In the former sense, the system "stairs-plus-perceiver" exhibits the property of climbability, which is an *affordance property*. Whereas in the latter the possibility of climbing is clearly *what* an object affords.

6.2.3 Affordances Are Ecological Facts

Concerning this point, Gibson argued that affordances are ecological facts. Consider, for instance, a block of ice. Indeed, from the perspective of physics a block of ice melting does not cease to exist. It simply changes its state from solid to liquid. Conversely, to humans a block of ice melting does go out of existence, since that drastically changes the way we can interact with it. A block of ice can chill a drink the way cold water cannot. Now, the point made by Gibson is that we may provide alternative descriptions of the world: the one specified by affordances represents the environment in terms of action possibilities. As Vicente put it, affordances "[...] are a way of measuring or representing the environment with respect to the action capabilities of an individual [...] one can also describe it [a chair] with respect to the possibilities for action that it offers to an organism with certain capabilities" [Vicente, 2003]. Taking a step further, we may claim that affordances are chances that are *ecologically rooted*. They are ecological rooted because they rely on the mutuality between an agent (or a perceiver) and the environment. As ecological chances, affordances are the result of a hybridizing process in which the perceiver meets the environment. The emphasized stress on the mutuality between the perceiver and the environment provides a clear evidence of this point.

[20] Cf. [Greeno, 1994; Stoffregen, 2003; Scarantino, 2003; Chemero, 2003].

6.2.4 *Affordances Imply the Mutuality of Perceiver and Environment*

Recently, Patel and Zhang [2006], also going beyond the ecological concept of affordance in animals and wild settings by involving its role in human cognition and artifacts, in an unorthodox perspective, connect the notion of affordance to that of distributed representation. They maintain that affordances can be also related to the role of distributed representations extended across the environment and the organism. These kinds of representation come about as the result of a blending process between two different domains: on the one hand the internal representation space, that is the physical structure of an organism (biological, perceptual, and cognitive faculties); on the other the external representation of space, namely, the structure of the environment and the information it provides. Both these two domains are described by constraints so that the blend consists of the allowable actions. Consider the example of an artifact like a chair. On one hand the human body constrains the actions one can make; on the other the chair has its constraints as well, for instance, its shape, weight, and so on. The blend consists of the allowable actions given both *internal* and *external* constraints.[21]

Patel and Zhang's idea tries to clarify that affordances result from a hybridizing process in which the environmental features and the agent's ones in terms of constraints are blended into a new domain which they call *affordance space*. Taking a step further, Patel and Zhang define affordances as allowable actions. If this approach certainly acknowledges the hybrid character of affordance I have described above and the mutuality between the perceiver and the environment, it seems however lacking with regard to its conceptual counterpart. As already argued, affordances are action-based opportunities.

6.3 Affordances as Eco-Cognitive Interactional Structures

Taking advantage of some recent results in the areas of distributed and animal cognition, we can find that a very important aspect that is not sufficiently stressed in literature is the dynamic one, related to designing affordances, with respect to their evolutionary framework: human and non-human animals can "modify" or "create" affordances by manipulating their cognitive niches. Moreover, it is obvious to note that human, biological bodies themselves evolve: and so we can guess that even the more basic and hardwired perceptive affordances available to our ancestors were very different from the present ones.[22] Of course different affordances can also be detected in children, and in the whole realm of animals.

---

[21] This idea can also be connected to the concept of cognitive fluidity argued by Mithen [1996]. From the perspective of cognitive palaeoanthropology, Mithen claimed that the modern mind is characterized by the capacity of applying to heterogeneous domains forms of thinking originally designed for specific tasks. He also contends that in hominids this change originated through a blend of two different intelligence domains, namely, that of internal representations entities and external artifacts [Magnani, 2006c]. Cf. also section 3.3, chapter three of this book.

[22] The term "wired" can be easily misunderstood, cf. the explanation illustrated at p. 226, footnote 4.

6.3.1 Pseudothoughts and Model-Based Thinking in Humans and Animals: Affordances as Chances

In section 5.4.4 of the previous chapter I have anticipated some of the aspects of Gibson's affordances in the perspective of the problem of the so-called model-based protothoughts (cf. [Bermúdez, 2003]), I have called pseudothoughts, which concern any kind of cognition far from the cognitive features granted by human language, typical of infants and of many animals. I described Gibsonian affordances as resources or chances that the environment presents to the organism, such as the availability of water or of finding recovery and concealment showing how the same part of the environment offers different affordances to different organisms.

Hence, it is clear that the concept of affordance can be extended to animals and artificial environments built by human and non-human animals. Deepening the problem of the interplay between direct and mediated perception, and of the interplay between proximal and distal environment will be of help to especially clarify the relationships between affordances and artifacts.

6.4 Direct and Mediated Perception, Proximal and Distal Environment

Some questions are still open. As briefly mentioned in section 6.2, an affordance is neither a property nor the result of an action. An affordance cannot be confused with the fact that one can sit down on a chair. And it is not the action of sitting either. Consider the following argument. We can say that the fact that a chair affords sitting means we can perceive some clues (robustness, rigidity, flatness) from which a person can easily say "I can sit down". Now, suppose the same person has another object O. In this case, the person can only perceive its flatness. He/she does not know if it is rigid and robust, for instance. Anyway, he/she decides to sit down on it and he/she does that successfully. Now, the question is: does the object O afford sitting? Do we directly access affordances? Can we say we have direct access to affordances?

6.4.1 Direct and Mediated Affordances

In his research Gibson basically referred to "direct" perception, which does not require the internal inferential mediation or processing by the agent. In this sense affordances express the complementary nature of an organism and its environment through the direct, effortless way in which an organism picks them up (cf. [Wells, 2002]). As already illustrated, they provide "a description of the distal structure of the environment" and are goal/action-relevant description of it, like Vincente says [2003, p. 248]. It has to be said that perception of affordances is in itself action: "Perception is an act, not a response, an act of attention, not triggered impression, an achievement, not a reflex" [Gibson, 1979, p. 149]. Perceiving, Gibson says, "does not have an end" but "goes on" (*ibid.*, p. 253), and nevertheless, "precedes predicating" (*ibid.*, p. 260).[23] Direct perception depends on the agent "picking up" the information that characterizes the affordances and it is related to the agent's sense faculties at a given moment: "Take, for example, a hidden door in a paneled room. The door affords passage to an appropriately sized individual even though there is no information to specify that passage is in fact an action possibility. Here, direct perception is clearly not possible" [McGrenere and Ho, 2000, p. 180]. Moreover, affordances can also be multifaceted, like for example in the case of a banana, which affords eating, but eating is composed by biting, chewing, and swallowing.

It is important to note that they are relative to organisms, and so can also be valued in ecology, and not merely in physics, as already illustrated in subsections 6.2.1 and 6.2.3. It is significant to note that Donald Norman modifies the original Gibsonian notion of affordance also involving mental/internal processing: "I believe that affordances result from the mental interpretation of things, based on our past knowledge and experience applied to our perception of the things about us" [1988, p. 14]. It appears clear that in this case affordances depend on the organism's experience, learning, and full cognitive abilities, i. e. they are not independent of them, like Gibson maintained. for example infants 12 to 22 weeks old already show complicated cognitive abilities of this type, as reported by Rader and Vaughn [2000]. These abilities allow them to lean on prior experience of an object and therefore detect what Rader and Vaughn call "hidden affordances". As argued by these authors, hidden affordances are those affordances specified by the information not available at the time of the interaction, but drawn from past experiences [2000, p. 539].[24] The same event or place can have different affordances to different organisms but also multiple affordances to the same organism. Following Donald Norman's perspective, affordances suggest a range of *chances*: given the fact that artifacts are complex things and their affordances normally require high-level supporting information, it is more fruitful to study them following this view. To give an example, perceiving

[23] As I will illustrate below in the following subsection, the Gibsonian relationships between objects and cues are replaced by Brunswik, in a symmetrical fashion, by the relationship between "proximal actions or habits and distal results". An example of distal variable is the distance of objects: Brunswik says: "Causal chains determined by distance will, on their way into the organism, exert certain proximal effects, or criteria, upon the sensory surface of the organism. The most important feature of the general relationship between distal and proximal stimulus variables is its lack of univocality" [1943, p. 255] .

[24] Norman's conception is supported by some empirical evidences related to patients suffering from visual agnosia. Visual agnosia is caused by damage of the ventral stream of one's visual system [Milner and Goodale, 1995], which impedes patients to consciously experience objects and access semantic knowledge related to them. Empirical studies reported that patients suffering from visual agnosia are still able to perform certain tasks that require the detection of simple affordances, for instance, holding and grasping pliers. However, the same patients showed the inability of understanding how to use instruments to accomplish more skilled tasks, for example, clasping the handles to manipulate jaws [Carey *et al.*, 1996]. As argued by Young [2006], that inability may be due to the lack of *functional knowledge* required to skillfully manipulate objects, which is provided by the ventral stream. This can support the claim according to which the exploitation of some affordances of an object involves mental/internal processing. Cf. also footnote 34 at p. 344.

the full range of the affordances of a door requires complex information about for example direction of opening or about its particular pull. Scarantino clearly explains this problem: affordances are perceivable because they are

> [...] invariants and disturbances in ambient-energy arrays that specify the threats and promises of items in the environment. For example, to say that the eat-ability of a given apple is perceivable or, that the being-hit-by-ability of a flying ball is perceivable is to say that there is a sensory appearance – a way to be visible/audible/tangible/odorous/tastable – typical respectively, of apples affording eating, and of flying balls affording being hit by. Gibson was very clear that we cannot establish a "priori" what affordances are specified in ambient energy. [...] Information is available for perceiving all and only those offerings of the environment that are associated with typical sensory appearances. In some cases, the organism will have to learn to perceive a perceivable affordance, that is, to learn to become attuned to the invariant or disturbance specifying it [Scarantino, 2003, pp. 953-954].

When ambient energy arrays, for example "optic arrays" – completely transparent in the case of many human and non human animals –, or "acoustic waves", become sources of affordances they immediately acquire the physical property values typical of an "ecological" physics. From this perspective stimuli are – stricto sensu – energies and not objects "[...] a chair and the light reflecting of a chair are qualitatively different entities. The human eye is a transducer for the latter and not for the former. In other words, distal stimulus is not a stimulus at all" [Vicente, 2003, p. 258].

As I have said before becoming attuned to invariants and disturbances often goes beyond the mere Gibsonian direct perception, and higher representational and mental processes of thinking/learning have to be involved.[25] This means that for example in designing an artifact to the aim of properly and usefully exhibiting its full range of affordances we have to clearly distinguish among two levels: 1) the construction of the utility of the object and 2) the delineation of the possible (and correct) perceptual information/cues that define the available affordances of the artifact. They can be more or less easily be undertaken by the user/agent (cf. [Gaver, 1991; Warren, 1995; McGrenere and Ho, 2000]): "In general, when the apparent affordances of an artifact match its intended use, the artifact is easy to operate. When apparent affordances suggest different actions than those for which the object is designed, errors are common and signs are necessary" [Gaver, 1991, p. 80]. In this last case affordances are apparent because they are simply "not seen". In this sense information arbitrate the perceivability of affordances, and we know that available information often goes beyond what it can be provided by direct perception but instead involves higher cognitive endowments. My study on affordances as chances, in term of abduction (cf. subsection 6.5 and section 6.6), will take advantage of this extended framework. Here a final note as to be added.

Vicente contends that it has to be said that of course it is impossible to think that direct perception can explain all psychological phenomena, like many Gibsonian

[25] Authors like Turvey and Shaw [2001] and Hammond et. al [1987] pointed out that high-level organisms' cognitive processes like those referred to language, inference, learning, and the use of symbols would have to be accounted for by a mature ecological psychology.

researchers seem to maintain. Moreover, according to Reed, the opinion that mediated perception or cognition is inconsistent with Gibson's view of ecological psychology is "simply mistaken" [1988, p. 305], like the following passage by Gibson would clearly illustrate:

> At least three separate levels [of theorizing] will be required: first, a theory of how we perceive the surfaces of objects [...]; second, a theory of how we perceive representations, pictures, displays, and diagrams; and third, a theory of how we apprehend symbols. There is no reason to suppose that the physiological concomitants of all these experiences will be the same; in fact, since pictures and symbols presuppose objects, their physiological explanations will probably have to be found at increasing levels of complexity [Gibson, 1951, p. 413].

Of course Gibson mainly preferred to study the first of the three categories of theories, related to relatively narrow psychological phenomena.

6.4.2 Proximal and Distal Environment

In the extended perspective above we can consider an artifact with its affordances in the framework of a distributed cognitive system where people interact with external cognitive representations.[26] Internal representations are the knowledge and structure in individual's brains, external representations are the knowledge and structure in the external environment, for example in a specific artifact (cf. [Zhang, 1997]). These external representations have many non-trivial properties (symbolic/semiotic patterns) depending on the kind of cognitive delegations operated when building them, the structure of the artifact itself (physical and chemical configurations), and the constraints intrinsic to its materiality.

I have already illustrated in chapter three, subsection 3.6.5, that following Clark's perspective on language as an external tool [Clark, 1997] Wheeler speaks of *on-line* – like in the case of manipulative abduction, which involves both internal and external representations – and *off-line* (also called *inner rehearsal*), based on internal representations alone. A true situation of distributed cognition is occurring in the case of on-line thinking, like in our case of manipulative abduction and in other less expert and less creative cases, where the resources are not merely inner (neurally-specified) and embodied but hybridly intertwined with the environment: in this case we face with an abductive/adaptive process produced in the dynamical inner/outer coupling where internal elements are "directly *causally locked onto* the contributing external elements" [Wheeler, 2004, p. 705].

As I have already noted in chapter one (section 1.5.2) my epistemological distinction between theoretical and manipulative abduction is certainly also based on

---

[26] Of course cognition can also be distributed across a collective of individuals. This is in line with the so-called "social manifestation thesis" put forward by Wilson. As he contends: "[...] individuals engage in some forms of cognition only insofar as they constitute part of a social group. [...] These psychological traits are not simply properties of the individual members of the group, but features of the group itself" [Wilson, 2005, p. 228]. Cf. also chapter one, section 1.6.1, p. 43.

the possibility of separating – at least in abstract – the two aspects in real cognitive processes, that resort to the distinction between off-line (theoretical, when only inner aspects are at stake) and on-line (manipulative, where the interplay between internal and external aspects is fundamental). Of course I have to repeat that various thinkers like Esther Thelen and Andy Clark have raised doubts about the on-line/off-line distinction. However, we can contend that, even if manipulative/on-line cases predominate, there are also cognitive processes that seem to fall into the class of off-line thinking.

In this perspective affordances can be considered as related to distributed representations extended across the environment and the organism. In the case of the affordances possibly offered by an artifact a basic problem is that the "cognitive" properties of the components of it cannot be inferred from the properties of those components themselves [Hutchins, 1995]. Following Zhang and Patel [2006, p. 336] we can say that

> The external and internal representation spaces can be described by either constraints or allowable actions. Constraints are the negations of allowable actions. That is, the allowable actions are those satisfying the constraints, and the constraints set the range of the allowable actions. If the external and internal representation spaces are described by constraints, then the affordances are the disjunction of the constraints of the two spaces. If the external and internal representation spaces are described by allowable actions, then the affordances are the conjunction of the allowable actions of the two spaces.

Brunswik's [1952] ecological theory (where agents and their environments are still seen, like for Gibson, as strongly intertwined) in terms of proximal and distal environment (cf. his famous lens-model), suitably combined with the theory of affordances, can further clarify the role in cognition of artifacts and of high-level cognitive processes, especially in human beings. Following Brunswik's theory an organism is not able to perceive distal stimuli directly but instead must *infer* what is going on in its ecological niche from the cues available (cues provided by proximal stimuli). The success (Brunswik says the *ecological validity*) of this "vicarious" inference[27] is of course jeopardized by the constitutive incompleteness (and unreliability, ambiguity, and equivocality) of the cues available and by their highly variable diagnostic character: Brunswik, implicitly professing an abductive attitude worthy of Peirce, says: "[...] ordinarily organisms must behave as if in a semierratic ecology" [1955, p. 209], given the intrinsic "ambiguity in the causal texture of the environment" [1943, p. 255].[28] In this sense he adds that the cues and the mediating inference are both "probabilistic", like in the case of abduction where it is always the case that: "Both the object-cue and the means-end relationship are relations between probable partial causes and probable causal effects" [Brunswik, 1943, p. 255].

---

[27] Here the word inference has to be intended in the expanded semiotic Peircean sense, i. e. the term must not collapse in the restricted logical meaning (cf. below section 6.5).

[28] A detailed illustration of these aspects of Brunswikian theory are given in Hammond and Steward [2001, part I].

As I will better describe in subsection 6.5 and in section 6.6 the inference above is of course abductive, and *mediates* the relationship between an organism's central desires and needs and the distal state of affairs, enabling it to (provisionally) *stabilize* the relationship itself by reaching goals. Of course this mediating process involves different means (for example selecting, combining, and substituting cues to overcome their limitations) [Rothrock and Kirlik, 2006] in different environmental circumstances [Goldstein, 2006, p. 12], where time is often a key element which strongly affects information incompleteness and the degree of success of the various means and inferential procedures used.

The organism's "end stage" of being afforded, which in the case of humans, involves all aspects of what is called *multimodal abduction* (cf. this book, chapter four), is of course reached, like Gibson contends, through *perception*, as a suitable collections of sensory information rather than through the organism's overt behavior (in this case perception resorts to a spontaneous abduction performed through various sensory mechanisms and their interplay, cf. chapter one, section 1.3).

It has to be said that recent research in human and not human animals has jeopardized the assumption concerning the basic cognitive impenetrability of perception (an analysis of this problem if presented in the previous chapter) in the framework of animal cognition). In sum, perception is informationally "semi-encapsulated", and also pre-wired, i. e., despite its bottom-down character, it is not insulated from plastic cognitive processes and contents acquired through learning and experience. Many results in the field of cognitive experimental psychology and of neuroscience corroborate the assumption above: 1) the role of emotions (anxiety) in perceiving affordances and the role of attentional narrowing mechanisms in the related changes [Pipers *et al.*, 2006]; 2) the nonconscious effects of perceptual accuracy and on consequent actions of tool use (reachability influences perceived distance [Witt *et al.*, 2006]); 3) the role of moral and social weights in people that carry babies or groceries [Godges and Lindheim, 2006]; 4) the role of motivational states (desires, wishes, preferences) [Balcetis and Dunning, 2006]; 5) the effect of the adoption of the actor's perspective while observing and understanding actions through mentalistic or motor schemes (for experimental results cf. [Lozano *et al.*, 2006] and [Anquetil and Jeannerod, 2007]; for neurobiological correlates cf. [Paccalin and Jeannerod, 2000] and [Decety and Grèzes, 2006]).

Thus affordances emerge through perceptions that are semi-encapsulated, which in turn grant the final (pragmatic) *action*.

In this light the Brunswikian concept of *ecological validity* can be seen in terms of abductive plausibility of the inference at play, given the available data/cues. The quality of the inferential abductive performance measures the degree of adaptation between an organism's behavior and the structure of the environment, i. e. the fitness of the behavior based on the particular adopted inference. When the cues are object of simple and immediate perceptual appraisal the situation reflects what I have illustrated in the case of the so-called "visual abduction" [Magnani, 2001b].[29] On the contrary, in the other cases, organisms inferentially abduce a "hypothesis/judgment"

[29] In this last case we could say there is a one-to-one mapping between proximal and distal structure [Vicente, 2003, p. 261].

about distal structure of the environment. It is worthy to quote again Gibson's intuition, which can better be grasped in this perspective: "Perceiving is the simplest and best kind of knowing. But there are other kinds, of which three were suggested. Knowing by means of instruments extends perceiving into the realm of the very distant and the very small; [...] Knowing by means of language makes knowing explicit instead of tacit" [Gibson, 1979, p. 263]. An example can be a forecast – usually probabilistic – about the behavior of wind based on the current wind speed measured at a ground station as illustrated in a computer screen – the "cue". It is important to say that in this last case the proximal perception affords the day-after action of wearing clothes suitable to the weather.

I have said above that in the mediating inferential process also cognitive delegations to the environment (for example to automated artifacts) can also play a role, for example in facilitating reliable action/decisions.[30] In both human and non-human animals artifacts can reduce the uncertainty of the relationship between organism and environment. Even in these cases, for example – in the case of humans – when technology is directly designed to respond with greater precision than people can do, a person's abductive judgment can still fail to correspond with the distal environment. Recent research based on Brunswikian tradition has emphasized the essential ecological character of the cognitive engineering enterprise in the framework of systems composed of the interaction among humans, mediating technologies, and tasks environments. Many results have shown in various interesting ways how many technological devices help humans to fulfill their adaptation to the environment by enhancing hypothesis generation/judgment and, consequently, decision-making. Often the technology itself fails to provide the correct judgment about a given situation, in other cases the gap in the proximal/distal relationship is embedded in the interaction with the user [Kirlik, 2006b].

For example [Byrne *et al.*, 2006] study the airline pilot's performance in landing and taxiing to the gate in foggy conditions by showing, thanks to an ecological analysis, how the cockpit artifacts (and their interfaces which provide proximal cues/chances for action) affect the overall cognitive performance. Often the problem is related to the fact that technologies exhibit a *discrete* ecology that does not sufficiently involve the approximation and convergence performed by *continuous* ecologies of natural environmental structures. This fact for example explains why it has been recently shown that current design of cockpit automation leaves pilots less supported in special uncertain (and more challenging) – unsafe – situations: control systems proximal to pilots are discrete, whereas the behavior of the distal controlled system – the aircraft – is continuous [Degani *et al.*, 2006]. Finally, in the mediating abductive inferences occurring through artifacts, time (and the so-called "time-stress" and "time-pressure" effects and their relationship to knowledge deficits and task simplification) is still a key element which strongly affects information incompleteness and the degree of success of various heuristic schemes.

The discreteness of the ecology of the artifacts above is related to the basic problem of the discretization of knowledge caused by the digital revolution, which relates

---

[30] The case of computer interfaces and time pressure is treated in [Adelman *et al.*, 2006].

to the profound epistemological status of concrete digital machines with respect to physical continuous systems.[31] [Longo, 2009] provides an insightful description of the discrete/continuous dichotomy, referring to the "mimetic" role of computational simulations in physics and engineering, the virtual digital world is far from providing a satisfactory theoretical model, or "knowledge", of the real physical system:

> Computational simulations cost a lot less than experiments leading many physicists to renounce experiments and work only on implementations. The simulation of turbulence, an extreme case of chaos, not only saves on wind tunnels, but its iterability is also an asset. The expert's qualitative judgment has easy access to as many iterations as necessary in order to appreciate the behavior of an airplane's wing or cockpit and the small variations induced provides a good picture of the particular dynamic's sensitivity (but it does not allow joint analysis of the wing and cockpit: virtuality in this case is too far removed from the phenomenon which is excessively complex).

6.4.3 Reconciling Direct and Mediated Perception: Ecological and Constructivist Approaches Intertwined

It is Joel Norman who, taking advantage of a wholly neuropsychological perspective, tries to account for a reconciliation of the two approaches above (ecological and constructivist). They resort to two cortical visual systems, the first of which he calls dorsal – hardwired, direct and active, less representational, without the recourse to memory, and so expressing Gibson's classical affordances – and the second ventral, which is more representational and judgemental, indirect, and related to mentalistic processes, and which basically performs different transformations of the available visual information.[32] Both systems perform different functions, and, present consciousness at different degrees (for example the ventral system brings the relevant information, picked up in a more unconscious way by the dorsal system, to conscious awareness).[33] Finally, both systems analyze the visual input, but the analysis is carried out for different purposes. Certainly both systems deal with object shapes, sizes, and distances, but

> The primary function of the ventral system would seem the *recognition* and *identification* of the visual input. Recognition and identification must depend on some comparison with some stored representations. In contrast, the primary function of the dorsal system is analysis of the visual input in order to allow *visually guided behavior* vis

[31] A more detailed analysis of this problem is presented in chapter eight, subsection 8.1.1.

[32] From a methodological perspective, the so-called "double-dissociation of function" constitutes a major evidence for the existence of neurologically distinct functional systems. Generally speaking, double dissociation is considered a strong neurological evidence when a lesion of structure *X* [ventral stream] will specifically disrupt function *A* [functional knowledge] while sparing function *B* [manipulation of an object], and a lesion of structure *Y* [dorsal stream] will specifically affect function *B* [manipulation of an object] while function A would remain intact [functional knowledge]. For a critical point of view on double-dissociation and the two visual systems, see [Pisella *et al.*, 2006].

[33] On the neural correlates of consciousness cf. [Vakalopoulos, 2005].

> à vis the environment and objects in it (e.g. painting, reaching, grasping, walking towards or through, climbing, etc.). [...] Thus, when one picks up a hammer, the control and monitoring of the actual movements is by the dorsal system but there also occurs intervention of the ventral system that recognizes the hammer as such and directs the movement towards picking up the hammer by the handle and not by the head [Norman, 2002, p. 84].

This perspective further clarifies that, when Gibson says that a postbox affords letter-mailing, it would be better to say that the slot in the mailbox affords inserting an object of appropriate size and shape (*ibid.*, p. 86).[34]

Recent discovery that different mental states concerning action (in the case of both intending an action and observing it performed by another agent) appear to share the same neural correlates, also stresses the intertwining between perception and action (cf. [Jeannerod and Pacherie, 2004]): in humans the signal produced by my mirror neurons appears to be the same as that of an action performed by the self and by another agent, and knowledge of other minds appears to be performed by exploiting both motor (devoted to grasping naked intentions) and inferential/mentalistic processes. The Cartesian gap, which maintains that there is an ontological fissure between perception and action, can be avoided. Jeannerod and Pacherie observe that the cortical network common to intending actions and preparing for execution, imagining actions, and observing actions performed by other people, is quite extended:

> My contention is that this cortical network provides the basis of the conscious experience of goal-directness – the primary awareness of intentions [naked] – but does not

[34] The debate (and criticisms) concerning Norman's so-called "dual-process approach" can be found in [Norman, 2002, pp. 96–137]. The recent discussion on the neurophysiology underlying affordances is illustrated in [Garbarini and Adenzato, 2004]. It is worth noting that the alignment of affordances with the dorsal stream argued by Norman is still problematic, as suggested by [Young, 2006]: "[...] only certain affordances are processed along the dorsal stream, and another neural pathway – the ventral system – may be implicated in the processing of other afforded properties" [Young, 2006, p. 136]. Several studies on patients diagnosed with optic ataxia (dorsal stream impaired) contributed to shed light on this issue. Patients affected by optic ataxia are able to name an object appropriately and recognize its function, but unable to grasp and locate it. However, if allowed a delay between target presentation and movement execution, they demonstrate skilled pantomimed action by relying on visual recollection of an object instead of its actual location [Young, 2006]. What they lack is therefore the ability to unconsciously adjust ongoing movements that seems to suggest the existence of an automatic pilot, which in this case results impaired (cf. [Himmelbach *et al.*, 2006, p. 2750]). This suggests that patients successfully interact with the object by retrieving past information stored in their memory [Milner *et al.*, 2001] instead of picking them up upon occasion. As maintained by Himmelbach and Karnath [2005, p. 633], "[...] the contribution of ventral system increases as the delay between target presentation and movement execution gets longer". This seems to suggest a division of labor between dorsal and ventral system in allowing people to interact with their environment. From this perspective affordances processed along the dorsal stream are merely picked up unconsciously, "[...] whereas those processed via the ventral stream constitute one aspect of the content of the subject's phenomenal experience, although I accept that this does not mean that the subject must be reflectively aware of such content" [Young, 2006, p. 141]. This ambivalence will be more extensively described taking advantage of the concept of abduction, cf. the following sections.

> by itself provide us with a conscious experience of self- or other-agency. This latter experience has its basis in the activation of cortical zones that do not overlap between conditions [Jeannerod and Pacherie, 2004, p. 140].[35]

Finally, Adolphs stresses the fact that perception in social cognition is "special", in both human and some non-human animals, and involves psychological and neural structures that other aspects of (non-social) cognition do not. Perception is not only affected by the inferences we make once sensory information has been perceived, but also "[...] by the possibility of discovering new information in the environment in the first place. We explore our environment, and we actively seek out social information" [Adolphs, 2006, p. 26].

From Joel Norman's perspective perception is first of all a "special" chance-seeker system, which presents pre-mentalistic encaspulated aspects. Examples are given from the field of facial recognition, hypothesized as domain-specific. The amygdala plays a role "much earlier than initially thought" [Adolphs, 2006, p. 27] in seeking salient information, so involving extra-neural enactive processes related to bodies, head, and eye muscles, and social environment, where movement is prior to sensation. The role on human mentalization in inferring other minds shows a propensity to infer mental states even in the presence of impoverished stimuli and to anthropomorphize stimuli that are not inherently social: again it seems that social cognition in this case is special or at least involves neural structures that organisms can only modulate for other limited cognitive performances, like decision-making or attention. Empathy, which notoriously involves simulation structures like mirror neurons and the insula – an interoceptive somatosensory cortex which provides representation of our own somatic states rendering them accessible to consciousness – also seems to be related to a wider cluster of neural mechanisms and to the use of the body as a substrate of the simulation, beyond the merely neural mechanism granted by mirror neurons.[36]

Interpreting perception and other forms of cognition as ways of seeking salient information that involves extra-neural enactive processes related to bodies and social environment rejoins what I have illustrated in the previous subsection about cognitive *action* and its "situatedness". I have said it is a way of getting more sensory data, compensating for their equivocality, and obtaining cognitive feedback, and/or a way of manipulating them, and also of exploiting *cognitive delegations* to the environment and to artifacts. In this sense brains do not need to form rich internal models of the environment and so they do not store "[...] all knowledge about the world in explicit form and do not hold comprehensive explicit models or representations of the environment. Rather, it has been argued [...] our brains

[35] I have already illustrated how the discovery of mirror neurons has given rise to speculation on various aspects of social cognition: intentions, action, empathy, mind-reading, emergence of language. Further details are illustrated above in chapter four, subsection 4.3.2.

[36] An interesting logico-epistemological analysis of the role of empathy in the so-called interpretation abduction, where linguistic understanding and understanding of other agent's inarticulate utterances is at play, is illustrated in [Gabbay and Woods, 2005, pp. 319–323]. They conclude that empathetic attachment "[...] is never a necessary ingredient *in* the successful negotiation of – especially intractable – interpretation problems."

contain recipes for seeking out information – often rather trivially by deciding where in the environment to look" in the interplay between internal and external (delegated) representation [Adolphs, 2006, p. 32].

6.4.4 *Attunement, Affordances, and Cognitive Artifacts: Extracting and Creating Affordances*

In subsection 5.4.4 of the previous chapter I treated the problem of attunement to the environment and the multifarious cognitive means the organisms exploit to reach this target, hardwired, pre-wired, and plastic. In the case of hard-wired capabilities, I contended that it is through this embodied process that affordances can arise both in wild and artificially modified niches and that Peirce was in some sense already aware of their cognitive role when speaking of abduction even in the case of the new born chick that picks up the right sort of corn (cf. section 5.2 of the previous chapter). I also stressed the fact that animals can act on a goal that they cannot perceive – the predator that waits for the prey for example – so organism's appraisal of the situation includes factors that cannot be immediately perceived. Animals occupy different environmental niches that "afford" their possibility to act, like Gibson's theory of affordances teaches, but in this interplay cognitive niches built and created by them also provide affordances: higher-level cognitive endowments either shaped by the evolution or plastically learnt are at play. Dishabituation experiments have shown how for example infants use high-level physical principles to relate themselves to the environment and to be afforded by it: in sum organisms already have affordances available because of their instinctive gifts, but also they can *dynamically* abductively "extract" natural affordances through "affecting" and modifying perception (which becomes semi-encapsulated). Finally, organisms can also "create" affordances by building artifacts and cognitive niches. These last affordances were not previously available taking advantage of both their instinctual and cognitive capacities. Like Gibson says: "[...] The human young must learn to perceive these affordances, in some degree at least, but the young of some animals do not have time to learn the ones that are crucial for survival" [Gibson, 1979, p. 406].

In subsection 5.4.4 of the previous chapter I contended that it can be hypothesized that in many organisms the perceptual world – in Brooks' words [1991] – is the only possible model of itself, and they can be accounted for in terms of a merely reflex-based organisms, and so no other internal representations are available. However, I also illustrated that in the case of affordances in various sensitive organisms the coupling with the environment is much more flexible because it is crucial in coupling with the niche to ascertain what environmental dynamics are currently the most important, among the several that afford and that are available. Of course, both affordances and the plastic cognitive processes of their selection in specific situations can be "stabilized", but they can also be modified, increased, and substituted with new ones: indeed in the same section I have illustrated the amazing "plastic" activity we can see in the management of information of bacteria.

6.5 Affordances and Abduction: The Plasticity of Environmental Situatedness

I have quoted above Gibson conviction that "The hypothesis that things have affordances, and that we perceive or learn to perceive them, is very promising, radical, but not yet elaborated" [Gibson, 1979, p. 403]. Let us deepen this issue: we can say that the fact that a chair affords sitting means we can perceive some clues (robustness, rigidity, flatness) from which a person can easily say "I can sit down". Now, suppose the same person has another object O. In this case, the person can only perceive its flatness. He/she does not know if it is rigid and robust, for instance. Anyway, he/she decides to sit down on it and he/she does that successfully. Again, we are faced with the problem of direct and indirect visual perception. It is thanks to the effect of action that we can detect and stabilize the new affordances.

Now, my point is that we should distinguish between the two cases: in the first one, the cues we come up with (flatness, robustness, rigidity) are *highly diagnostic* to know whether or not we can sit down on it, whereas in the second case we eventually decide to sit down, but we do not have any precise clue about. How many things are there that are flat, but one cannot sit down on? A nail head is flat, but it is not useful for sitting. This example further clarifies two important elements: firstly, finding/constructing affordances certainly deals with a (semiotic) inferential activity (cf. [Windsor, 2004]); secondly, it stresses the relationship between an affordance and the information that specify it that only arise in the *eco-cognitive interaction* between environment and organisms. In this last case the information is reached through a simple action, in other cases through action and complex manipulations. I maintain that the notion of abduction can further clarify this puzzling problem (cf. below section 6.6).

The term "highly diagnostic" explicitly refers to the abductive framework. In section 1.3 of chapter one I have said that abduction is a process of *inferring* certain facts and/or laws and hypotheses that render some sentences plausible, that *explain* or *discover* some (eventually new) phenomenon or observation. The distinction between theoretical and manipulative abduction extends the application of that concept beyond the internal dimension. I repeatedly said that from Peirce's philosophical point of view, all thinking is in signs, and signs can be icons, indices or symbols: moreover, all inference is a form of sign activity, where the word sign includes "feeling, image, conception, and other representation" [Peirce, 1931-1958, 5.283], and, in Kantian words, all synthetic forms of cognition. That is, a considerable part of the thinking activity is "model-based" and consequently non sentential. Of course model-based reasoning acquires its peculiar creative relevance when embedded in abductive processes, so that we can individuate a *model-based abduction*. In the case of diagnostic reasoning in medicine, a physician detects various symptoms (that are signs or clues) in a multimodal way, for instance, cough, chest pain, and fever, *then* he/she may infer that it is a case of pneumonia.

The original Gibsonian notion of affordance (cf. section 6.2) especially deals with those situations in which the "perceptual" signs and clues we can detect prompt or suggest a certain action rather than others.[37] They are already available and belong to the normality of the adaptation of an organism to a given ecological niche. Nevertheless, if we acknowledge that environments and organisms' instinctual and cognitive plastic endowments change, we may argue that affordances can be related to the variable (degree of) *abducibility* of a configuration of signs: *a chair affords sitting* in the sense that the action of sitting is a result of a sign activity in which we perceive some physical properties (flatness, rigidity, etc.), and therefore we can ordinarily "infer" (in Peircean sense) that a possible way to cope with a chair is sitting on it. So to say, in most cases it is a spontaneous abduction to find affordances because this chance is already present in the perceptual and cognitive endowments of human and non-human animals.

As I have already anticipated in the previous chapter (subsection 5.4.4) I maintain that describing affordances that way may clarify some puzzling themes proposed by Gibson, especially the claim concerning the fact that we directly perceive affordances and that the value and meaning of a thing is clear at first glance: organisms have at their disposal a standard endowment of affordances (for instance through their hardwired sensory system), but at the same time they can extend and modify the scope of what can afford them through the suitable cognitive abductive skills. I also stressed the important fact that environments change but also perceptive capacities enriched through new or higher-level cognitive skills change, those capacities which go beyond the ones granted by the merely instinctual levels: if affordances are usually stabilized this does not mean they cannot be modified and changed and that new ones can be formed.[38]

First of all, affordances appear durable in human and animal behavior, like kinds of habits, as Peirce would say [Peirce, 1931-1958, 2.170]. For instance, that a chair affords sitting is a fair example of what I am talking about. This deals with what we may call *stabilized affordances*. That is, affordances that we have experienced as highly successful. Once evolutionarily formed, or created/discovered through cognition, they are stored in embodied or explicit cognitive libraries and retrieved upon occasion. Not only, they can be a suitable source of new chances, through analogy. We may have very different objects that equally afford sitting. For instance, a chair has four legs, a back, and it also stands on its own. The affordances exhibited by a traditional chair may be an analogical source and transferred to a different new artifact that presents the affordance of a chair for sitting down (and that to some extent can still be described as a chair). Consider, for instance, the variety of objects that afford sitting without having four legs or even a back. Let us consider a stool: it does not have even a back or, in some cases, it has only one leg or just a pedestal, but it affords sitting as well as a chair.

---

[37] In the original Gibsonian view the notion of affordance is mainly referred to proximal and immediate perceptual chances, which are merely "picked up" by a stationary or moving perceiver. I maintain that perceiving affordances also involves evolutionary changes and the role of sophisticated and plastic cognitive capacities.

[38] Thom nicely says that "Nature (φύσις) is present in the behavior of inanimate beings. But the animate being is able to exploit natural regularities in order to stabilize connections that would be accidental, not generic, in the inanimate world" [Thom, 1988, p. 217]. On Thom's catastrophe theory and its consequences for the concept of abduction cf. chapter eight.

Second, affordances are also subjected to changes and modifications. Some of them can disappear, because new configurations of the cognitive environmental niche (for example new artifacts) are invented with more powerful offered affordances. Consider, for instance, the case of blackboards. Progressively, teachers and instructors have partly replaced them with new artifacts which exhibit affordances brought about by various tools, for example, slide presentations. In some cases, the affordances of blackboards have been totally re-directed or re-used to more specific purposes and actions. For instance, one may say that a logical theorem is still easier to be explained and understood by using a blackboard, because of its affordances that give a temporal, sequential, and at the same time global perceptual depiction to the matter.

Of course – in the case of humans – objects can afford different persons in different ways. This is also the case of experts: they take advantage of their advanced knowledge within a specific domain to detect signs and clues that ordinary people cannot detect. For instance, a patient affected by pneumonia affords a physician in a completely different way compared with that of any other uncultured person. Being abductive, the process of perceiving affordances mostly relies on a continuous activity of hypothesizing which is cognition-related. That A affords B to C can be also considered from a semiotic perspective as follows: A signifies B to C. A is a sign, B the object signified, and C the interpretant. Having cognitive skills (for example knowledge contents and inferential capacities but also suitable pre-wired sensory endowments) about a certain domain enables the interpretant to perform certain abductive inferences from signs (namely, perceiving affordances) that are not available to those who do not possess those apparatuses. To ordinarily people a cough or chest pain are not diagnostic, because they do not know what the symptoms of pneumonia or other diseases related to cough and chest pain are. Thus, they cannot make any abductive inference of this kind and so perform subsequent appropriate medical actions.

6.6 Innovating through Affordance Creation

Consider, for instance, a huge stone and a chair. Indeed, both these objects afford sitting. The difference is that in the case of a stone affordances are simply already, so to say, *given*: we find a stone and we ordinarily "infer" (in Peircean sense) that it can be useful for sitting. In contrast, those of a chair are somehow *manufactured* from scratch. In the case of a chair, we have entirely created an object that displays a set of affordances. This process of building affordances can be made clearer taking advantage of the abductive framework we have introduced above.

That an object affords a certain action means that it embeds those *signs* from which we "infer" – through various cognitive endowments, both instinctual and learnt – that we can interact with it in a certain way. As already said, in the case of a stone, humans exploit a pre-existing configuration or structure of meaningful sign data that are in some sense already established in the interaction organism/environment shaped by past evolution of the human body (and in part "material cultural" evolution, for example when hominids exploited a stone/chair to sit down in front of a rudimentary altar). In the case of a chair, this configuration is invented. If this perspective is correct, we may argue that building "artificial" affordances means configuring signs in an external environment expressly to trigger new proper inferences of affordability. In doing this, we perform smart manipulations and actions that – I conjecture – can produce new (and sometimes "unexpected") affordances. Accordingly, creating affordances is at the same time making new ways of inferring them feasible.[39]

6.6.1 Latent Constraints

The organism's "end stage" of being afforded, in the Brunswikian interplay proximal/distal, not only involves perception, but also a more complicated *inferential processing* which in turn occurs through either

1. a mere *internal* cognitive processing (expressed by model-based abduction and, in the case of human beings, also by sentential abduction, cf. chapter one, section 1.3)
   or
2. a (cognitive) composite *action* with the aim of getting more sensory data, compensating for their equivocality, and obtaining cognitive feedback (confirmatory and disconfirmatory), and/or with the aim of manipulating them (manipulative abduction), also exploiting *cognitive delegations* to the environment and to *artifacts* (cf. chapter one, section 1.3).

   [Indeed these high-level inferential processes affect perception in various ways, like its is shown by the evidence coming from the studies illustrated above].

I have consequently said that affordances emerge through perceptions that in humans are basically semi-encapsulated, and also affected by the two processes above and that grants the subsequent (pragmatic) *action*.

Consequently, two kinds of "actions", cognitive and pragmatic (performatory) are at play. Kirlik [2006a], also resorting to a distinction between pragmatic and epistemic action offered by Kirsh and Maglio [1994],[40] offers an analysis of the problem that is extremely interesting to further clarify the dichotomy. In Kirlik's words, the first type of action I have just indicated plays an "epistemic" or

[39] On this note, see [Magnani and Bardone, 2005] about the role of abduction in designing computer interfaces.

[40] Kirsh describes everyday situations (e. g. grocery bagging, salad preparation, where people adopt actions to simplify choice, perception, and reduce the amount of internal computation through the use of suitable cognitive delegations to the environment and to artifacts [Kirsh, 1995].

"knowledge-granting" role and the second a "performatory role in the execution of interactive tasks" [Kirlik, 2006a, p. 214]. It is well-known that traditionally, cognitive scientists have seen action systems in a purely performatory/pragmatic light, thus lacking any immediate "cognitive" features other than ability to execute commands.

Recent research, taking an explicit ecological approach to the analysis and design of human-machine systems [Kirlik, 1998], has shown how expert performers use action in everyday life to create an *external* model of task dynamics that can be used in lieu of an internal model: "Knowing of not, a child shaking a birthday present to guess its contents is dithering, a common human response to perceptually impoverished conditions". Not only a way for moving the world to desirable states, action performs an epistemic and not merely performatory role: people structure their worlds to simplify cognitive tasks but also in the presence of incomplete information or faced with a diminished capacity to act upon the world when they have less opportunities to gain knowledge. Epistemic action can also be described as resulting from the exploitation of latent constraint in the human-environment system. This additional constraint grants additional information: in the example of the child shaking a birthday present she is taking action that will cause variables relating to the contents of the package to covary with perceptible acoustic and kinesthetic variables. Epistemic actions result from exploiting *latent constraints* in the human-environment system as chances for further inferences. "Prior to shaking, there is no active constraint between these hidden and overt variables causing them to carry information about each other". Similarly, "[...] one must put a rental car 'through its paces' because the constraints active during normal, more reserved driving do not produce the perceptual information necessary to specify the dynamics of the automobile when driven under more forceful conditions" [Kirlik, 1998, p. 24] . In this light [Powers, 1973] studied behavior considering it as a *control of perception* and not only as controlled by perception. [Flach and Warren, 1995] used the term "active psychophysics" to illustrate that "[...] the link between perception and action [...] must be viewed as a dynamic coupling in which stimulation will be determined as a result of subject actions. It is not simply a two way street, but a circle" (p. 202).

Moreover, a very interesting experiment is reported concerning short-order cooking at a restaurant grill in Atlanta: the example shows how cooks with various levels of expertise use external models in the dynamic structure of the grill surface to get new information otherwise inaccessible.

In Brunswikian terms some variable values are proximal because they can be perceived and others cannot because they are distal. This distinction, Kirlik observes, is relative:

> A particular variable (e.g. the velocity of an automobile) could be described as proximal if the purpose of a study was to understand how a driver might use velocity to infer less readily available information relating to a vehicular control task (e.g. whether he or she can take a given turn safely). In other cases, velocity could be considered a distal variable if the goal of the effort was to understand how velocity itself was estimated on the basis of even more immediate perceptual information (e.g. optical flow patterns, speedometer readings, etc.) [Kirlik, 2006a, p. 216].

Of course from the perspective of action the values of proximal variables can directly be manipulated. On the contrary distal variables can be changed (and so inferred) only by manipulating proximal variables. On this basis Kirlik illustrates a generalized framework that provides an ontology for describing an environment of a performer who is undertaking an interactive activity. Variables exhibited to an agent in a world of surfaces, objects, events, and artifacts present various values:

1. Type 1 [PP.PA]: variables are proximal from the perspective of both perception and action. "The location of the book you are now reading can most likely be represented by variables of this type: most of us can directly manipulate the values of these variables by hand and arm movements, and we can also perceive the location of the book directly. One can think of these variables as representing Gibson's directly perceptible affordances" [Kirlik, 2006a, p. 216].
2. Type 1 [PP.DA]: variables proximal for perception but distal for action. Variables are directly available to perception but we can change their values by manipulating proximal variables that cause changes in the values of the distal action variable. You feel it is cold in the room but you need to manipulate the thermostat to change the temperature and consequently your feeling, but you cannot directly change the temperature.
3. Type 1 [DP.PA]: variables distal for perception and proximal for action. It is the case that many of my actions through for instance an artifact can change the values of environment, which I cannot directly perceive. A common outcome of the manipulation of many artifacts is that they for example have moral consequences, perhaps even unexpected ones, for human beings and objects very distant and remote from us, which might last forever. "When posing for a photograph I change the location of my image in the viewfinder of a camera without being able to perceive how it has been changed" [Kirlik, 2006a, pp. 217–218].[41]
4. Type 1 [PP.PA]: variables that are distal from both perception and action, very common in the case of technology. "We infer the values of these variables from interface displays, and we change the value of these variables by using interface controls" (*ibid.*).

In the framework above, which stresses the importance of inner-outer interaction in proximal-distal dynamics, it is very easy to interpret artifacts built by humans as ways of adaptation that, through the construction of suitable cognitive feasible niches, mediate and augment our interaction with the distal world. They aim at enhancing intellectual functioning by offering suitable differentiated affordances/proximal-cues, which are easier to perceptually detect and present new opportunities for action. This strategy, which modifies previous available human

[41] I have illustrated a related issue in the recent [Magnani, 2007d]: many artifacts play the role of "moral mediators". This happens when macroscopic and growing phenomenon of global moral consequences and collective responsibilities result from the "invisible hand" of manipulations of artifacts and interactions among agents at a local level, like for example in the case of the effect of the Internet on privacy or of the derivatives in the international economical crisis: for example my manipulations on the net can affect the privacy of people with effects that I cannot perceive.

"cognitive ecology", can offload some cognitive demands to the world through a better exploitation of artifacts.[42]

6.6.2 *Creating Chances through Manipulating Artifacts and External Representations*

It is now clear that the history of culture, artifacts, and various tools can be viewed as a continuous process of building and crafting new affordances upon pre-existing ones or even from scratch. From cave art to modern computers, there has been a coevolution between humans and the environment they have built and they live in. Indeed, what a computer can afford embraces an amazing variety of opportunities and chances comparing with the ones exhibited by other tools and devices. More precisely, a computer as a Practical Universal Turing Machine (cf. [Turing, 1992] and chapter three of this book) can mimetically reproduce a considerable part of the most complex operations that the human brain-mind systems carry out (cf. [Magnani, 2006c]). For instance, computers even result in many respects more powerful than humans in memory capacity and in mathematical reasoning. From a semiotic perspective, computers bring into existence new artifacts that present "signs" (in Peircean sense) – for exploring, expanding, and manipulating our own brain cognitive processing (and so they contribute to "extend the mind beyond the brain"), that is, by offering and creating new affordances. As just argued above in section 6.1, building affordances deals with a semiotic activity, mainly abductive, in which signs are appropriately scattered all around in order to prompt (or suggest) a certain interaction rather than others.

In order to clarify this point, consider the simple diagrammatic demonstration that the sum of the internal angles of any triangle is 180° illustrated in chapter three (cf. Figure 3.5, p. 176.) as a prototypical case of manipulative abduction. This is also an example of construction of affordances taken from the field of elementary geometry. In this case a simple manipulation of the – suitably externally depicted – triangle in Figure 3.5(a) (cf. p. 176) gives rise to another external configuration (Figure 3.5(b)) that uncovers new visual affordances. The process occurs through construction and modification of the initial external diagram, so as we can easily arrive at the generalized result that the sum of the internal angles of any triangle is

[42] Kirlik [2007] describes various interesting examples on the use of artifacts to enhance cognitive skills in technological disciplines like architecture and aeronautics, but also in expert everyday performances like cooking. A negative example of proximal-distal dynamics has to be noted in many contemporary artifact devices (for example in cars) which offer signs that mainly appeal to the visual sense. This involves learning to answer to visual rather than to auditory, olfactory – for example motor smell is extinct in new cars – and tactile proximal cues so loosing out on a lot of important body-mind sensorimotor resources [Bauters, 2007]. We can say, in Peircean terms, that in these cases the semiotic ground active in perception and in which the actions can emerge seems poor, allowing bad and unsatisfactory habits to be afforded, promoted, and stabilized [Vandi, 2007]: indeed Peirce maintains that habits are "[. . .] a general law of action, such that on a certain general kind of occasion a man will be more or less apt to act in a certain general way" [Peirce, 1931-1958, 2.148].

180°. The process happens in a distributed interplay between a continuous externalization through cognitive acts, its manipulation, and re-internalization that acknowledges what has been discovered outside, picking up the result and re-internalizing it. In the action new affordances arise and lead to the result whereby, the sum of the internal angle is found.

From the epistemological point of view this example is a typical example of the so-called manipulative abduction (cf. [Magnani, 2001b] chapter three, and [Magnani, 2007c]). Reframed in terms of affordances this is a cognitive manipulation (entirely abductive, the aim is to find a incontrovertible geometrical hypothesis) in which an agent, when faced with merely "internal" representational geometrical "thoughts", from which it is hard or impossible to extract new meaningful features of a concept, organizes epistemic action that structures the environment in such a way that unearths new affordances as chances which favor new outcomes. As already mentioned, the detection of affordances is hypothesis-driven: it is not said that just anybody can detect it. In the example of the triangle, only a person that has been taught about geometry can infer the affordances within the manipulated construction built upon the original triangle. In this sense, "artificial" affordances are intimately connected to culture and knowledge available in specific human cognitive niches but also to their availability to the suitable individuals involved in the epistemic process.

We can say that in a given geometrical diagram various affordances still exist regardless of correct interpretation or even perception by the agent: indeed the diagram embeds geometrical knowledge that potentially means something to someone. To correctly grasp the affordances

1. the agent has to appreciably "know" geometry,
2. the particularly expert agent – through suitable cognitive manipulations in the internal/external representations interplay – can also extract/find/discover "new" affordances like chances offered to possibly increase of knowledge.

The diagram offers nested affordances:[43]

1. it is a simple picture, which almost everybody (and many animals) can see and understand as a perceptual picture that afford potential colors and shapes, depending on the perceptive hardwired endowments of the organism in front of it – for example expert and uncultured persons but also cats – (strict Gibsonian case);

[43] [Turner, 2005] clearly acknowledges the character of affordances as nested. They are seen as "complex" affordances expressed by "significances" delegated to the natural and artificial environment – in the perspective of an interesting semiotic and Heideggerian approach to distributed cognition. On the role of the "extra-bodily world" – an artifact, for example – that becomes poised itself to the user not just as a problem-space but also as a problem solving resource via "inhabited interaction", which grants fluent acts on ways that simplify or transform the problems, cf. [Clark, 2008, p. 10]. Clark usefully stresses that in case of the inhabited interaction the body has become "transparent equipment" and tools lead to the alterations of the local sense of embodiment: a true incorporation rather than mere use is at play, like Heidegger already contended.

2. it is a picture which can be seen and understood as a geometrical diagram with all its technical properties and features (in this case higher cognitive endowments are needed in the organism at stake: a human being of course, other animals do not understand geometry),
3. it is an artifact which can offer – through even higher creative cognitive manipulations – new affordances to be picked up and possibly inserted in the available encyclopedia of geometrical knowledge. Imagine a child who must "demonstrate" a theorem of the elementary geometry (the sum of the internal angles of a triangle). The child does not have to demonstrate this theorem for the first time, in the sense that it has already been discovered (demonstrated) historically and reported in every manual of Euclidean geometry. However, excluding the case when he passively repeats by rote, he can achieve this demonstration by using the series of elementary extracted affordances illustrated above, based on the appropriate use of basic geometric concepts which are already available to him. We can also say that the child used a heuristic, that is an advanced procedure of discovery. And this heuristic, naturally, considered from the viewpoint of an already available manual of geometry (as an abstract and static system of knowledge) is a true "demonstration", and obviously does not lead to a discovery. Rather, it is a kind of re-discovery. Also from the viewpoint of the child-subject, it is a re-discovery, because he re-discovers a property that is already given at the beginning). Instead the inferences that were employed at the moment of the first historical discovery (maybe Greek) of that property of the triangles (and the assessment of the respective theorem) established a kind of creative achievement (as I have said, a creative manipulative abduction). Moreover, both types of reasoning are also mainly model-based, considering the fact they are performed with "hybrid" forms of representation, including considerable non-verbal devices (like the geometrical diagrams).[44]

The case indicated in item 1) also explains why an affordance can be grasped at the same time by animals, infants, and adult human beings, despite their cognitive differences: all of them can perceive "the brink of a cliff as *fall-off-able* according to a common perceptual process" [Scarantino, 2003, p. 960], even if they have different perceptual endowments. "This is much the same as we would describe a piano as having an affordance of music playability. Nested within this affordance, the piano keys have the affordance of depressability" [McGrenere and Ho, 2000, p. 340]. We can add that the piano also offers the chance, in the cognitive interplay artifact/agent, of providing new affordances of *new* good melodies, not previously invented in a merely internal way, in the agent's mind, but found over-there, in the hybrid interplay with the artifact. Of course both artifacts, the diagram and the piano, offer in themselves various constrained conditions for affordances, depending

[44] Of course the agent can modify and change in a more or less creative way the features of the artifacts just to make the available affordances visible or more exploitable, or to build new ones that are offered as options. This is for example the case of a user who customizes a computational interface making an alias for a long command string. He/she can gain easier use of the tool by simply invoking a single key or by the simultaneous pressing of multiple keys instead of writing a long string of characters [McGrenere and Ho, 2000].

on their properties, quality, materiality, and design, and of course various degrees of affordances. From Kirlik's perspective in terms of variables and proximal/distal distinctions the example above can be illustrated in this way: the agent builds a diagram in which he/she can work on the surface by taking advantage of the constraints that guarantee that latent variables intrinsic to the materiality of the artifact at hand "to take care of themselves, so to speak" [Kirlik, 2006a, p. 221].

The need of having a rich internal model of the depicted triangle is weakened because various aspects are discharged from the agent and delegated to the external representation, which offers a proximal perceptual and a manipulative environment containing all the resources needed to effectively perform the creative task of finding new properties of a triangle. The result is immediately perceived and consequently can be picked up and internally re-represented. The diagram itself is a model in the dynamics of the geometrical reasoning and its manipulation expresses an example of situated cognitive action completely intertwined with perception.

From a semiotic perspective, at first glance we do not have sufficient cognitive capacities to internally infer what the sum of internal angles is. Manipulating the externalized configuration (the external diagram of a triangle) we come up with a new configuration of perceptual signs that adds properties not contained neither in the initial internal representation of a triangle nor in the initial external depiction (cf. above chapter three). This new configuration of signs gives birth to a new set of affordances that allow us to find the solution. In a Euclidean sense it is a way of "demonstrating" a new theorem. This example furnishes an epistemological example of the nature of the cognitive interplay between internal neuronal semiotic configurations that permit the representational thought about a triangle (together with the help of various embodied "cognitive" kinesthetic abilities) and external representations: also for Peirce, more than a century before the new ideas derived from the field of distributed reasoning, the two aspects are pragmatically intertwined.

Indeed, the human possibility of creating affordances is constrained in this interplay. More precisely, this possibility depends on:

- the specific cognitive delegation to externalities suitably artificially modified, and
- the particular intrinsic constraints of the manipulated materiality at play, which usefully provide to us new – so to say – "insights".

The first aspect is closely related to the idea of humans as chance-seekers illustrated above. As extensively argued throughout this chapter, securing meaning to the environment relies on smart manipulations that are always hypothetical and therefore withdrawable. That is, it depends on our ability to exploit latent chances. Regarding the second aspect, the environment has constraints of its own. When designing new affordances, we do not work on a blank slate. Conversely, we are in a network of pre-existing ones that constrain our possibilities. On one hand, the possibility of uncovering new latent chances depends on human agency. On the other hand, this agency is mediated/constrained by what the environment offers in terms of pre-existing affordances and its potentiality to provide new ones, as maintained.

As already mentioned in the previous chapter, this "hypothetical" status of affordances reminds us that it is not necessarily the case that just anybody can detect it. Affordances are a mere potentiality for organisms. First of all perceiving affordances results from an abductive activity in which we infer possible ways to cope with an object from the signs and cues available to us. Some of them are more or less unchanging and in some cases they are neurally hardwired or pre-wired (pre-specified) in the perceptual system – they are "invariants", to exploit a term derived from physics, also used by Gibson : this is the case of affordances that have a high cognitive valence.[45] We said the differences that we can appreciate are mostly *inter-species* – so to speak. In the high-level cognitive performance on triangles there is something different, since *intra-species* differences seem to be strongly involved. For instance, only a person that has been taught about geometry can infer (and so "perceive") the affordances "inside" the new manipulated construction built on the original triangle I have illustrated above. This concerns the problem of "expertise" I already noted. First of all, artificial affordances are intimately connected to culture and the social settings in which they arise and the suitable availability of knowledge of the individual(s) in question. Secondly, affordances deal with learning. Perceiving some affordances like those of a triangle is not a *built-in* activity, indeed, once manufactured, they can be learnt and taught. Of course acknowledging this last fact places much more emphasis upon the dynamic character of affordances in organisms's plastic cognitive life, beyond the their evolutionary character.

Hence, the capacity to perform smart manipulations is concerned with the activity of creating external representations. Humans are continuously engaged in cognitive mimetic and creative processes (cf. [Donald, 2001] and chapter three of this book) in which they represent thoughts, private ideas, solutions, into external structures and suitably constructed artifacts. In doing so they create external counterparts of some internal, already available propositional and model-based representations, suitably stored in their brains. In some cases these external counterparts, which initially just simply mirror ideas or thoughts already present in the mind, later on can be creatively used to find room for finding new concepts and new ways of inferring that cannot be exhibited by the mere "internal" representation alone [Magnani, 2006c]. In building these external representations (that – I repeat – can be hold as merely mimetic, but that can become "creative"), people manipulate the world in such a way that new cognitive *chances* are uncovered. In doing this, new affordances are thus manufactured and made "socially" available. More generally, we can reiterate, also from this perspective, that abduction also deals with the continuous activity of manipulating the environment in order to create new chances and opportunities to action, namely, affordances.[46]

[45] Perceiving the affordances of a chair is indeed rooted and "stabilized" in our cultural evolution because for human beings it is easy – and *possible*, given our cognitive-biological configuration – to gain the corresponding cognitive ability as a "current" and "reliable" ability (cf. [Scarantino, 2003, p. 959]).

[46] An interesting application of the relationship between affordances and abduction I have proposed in this chapter is given by [Abe, 2009]. In order to support dementia persons lives, a model based on the concept of affordance, abduction, and chance discovery, which implements a (semantic) dementia care under the concept of affordance, is proposed.

Summary

In this chapter I have argued that cognitive niche construction is one of the most distinctive traits of human and animal cognition. As a matter of fact, humans and many other organisms continuously manipulate the environment in order to exploit its offerings. In so doing, they are engaged in a process of altering or even creating external structures to lessen and overcome their limits. New ways of coping with the environment, through both evolution and cultural evolution (i.e "cognitive niche construction") are thus created.

In depicting the intricate relationship between humans (and other animals) and the environment I have argued that the concept of affordance may provide new valuable insights. In my view, affordances can be considered as chances that are either already present and available or manufactured upon occasion via niche construction. I have pointed out that the notion of affordance is to some extent ambiguous. One of the most puzzling questions is whether the process of affordance in visual perception is or is not mediated. To address this problem I have taken advantage of some recent neurocognitive results and focused attention on the evolutionary aspects of affordances. The key concept of abduction has played the role of unifying the whole cognitive perspective.

First of all the relationship between humans and environments is highly dynamic and evolution has provided humans and other animals with a set of hardwired ways to be afforded. Nevertheless, if humans and other animals have at their disposal a standard endowment of hardwired capacities to be afforded by the environment, they can also enrich, manipulate, and transform it through new and plastic cognitive skills by enhancing their cognitive niches. In the case of humans it is easy to see that they can create, modify and often stabilize affordances that, in turn, form a great part of what it is called the eco-cognitive inheritance system. Once this system is made available, taking advantage of certain affordances it can be learnt, as Gibson himself acknowledged, and passed on through generations. This is clearly captured by the idea that the perception of affordances is "semi-encapsulated", i.e. it is not insulated from plastic cognitive processes that lean on learning and experience. I have argued that this perspective develops some of Gibson's intuitions clearly stated in his writings.

I have also argued that abduction can shed light on the process of detecting and building new affordances as fundamental cognitive "chances". I have maintained that abduction can clarify the process through which humans design external representations and artifacts, thus providing new ways of being afforded. Taking advantage of Peircean insights on abductive reasoning, I have stated that both finding and constructing affordances refers to a (semiotic) inferential activity: we come up with an affordance insofar as an object exhibits those signs from which we infer a possible way to interact with it, performing suitable actions. This inferential process relies on various cognitive endowments, both instinctual and learnt, and occurs in the eco-cognitive interaction between environment and organisms.

It is worth noting that the changes and modifications made to the environment can be passed on, and they become socially available to other organisms. This introduces an interesting issue for future development. As argued above, the evolutionary impact of a niche depends on its persistence and stability. In turn, its persistence is related to the possibility of transmitting the various resources made ecologically available through cognitive niche construction. Now, given the tremendous success of humans as ecological engineers, it follows that humans should be equipped with a mechanism that facilitates the exploitation and transmission of those ecological resources. That mechanism is, indeed, connected to information-gaining ontogenetic processes. Consequently, from an evolutionary perspective cognitive niche construction cannot be separated from various brain-based learning processes, which are also prominently social. This social dimension is intimately connected to the capacity of people to exploit the construction of a "social" medium as the basis for creating chances, solving problems or making decisions.

Chapter 7
Abduction in Human and Logical Agents
Hasty Generalizers, Hybrid Abducers, Fallacies

After having illustrated in the previous chapters the main features of my cognitive-epistemological analysis of abduction, it is necessary to further stress the dynamics involved in the interplay between internal and external representations in the case of logic. This will provide a tool for exploring the relationship between human and logical agents in section 7.1. First of all I will further develop my distinction between creative and mimetic artifacts. As I explained in chapter three these artifacts play the role of external objects (representations) active in what I have called disembodiment of mind. Mimetic external representations mirror concepts and problems that are already represented in the brain and need to be enhanced, solved, further complicated, etc. so they can sometimes creatively give rise to new concepts and meanings. From this perspective the expansion of the minds is, in the meantime, a continuous process of disembodiment of the minds themselves into the material world around them so that the evolution of the mind is inextricably linked with the evolution of many kinds of large, integrated, material cognitive systems. This chapter illustrates some features of this extraordinary interplay by focusing on the construction of *logical cognitive systems* and its consequences for abductive reasoning.

I will exploit the so-called *agent-based reasoning* framework, which adopts the perspective of a cognitive agent, that can naturalistically be seen in the perspective of the role of manipulations and of the interplay between internal – neural – representations and external ones, where both the conscious and unconscious are at work. I acknowledge that intellectual artifacts like *logical agents* are "ideal" tools for thoughts, as is language, they are tools for exploring, expanding, and manipulating our own minds. Logical systems can be considered mimetic, in the sense of the "mimetic representations" I have introduced in the previous chapters, (an example being that nonmonotonic systems seem to "mime" human reasoning performances better than classical logic – they are more psychologically adequate). However, they can be seen as creative when some "new ways of inferring" performed by the biological human agents arise in an unexpected and distributed interplay between brains (and their internal representations) and external representations which then leads to the creation of new ideal logical systems. I call this process – creative – externalization in "demonstrative environments" (section 7.3): I argue that it is central to the

L. Magnani: Abductive Cognition, COSMOS 3, pp. 361–416.
springerlink.com

creation of logical models of abduction, such as externalization in objective logical systems, it is communicable and sharable and able to grant stable perspectives endowed with symbolic, abstract, and rigorous cognitive features.

The proposed perspective allows us to see that deductive reasoning also means the employment of logical rules in a heuristic manner, maintaining the truth preserving character. Application of the rules is organized in such a way that one particular course of action can be recommended over another one. Consequently, very often the heuristic procedures of deductive reasoning are themselves performed by means of "in-formal" abductions, which often show model-based aspects. I illustrate this topic taking advantage of some classical research by Hintikka and his collaborators concerning the analysis of the so-called "singular terms" and of the Kantian idea of "construction" in logical and geometrical thinking. The distinction between strategic and definitory rules is also introduced, giving further insight into abductive cognition.

As already described in chapter three (section 3.6.2) *epistemic actions* in the environment often "mediate" the formation of new meanings. This framework also provides a useful perspective when studying the role of externalization of the mind in producing logical representations and a better understanding of the distinction between *human* (practical) and *ideal* (theoretical, or institutional) logical agents. I further stress how demonstrative ideal systems can be described not only in terms of symbolic, abstract, and rigorous terms, but also as endowed with what I call a *maximization of memorylessness*. The example of logic programs as ideal agents will be addressed, which is particularly useful in demonstrating their main epistemological difference with respect to the classical logical systems. Logic programs opened up a new perspective on the logic of abduction, in which sensitivity to the growth of information [and the suitable extension of logical language] is fundamental. It is considered fundamental to the whole logic itself by some contemporary logicians.

The close relation between *fallacies* and abduction will also be studied. As regards classical logic (and informal logic) abduction and inductive reasoning can be defined as fallacious. I will describe how in agent-based reasoning these and other kinds of the so-called fallacious reasoning can in some cases be redefined and considered as a good way of reasoning. In the light of the agent-based framework the fallacious character of abduction and induction can be clarified: abduction, that in chapter one was described in terms of the well-known fallacy of affirming the consequent, will be recognized – in the extended agent-based perspective – as very precious method of explanation and discovery in science and in everyday reasoning. Similarly, bad inductions – hasty generalizations – will be studied both from the perspective of their possible fallacious character and regarding their usefulness in reasoning. What has been called manipulative abduction in the previous chapters will be re-interpreted as a form of practical reasoning, a better understanding of which can furnish a description of human beings as *hybrid reasoners* in so far they are users of ideal (logical) and computational agents. I think that the issue could invite further research on the role played by symbolism, abstractness, and rigor, regarding their capacity to characterize externalized demonstrative systems.

Taking advantage of the agent-based analysis of the interrelations between fallacies and abduction, the last part of the chapter deals with three basic eco-logical issues:

1. abductive reasoning is involved in dialectic processes, which are at play in both agent-based everyday and scientific settings;
2. abductive reasoning is strictly linked to so-called smart heuristics and to the fact that very often less information gives rise to a better performance;
3. heuristics like "following the crowd", or social imitation, more or less linked to fallacious aspects which involve abductive steps, are often very effective. For example, these and other heuristics play an argumentative role in so far as they are linked to what I call "military intelligence", exploiting Réné Thom's expression, which relates to the role of language in coalition enforcement, I will introduce in the following chapter (section 8.6). As regards the agent-based approach of this chapter we can clearly see that various arguments, linked to the so-called fallacious aspects, intertwined as they are with their constitutive concealedness of error, are usually fruitfully exploited in a *distributed* cognitive framework, where moral conflicts, violence, and negotiations are normally at play. The specific role of abduction in argument evaluation and assessment is illustrated.

7.1 Beyond Peirce: Human Agents, Logical Agents

When searching for a satisfying logic of abduction (see above section 1.4, chapter one) [Gabbay and Woods, 2005] disagree with the Peircean rejection of the possibility of a practical logic and consequently of a logic of abductive reasoning in practical contexts. We have to overcome Peirce's limitations that are stated in the following passage I have already quoted in chapter five, section 5.2.4:

> My proposition is that logic, in the strict sense of the term, has nothing to do with how you think [...]. Logic in the narrower sense is that science which concerns itself primarily with distinguishing reasonings into good and bad reasonings, and with distinguishing probable reasonings into strong and weak reasonings. Secondarily, logic concerns itself with all that it must study in order to draw those distinctions about reasoning, and with nothing else [Peirce, 2005, p. 143].

Gabbay and Woods clearly maintain that Peirce's abduction, depicted as both a) a surrender to an idea, and b) a method for testing its consequences, perfectly resembles practical reasoning. Anyway, Peirce says "In everyday business reasoning is tolerably successful but I am inclined to think that it is done as well without the aid of theory as with it" [Peirce, 2005, p. 109].

Facing the problem of "reaching" a logic of abduction Woods contends it is just a formal and somewhat idealized description of an abductive agent.[1] Following my

[1] The author also maintains it is necessary to construct an "empirically sensitive logic" (ESL): "ESL is a part of the natural science of human reasoning. It is an approach to reasoning that pays attention to what people are like, to how they are put together and what they get up to when they reason" [Woods, 2010].

cognitive description of abductive reasoning illustrated above, an abductive human agent can naturalistically be seen in the perspective of the role of manipulations and of the interplay between internal – neural – representations and external ones. We have also seen that in this framework both conscious and unconscious inferences are important. Gabbay and Woods [2005, p. 27–29] seem to agree with me: they indeed stress the function of consciousness and indicate both its narrow bandwidth and its slow processing of information, an extraordinary quantity of information processed by the human system cannot be accessed by consciousness. Analogously, the bandwidth of natural language is narrower than the bandwidth of sensation, a great quantity of what we know we are not able to communicate one to another – linguistic intercourse is a series of exchanges whose bandwidth is 16 bits per second. Moreover, a side effect of consciousness is that it suppresses information.

The importance of abduction from the evolutionary point of view is clear: as Dennett [2003, p. 248] points out, it is better to kill a hypothesis in the mind than to be killed ourselves because we precipitously experienced the real environment. It is in this way, he says, that we are "Popperian creatures". I would think more appropriate to say we are "Peircean creatures", but what it is important here is that we do not have to know we are Popperian creatures to be one, as is the case with my consciousness-free – but hypotheses-making cat – as well as with computers programmed, for example, to simulate planning and mean-ends reasoning.

Clearly logic is historically related to conscious and propositional thinking and it seems to disregard the subconscious and prelinguistic levels of thinking. This fact leads to the following dilemma: rules of logic are thought of as having something to do with how human beings actually think as practical agents, then by and large they are too complex for conscious deployment; on the other hand, unconscious performance or tacit knowledge is a matter of certain things happening under the appropriate conditions and the right order, but it is unlikely to suppose that this is a matter of following rules (an inclination which seems embedded in contemporary computer science): "Given the cognitive goals typically set by practical agents, validity and inductive strength are typically not appropriate (or possible) standards for their attainment. [. . .] This, rather than computational costs, is the deep reason that practical agents do not in the main execute systems of deductive or inductive logic as classically conceived" [Gabbay and Woods, 2005, p. 25].

In the following sections I plan to offer a new clarification of this dilemma. Frequent complaints claim that logic provides rules which humans cannot conform their conscious thinking to except for very few exceptions, so that logicality cannot be considered just as a simple matter of following rules: indeed the intervention in reasoning of implicit knowledge sometimes renders performances relatively effortless, as I have illustrated in the first three chapters.

If, as Gabbay and Woods contend, a logic is a formalized *idealization* of a type of agent, a *logical agent*, which features of a human agent will a logic of abduction describe? They maintain that a real human agent is a kind of biological realization of a nonmonotonic paraconsistent base logic and surely the strategies provided by classical logic and some strictly related non standard logics form a very small part of the individual cognitive skills given the fact that human agents are not in general dedicated to error avoidance like "classical" logical agents. They say: "A formal model is an idealized description of what an abductive agent does. As such, it reflects some degree of departure from empirical accuracy. Thus an ideal model I is distinct from a descriptive model D" (p. 22). Immediately they add that in this logic of a practical agent basic questions of *relevance* and *plausibility* are central.[2]

Gabbay and Woods describe the features of "real" human thinking agents in the following way, consistent with the cognitive framework I have illustrated in the previous section: a human abductive agent certainly operates at two levels, conscious and unconscious, and at *both* levels it engages (or it is influenced by) truth conditions on propositional structures, state conditions on belief structures and their fixation, and sets of rules defined for various argumentative structures, for instance for evaluating arguments. These three capacities cut across explicit and implicit thinking. I think that many of the most important inference skills of a human agent are endowed with a *story* which varies with the multiple propositional relations she finds in her environment and which she takes into account, and with various cognitive reasons to change her mind or to think in a different way, and with multiple motivations to deploy various tactics of argument (see below section 7.3).

Facing the problem of logical modeling this kind of practical (abductive) human agent, logical systems can be considered *mimetic*, in the sense of the "mimetic representations" I introduced above in chapter three, (to make an example, nonmonotonic systems seem to "mime" better – they are more psychologically adequate – human beings' reasoning performances).

First of all a good mimetic abductive ideal logical agent is embedded in a situation of *nescience* and it is characterized by an *ignorance preserving* (but also *ignorance mitigating*) character [Gabbay and Woods, 2005][3] and by the following general distinct levels

- a *base logic* L_1 with proof procedures Π;
- an *abductive algorithm* which deploys Π to look for missing premises and other formulas to be abduced;
- a *logic* L_2 for deciding which abduced formulas to choose, which criteria of selection apply, etc. This logic is related to the specification of suitable problems of plausibility, relevance (topical, full-use, irredundancy-oriented, probabilistic) and economy, making the ideal agent able to discount and select information which does not resolve the task at hand.

The second and the third component together – endowed with what it is called a *filtrating* power – form *the logic of discovery*. It is in this new formal framework (GW model) that Gabbay and Woods criticize the so called AKM model of abduction[4] and also present the so-called *non-explanatory* abduction, considered as not intrinsically consequentialist.

[2] Cf. chapter two of this book, section 2.3.

[3] Section 2.1 – chapter two of this book – illustrates this role of ignorance in abductive reasoning. Cf. also below section 7.5.2.

[4] Among its supporters are [Aliseda, 1997; Kuipers, 1999; Magnani, 2001b; Meheus *et al.*, 2002; Meheus and Batens, 2006]. See also chapter, two section refakm-model.

7.2 Logical Agents as Mimetic and Creative Representations and Mediators

Following our previous considerations and the same scheme I have already illustrated in chapter three (subsection 3.4.1) it would seem that logical systems can be fruitfully seen as *external representations* expressed through artificial languages (in part mathematical) and through ordinary language, to the aim of *mimicking* various human reasoning performances in an idealized and rigorous way.

I think this view is partial and does not reveal other important features of logical agents. Indeed it is important to stress that logical agents can also play the role of *creative* representations human beings externalize and manipulate not just to mirror the internal ways of thinking of human agents but to find room for concepts and new ways of inferring which cannot – at a certain time – be detected internally "in the mind".

In summary, we can say that

- logical systems as *external representations* are formed by "external materials" that either *mimic* (through reification) concepts and problems already internally present in the brain or *creatively* express concepts and problems that do not have a "natural home" in the brain;[5]
- *internalized logical representations* are internal re-projections, a kind of recapitulations, (learning) of external logical representations in terms of neural patterns of activation in the brain. In some simple cases – cf. for example some precise rules of inferences of classical logic in the case of ordinary people or more complex logical rules and structures in the case of expert logicians – can be "internally" manipulated like external objects and can originate new internal reconstructed representations through the neural activity of *transformation* and *integration*.

I have already stressed in chapter three (subsection 3.4.1) that this process explains – from a cognitive point of view – why human agents seem to perform both computations of a *connectionist* type such as the ones involving representations as

- (I Level) *patterns of neural activation* that arise as the result of the interaction between body and environment (and suitably shaped by the evolution and the individual history): pattern completion or image recognition,
  and computations that use representations as
- (II Level) *derived combinatorial syntax and semantics* dynamically shaped by the various artificial external representations and reasoning devices found or constructed in the environment (in this case logical systems); they are – more or less completely – neurologically represented contingently as patterns of neural activations that "sometimes" tend to become stabilized structures and to fix and so *to permanently belong to the I Level* above.

[5] Also Gabbay and Woods consider mimetic the logical agent: "An abductive logic describes or mimics what an abductive agent does" [Gabbay and Woods, 2005, p. 346].

The I Level originates those *sensations* (they constitute a kind of "face" we think the world has), that provide room for the II Level to reflect the structure of the environment, and, most important, that can follow the computations suggested by the *logical external structures* available. It is clear that in this case we can conclude that the growth – so to say – of the "logical" brain and especially the synaptic and dendritic growth are deeply settled by the environment. Consequently we can hypothesize a form of coevolution between the logical brain and the development of the external logical systems. Brains build logic manipulating external symbols and structures and learn from them after having manipulated them to discover new concepts and features.

When the fixation is reached – imagine for instance the precise rule *modus tollens* as an internally explicit and stabilized form of thinking – the pattern of neural activation no longer needs a direct stimulus from the external logical representation in the environment for its construction. It can be viewed as a *fixed internal record* of an *external structure* that *can exist* also in the absence of such external structure. The pattern of neural activation that constitutes the I Level Representation has kept record of the experience that generated it and, thus, carries the II Level Representation associated to it, even if in a different form, the form of *memory* and not the form of the vivid sensorial experience of *modus tollens* written externally, over there, for instance in a blackboard. Now, the human agent, via neural mechanisms, can retrieve that II Level Representation and use it as an *internal* representation (and can use it to construct new internal representations for example less complicated than the ones previously available and stored in memory).

At this point we can easily understand the particular mimetic and creative role played by external logical representations:

1. some "new ways of logical inferring" performed by the biological human agents appear hidden and tacit (see the considerations on unconscious thinking above) and can be rendered explicit by building external logical *mimetic* models and structures; later on the agent will be able to pick up and use what suggested by the constraints and features intrinsic to their external materiality and the relative established conventionality: artificial languages, proofs, examples, etc.;[6]
2. some inferences can be *discovered* only through a problem solving process occurring in a distributed interplay between brains and external representations. In chapter three I have called this process *disembodiment of the mind*: the representations are *mediators* of results obtained and allow human beings
    a. to re-represent in their brains new logical reasoning devices picked up outside, externally, previously absent at the internal level and thus impossible: first, a kind of alienation is performed, second, a recapitulation is accomplished at the neuronal level by re-representing internally that which has

[6] A simple mathematical example I have previously already quoted in this book: it is relatively neurologically easy to perform an addition of numbers by depicting in our *mind* – thanks to that brain device that is called visual buffer – the images of that addition *thought* as it occurs concretely, with paper and pencil, taking advantage of external diagrammatic representations. Cf. also chapter three, subsection 3.6.5, this book.

been "discovered" outside. We perform cognitive logical operations on the structure of data that synaptic patterns have "picked up" in an analogical way from the explicit logical representations in the environment;

b. to re-represent in their brains portions of reasoning devices which, insofar as explicit, can facilitate logical inferences that previously involved a very great effort because of human brain's limited capacity. In this case the thinking performance is not completely processed internally but in a *hybrid* interplay between internal (both tacit and explicit) and external logical representations. In some cases this interaction is between the internal level and a computational tool which in turn can exploit logical representations to perform inferences (cf. above chapter one, subsection 1.6.2).

In this logical case too, like in case concerning the mythical being illustrated in chapter three (subsection 3.3.3), an evolved mind is unlikely to have a *natural home* for complicated concepts like the ones a logical ideal agent introduces: so whereas evolved minds could perform some trivial deductions or abductions in a more or less tacit way by exploiting modules shaped by natural selection,[7] how could one think exploiting explicit concepts without having picked them up outside, after having produced them? The only way is *to extend* the mind into the *material world*, exploiting paper, blackboards, computers, symbols, artificial languages, and other various semiotic tools, to provide anchors for ways of logical inferring that have no natural home within the mind.

Hence, we can hypothesize – for example – that many truth preserving reasoning habits which in human agents are performed internally have a deep origin in the past experience lived in the interplay with logical systems at first represented in the environment. As I have just illustrated other recorded thinking habits only partially occur internally because they are hybridized with the exploitation of already available external logical artifacts.

7.3 Externalization in Demonstrative Environments

7.3.1 Model-Based Abduction in Demonstrative Frameworks

In her book *Abductive Reasoning* Atocha Aliseda [2006] stresses the attention to the logical models of abduction, centering on the semantic tableaux as a method for extending and improving both the whole cognitive/philosophical view on abduction and on other more restricted logical approaches. I think this perspective wonderfully achieves many results. Logic is definitely offered to the appreciation of an interdisciplinary cognitive audience in both its plasticity and rigor in modeling various kinds of reasoning, beyond the rigid character of many of the traditional perspectives.

[7] Logicians express this idea in terms of the first of two basic low resources principles: "[...] 1. low resources individual agents perceive *natural kinds* around them, 2. low resources individual agents are *hasty generalizers* of *generic rules*" [Gabbay and Woods, 2005, p. 352]. On hasty generalizations see also [Walton, 2004, p. 138–142]. Indeed Gabbay and Woods characterize human agents as those that command few cognitive resources, cf. also my considerations on the main features of practical and theoretical agents below in the following section 7.3.

Two aspects have to be pointed out. The first is related to the importance of increasing logical knowledge on abduction: Aliseda clearly shows how the logical study on abduction in turn helps us to extend and modernize the classical and received idea of *logic*. The second implicitly refers to some issues coming from the so-called *distributed cognition* approach and concerns the role of logical models as forms of cognitive externalizations of preexistent in-formal human reasoning performances. The logical externalization in objective systems, communicable and sharable, is able to grant stable perspectives endowed with symbolic, abstract, and rigorous cognitive features. Indeed, Aliseda says, this character of stability and objectivity of the logical achievements is not usually present in models of abduction that are merely cognitive and epistemological and, moreover, they are central to computational implementation.

As I described in my previous book on abduction [Magnani, 2001b] and in chapter one above Peirce clearly indicated the importance of logic (first order syllogism) to grasp the inferential status of abduction, also creating a wonderful new broad semiotic view at the same accompanied by the well-known philosophical commitment to the new vision on pragmatism.[8] Given the restricted scope – classical, in terms of first order syllogisms – of the logical tools that were available to him the logical framework just depicted abduction as the well-known "fallacy of affirming the consequence". Aliseda's work belongs to this fundamental Peircean "logical" tradition and it presents plenty of recent logical models of abduction which are clearly illustrated in their rigorous "demonstrative" frameworks.[9]

7.3.2 *Model-Based Heuristic and Deductive Reasoning*

I have already illustrated in chapter one that the kind of reasoned inference that is involved in selective and creative abduction[10] goes beyond the mere relationship that there is between premises and conclusions in valid "classical" deductions, where the truth of the premises guarantees the truth of the conclusions, but also beyond the relationship that there is in probabilistic reasoning, which renders the conclusion just more or less probable. On the contrary, we can usefully see selective and creative abduction as formed by the application of "heuristic procedures" that involve all kinds of good and bad inferential moves, and not only the mechanical application of rules. It is only by means of these heuristic procedures that the acquisition of *new* truths is guaranteed. Also Peirce's mature view on creative abduction as a kind of inference seems to stress the strategic component of reasoning.

---

[8] [Aliseda, 2006, chapter seven] clearly shows how abduction is at the basis of Peirce's pragmatism.

[9] A recent issue of the journal *Theoria* – 22-23(60) 2007 – collects papers by, among others, Van Benthem, Magnani, Nepomuceno, Meheus, and Woods, devoted to discuss Aliseda's logical approach to abduction in terms of semantic tableaux.

[10] Epistemologically selective abduction occurs when we reach a hypothesis among – to use a word of the logical tradition also exploited by Aliseda – already available "abducibles" hypotheses (like for instance in the case of medical diagnosis). Creative abduction occurs when, through our reasoning processes, we are able to create "new" abducibles, which can be tested and added to [or which can replace] the available ones.

Many researchers in the field of philosophy, logic, and cognitive science have maintained that deductive reasoning also consists in the employment of logical rules in a heuristic manner, even maintaining the truth preserving character: the application of the rules is organized in a way that is able to recommend a particular course of actions instead of another one. Moreover, very often the heuristic procedures of deductive reasoning are performed by means of an "in-formal" (often model-based) abduction.[11] So humans apply rudimentary abductive/strategic ways of reasoning in deduction too.

Hence we have to say that theoretical model-based abductions also operate in deductive reasoning. Following Hintikka and Remes's analysis [1974] proofs of general implication in first order logic need the use of instantiation rules by which "new" individuals are introduced, so they are "ampliative". In ordinary geometrical proofs auxiliary constructions are present in term of "conveniently chosen" figures and diagrams. The system of reasoning exhibits a dual character: deductive and "hypothetical". The story of *Meno* dialogue I have illustrated in [Magnani, 2001b, chapter one] shows the role of these strategic moves and their importance in hypothesis generation. These strategic moves correspond to particular forms of abductive reasoning. In Beth's method of semantic tableaux the strategic "ability" to construct impossible configurations is undeniable [Hintikka, 1998; Niiniluoto, 1999].[12]

This means that also in many forms of deductive reasoning there are not trivial (and mechanical) methods of making inferences but we have to use "models" and "heuristic procedures" that refer to a whole set of strategic principles. All the more reason that Bringsjiord [2000] stresses his attention on the role played by a kind of "model based deduction" that is "part and parcel" of our establishing Gödel's first incompleteness theorem, showing the multimodal character of this great abductive achievement of formal thought.[13]

---

[11] We have already said in chapter four (section 4.1) that multimodal abduction takes advantage of internal (or of external models suitably re-internalized) that are not merely symbolic/propositional but which exploit for example diagrams, visualization, configurations, schemes, thought experiments, and so on [Magnani, 2001b].

[12] Also Aliseda [1997; 2006] provides interesting use of the semantic tableaux as a constructive representation of theories, where abductive expansions and revisions, derived from the belief revision framework, operate over them. The tableaux are so viewed as a kind of reasoning (non-deductive) where the effect of "deduction" is performed by means of abductive strategies.

[13] Many interesting relationships between model-based reasoning in creative settings and the related possible deductive "dynamic" and "adaptive" logical models are analyzed in [Meheus, 1999; Meheus and Batens, 2006]. Dynamic logic is also related to the formal treatment of inconsistencies. [Batens, 2006] proposes and interesting new formal diagrammatic proof procedure which takes advantage of the use of icons – derived from the premises – to build an algorithm for derivability that leaves choices to the user and is more efficient than tableaux method. Recently [Gabbay, 2007] has proposes a new perspective on various logical approaches, both classical and non classical, which takes advantage of the diagrammatic exploitation of graphical networks and the so-called "network modalities"; it is interesting to note that in this perspective a network can be interpreted as an "ecology", where nodes represent species.

7.3.2.1 Intuition, Construction, and the Logic of Singular Terms

As I have already pointed out in chapter three an example concerning the important role played in abduction by the manipulation of model-based aspects of reasoning is clearly given by Kant when he illustrated the synthetic character of mathematics. Kant notes in the "Transcendental Doctrine of Method" (cf. [Magnani, 2001c, chapter 2, sections 4 and 6]): "*Philosophical* knowledge is *the knowledge gained by reason from concepts*; mathematical knowledge is the knowledge gained by reason from the *construction* of concepts" [Kant, 1929, A713-B741, p. 577]. An important interpretative perspective which examined this fundamental passage of Kantian philosophy is represented by Hintikka and Beth's investigations (cf. [Hintikka, 1973, chapters 5-9], and also [Beth, 1956–1957; Beth, 1965]).

As an *intuition* [*Anschauung*] represents a "single object" [Kant, 1929, A713-B741, p. 577], in contrast with the *concept* ("The former relates immediately to the object and is single, the latter refers to it mediately by means of a feature which several things may have in common" [Kant, 1929, A320-B377, p. 314]), similarly, in Hintikka's interpretation intuitions are considered as *singular terms*, that is symbols which refer to singular objects" "[...] an intuition is almost like a 'proper name' in Frege's unnaturally wide sense of the term, except that it did not have to be a linguistic entity, but could also be anything in the human mind which 'stands for' an individual" [Hintikka, 1973, p. 207]. Thus, Brittan appropriately observes [1978, p. 51]:

> In constructing a geometrical figure, a triangle say, we often represent it by a figure drawn on a blackboard. In the same way, we "construct" arithmetical or algebraic concepts when we represent individual quantities, perhaps by the finger of a hand, perhaps by numerals or letters. What is important is not so much the "representative" – that is, the "pictorial" character of such "constructions" – but the fact that in using them we *refer to* or *talk about* individual mathematical objects, whatever these might be.

Moreover, Kant observes in the *Prolegomena*: "Intuition is a representation, such as would depend directly on the presence of the object" [Kant, 1966, §8, p. 37]. Kant clearly affirms in the "Transcendental Doctrine of Method" that a method is synthetic if constructions are made, that is, new geometric entities are introduced in the argument. Thus, Kant returns to the tradition which envisions processes of analysis and synthesis operating in geometrical reasoning.[14] In proving geometrical theorems, it suffices to consider the given figure, i.e., the figure of which the antecedent of the proposition speaks. Then, it is necessary to provide a "preparation" or a "construction" – or "machinery" – (in Greek: κατασκευή) in order to be able to conduct the proof (that is to complement the given figure by drawing new lines, circles, and other diagrams) [Hintikka, 1973, p. 202].[15]

[14] On the relationship between analysis and synthesis in the history of geometry and in particular in the history of philosophy of geometry refer to the many remarks given in [Lakatos, 1976] and [Otte and Panza, 1999]. Cf. also [Angelis, 1964].

[15] On the division of the figures in various parts in Euclid's *Elements* cf. [Heath, 1992].

Hintikka noted that Leibniz too was already aware of the central role of constructions: "Geometers start their demonstrations with the 'proposition' which is to be proved, and then prepare the way for demonstration of it by offering the 'exposition', as it is called, in which whatever is *given* is displayed in a diagram; after which they proceed to the 'preparation', drawing in further lines which they need for reasoning – the finding of this preparation often being the most skilful part of the task" [1690] [Leibniz, 1949, IV, chapter 17, section 3, p. 476]. Hintikka continues affirming that "When Pappus says that in a theoretical analysis 'we assume what is sought to exist and to be true', this can be understood as requesting us to assume that the proposition in question is true and that the auxiliary constructions which are needed for the proof have already Hintikka been carried out" [Hintikka, 1973, p. 202].

Moreover, Hintikka recalls, Socrates himself traced figures on the basis of *Meno*'s reasoning and these figures are the same starting points of the slave's analysis:[16] "To some extent, Kant also seems to have thought of another part of the proposition as being synthetic, namely, the setting-out or *ecthesis* (ἔκθεσις) which immediately follows the general enunciation of the proposition in question and in which the geometrical entities with which the general enunciation deals are 'set out' or 'exposed' in the form of a particular figure" (*ibid.*, p. 209). Indeed Kant considers the true demonstration (ἀπόδειξις) to be analytic "[...] which follows the auxiliary construction and in which no new geometrical objects are introduced. In this ἀπόδειξις we merely analyse, in a fairly literal sense of the word, the figure introduced in the ecthesis and completed in the 'construction' or 'machinery' " [Hintikka, 1973, p. 209]. The presence in this process of those manipulative aspects of cognition I have stressed in the previous chapter is evident.

The ἔκθεσις (in this case intended in the sense of Aristotle, as a form of syllogistic reasoning which approximates the rule of existential instantiation), represents the situation in which Euclid firstly introduces a new single individual: for example, "Let ABC a triangle. I say that in the triangle ABC, two sides taken together in any manner are greater than the remaining one, etc. Therefore, in any triangle, etc." Kant seems to have this situation in mind when he affirms that the geometer, contrary to the philosopher, "at once begins by constructing a triangle", and then, by elaborating an argumentation related to this construction, "[...] through a chain of inferences guided throughout by intuition, he arrives at a fully evident and universally valid solution of the problem" [Kant, 1929, A716/717-B744/745, p. 579].

Following this point of view, both Kant's conception of a construction as the introduction of a new "intuition", and the rule of existential instantiation, can be considered as developments of the notion of ἔκθεσις. The importance of the "constructive" procedures (of which the rule of existential instantiation is a simple example) in modern quantification theory is a tribute to the acumen of those earlier philosopher of mathematics who considered the constructions (the "machinery" part of a Euclidean proposition) as the essence of geometrical proofs.

[16] On this Plato's dialogue cf. [Magnani, 2001b, chapter six, section one].

It is important to note that in this type of demonstration, which can lead to a general conclusion, is the fact that nothing is assumed in regard to a particular individual or singular term introduced beyond what has already been stated in the premises. Kant, in fact, insists in declaring that the geometer, in order to definitely know something *a priori*, should not attribute qualities to the figure unless it results to be necessary from "what he has himself set into it in accordance with his concept":

> A new light flashed upon the mind of the first man (be he Thales or some other) who demonstrated the properties of the isosceles triangle. The true method, so he found, was not to inspect what he discerned either in the figure, or in the bare concept of it, and from this, as it were, to read off its properties; but to bring out what was necessarily implied in the concepts that had himself formed *a priori*, and had put into the figure in the construction by which he presented it to himself. If he is to know anything with *a priori* certainty he must not ascribe to the figure anything save what necessarily follows from what he has himself set into it in accordance with his concept [Kant, 1929, Bxii, p. 19].

According to Hintikka, the Kantian principle is equivalent to the affirmation which states that we can have a conceptual guaranteed knowledge about certain things if and only if this knowledge is based on the ways we acquire such knowledge, and "reflect the form (structure) of our sensibility" [Hintikka, 1973, p. 117]. In fact, for Kant "Objects are given to us by means of sensibility, and it alone yields us *intuitions*" [Kant, 1929, A19-B31, p. 65], that is representations of individual objects. Hintikka explains:

> Hence the relations and properties with which mathematical reasoning deals and which are anticipated in the *a priori* intuitions which a mathematician uses must have been created by ourselves in the process of sense perception. Only in this way Kant thinks that one can explain the possibility of the use of *a priori* intuitions (introduction of the new individuals) in mathematical arguments. If the existence and properties of a triangle had no "relation to you, the subject", then "you could not add anything new (the figure) to your concepts (of three lines) as something which necessarily must be met with in the object, since this object is given antecedently to your knowledge and [on that view] not by means of it". From this Kant concludes that the knowledge we gain through mathematical reasoning applies to objects only in so far the relations with which mathematical reasoning is concerned merely reflect the form (structure) of our sensibility [Hintikka, 1973, p. 117].

The importance of Hintikka's interpretation lies in ascertaining the fact which, granted that mathematics can be reduced to the general theory of quantification, nevertheless it cannot cease to be *synthetic*, from the moment that also that part of logic is synthetic, that is dependent on non-logical "intuitive" methods, in Kantian sense. We could say that for Kant, border-line between logic and mathematics, is given by the logic of monadic predicates.

In this perspective, Hintikka can consider Russell's interpretation [Russell, 1919, p. 175] to be incorrect: according to Russell, the intuitive and synthetic factors are in fact extraneous to the axiomatic and deductive structure and are connected in an irrelevant way with our geometric "imagination". But for Kant, is the use itself of constructions (intended to introduce new geometric entities) *as such*, and not their use as an aid for the imagination (in Russell's sense), which renders a geometric concept synthetic. What made mathematics synthetic is the introduction of such singular terms to represent the individuals, in the same way that geometrical entities are introduced by geometrical "constructions" into the figure.

In the "Transcendental Doctrine of Method" of the *Critique of Pure Reason* Kant illustrates the question involving the constructive character of geometry by presenting the *demonstration* (which appears as a *construction*) of the relation between the angles of the triangle and the right angle [Kant, 1929, A716-B744, pp. 578-579]. He then indicates that the introduction of new geometric entities in constructions produced in the course of geometric demonstrations is governed, in the Euclidean system, by the assumptions of *postulates*, which consequently appear as eminently synthetic, and referred to as the "intuition". Without the constructive procedure, guaranteed by postulates, despite attempts to change or combine different points of view, we can never directly perceive, by means of the simple observation of a triangle, that the sum of its angles is equal to straight lines. When producing, however, a suitable construction, the conclusion becomes obvious.

The epistemological significance of the notion of construction is also very broad and goes beyond the mere application of the geometric method. Hintikka states that:

> When Ernst Cassirer made his illuminating comparison between the geometrical method of analysis and the methods of the first great modern scientists, he had in mind just the analogy between the experimental "analysis" of a given physical configuration and the "problematical" analysis of the interrelations of the different parts of a geometrical figure, an analogy which in fact seems to have been one of the basic methodological ideas of such scientists as Galileo (who still employed the medieval terms "composition" and "resolution" for "synthesis" and "analysis") and Newton [Hintikka, 1973, p. 205].

We have to clearly illustrate how Kant's notion of construction can be interpreted from a logical point of view. On the basis of the logical re-interpretation of the Kantian notion of intuition, Hintikka can in fact conclude that "constructions" (in this book I have called them "cognitive manipulations", involving the case of manipulative abduction) are central because many inferences deal with the introduction of "individual" representations, as we have seen. When this occurs, the inferences have a synthetic nature. Hintikka notes that for Kant, the fundamental condition of possibility of mathematical knowledge is the condition according to which mathematics must be capable of representing its objects concretely as well as *a priori* (that is representing *a priori* the corresponding "intuition", corresponding to the concept). This definition of construction means that Kant saw the central aspect of mathematics in the introduction of representatives of individuals (intuitions, that is free individual terms) which instantiate general concepts. The rule of existential instantiation offers a very typical case.

Thus, Brittan appropriately notes: "On Hintikka's reconstruction, the paradigmatic synthetic method in quantification theory is the natural deduction rule of existential instantiation, that is, a rule that permits us to move from an existentially quantified sentence $(\exists x)(Fx)$ to a sentence instantiating or specifying it, for example $F(a|x)$, where a is a free individual symbol (or argument constant) and $F(a|x)$ the result of replacing x by a in F" [Brittan, 1978, pp. 52–53].

Hintikka's problem is therefore the same as Kant's, translated in logical terms, that is "to anticipate the existence of an individual before experience has provided us with one" [Hintikka, 1973, p. 132]. Kant expresses himself in *Prolegomena*:

> But with this step the difficulty seems rather to grow than to decrease. For now the question runs: *How is it possible to institute anything* a priori? Intuition is a representation, such as would depend on the presence of the object. Hence it seems impossible to intuit anything *a priori originally* [*ursprünglich*], because the intuition would then have to take place without any object being present, either previously or now, to which it could refer, and so could not be an intuition [Kant, 1966, §8, p. 37].

For Hintikka, mathematical demonstrations can be seen as procedures of quantification in which free singular terms are introduced. Thus, Hintikka affirms that it is not possible, using the "analytic method", to pass from assertions on the existence of an individual to an assertion regarding a different individual:

> III (a) An analytic step of argument cannot carry us from the existence of an individual to the existence of a different individual. Kant's writings show that inferences violating III (a) were in his view virtually paradigmatic instances of synthetic modes of reasoning. Time and again he returns to the problem which he takes to be a generalization of the Humean problem of causality and which he once formulates as follows: "Wie soll ich es verstehen, daß weil etwas ist, etwas anderes sei?" [Kant, 1902–1983, vol. II, pp. 202–203]. He goes on to point out that this cannot take place "durch den Satz des Widerspruchs", that is, analytically. From the *Prolegomena* it appears that Kant took his own main problem of the justifiability of synthetic *a priori* truths to be a generalization of the problem of justifying inferences synthetic in sense III (a) [Hintikka, 1973, pp. 137–138].

Hintikka also quotes Mach [1905] who affirms that a logical inference cannot offer us anything new. Moreover, Kant affirms: "For through mere concepts of these things, analyse them as we may, we can never advance from one object and its existence to the existence of another or to its mode of existence" [Kant, 1929, A217-B264, p. 238]. On the other hand, an inference from the existence of a certain individual to the non-existence of another individual is equally as problematic of the inference to its existence. Therefore, according to Hintikka, when one ascertains an *increase* in the number of individuals considered in their reciprocal relations in a procedure, then it can be said that it constitutes a synthetic argument: "Syntactic steps are those in which new individuals are introduced into true argument; analytic ones are those in which we merely discuss the individuals which we have already introduced" [Hintikka, 1973, p. 210].

Hence, as we have already observed, Kant was aware of the fact that the introduction of *new* entities as occurs in the course of geometric demonstrations is governed, in the Euclidean system, by the so-called postulates. Consequently, it is erroneous to think that the principles of mathematics can be proved according to the principle of contradiction: "[...] it was supposed that the fundamental propositions of the science can themselves bp known to be true through that principle. This is an erroneous view. For though a synthetic proposition can indeed be discerned in accordance with the principle of contradiction, this can only be if another synthetic proposition is presupposed, and if it can then be apprehended as following from this other proposition; it can never be so discerned in and by itself" [Kant, 1929, A20-B14, p. 52].

If we overestimate the analytical aspects, which also constitute the most conspicuous part of mathematical demonstrations, then we imagine to be able to demonstrate the same principles with purely analytical means so that, following Leibniz, mathematical existence is limited to the mere logical existence.

Kant concludes the argumentation in this way:

> But since the mere form of knowledge, however completely it may be in agreement with logical laws, is far from being sufficient to determine the material (objective) truth of knowledge, no one can venture with the help of logic alone to judge regarding objects, or to make any assertion. We must first, independently of logic, obtain reliable information; only then are we in a position to enquire, in accordance with logical laws, into the use [Kant, 1929, A60-B85, pp. 98-98].

As far Hintikka's rule, Brittan observes that: "[...] the soundness [...] is that the instantiating symbol must be different from all the free individual symbols occurring earlier in the proof. Hence use of the rule inevitably introduces new representatives of individuals, what Hintikka takes Kant to mean by 'intuitions', into the argument. For this reason, mathematical proofs that can be reformulated as quantificational arguments in which the rule of existential instantiation is applied are synthetic" [Brittan, 1978, p. 53].[17]

7.3.2.2 Definitory and Strategic Rules

Hence, we have to say that a kind of "in-formal" model-based abduction also operate in deductive reasoning performed by humans who use logical systems. Hintikka's approach is both game-theoretic and erotetic. I have already pointed out that, following Hintikka and Remes's analysis [1974] proofs of general implication in first order logic need the use of instantiation rules by which "new" individuals are introduced, so they are "ampliative". In ordinary geometrical proofs auxiliary constructions are present in term of "conveniently chosen" figures and diagrams. In Beth's method of *semantic tableaux* the "strategic ability" performed by humans to construct impossible configurations is striking [Hintikka, 1998; Niiniluoto, 1999]. I have already noted that Aliseda's approach provides interesting uses of the semantic tableaux as constructive representations of theories, where for example, abductive expansions and revisions, derived from the belief revision framework, operate over them. In the

[17] Hintikka's perspective is also related to Beth's previous works [Beth, 1956–1957; Beth, 1965] that had already focused on the use of free variables in mathematical proofs so that general affirmations can be obtained for all the individuals of a given type. On this topic cf. also [Friedman, 1992, chapter 1].

case of tableaux, their symbolic character is certainly fundamental, but it is also particularly clear they also are configurations – model-based – of proofs externalized through suitable notations.[18]

It is well-known that the formulation and the proving of theorems in an established mathematical theory, of the kind that creativity theorists typically consider, takes place by means of well established rules which are the same for all theorems [Hintikka, 1997]. There is no distinction between the rules that permit the derivation of "merely novel" theorems or "genuinely original" ones from the axioms. Creativity theorists like Boden [1992] operate the distinctions in terms of the mental operations associated with the different steps of the logical argument, instead of in terms of the kind of deductive rule at stake in the reasoning step in question.

Hintikka clearly finds the above perspective misleading and proposes a distinction between *definitory* and *strategic* rules. This distinction cuts accross the deduction-induction-abduction trichotomy [Gabbay and Woods, 2005, p. 139]. The first ones are merely permissive and do not say anything as to what the player should do or about which moves are good or bad or better than others. In the case of chess, if "you only know the definitory moves" [Hintikka, 1997, p. 68], you cannot say that you know how to play chess. You need some appropriate knowledge of what counts as a good or bad move. These are called strategic rules: "Strategies in this game-theoretical sense are rules that specify what a given player should do in every possible situation that can arise in the course of a play of the game", and, Hintikka concludes "Creativity is a matter of strategic rules" (*ibid.*) that of course involve long sequences of strategic moves. It is at this point clear that I was referring to this kind of rule when in section 5.2 of chapter five I illustrated the importance in abduction of what I called "heuristic strategies". Strategic rules are smart rules, even if they fail in individual cases, and show a propensity for cognitive success. In the case abduction, they tacitly fulfil the ignorance condition I have illustrated in chapter two, subsection 2.1, thus abduction aims at neither truth-preservation not probability-enhancement, as Peirce maintained.[19] Moreover, definitory rules are recursive but in several important cases strategic rules are not: therefore, playing a game strategically requires some kind of creativity.

An initial consequence is that the rules of inference of deductive logic are merely definitory rules, they define the rules of the game and certainly cannot be considered rules that reflect the way people actually reason or how they should reason. At this point an important question concerns the clarification of creativity in deductive reasoning.

Following Hintikka, we can say that the "ground floor" of deductive reasoning, the first-order logic, is nothing but operating with certain models or approximations of models, as is just simply demonstrated by some fundamental techniques such as

[18] It is worth to be noted that semantic tableaux method provides further insight on the problem of theory evaluation, intrinsic in abductive reasoning. Semantic tableaux can deal with "causal" aspects of abductive reasoning that cannot easily be considered with the only help of the logic programming tradition [Aliseda, 2006, chapters six and seven].

[19] In a strict sense, Hintikka's heuristic rules are strategic rules, even if merely tentative and partial [Hintikka, 1997, p. 69].

Beth's semantic tableaux. It is important to note that Hintikka is perfectly aware of the double character of these "models", *internal* and/or *external*:

> These models can be thought of as being mental, or they can be taken to consist of sets of formulas on paper – or in this day and age perhaps rather on the screen and in the memory of a computer. In fact, in this perspective all rules of "logical inference" obviously involve "mental models". Johnson-Laird's discovery hence does not ultimately pertain to the psychology of logic. It pertains, however confusedly, to the nature of logic itself. *The most basic deductive logic is nothing but experimental model construction* [Hintikka, 1997, pp. 69–70].

In this way Hintikka implicitly rejoins the distributed cognition approach to logic I have stressed in this book, where the interplay between internal and external – as kinds of "semiotic anchors", *symbolic*, in this case [Magnani, 2006d] – aspects of logical reasoning are illustrated. For example the role in logical deduction of the strategies of experimental (counter) model-construction is stressed, but also the importance of the introduction of the right new individuals by means of existential instantiation to be introduced in the model. The most important "strategic" question – in deductive reasoning – is to determine in what order the instantiations are to be treated. In geometrical reasoning the role of existential instantiation is obvious and occurs through the iconic so-called "auxiliary constructions", where conceptually manipulating a configuration of geometrical objects and extending it by introducing new individuals is at stake. The possible creative character is for example reflected in the fact that there is not always a mechanical (recursive) method for modeling these human deductive performances: in this case the role of manipulative abduction is implicitly at play. Of course, as Aliseda [2006] shows in chapter four "Abduction as computation" of her book, a suitable computational counterpart can take advantage of algorithms which render mechanical the suitably chosen reasoning processes, and so suitable to be implemented in a computational program.

The logical tradition of Frege and Russell rejected all reasoning that had been made in terms of geometrical icons as being responsible for introducing an appeal to intuition. On the contrary, I am very inclined to agree with Hintikka, who maintains that the traditional idea of logical reasoning as a discursive process is wrong, it is an "optical illusion", because all deduction is a form of "experimental model construction" which follows that interplay between internal and external representations I have already indicated. It is important instead to note that for instance already at the level of elementary geometry:

> [...] geometrical figures are best thought of as a fragmentary notation for geometrical proofs alternative to, but not necessarily intrinsically inferior to, the "purely logical" notation of formalized first order logic. [...] They are intrinsic features of certain deductive methods. They are part of the semantics of logical reasoning, not only of its psychology or its heuristics. If it is suggested that heuristic ways of thinking are needed to make mathematical reasoning intuitive, I will borrow a line from Wittgenstein's *Tractatus* 6.233 and say that in this case the language (notation) itself provides the intuitions [Hintikka, 1997, p./ 73].

Moreover, in the case of human performances, as I have observed above in section 7.3.2, in many forms of deductive reasoning there are not trivial and mechanical methods of making inferences but we have to use "models" and "heuristic procedures" that refer to a whole set of strategic abductive principles.[20]

7.3.3 *Ideal Logical Agents*

In section 3.7 of chapter three I have said that there are external representations that are representations of other external representations. In some cases they carry new scientific knowledge. I provided the example of Hilbert's *Grundlagen der Geometrie*, as a "formal" – diagram-free – representation of the geometrical problem solving through diagrams: in Hilbertian systems solutions of problems through diagrams typical and ancient Euclidean axiomatics become proofs of theorems in terms of a modern logical/axiomatic model. Similarly, a calculator can re-represent (through an artifact) (and perform) those geometrical proofs with diagrams already performed by human beings with pencil and paper (cf. chapter two, section 2.12). In this case we have representations that *mimic* particular cognitive performances that we usually attribute to our minds.

In the previous chapters I have contended that our brains delegate cognitive (and epistemic) roles to externalities and then can "adopt" and recapitulate what they have checked occurring outside, over there, after having manipulated – in some cases with creative results – the external invented structured model. I have said that mind representations are also over there, in the environment, where mind has objectified itself in various structures that *mimic* and *enhance* its internal representations. One of the clearest example of the externalizations of cognitive abilities is furnished by the Turing machines, as abstract tools (the – Universal – Logical Computing Machine, LCM) endowed with powerful mimetic properties. Turing's LCM, which is an externalized device that can be realized in the so-called , is able to *mimic* human cognitive operations that occur in that interplay between the internal mind and the external one: I have already quoted the Turing's passage where he says that, taking advantage of the existence of both the (Universal) Logical Computing Machine (LCM) (the theoretical artifact) and the (Universal) Practical Computing Machine (PCM) (the practical artifact): LCMs "can in fact mimic the actions of a human computer very closely" [Turing, 1950]. It is in this sense I said digital computers

[20] [Grialou and Okada, 2005], quoting Hilbert of the late 1910s and the 1920s, who goes beyond a nominalistic/formalist view of logic in terms of rule-based syntax, also emphasize the importance of studying logical and mathematical proof not only at the level of models or mathematical object domains but focusing on the properties of formal proof "figures", considering them as concrete objects whose structure is determined by a set of formal inferences rules. From this perspective Grialou and Okada speak of a Hilbertian "finitist evidence theory", in which perceptual/sensitive-based finitist *intuitions* of proof play a fundamental role related to model-based spatial/figural cognitive processes. Also the notion "of proof-nets" (a geometry of proofs), such as graphical representations of proof in Girard's linear logic [1987] can be related to this perspective resorting both to Hilbert and Gentzen . [Grialou and Okada, 2005] also provide neurological evidence of the involvement of both language and visual/diagrammatic processing systems in logical reasoning. On the role of errors, due to the context, in spontaneously performed logical reasoning in humans and on their rudimentary deductive competence see [Evans, 2002b].

are whole *mimetic minds*, as a case of a further extremely fruitful step of the disembodiment of the mind I have described in chapter three.

I have said that PCMs can be considered mimetic minds (they are ideal – even if "practical", in Turing's sense – agents): what is in turn the cognitive status of "logical agents" (as theoretical agents) from the point of view of their *demonstrative* aspect? It is at the level of the external representations originating the logical ideal agent that we first realize the *demonstrative* character of a reasoning process. For example this is the case in Hilbertian formal systems, in Gentzen's proofs and in semantic tableaux but also in heterogeneous systems. Let us consider the amazing case of the heterogeneous systems: they produce representations in a demonstrative framework which originate from a number of different representation systems, sentential, but also model-based, diagrammatic, usually considered non-demonstrative: the advantage is that they allow a reasoner to bridge the gaps among various formalisms and to construct threads of proof which cross the boundaries of the systems of representations [Swoboda and Allwein, 2002].

In doing this, heterogeneous systems allow the reasoner to take advantage of each component system's ability to express information in that component's area of expertise. For example "recast" rules are clearly elicited as rules of inference that allow the exchange of information between the various representations (cf. Figure 7.1). We have two ways to use them: one for the extraction of information from a diagram to be expressed in a sentential form and another that allows the extraction of information from a formula to be incorporated into a diagram [Swoboda and Allwein, 2002].[21]

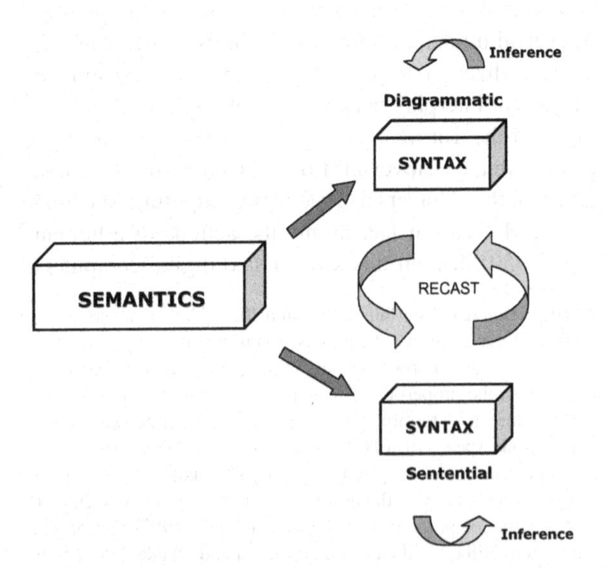

Fig. 7.1 A heterogeneous logical system (cf. [Swoboda and Allwein, 2002])

[21] Clarifications of exact processes and semantic requirements of manipulative inferences and distributed cognition are given in [Shimojima, 2002].

Gabbay and Woods propose a hierarchy of cognitive agents: "An agent is a practical agent to the extent that it commands comparatively few cognitive resources in relation to comparatively modest cognitive goals. [...] An agent is theoretical to the extent that it commands comparatively much in the way of cognitive resources, directed at comparatively strict goals" [Gabbay and Woods, 2005, pp. 12-13]. They also add that it is typical of human individuals to function as practical agents and that it is typical of what they call "institutions" to function as theoretical agents (p. 14); moreover, agents tend toward enhancement of cognitive assets when this makes possible the achievement of cognitive goals previously unaffordable or unattainable.[22] Of course the ideal logical agents I am considering here are theoretical agents.

The externalization of inferential skills in demonstrative systems presents interesting cognitive features (cf. also [Longo, 2005]) which I believe deserve to be further analyzed and which can further develop the distinction above between theoretical and practical agents:

1. *symbolic*: they activate and "anchor" meanings in material communicative and intersubjective *mediators* in the framework of the phylogenetic, ontogenetic, and cultural reality of the human being and its language. It seems these logical agents originated in embodied cognition, gestures, and manipulations of the environment we share with some mammals but also non mammal animals (cf. the case of monkeys' knots and pigeons' categorization, in [Lestel and Herzfeld, 2005; Aust *et al.*, 2005]).[23]
2. *abstract*: they are based on a *maximal independence* regarding sensory modality; they strongly stabilize experience and common categorization. The maximality is especially important: it refers to their practical and historical invariance and stability;
3. *rigorous*: the rigor of proof is reached through a difficult practical experience. For instance, in the case of mathematics, as the maximal place for convincing and sharable reasoning. Rigor lies in the stability of proofs and in the fact they can be iterated. Following this perspective mathematics is the best example

[22] Interesting research about the interplay between practical (human) and ideal (logical) agents is illustrated in part one of the book [Andler *et al.*, 2006], which shows puzzling cognitive effects generated by the interplay of the two sides: 1) the role of premise presentation in conditional reasoning and in the recognition of the fallacy of affirming the consequent [Henst *et al.*, 2006]; 2) the role of rationality assumptions in the agents' recognition of their own inferential performances in the case of errors caused by misleading perceptual saliency in data problems [Burgeois-Gironde, 2006]; 3) the role of proto-logical and illogical performances related to the equivalence relations humans share (or do not share) with a particular kind of practical agent devoid of "linguistic knowledge": animals like squirrels and resus, monkeys, chimpanzees, rats, and pigeons) [Yamazaki *et al.*, 2006]. Humans seem to differ from animals by showing an early capacity to recognize stimulus equivalence, "That is, though illogical, the ability to establish stimulus equivalence, which requires minimum amount of direct training and is highly productive in acquisition of stimulus relations, would support the symbol acquisition in humans from early life, hence is a kind of cognitive module that can be seen only in humans" [Yamazaki *et al.*, 2006, p. 71].

[23] Cf. also the cognitive analysis of the origin of the mathematical continuous line as a preconceptual invariant of three cognitive practices [Theissier, 2005], and of the numeric line [Châtelet, 1993; Dehaene, 1997; Butterworth, 1999].

of maximal stability and conceptual invariance. Logic is in turn a set of proof invariants, a set of structures that are preserved from one proof to another or which are preserved by proof transformations. As the externalization and result of a distilled praxis, the praxis of proof, it is made of maximally stable regularities;

4. I also say that a *maximization of memorylessness*[24] "variably" characterizes demonstrative reasoning. This is particularly tangible in the case of the vast idealization of classical logic and related approaches. The inferences described by classical logic do not yield sensitive information – so to say – about their real past life in human agents' use, contrarily to the "conceptual" – narrative – descriptions of human non-demonstrative processes, which variously involve a remarkable amount of "historical", "contextual", and "heuristic" memories and constraints. Indeed many thinking behaviors in human agents – for examples abductive inferences, especially in their generative part – are context-dependent. As already noted their *stories* vary with the multiple propositional relations the human agent finds in her environment and which she is able to take into account, and with various cognitive reasons to change her mind or to think in a different way, and with multiple motivations to deploy various tactics of argument:

> Good reasoning is always good in relation to a goal or an agenda which may be tacit. [...] Reasoning validly is never *itself* a goal of good reasoning; otherwise one could always achieve it simply by repeating a premiss as conclusion, or by entering a new premiss that contradicts one already present. [...] It is that the reasoning actually performed by individual agents is sufficiently reliable not to kill them. It is reasoning that precludes neither security not prosperity. This is a fact of fundamental importance. It helps establish the fallibilist position that it is not unreasonable to pursue modes of reasoning that are known to be imperfect "Given the cognitive goals typically set by practical agents, validity and inductive strength are typically not appropriate (or possible) standards for their attainment" [Gabbay and Woods, 2005, pp. 19-20, p. 25].

As I will illustrate in subsection 7.5.2 human agents, as practical agents, are hasty inducers and bad predictors, unlike ideal (logical and computational) agents. In conclusion, we can say abductive inferences in human agents have a memory, a story: consequently, an abductive ideal logical agent has to variably weaken many of the aspects of classical logic and to overcome the relative demonstrative limitations.

In the Peircean semiotic perspective I have illustrated in section 3.5 of chapter three, where the so-called "logical interpretants" can lead to relatively stable cognitive or intellectual "habits" and "belief" changes (often made possible thanks to conceptual creative abductions), the externalizations of the inferential skills I have just described can be seen as resulting from the formation of a kind of ultimate "meaning". As objectified (symbolic, abstract, rigorous, memoryless), cognitive logical tools, available and crystallized over there, in the environment, they can be picked up to

[24] I derive this expression from [Leyton, 2001], cf. also footnote 96 at p. 210.

generate habit-changes in individuals' rational attitudes and actions "[...] the ultimate logical interpretant of the concept [...] that is not a sign but is of a general application is a *habit-change*; meaning by a habit-change a modification of a person's tendencies toward action" [Peirce, 1931-1958, 5.476]. Of course, even if logical interpretants provide the widest scope or "general meaning" of a class of signs, so justifying their cognitive stability and – so to say – universality, they are never final, because they are always in relation to other logical interpretants, to which they are in turn related or under which they are subsumed, in a situation of infinite progression, that in fact impedes their ever actually having an ultimate "meaning". Even in the case of logical systems, which certainly are strongly made stable and "institutional" (in Gabbay and Woods' sense) thanks to their artificial and abstract languages.[25]

I think that a great contribution given to logic by Gabbay is the creation of the *labelled deductive systems* (and their application to the logic of abduction), where data is structured and labelled and different insertion policies can be formulated [Gabbay, 2002; Gabbay and Woods, 2006]. The labelled deductive systems fulfill the request of weakening the rigidity of classical logic but also of many non standard logics strictly related to it, opening a new era in logic: the attention to the role of *meta-levels* – for instance in the logic of abduction – formalizes the flexibility and "historicity" of many kinds of human thinking which are meaningful in certain application areas they address. Gabbay and Woods' conclusion about psychologism is clear and leads to a new conception of logic:

> If [...] it is legitimate to regard logic as furnishing formal models of certain aspects of the cognitive behavior of logical agents, then not only do psychological considerations have a defensible place, they cannot [Gabbay, 2002; Gabbay and Woods, 2006] reasonably be excluded [Gabbay and Woods, 2005, p. 2].[26]

We can conclude by stressing the fact that human non-demonstrative inferential processes are more and more externalized and objectified at least in three ways:

1. through Turing's Universal Practical Computing Machines we can have running programs that are able to mimic "the actions of a human computer very closely" in Turing's classical sense [Turing, 1950], but also to mimic – amazingly – those human agents' "actions" that correspond to the complicated inferential performances like abduction (cf. the whole area of artificial intelligence);
2. human non-demonstrative processes are more and more externalized and made available in a multiplicity of explicit narratives and learnable templates of behavior (cf. also the study of fallacies as important tools of the human "kit" that

[25] Woods provides other examples of institutional agents, such as Nato, M15 or an university. In this perspective individual organic agents possess "fewer" cognitive assets than institutional agents [Woods, 2010, chapter eight].

[26] An analogous example of the new modeling flexibility of recent logic is represented by the work in the dynamic logics of reasoning of van Benthem [1996]. This logic offers a distinction between inferences that are dependent on short term representations and those that depend on long-term memory, which involves the processing of representations of greater abstraction. In this way it is possible to formally and flexibly reproduce the interplay that occurs in human agents' thinking both at the level of short-term memory – more inclined to be damaged by inconsistencies – and at the level of the long-term memory, where inconsistencies can be inert.

provides evolutionary advantages, in this sense a fallacy of the affirming the consequent – which depicts abduction in classical logic – is better than nothing [Woods, 2004]).[27]

3. new demonstrative systems – ideal logical agents – are created able to model in a deductive way many non-demonstrative thinking processes, like abduction, analogy, creativity, spatial and visual reasoning, etc.[28]

7.4 Hasty Generalizers and Hybrid Abducers in Agent-Based Reasoning

In this section of the chapter I will try to describe the close relation between fallacies and abduction. Then I will analyze some types of reasoning that in the perspective of classical and informal logic are defined fallacies. First of all I would like to describe inductive and abductive reasoning in the light of the agent-based framework to the aim of clarifying their fallacious character and the role of the so-called ideal systems (logical and computational). Then I will analyze some inductive and abductive types of reasoning that in the perspective of classical and informal logic are defined *fallacies*: we will recognize in this new perspective abduction is a very precious method of guessing hypotheses and of discovery in science and in everyday reasoning (for example abduction is able to govern inconsistencies that arise in scientific domains, when anomaly is treated through the generation of new hypotheses (cf. chapter two, this book). Then, I will describe how in an agent-based reasoning this kind of *fallacious* reasoning can in some cases be *redefined* and considered as a good way of reasoning. Finally, I will illustrate how what I call *manipulative abduction* can be interpreted as a form of practical reasoning a better understanding of which can furnish a description of human beings as *hybrid reasoners* in so far they are users of ideal and computational agents.

7.4.1 Agent-Based Reasoning, Agent-Based Logic, Abduction

First of all I would like to describe inductive reasoning in the light of the so-called agent-based framework. This analysis will permit us to explain the traits of the fallacious character of induction (and abduction) and the role of the "idealized" logical systems.

[27] Cf. also [Gabbay and Woods, 2005, pp. 33-36];

[28] A skeptical conclusion about the superiority of demonstrative over non-demonstrative reasoning is provided by the following philosophical argumentation of Cellucci [2005] I agree with, which seems to emphasize the role of *ignorance preservation* in logic: "To know whether an argument is demonstrative one must know whether its premises are true. But knowing whether they are true is generally impossible", as Gödel teaches. So they have the same status of the premises of non-demonstrative reasoning. Moreover: demonstrative reasoning cannot be more cogent than the premises from which its starts; the justification of deductive inferences in any absolute sense is impossible, they can be justified as much, or as little, as non-deductive – ampliative – inferences. Also checking soundness is a problem.

It is well-known that in classical logic a good argument is a sound argument and, from a semantic point of view, it is a valid argument based on true premises. Even if this conception of good inference is usually able to model many kinds of argumentation of real human beings, its appeal to true premises is ill suited to many contexts which are often characterized by the presence of hypothetical and uncertain beliefs, by great disagreement about what is true and false, by ethical and aesthetic claims which are not easily categorized as true or false, and, finally, by variable contexts in which dramatically different assumptions may be accepted and rejected.

I share with Gabbay and Woods [2005] the idea that logic can be considered a formalization of what is done by a cognitive agent. Starting from this perspective, logic is *agent-based* [Gabbay and Woods, 2001]. From this point of view logic should also give an account of the cognitive agent's concrete ways of reasoning and if reasoning is an aid to cognition, a logic, when conceived of as a theory of reasoning, must seriously take this cognitive orientation into account.

According to Norman [1993] and Hutchins's [1995] view, we can say that a *cognitive system* is a triple of a cognitive agent, cognitive resources, and cognitive target performed in real time. Correspondingly, a logic of a cognitive system, or an *agent-based logic*, is a principled description of conditions under which agents deploy resources in order to perform cognitive tasks. The more general notion involved in this conception is that reasoning is target-motivated and resource-dependent. That is, agent-based logic deals with the problems, limitations, and ways of reasoning of cognitive agents.

In this perspective the problem of describing how agents perform reasoning is constrained in three crucial ways: in what they are disposed towards doing or have it in mind to do (i.e., their *agendas*); in what they are capable of doing (i.e., their *competence*); and in the means they have for converting competence into performance (i.e., their *resources*). Loosely speaking, agendas are plans of action, for example belief-revision and belief-update, decision-making and various kinds of case-making and criticism transacted by argument [Gabbay and Woods, 2005].

A description of the cognitive practice should include a description of the *type of cognitive agent* involved. Agency-type is set by two complementary factors. One is the degree of command of resources an agent needs to advance or close his (or its) agendas. In this perspective agent-based reasoning consists in describing and analyzing the reasoning occurring in problem solving situations where the agent access to cognitive resources encounters limitations such as

- bounded information,
- lack of time,
- limited computational capacity.

Hence, the "beings-like-us" that Woods describes in his "Epistemic bubbles" [Woods, 2005] discharge their cognitive agendas under press of incomplete information, lack of time, and limited computational capacity. We can consequently say that cognitive performances depend on information, time, and computational capacity. An *agent-based logic*, as a discipline that furnishes ideal descriptions of *agent-based reasoning*, returns to be thought of as a science of reasoning and considered agent-centered, task-oriented, and resource-bound. Woods says:

> So, then, a principal function of reasoning is to facilitate cognition, this means the reasoning agent is also cognitive agent. If logic is to press forward as a renewed science of reasoning, it would do well to reflect on what cognitive agency is like, on what it is like to be a knower [Woods, 2005, p. 732].

In dealing with these features we arrive to what has been called the "Actually Happens Rule" [Woods, 2005] that states that "[...] to see what agents should do we should have to look first at what they actually do and then, if there is particular reason to do so, we would have to repair the account". This rule is a particular attractive assumption about human cognitive behavior mainly for two reasons. The first is that beings like us make a lot of errors; the other is that cognition is something that we are actually very good at.

In the following subsection I will discuss the case of "fallacies" as errors that people make. These errors occur in ways of reasoning and acting that from some perspectives are good and from others are bad. In dealing with this matter I will try to give an account of fallacies seen from the viewpoint of agent-based reasoning. I will try to give some examples of fallacious reasoning treating both informal fallacies (such as the inductive ones like "hasty generalization") and formal fallacies (such as abduction). I will treat induction and abduction as ways of reasoning that, in spite of the fact that they can be seen as fallacious from ideal perspectives, are fruitful for the cognitive agent: so to say, a way of being rational through fallacies. I have to note that my discussion of hasty generalization and of abduction as a fallacy does not aim at warranting general conclusions about the nature of fallacies. My only target is to stress the interplay between "logical" and "eco-logical" aspects of some kinds of hypothetical cognition, that can be further enlightened taking advantage of the analysis of the appropriate features of few fallacious inferences (also stressing their role at the level of a kind of "social epistemology"– cf. below section 7.8).

Abduction can be easily considered in the perspective of agent-based reasoning because in abductive reasoning [Magnani, 2001b] both the activity of guessing new explanatory, non-explanatory, and instrumental hypotheses and the activity of selecting already existing ones, is based on incomplete information. In this case we deal with "nonmonotonic" inferences: we draw defeasible conclusions from incomplete information.[29] In this perspective, abductive reasoning also represents a prototypical case of practical reasoning: we adopt deliberations based on incomplete information and on particular abduced hypotheses – guesses – that serve as "reasons."

---

[29] It has to be noted that an explanatory reasoning can be causal, but explanations are also based on other aspects. Thagard [1992, chapter five] illustrates various types of explanations (deductive-nomological – so-called in the neopositivist tradition of philosophy of science – statistical, causal, analogical, schematic).

7.4.2 *Beings-Like-Us as Hasty Generalizers: Induction as a Fallacy*

As already noted, people make errors in reasoning. This means that in analyzing the beings-like-us argumentations we have to face problems regarding agent's access to cognitive resources such as information, time, and computational capacity, and logical attributes such as truth-preservation. It is in this sense that I have previously said that agents discharge their cognitive agendas under press of bounded information, lack of time and limited computational capacity.

The successful use of inferences that from an intellectual perspective can be clearly seen as fallacies into many kinds of hypothetical reasoning can be fruitfully accounted for in the framework of agent-based reasoning. The peculiarity of the so-called fallacies seen in the perspective of agent-based reasoning is that mistakes that are actually committed are mistakes that do not seem to be mistakes to those who commit them. In some sense we can say that they are ways of reasoning that are felt truth preserving for the reasoner but are not considered truth preserving for the logicians!

A fallacy can be considered a pattern of poor reasoning which appears to be a pattern of good reasoning [Hansen, 2002]. The so-called fallacies are forms of reasoning and argumentation typical of organic agents and in this sense we can say they are suitably shaped by that coevolution between brain and artificial cognitive niches, built by both humans and animale, I have illustrated in chapter three. Simple inductions and abductions performed more or less consciously by humans and implicitly by animals are surely two great results of this coevolutionary process. Two main disciplines respectively clearly illustrate different kinds of fallacies: formal logic, which recognizes and explains "formal fallacies", and informal logic, that describes the so-called "informal fallacies". First of all, we can say that the validity of a deductive argument depends on its form, consequently, formal fallacies are arguments which have an invalid form and are not truth preserving (for example the fallacy of the "affirming the consequent" and of "denying the antecedent"). On the other hand, informal fallacies are any other invalid modes of reasoning whose failing is not strictly based on the shape of the argument (for example the "*ad hominem* argument" or the "hasty generalization").

Even if there is no agreement upon an established taxonomy, the fallacies discussed in informal logic contexts typically include formal fallacies such as the famous fallacy of affirming the consequent[30] and the fallacy of denying the antecedent, but also informal fallacies such as "*ad hominem*" (against the person), "slippery slope", "*ad baculum*" (appeal to force), "*ad misericordiam*" (appeal to pity), and "two wrongs make a right". Some authors use the fallacy nomenclature as a way for designing the properties of particular kinds of fallacious arguments. Anyway, we will see that in the perspective presented in this chapter the distinction between formal fallacies and informal fallacies is not crucial so, in dealing with agent-based reasoning, can be disregarded.

[30] We already know that in the light of classical logic abduction coincides with the fallacy of affirming the consequent (see the following section).

From the point of view of classical logic a fallacy is a bad argument that in general looks good (cf. Figure 7.2). From the point of view of agent-based reasoning a fallacy is not an argument that looks good but is bad, but an argument that is bad in some aspects and good in some others. Finally, as I will illustrate in the last sections of this chapter, from a wide aco-cognitive perspective the (supposed to be) fallacies are not fallacies at all. We can say that a fallacy is a mistake made by an agent. It is a mistake that seems not to be a mistake. Correspondingly, it is a mistake that is naturally made, commonly made, and not easy to repair (i.e., to avoid repeating) [Woods, 2004, chapter one].

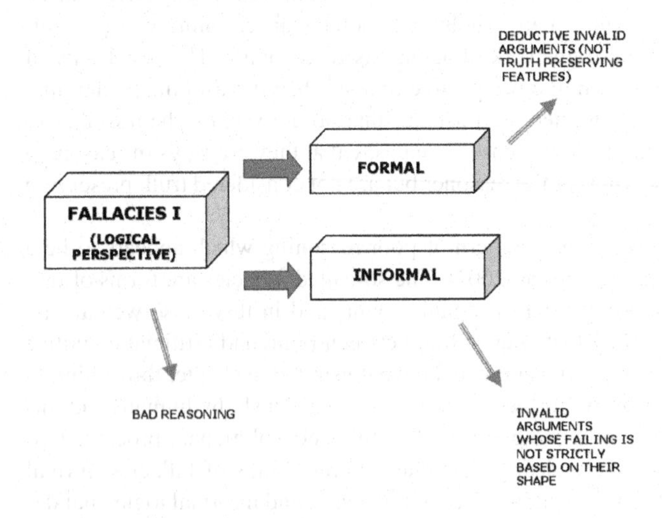

Fig. 7.2 Fallacies from the point of view of classical logic

Let us consider the inductive case of the so-called "hasty generalization", that can lead the cognitive agent – in spite of the fact that, from the perspective of "ideal" (Millian) inductive models of reasoning, it can be considered fallacious – to fruitful outcomes.

This fallacy occurs when a person (but there is evidence of it also animals, for example in mice, where the form of making hypotheses can be ideally modeled as a hasty induction) infers a conclusion about a group of cases based on a model that is not large enough. Of course in the case of animals we cannot imagine hasty generalization is occurring thanks to the "mental" exploitation of propositions (see above chapter five). The fallacy can be reconstructed in the following form:

- Sample S, which is too small, is taken from the group of persons P.
- Conclusion C is drawn about the group P based on S.

It could take also the form of:

- The person X performs the action A and has a result B.
- Therefore all the actions A will have a result B.

The fallacy is committed when not enough A's are observed to warrant the conclusion. If enough A's are observed then the reasoning is not fallacious, at least from the *informal* point of view. Males, driving their cars, have probably quarreled with a woman driving her car and, while quarreling, they have argued (when not shouted) "all woman are bad drivers!" That's our case of fallacious reasoning.

Insofar, small samples will be likely to be unrepresentative. Another simple case is the following. If we are asking one person that met a lot of Italians what those Italians in general thinks about the recently new established Italian proportional-oriented electoral system, his answer clearly would not be based on an adequate sized sample for determining what Italians in general think about the issue. This is because the answer given is based only on a reduced experience and that judgment can not be relevant in dealing with a generalization about the matter in question. This means that this fallacious argument implies that small samples are less likely to contain numbers proportional to the whole group of cases.[31]

For example, if a container has plenty of blue, red, green and orange coins, then a sample of three coins cannot possibly be representative of the whole amount of coins. As the sample size of coins increases the more likely it becomes that coins of each color will be selected in proportion to their numbers in the whole amount. The same holds true for things others than coins, such as people and their political views.

People often commit hasty generalizations because of bias or prejudice (biased evidence can implicitly influence reasoning and certain emotions can furnish an immediate positive evaluation of the reached hypothesis, cf. the previous footnote). For example, someone who is a sexist might conclude that all women are unfit to fly jet fighters (or to drive a car) because one woman crashed in either case. People also commonly commit hasty generalizations because of laziness or sloppiness. It is very easy to simply jump to a conclusion and much harder to gather an adequate sample and draw a justified conclusion. Thus, avoiding this fallacy requires minimizing the influence of bias and taking care to select a sample that is large and meaningful enough.

7.4.2.1 Casual Truth-Preserving Inferences

Moreover, we can recognize another important occurrence. I have said that people commit errors and are hasty generalizers because of prejudice, mindlessness, bias, and so on. What I am trying to underline is that the hasty generalization is not always a bad generalization for two reasons. The first is that, getting true conclusions, hasty generalization might be good if the result of the generalization we made coincides with the result of a good generalization in the philosophical – for example Millian – sense of induction (or in the sense of inductive logics). We can call this case *"casual" truth preserving* feature of hasty generalization. The second reason is that,

[31] [Corruble and Ganascia, 1997; Ganascia, 2008] illustrate a computational program that has been built expressly to simulate erroneous/prescientific and commonsense inductions – due to the lack of facts but also to the poor description of existing facts and to implicit knowledge that is transmitted socially.

in some sense, even if we do not reach good conclusions, not exploiting the casual truth preserving feature, we can say that hasty generalization is good in some sense. We will now try to understand what it can be.

Think of a toddler that for the first time touches a stove in his kitchen [Woods, 2004, pp. 314–316]. His finger is now burned because the stove burns. Starting from this evidence, the hasty generalizer toddler thinks that all the stoves are hot and decides not to touch stoves anymore. I do not think so, as I will illustrate in the last sections of this chapter. This is obviously a hasty generalization:

- X of observed A are B (The stove *touched* burns).
- Therefore all A are B (*All* the stoves burn.)

Or:

- Sample S, which is too small, is taken from the group of persons P. (The toddler touches the stove and at a first touch the stove burns).
- Conclusion C is drawn about the group P based on S. (Whenever the toddler will touch the stove, it will burn).

Given its eco-cognitive fruitful outcome, can we still call this inference fallacious?

I have to add that this kind of reasoning is here ideally depicted and reconstructed as a case of hasty generalization, in the framework of a sentential inferential framework. An objection could be provided by the psychologically realistic observation that an actual child does not really perform a propositional inference as become averse to hot stoves, but instead he is basically directed by some emotional abductive reactions. I have already illustrated this kind of emotional abduction in this book and especially in chapter four, taking advantage of the analysis of the model-based ways of abducing in human and non-human animals. It is true, the account in terms of emotions is more psychologically realistic, given the fact we are not dealing with an adult human being. However, this does not impede us to choose the other perspective of "reconstructing" in a non psychologically way but in an inferential/propositional scheme the hypothetical inference performed, so revealing its fallacious character.

7.4.2.2 Strategic Rationality

We can also say that hasty generalization is a case of bad argument also from the formal point of view because it is not truth preserving, in the light of classical logic. However, in the perspective of agent-based reasoning the problem now is: can we say that this argument is good from some perspective? Indeed the hasty generalization is sometimes a "prudent" strategy. It also presents a cognitive economy: given the task of not being burnt for a second time, the hasty generalization is a kind of reasoning that is fruitful because, being a prudent strategy, it embeds the canons of *strategic rationality* in the sense of the "strife for survival". Moreover, it also involves a *cognitive success*.

First, fallacies (hasty generalization in this case) have some relevant relations with strategic rationality. However, the prudent strategy of "not touching the stove" is obviously incorrect for at least two reasons. 1) The first reason is that it is not good to generalize from only one sample available and 2) from applied natural physics, we can say that it is a state of affairs that a stove does not burn because a stove is made of iron or some other metals and metals burn only if they are overheated. So there is something "bad" in this kind of reasoning both from an informal logic point of view and from the perspective of natural physical principles of heat. But even if we recognize these wrong steps, there is an idea of some rationality embedded in this example due to the fact that the toddler prevents himself from being burned again. It seems that hasty generalization (like in the case of other fallacies, too, like the fallacy of affirming the consequent – abduction) can be considered – from an eco-cognitive perspective – a resource that enters in a sort of *human survival kit* [Woods, 2004, p. 7]. As some unconditioned reflexes, hasty generalization is a response (in the form of a reasoning and then of an action) to something that the toddler is involved to. The cognitive result of a hasty generalization is bad but only in the sense that it does not *explain* the burnt stove. It is instead a form of good reasoning because it preserves the toddler from being burnt another time.

Second, hasty generalization also allows the toddler to produce a new successful cognitive information and knowledge. In the perspective of the logical tradition and of the "ideal" models of good induction, this piece of information is "bad" because obtained through fallacious reasoning, but in agent-based terms we notice that the same information contributes to solve the toddler's problem and, in this sense, can be endowed with "good" cognitive relevance.

The goal reached by the toddler through the hasty induction is in some sense a success. In describing agent-based reasoning, we have said that agent-based logic (as a matter analyzing agent-based reasoning) returns to be thought of as science of reasoning and considered agent-centered, task-oriented and resource-bound. The task of the toddler was to prevent himself from being burnt: the task is reached with a performance that is considered bad if seen in the perspective of ideal models of reasoning. We can say that his inferential strategy "satisfies" his purpose of not being burnt; we can also say not only that hasty generalization halts upon the attainment of the good-enough, but also reaches a specific target that responds to a survival strategy. In some sense, the toddler's reasoning in hasty generalization terms can be seen as a very good strategy that satisfies his purpose in an optimizing way. From this point of view, the study of fallacies allows us to think about the interplay between reasoning, strategy, goal, and evaluation of goals.

I have contended above that the so-called fallacies can play the role of forms of hypothetical reasoning/argumentation typical of organic agents and in this sense we have concluded they are part of a "survival kit" suitably shaped by the coevolution between brains and culture/cognitive niches ("natural culture", in the case of animals, to use the Jungian expression I adopted in chapter three, footnote 95, p. 210). I have also added that induction and abduction performed more or less consciously by humans and implicitly by animals are surely two great results of this coevolutionary process. We know that in the last centuries humans have also characterized induction and abduction in various "ideal" philosophical and logical ways, so intellectually amending the spontaneous use of those kinds of hasty thinking I have just illustrated. Already Mill provided "Methods" for Induction and Peirce integrated abduction and induction through the famous syllogistic framework where the two non-deductive inferences can be clearly distinguished: it has to be noted that Mill also said that what he called "institutions" rather than individuals are the real embodiment of "inductive logics". As I have already pointed out in section 7.3.3 above, following this Millian perspective Gabbay and Woods also add that it is typical of human individuals to function as *practical agents* and that it is typical of "institutions" to function as *theoretical agents* (or ideal agents) [Gabbay and Woods, 2005, p. 14]; moreover, agents tend toward enhancement of cognitive assets when this enables the achievement of cognitive goals previously unaffordable or anattainable.

In summary, organic agents like human beings are hasty generalizers and more or less naive inducers and abducers but are also the creators of external cognitive and logical representations that for example provide deductive and/or computational realizations of those reasoning performances in an institutionalized way. The interplay between these "external" tools and the already "internalized" templates of reasoning certainly realizes a continuous improvement of the internal templates themselves but also expresses the centrality of the organic agents' hybrid exploitation of both levels in reasoning. The ideal agents (logical and computational) I will describe in the following sections are theoretical agents, that *mimic* "institutions", in Millian sense, more than individuals' reasoning performances.

To clarify the process that underlies the formation of ideal inductive and abductive agents I have to briefly recall in the following section the distinction between internal and external representations.

7.5 External and Internal Representations in Hybrid Abducers and Inducers

7.5.1 Logic Programs as Agents: External Observations and Internal Knowledge Assimilation

As I have illustrated in chapter three it is in the area of distributed cognition that the importance of the interplay between internal and external representations has recently acquired importance (cf. for example [Clark, 2003] and [Hutchins, 1995]). This perspective is particularly coherent with the agent-based framework I have introduced above, as we will see. It is interesting to note that a clear attention to the agent-based nature of cognition and to its interplay between internal and external aspects can be found in the area of logic programming which Aliseda [2006] describes as one of the two main ways – we already said that the other is the semantic tableaux method – of logically and computationally dealing with abduction. I think in logic programming a new idea of logic – contrasted with the classical one – arises, which certainly opens to abduction the door that grants access to its full treatment through logical systems. Indeed, logic programs can be seen in an agent-centered, computationally-oriented and purely syntactic perspective. Already in 1994 Kowalski [1994] in "Logic without model theory" introduced a knowledge assimilation framework for rational abductive agents, to deal with incomplete information and limited computational capacity.

"Knowledge assimilation" is the assimilation of new information into a knowledge base, "[...] as an alternative understanding of the way in which a knowledge base formulated in logic relates to externally generated input sentences that describe experience". The new pragmatic approach is based on a proof-theoretic assimilation of observational sentences into a knowledge base of sentences formulated in a language such as CL.[32] Kowalski proposes a pragmatic alternative view that contrasts with the model-theoretic approach to logic. In model theory notions such as *interpretation* and *semantic structures* dominate and are informed by the philosophical assumption that experience is caused by an independent existing "reality composed of individuals, functions and relations, separate from the syntax of language".

On the contrary logic programs can be seen as agents endowed with deductive databases considered as *theory presentations* from which logical consequences are derived, both in order to *internally* solve problems with the help of *theoretical sentences* and in order to assimilate new information from the *external* world of observations (*observational sentences*). The part of the knowledge base, which includes observational sentences and the theoretical sentences that are used to derive conclusions that can be compared with observations sentences, is called *world model*, considered a completely syntactic concept: "World models are tested by comparing the conclusions that can be derived from them with other sentences that record inputs, which are observational sentences extracted – *assimilated* – from experience". The agent might generate outputs – that are generated by some plan formation process in the context of the agents's "resident goals" – which affect its environment and which of course can affect its own and other agents' future inputs. Kowalski concludes "The agent will record the output, predict its expected effect on the environment using the 'world model' and compare its expectations against its later observations".

The epistemological consequence of this approach is fundamental: in model theory truth is a static correspondence between sentences and a given state of the world. In Kowalski's computational and "pragmatic" theory, the important is not the correspondence between language and experience, but the appropriate assimilation of an inevitable and continuous flowing input stream of "external" observational sentences into an ever changing "internal" knowledge base (of course the fact that computational resources available are bounded suggests to the agent to make the best use of them, for instance avoiding redundant and irrelevant derivation of consequences). The correspondence (we can say the "mirroring") between an input sentence and a sentence that can be derived from the knowledge base is considered by Kowalski only a limiting case. Of course the agent might also generate its own hypothetical inputs, as in the case of abduction, induction, and theory formation.

[32] CL, computational logic, refers to the computational approach to logic that has proved to be fruitful for creating non-trivial applications in computing, artificial intelligence, and law.

Aliseda seems to acknowledge this fact and further improves this perspective. The task is accomplished with the help of the semantic tableaux framework which can control in several ways various and meaningful logical and computational abductive strategies, some of them reflecting types of abducing already present in "actual" human performances: "That is, we must provide the automatic procedures to operate a logic, its control strategy, and its procedures to acquire new information without disturbing its coherence and hopefully achieve some learning in the end" [Aliseda, 2006, p. 23]. It is important to stress that thanks to this perspective in the logic of abduction the sensitivity to the growth of information [and the suitable extension of logical language] is definitely considered fundamental for the whole logic itself. Her analogy with the non-Euclidean revolution is striking and conclusive: "Whether non classical modes of reasoning are really logical is like asking if non-Euclidean geometries are really geometries" [Aliseda, 2006, p. 92]. I have described in the last sections of chapter two that that discovery represented an irreversible extension of geometry and mathematics beyond the restricted "intuitive" areas of the elementary perspectives.

The conceptual framework above, that is derived from a computationally-oriented logic approach that strongly contrasts with the traditional one in terms of model theory, is extremely interesting. It stresses the attention on the flowing interplay between internal and external representations/statements, so *epistemologically* establishing the importance of the agent-based character of cognition. In the following subsection I will illustrate that an analogous perspective is convenient also for depicting human beings' cognition so far as we are interested in studying its essential distributed dynamics.

7.5.2 Hybrid Inducers and Abducers

In the previous chapters of this book I have repeatedly contended that even if we can say that a large portion of the complex environment of a thinking agent is internal, we have to recognize that "human" cognitive systems are composed by distributed cognition among people and some "external" objects and technical artifacts. It is the case of the human use of the construction of external diagrams in geometrical reasoning, useful to make observation and experiment to transform one cognitive state into another for example to discover new properties and theorems. I have described in detail this dynamics in section 7.2 above. In this perspective, I have concluded, mind, classically conceived as "internal", is limited, both from a computational and an informational point of view: the act of extending it by delegating some aspects of cognition becomes necessary. In is in this sense that we have stressed in chapter four that abductive cognition is essentially multimodal.

In addition, I have also stressed that, adopting this perspective, we can give an account of the complexity of the whole human cognitive systems as the result of an interplay and *coevolution* of states of mind, body, and external environments suitably endowed with cognitive significance. I acknowledge that material artifacts like for example *inductive and abductive logical and computational agents* are tools for thoughts as is language: tools for exploring, expanding, and manipulating our own minds. The "agent-based" view illustrated in this chapter aims at analyzing the features of "real" human thinking agents by recognizing the fact that a being-like-us agent functions "at two levels" and "in two ways". I define the two levels as *explicit* and *implicit* thinking. Classic perspective on logic only exploits the first of these two features; *agent-based* perspective in logic has the power of recognizing the importance of both levels. The two ways mentioned above are the *external way* and the *internal way*. In fact inductive and abductive logical and computational systems can be seen as external representations and tools expressed through artificial (in part mathematical) and ordinary language and the use of suitable artifacts. These ideal systems not only mirror and mimic the internal ways of inferring of the *being-like-us* reasoners we have illustrated above; they can also play a *creative* role. The activities of externalizing play a central role not just in mirroring the internal ways of thinking but also in finding room for concepts and new ways of inferring which cannot be found internally "in the mind".

In summary, organic agents like human beings are hasty generalizers and more or less naive inducers and abducers but are also the creators of sophisticated external cognitive representations that for example provide *demonstrative/deductive* and *computational* representations of those reasoning performances. The interplay between these "external" tools and the already "internalized" templates of reasoning certainly realizes a continuous improvement of the internal templates themselves but also expresses the centrality of the hybrid exploitation of both levels in reasoning.

Let us reconsider the case of abduction, I have indicated above that abduction appears to be a formal fallacy that can be recognized from the classical logic point of view: the fallacy of affirming the consequent. However, from the point of view of both everyday and scientific knowledge, abduction is an important kind of hypothetical inference used for example to explain facts and invent hypotheses and theories [Magnani, 2001b].

Abduction starts from the point that information is incomplete and its retroductive method reminds us that, in some respects, classical deduction cannot be used. In other words, in a classical logic system, adding *new* information to the old one does not let us to revise previous assumed hypotheses. But if we start from incomplete information we will have to face the problem of inferring conclusion that are not foreseeable or already stated. As a consequence of this we have that a conclusion can be invalidated and modified in presence of a new information. Limited information reminds us that abductive reasoning cannot be anything but a fruitful *ignorance preserving reasoning* [Gabbay and Woods, 2005], in which the need of a deductive reasoning is underlined by the importance of a validation (cf. section 2.1, chapter two of this book). An example is of course *diagnostic reasoning*, in which abduction is useful for the diagnosis and deduction for its validation.

This schema is built by a loop in which inferring explanations with abductive reasoning remarks the need of abductive fallacy in reasoning. In saying this we can give abduction an economizing feature: it can help us in building new hypotheses for explanation in force of its fallacious "formal" features. Hence, we can outline two different ways of thinking of abduction: 1) from the point of view of classical logic, abduction is a formal fallacy, not truth preserving; 2) from the point of view of epistemology, cognitive science, and non standard logics, abduction is an important kind of reasoning able to discover new hypotheses and give explanation to scientific facts.

I have already noted that in delineating the structure of a their agent-based perspective of logic Gabbay and Woods state that logic has to be considered an "account of how thinking agents reason and argue" [2005, p. 1]. Their idea is that logic has to be defined as the disciplined description of the behavior of real-life of logical agents. Logic has to be thought of as an *agent-based logic*. From this viewpoint, abduction can be rendered as that kind of logical reasoning in which the fact of not being truth preserving (but *ignorance-preserving*, as they contend) has to live together with the fact that it is fruitfully used by real logical agents. In this framework induction is seen as *probability-enhancing* and deduction as *truth-preserving*. The definition of this innovative logical perspective is based on the assumption that the reasoner reasons in a bounded information set and on the redefinition of fallacious reasoning (as has been shown in the case of abduction, but also of hasty generalization) as a kind of fruitful reasoning.

To conclude, the use of abduction is good for at least two reasons. Abduction is not only a simple formal fallacy, but also a specific case of ignorance-preserving reasoning that can be fruitfully *idealized* in theoretical logical agents; on the applicative side, abduction is a good process able provide new hypothesis and govern inconsistencies.

At this point I hope it is clear that organic agents are spontaneous inducers and abducers and that they also construct logical and computational (ideal) systems both able to *mimic* human inductions and abductions and to *create* new and more"rational" ways of inducing and abducing. These systems are in turn used by organic (human) agents: they consequently have to be seen as *hybrid reasoners*. In the following section I will illustrate how what I called manipulative abduction can furnish a perfect example of this hybridity of human logical reasoning.

7.6 Manipulative Abduction, Hybrid Reasoning, Fallacies

I have introduced the concept of *manipulative abduction* – contrasted with theoretical abduction – in chapter one (cf. also [Magnani, 2001b]) – to illustrate situations where we are thinking through doing and not only, in a pragmatic sense, about doing. It represents a clear example of hybrid reasoning.

In our logico/computational and epistemological cases we face with at least two cases of manipulative abduction:

1. the first one refers to the exploitation of external logical and computational abductive – but also inductive – systems/agents to form hybrid and multimodal representations and ways of inferring in organic agents. Doing this they are able to enhance their "rational" performances (see below subsection 7.6.1);
2. the second case refers to the role of manipulative abduction at the level of scientific experiment and of the so-called *thinking through doing* that in turn can

improve our knowledge of induction, and its distinction from abduction: manipulative abduction can be considered as a kind of basis for further meaningful inductive generalizations.[33]

7.6.1 *Merely Successful and Successful Abductive and Inductive Strategies*

In the perspective I have illustrated above in section 7.5, resorting to the distinction between internal and external inducers and abducers a novel perspective on external ideal logical agents can be envisaged.

Human beings spontaneously (and also animals, like already Peirce maintained) perform more or less rudimentary abductive and inductive reasoning. Starting from the low-level inferential performances of the kid's hasty generalization that is a strategic success and a cognitive failure human beings arrive to the externalization of "theoretical" inductive and abductive agents as *ideal agents*, logical and computational. It is in this way that *merely successful strategies* are replaced with *successful strategies* that also tell the "more precise truth" about things. Human informal nondemonstrative inferential processes of abduction (and of induction) are more and more externalized and objectified: these external representations can be usefully rerepresented in our brains (if this is useful and possible), and they can originate new improved organic (mentally internal) ways of inferring or suitably exploited in a hybrid manipulative interplay, as I have said above.

In this perspective human beings are hardwired for survival and for truth alike so best inductive and abductive strategies can be built and made explicit, through self-correction and re-consideration (since for example the time of the inductive Mill's methods). Furthermore human beings are agents that can cognitively behave as *hybrid agents* that exploit in reasoning both internal representations and externalized representations and tools, but also the mixture of the two. Let 's consider the example of the externalization of some inferential skills – explanation and generalization – in ideal logical demonstrative systems aimed at integrating abduction and induction, like for example the ones that are at the basis of logic programming.[34] The following distinction is introduced between explanation and generalization:

- *explanation*: hypothesis does not refer to observables – it is for example the case of selective abduction [but we have to remember that abduction also creates new hypotheses], indeed we have to also consider;

[33] In chapter one, subsection 1.6.2 on manipulative abduction, I have described how Gooding [1990] refers to this kind of concrete manipulative reasoning when he illustrates the role in science of the so-called "construals" that embody tacit inferences in procedures that are often apparatus and machine based.

[34] A survey on perspectives in logic programming about induction and abduction is given in Flach and Kakas [2000b], who also furnish a useful classical perspective on integration of abduction and induction.

- *generalization*: it is the introduction of a genuinely new hypothesis, which in turn can entail additional observable information on *unobserved* individuals, extending the theory T.

Imagine we have a new abductive theory $T' = T \cup H$ constructed by induction: an inductive extension of a theory can be viewed as a set of abductive extensions of the original theory T.

In sum, the ideal logico-computational systems present interesting cognitive features (cf. also [Longo, 2005]) which I believe deserve to be further analyzed and which can further develop the distinction above between theoretical and practical agents: as I have indicated in subsection 7.3.3, they are *symbolic*, *abstract*, *rigorous*, and *characterized* by what I have called *maximization of memorylessness*.

Finally, we have already noted the contribution given to logic by Aliseda [2006] is the improvement of the *semantic tableaux method* (and its application in the logic of abduction and in the enhancement of other logical models of abduction, like for example the belief-revision framework). The new extended semantic tableaux method fulfills the request of "weakening" the rigidity of classical logic but also of many non standard logics strictly related to it, helping to further open the new era of logic and focus attention on the role of *meta-levels* – for instance in the logic of abduction. It formalizes the flexibility and "historicity" of many kinds of human thinking which are meaningful in certain of the application areas they address. Aliseda's conclusion is clear, and by implicitly reflecting the four aspects I have just illustrated, also leads not only to a new perspective on abduction but also to a new conception of logic:

> The various types of abductive explanatory styles in a larger universe of other deductive and inductive systems of logic naturally commit us to a global view of logic [...] in which there is a variety of logical systems which rather than competing and being rival to each other, they are complementary in that each of them has a specific notion of validity corresponding to an extra-systemic one and a rigorous way for validating arguments, for it makes sense to speak of a logical system as correct or incorrect, having several of them. And finally, the global view states for abduction that it must aspire to global application, irrespective of subject matter, and thus found in scientific reasoning and in common sense reasoning alike [Aliseda, 2006, p. 89].

7.6.2 Abduction, Fallacies, Rhetoric, and Dialectics

Not only is abductive reasoning (especially explanatory) related to fallacies from many perspectives, as I have illustrated above, but it is also useful in shedding new light on the contextual "dialectic" processes in which a responder conveys greater understanding to a dialogue partner who has sought it through questioning (for example the so-called why "questions"), looking for an explanation [Walton, 2004]. In this process it is easy to see that process of comprehension at play [the *Verstehen* of the philosophical tradition], which resorts to the understanding of the explanation given. This kind of research aims at illustrating a dialogic model of explanation alternative to the DN-Model (or "law covering model of scientific explanation"

[Magnani, 2001b, pp. 38–39]), where the explanation is basically seen as based on a dynamic chain of reasoning/arguments. It also aims at depicting some dialectic aspects of scientific reasoning and of the processes of teaching and learning [Walton, 2004, p. 82–92].

Walton also claims for a procedural – agent-based – model of rationality, as an interplay of two or more agents who act in a coordinated way by communicating with each other as they act, a model that is also associated with the new field of computational dialectics [Hage, 2000].[35] In typical forensic and testimonial evidence used in a trial, reasoning embedded in arguments is based on abductive inferences[36] in which attempts are made to offer competing explanations of what the law calls the "facts" or the accepted findings of the case at hand [Walton, 2008]. In this kind of reasoning the so-called fallacies can play the role of deceptive tactics and enthymemes are common, something unlikely to happen in science, where the dialogue does not basically deal with a conflict of mere opinions. The analysis of these issues in various cognitive situations is certainly important in order to depict a complete theory of abduction.

For example, in the case of abduction in legal reasoning it has to be noted, following [Gabbay and Woods, 2005, pp. 256] that non-epistemic requirement of avoidance of injustice takes precedence over attainment of truth, so determining a situation of epistemic deficit. Standard epistemic relevance criteria I have illustrated in chapter two (subsection 2.3.1) are distorted:

> For these reasons, it is easily seen that legal systems such as those that evolved in Anglo-American jurisprudence have something of the same basic structure as ideologies and dogmatic religions. All are systems that impose prior constraints on what the evidence is allowed to show. In purely operational terms, the best way of disarming such evidence is to refuse to hear it. This, too, is often the counsel of the religious leader or the ideologue: stay away from considerations that may tend to discredit the requirements of orthodox belief. Thus some Christians advise against a secular education as an "occasion of sin"; and some ideologues recommend the prosecution or expulsion of those who consort with non-believers. In all three cases, there is known to be considerable potential for epistemic distortion, yet in only one is this knowledge given much formal recognition. It can safely be said, however, that of the three, the law's epistemic triflings have a non-epistemically coherent motivation.

Plausible and presumptive arguments (for instance in legal reasoning) embed inferences to hypotheses – just like abduction – which are those instantiating schemes that have traditionally been classified as fallacies. Walton maintains that presumptive[37] and plausible reasoning (where there is a selection among a set of alternatives whose evidence is often considered dubious) share a pragmatic nature with abduction. Nevertheless they differ from classical Peircean abductive reasoning because

[35] Given the multiplicity of agents at play, all endowed with a cognitive role, this approach is naturally in tune with many aspects of the tradition of distributed cognition that I have adopted throughout this book.

[36] On the constitutive link between inferences and arguments cf. below footnote 42.

[37] Presumptive reasoning is typically dialogic: it shows an interplay among intermediate assumptions and "reasons", often associated with the so-called fallacious *argumentum ad ignorantiam*.

they are often provided in the absence of *firm* evidence (for example empirical) or knowledge and (moreover, they do not necessarily look for explanations). It is in this sense that these reasoning processes (and the arguments that embed them) are defeasible, and often take the form of various fallacies: arguments *ad hominem*, from sign, consequences, witness testimony, analogy, precedent, slippery slope, gradualism, appeal to popular opinion, to pity, and to expert opinion. Their standard of rationality differs from the deductive and inductive one and resorts to that kind of contextual interplay I have indicated above. Walton also describes a computational system devoted to the construction of an argument diagram from the text of a given argument applied to presumptive legal argumentation [pp. 43–50], that often embeds reasoning steps that can be classified as abductive. I think it is interesting to study how abduction as a classical Peircean form of explanatory hypothetical reasoning can be encountered as intertwined in all these kinds of reasoning and not as only active in most cases of the argument from sign [Walton, 2004, p. 42]. I will provide some examples below in subsection 7.8.3.

Let us now briefly reiterate the fallacious character of abduction. As I have already illustrated in section 1.4 of chapter one the abductive inference rule corresponds to the well-known fallacy called *affirming the consequent*. It is now clear that in the light of classical logic abduction is a fallacy. Despite this fallacious character it is of fundamental importance in many agent-based reasoning situations like scientific explanation, scientific discovery, etc. and also in moral deliberation, as I have illustrated in this book. We can furnish another reason that stresses the fruitfulness of abduction in agent-based reasoning: it is a powerful inferential process able to govern inconsistencies (cf. chapter two, this book). For example, in the case of the formation of new scientific hypotheses and theories epistemologists have always recognized the role played by inconsistencies and anomalies that violate the paradigm-induced expectations derived from previously established conceptual frameworks. Logicians have in turn shown that inconsistencies generated by anomalies are difficult to be managed in deductive frameworks: they are unexpected facts that the rules of classical logic are not able to manage.

In fact, if we consider *deduction* in scientific reasoning, an anomaly in the system is a problem that logical hypotheses will explain with difficulty: *classical* deduction has no sense in giving reasons for unexpected or unknown facts. In a deductive system, all premises are given and new information that contradicts the previous one clashes against the classical law of *ex falso sequitur quodlibet*. However, we know that abduction can be rendered as that kind of reasoning in which the fact of not being truth preserving (but *ignorance preserving*), has to live together with the fact that it is fruitfully used by real cognitive agents. In this sense the study of abduction as a fallacy, and the study of all other fallacies, allows us to maintain that, in some sense, a fallacious form of reasoning is better than nothing [Woods, 2004]. In the case of abduction, we can go beyond this definition and say that fallacious reasoning give us even more: in fact, the use of abduction is good for at least two reasons. Abduction can play the role of a bad fallacy, but it can also perform a specific and complex case of ignorance-preserving *agent*'s reasoning that can be fruitfully modeled and idealized in theoretical logical agents; on the applicative side, abduction obviously can be a good process able to provide new hypothesis and govern inconsistencies.

7.7 Intelligence as Smart Heuristic: Ecological Thinking vs. Logical Reasoning

This section is devoted to further illustrate some aspects of heuristics to the aim of stressing the double logical and eco-logical aspect of hypothetical reasoning in general, and so of abductive cognition. Outside the realm of logical models, a view of hypothetical intelligence as consisting in "fast and frugal heuristic" [Gigerenzer and Todd, 1999; Gigerenzer and Selten, 2002; Raab and Gigerenzer, 2005; Gigerenzer and Brighton, 2009], still confirms that the perspective on strategic rationality illustrated in the previous sections is essential to understanding human (but also some aspects of animal) behavior. In this perspective the heuristics are devices that can solve a class of problems in situations with limited knowledge and time. In this case intelligence is sometimes viewed in a Darwinian way as an *adaptive toolbox* that involves an ecological and social view of rationality.

Many cases are described. For example *recognition heuristic*, when in front of a questions like "Which city has a larger population: San Diego or San Antonio?" people that know less are more likely to answer correctly, in an experiment 100% of German students got the answer right, but only two third of American students got the answer right: Americans cannot use this heuristic because they know too much. Limited knowledge makes this strategy advantageous. When wild Norwegian rats choose between two foods, one that they recognize from the breath of a fellow rat and one that they do not recognize, they are inclined to select the recognized one [Raab and Gigerenzer, 2005, p. 190]. "Recognition heuristic" is often accompanied by the *take-the-best* strategy, when for example, looking to buy a laptop, if you recognize one company label, and not the other, you will choose the one that you know, but, if more laptops of the same company are available, you continue to search for more information to find properties that favor the best choice.

Gaze heuristic is illustrated in the case of some baseball players: when the ball comes in high, the player fixates on the ball and starts running, the heuristic is to adjust the running speed so that the angle of gaze, that is the angle between the eye and the ball, remains constant (or within a certain range). Only one variable is considered: the information complexity of the environment is transformed into the input given a in a linear line. In time-to-contact heuristics pigeons use this type of information reduction and simplification, not to collide but to avoid collision in the air.

Social and distributed aspects are also important in intelligence as in smart heuristics. This is the case of *tit-for-tat* heuristic:

> Two people play a game: each has two behavioral options, to cooperate with the other, or to "defect". If one cooperates and the other defects, the first is exploited by the second, a situation that can be represented in monetary terms, for example, by stating that the first loses $1 whereas the second gains $3. If nether cooperates, nobody loses and

> nobody gains anything. If both cooperate, each gains 2$. Such a situation is known as the prisoner's dilemma. Standard rational choice theory says that the optimal behavior for both sides is to defect, because whatever the other person does, it is always an advantage to defect. There is, however, a fast and frugal heuristic, called tit-for-tat that can outperform the "optimal" strategy. In the first round, tit-for-tat always cooperates, that is, it trusts the partner. Thereafter it searches in memory for the partner's response (search rule), memorizes only the last move of partner (stopping rule), and reciprocates, that is imitates the partner's behavior (decision rule) [Raab and Gigerenzer, 2005, p. 192–193].

In social environments where mostly tit-for-tat players are present the heuristic is good, leads to efficient results, and can be combined with other heuristics such as detecting cheaters in social contracts. In summary, in this perspective the heuristics, like the so-called fallacies – for instance hasty generalization – I have illustrated in the previous sections, are not good (rational) or bad (irrational) in themselves, but context-dependent, domain-specific, this explains why they are called eco-logical. Even if they are not the best solution to a problem, if compared to some ideal models, based on omniscience and optimization, often being bad from the perspective of "flourishing" and/or"surviving". As is shown in the quotation above, heuristics are intertwined with search rules, stopping rules, and decision rules. For example: "searching for the cues with highest validity" is basically good, but also searching randomly can be advantageous, for example when searching for spouses! But they have to be stopped at some point because deciding is compelling in practical reasoning, where the costs of a further search can exceed its benefits [Simon, 1955].

7.7.1 Reducing Information

Animal and human sensorial endowments are domain-specific in themselves: for instance eyes process visual but not acoustic information. Abductively going beyond available data and performing suitable inferences is intrinsic to the adaptive toolbox idea of intelligence, and often these performances require multimodal processes. This is the case of catching a ball, that needs sensory intelligence but also motor abilities, but also of manipulative abduction, which certainly involves many hardwired – and thus implicit – schemes. Social intelligence, also called Machiavellian intelligence,[38] is typical of *homo sapiens* and takes advantage of smart heuristics as well as of all the more complicated ideal tools invented during the history of knowledge, such as logic, science, artifacts, and so on that humans can suitably pick up and use when available. In this perspective "following the crowd", or social imitation, is good too: an analogous behavior is detectable in many animals.[39]

It is important to note that in many reasoning performances, abduction included, in animals and humans, less information – and the act of reducing information – gives

[38] Cf. subsection 3.3.2, chapter three of this book.

[39] The context-dependency of this heuristic is clear for example when we acknowledge the evidence that, even if advantageous from the perspective of some individuals in a group, a "follow the crowd" approach can originate bad moral outcomes, such as *mobbing* an individual, with further possible disadvantage for an entire system, for example a business firm.

rise to better performance. This is present in gaze heuristic but also in the so-called *take-the-first* heuristic, which describes how options are generated from memory, for example in selective abduction in merely non-social internal inferences: "For instance, in chess it is known that an expert can generate a large number of options, but that the first ones generated are often the best options. The options are generated by order of their appropriateness to a specific situation. Like *take-the-best* (where cues are searched for in order of cue validity), take-the-first looks up alternative options by option validity (search rule)" (*ibid.*, p. 200). This heuristic is good when the environment is the one in which the person is trained by feedback, that is when options are simply generated from memory in the order of validity.

Intelligence in terms of smart heuristics completely acknowledges the hybrid character of cognition (and of hypothetical cognition) that I am emphasizing, considering both the mind and world-contexts, internal and external representations (in external objects and – socially – in other humans) and their interplay, that is the structure of the environment and the cognitive delegation it embodies. This perspective further and indirectly emphasizes that hypothetical thinking can be usefully characterized from both a logical and an eco-logical perspective.

7.8 Fallacies as Distributed "Military" Intelligence

In the sections above I have illustrated the main relationships among fallacies, abduction and logic. In the third part of chapter eight, a part which plays the role of a sort of appendix, section (8.6), I will describe the general framework of the so-called *coalition enforcement hypothesis*, a perspective which especially stresses the intrinsic *moral* (and at the same time *violent*) nature of language (and of abductive and other hypothetical forms of reasoning and arguments intertwined with the propositional/linguistic level). In this perspective language is basically rooted in a kind of *military intelligence*, as maintained by Thom [1988].[40] Indeed we have to note that many kinds of hypothesis generation (from abduction to hasty generalization, from *ad hominem* to *ad verecundiam*) are performed through inferences that embed formal or informal fallacies.

Of course not only language carries morality and violence, but also motor actions and emotions: it is well-known that overt hostility in emotions is a possible trigger to initiate violent actions. I have already discussed the "moral" role of emotions in subsection 4.4.2 of chapter four and I think the potential "violent" role of them is out of discussion.[41]

Essentially, language efficiently transmits *vital* pieces of information about the fundamental biological opponents (life, death – good, bad) and so it is intrinsically involved in moral/violent activities. This conceptual framework, together with the

[40] The section on coalition enforcement illustrates some theoretical issues, which offer a background that I think can be of help to better understand the content of this section and of the second part of the chapter eight, devoted to the description of abduction and its relationship with the concept of pregnance in the catastrophe theory.

[41] I have treated the moral character of actions, both spontaneous/immediate and planned, in my book [Magnani, 2007d, chapter six].

eco-logical approach illustrated in the previous sections, also sheds further light on some fundamental dialectical and rhetorical roles played by the so-called fallacies, which are of great importance to stress some basic aspects of human abductive cognition. In the following section I will consider some roles played by fallacies that have to be ideally related to the intellectual perspective of the coalition enforcement hypothesis quoted above and which I will illustrate in the following chapter.

Of course this "military" nature is not evident in various aspects and uses of syntactilized human language. It is hard to directly see the coalition enforcement effect in the many *epistemic* functions of natural language, for example when it is simply employed to transmit scientific results in an academic laboratory situation, or when we pick up some information through the Internet – expressed in linguistic terms and numbers – about the weather. However, we cannot forget that even the more abstract character of the knowledge packages embedded in certain uses of language (and in hybrid languages, like in the case of mathematics, which involves considerable symbolic parts) still plays a significant role in changing the moral behavior of human collectives. For example, the production and the transmission of new scientific knowledge in human social groups not only transmits information but also implements and distributes roles, capacities, constraints, and possibilities of actions. This process is intrinsically moral because in turn generates precise distinctions, powers, duties, and chances, which can create new inter- and intra-group more or less moral and/or violent conflicts or reshape the old ones.

New abstract biomedical knowledge about pregnancy and fetuses usually has two contrasting moral/social effects, 1) a better social and medical management of childbirth and related diseases; 2) the potential extension or modification of conflicts surrounding the legitimacy of abortion. In sum, even very abstract bodies of knowledge and more innocent pieces of information enter the semio/social process which govern the identity of groups and their aggressive potential as coalitions: deductive reasoning and declarative knowledge are far from being exempt from involving argumentative, deontological, rhetorical, and dialectic aspects. For example, it is hard to distinguish, in an eco-cognitive setting, between a kind of "pure" (for example deductive) inferential function of language and the argumentative or deontological one.[42] However, it is in the arguments traditionally recognized as fallacious, that we can more clearly grasp the military nature of human language and especially of some hypotheses reached through the so-called fallacies.

Woods contends that a fallacy is by definition considered "a mistake in reasoning, a mistake which occurs with some frequency in real arguments and which is

[42] As I have already stressed in chapter four Searle considers "bizarre" that feature of our intellectual tradition, according to which true statements – so to say, fruit of "pure" inferences – that describe how things are in the world can never imply a statement about how they ought to be: he contends that to say something is true is already to say you ought to believe it, that is other things being equal, you ought not to deny it. This means that normativity is more widespread than expected. It is in a similar way that Thom clearly acknowledges the intrinsic "military" (and so moral, normative, argumentative, etc.) nature of language, by providing a justification in terms of the catastrophe theory, as I will describe in the following chapter.

characteristically deceptive" [Woods, 2010].[43] As I have already illustrated, traditionally recognized fallacies like hasty generalization and *ad verecundiam* are considered "inductively" weak inferences, while affirming the consequent is a deductively invalid inference. Nevertheless, when they are used by actual reasoners, "beings like us", that is in an eco-logical[44] and not merely logical – ideal and abstract – way, they are *no longer* necessarily fallacies. Traditionally, fallacies are considered mistakes that appear to be *errors*, *attractive* and *seductive*, but also *universal*, because humans are prone to committing them. Moreover, they are "usually" considered *incorrigible*, because the diagnosis of their incorrectness does not cancel their appearance of correctness: "For example, if, like everyone else, I am prone to hasty generalization prior to its detection in a particular case, I will remain prone to it afterwards" (*ibid.*).

Woods calls this perspective the traditional – even if not classical/Aristotelian – "EAUI-conception" of fallacies. Further, he more subtly observes

> [...] first, that what I take as the traditional concept of fallacy is not in fact the traditional concept; and, second, that regardless whether the traditional concept is or is not what I take it to be, the EAUI-notion is not the right target for any analytically robust theory of fallacyhood. [...] But for the present I want to attend to an objection of my own: whether the traditional conception or not, the EAUI-conception is not even so a sufficiently clear notion of fallacyhood. [...] If the EAUI-conception is right, it takes quite a lot for a piece of reasoning to be fallacious. It must be an error that is attractive, universal, incorrigible and bad. This gives a piece of reasoning five defences against the charge of fallaciousness. Indeed it gives a piece of erroneous reasoning four ways of "beating the rap". No doubt this will give some readers pause. How can a piece of bad reasoning not be fallacious? My answer to this is that not being a fallacy is not sufficient to vindicate an error of reasoning. Fallacies are errors of reasoning with a particular character. They are not, by any stretch, all there is to erroneous reasoning [Woods, 2010, chapter three].

In the sections above I have depicted a sharp distinction between strategic and cognitive rationality, and many of the traditional fallacies - hasty generalization for example - call for an equivocal treatment. They are sometimes cognitive mistakes and strategic successes, and in at least some of those cases, it is more rational to proceed strategically, even at the cost of cognitive error. On this view, hasty generalization instantiates the traditional concept of fallacy (for one thing, it is a logical error), but there are contexts in which it is smarter to commit the error than avoid it.

According to Woods' last and more recent observations the traditional fallacies - hasty generalization included - do not really instantiate the traditional concept of fallacy (the EAUI-conception). In this perspective it is not that it is sometimes "strategically" justified to commit fallacies (a perfectly sound principle, by the way), but rather that in the case of the Gang of Eighteen traditional fallacies they simply are not fallacies. The distinction is subtle, and I can add that I agree with it

[43] Of course deception – in so far as it is related to *deliberate* fallaciousness – does not have be considered to be a part of the definition of fallacy [Tindale, 2007].

[44] That is when fallacies are seen in a social and real-time exchange of speech-acts between parties/agents.

in the following sense: the traditional conception of fallacies adopts – so to say – an *aristocratic* (ideal) perspective on human thinking that disregards its profound eco-cognitive character, where the "military intelligence" I have quoted above is fundamentally at play. Errors, in an eco-cognitive perspective, certainly are not the exclusive fruit of the so-called fallacies, and in this wide sense, a fallacy is an error – in Woods' words – "that virtually everyone is disposed to commit with a frequency that, while comparatively low, is nontrivially greater than the frequency of their errors in general".

My implicit agreement with the new Woods' "negative thesis" will be clearer in the light of the illustration of the general military nature of language I will give in the following sections and I will deepen in the following chapter: 1) human language possesses a "pregnance-mirroring" function, 2) in this sense we can say that vocal and written language is a tool exactly like a knife; 3) the so-called fallacies, are linked to that "military intelligence", which relates to the problem of the role of language in the so-called *coalition enforcement*; 4) as I have illustrated above, this "military" nature is not evident in various aspects and uses of syntactilized human language. As already stressed above, it is hard to directly see the coalition enforcement effect in the many (too easily and quickly supposed to be universally "good") *epistemic* functions of natural language. However, we cannot forget that even the more abstract character of the knowledge packages embedded in certain uses of language – for example truth-preserving – still plays a significant role in changing the moral behavior and it is still error prone fom many perspectives. In this sense it is hard to distinguish, in an eco-cognitive dimension, between a kind of "pure" (for example deductive) inferential function of language and the argumentative or deontological one, with respect to their error-making effects. So to say, there are many more errors of reasoning than there are fallacies. What I can bashfully suggest is that in the arguments traditionally recognized as fallacious, it is simply much more easier than in other cases, to grasp their error-proneness, for example when they constitute of assist hypothetical cognition. In sum, from an eco-cognitive perspective, when language efficiently transmits positive *vital* pieces of information through the so-called fallacies and fruitfully acts in the cognitive niche from the point of view of the agent(s), it is hard to still label the related reasoning performance as fallacious.

As I have already illustrated (cf. above section 7.4.1), if we see the so-called fallacies in practical *agent-based* and *task oriented* reasoning occurring in actual interactive cognitive situations, some important features immediately arise. In agent-based reasoning, the agent access to cognitive resources encounters limitations such as

- bounded information
- lack of time
- limited computational capacity.[45]

[45] We have seen in the previous section that, on the contrary, logic, which Gabbay and Woods consider a kind of theoretical/institutional ("ideal", in my terms) agent, is occurring in situations characterized by more resources and higher cognitive targets.

7.8.1 Distributing Fallacies

It is now necessary to stress that the so-called fallacies and hypothetical reasoning which embeds them are usually exploited in a *distributed* cognitive framework, where moral conflicts and negotiations are normally at play. This is particularly clear in the case of *ad hominem*, *ad verecundiam*, and *ad populum*. We can see linguistic reasoning embedded in arguments which adopt fallacies, as distributed across

- human agents, in a basically visual and vocal dialectical interplay,
- human agents, in an interplay with other human agents mediated by various artifacts and external tools and devices (for example books, articles in newspapers, media, etc.)

From this perspective the mediation performed by artifacts causes additional effects: 1) other sensorial endowments of the listener – properly excited by the artifact features – are mobilized; these artifacts in turn 2) affect the efficacy of the argument: in this sense artifacts (like for example media) have their own highly variable cognitive, social, moral, economical, and political character. For example, the same *ad hominem* argumentation can affect the hearer in a different way depending on whether it is watched and heard on a trash television program or instead listened to in a real-time interplay with a friend. It is obvious that in the second case different cognitive talents and endowments of the listener are activated:

- positive emotional attitudes to the friend can be more active and other areas of the information and knowledge at disposal of the arguer – stored in conscious but also in unconscious memory – are at play. Both these aspects can affect the negotiatory interaction, which of course can also acquire
- the character of a dialectical process full of feedbacks, which are absent when an article in a newspaper is simply passively read (in this case the "rhetorical" effect prevails).

7.8.2 Military Intelligence through Fallacies

Some aspects are typical of agent-based reasoning and are all features which characterize fallacies in various forms and can consequently be seen as good scant-resource adjustment strategies. They are:

1. Risk aversion (beings like us feel fear!),
2. guessing ability and generic inference,
3. propensity for default reasoning,
4. capacity to evade irrelevance, and
5. unconscious and implicit reasoning

Gabbay and Woods also contend that in this broader agent-based perspective one or other of the five conditions above remain unsatisfied, for example: i) fallacies are not necessarily errors of reasoning, ii) they are not universal (even if they are frequent), iii) they are not incorrigible, etc. Paradoxically, fallacies often *successfully* facilitate cognition (and hypothetical thinking) (*Abundance thesis*), even if we obviously acknowledge that actually beings like us make mistakes (and know it) (*Actually Happens Rule*). In sum, if we take into account the role of the so-called fallacies in actual human behavior, their cognitive acts show a basic, irreducible, and multifarious argumentative, rhetorical, and dialectic character. These cognitive acts in turn clearly testify that cognition can be successful and useful even in the presence of bounded information and knowledge and, moreover, in the absence of sound inferences. In this perspective deeper knowledge and sound inferences loose the huge privileged relevance usually attributed to them by the intellectual philosophical, epistemological, and logical tradition. "Belief" seems sufficient enough for human groups to survive and flourish, as they do, and indeed belief is more "economical" than knowledge as at the same time it simulates knowledge and conceals error.

The anti-intellectual approach to logic advanced by Woods' agent-based view is nicely captured by the *Proposition 8* (*Epistemic Bubbles*): "A cognitive agent *X* occupies an epistemic bubble precisely when he is unable to command the distinction between his thinking that he knows *P* and his knowing *P*" and *Corollary 8a*: "When in an epistemic bubble, cognitive agents always resolve the tension between their thinking that they know *P* and their knowing *P* in favour of knowing that *P*" [Woods, 2010]. Hence, we know a lot less than we think we do. Moreover, it is fundamental to stress that, when epistemic bubbles obviously change, the distinction between merely apparent correction and genuinely successful correction exceeds the agent's command and consequently the cognitive agent from his own first-person perspective favors the option of genuinely sound correction. In sum, detection of errors does not erase the appearance of goodness of fallacies.

This Humean skeptical conclusion is highly interesting because it shows the specific and often disregarded very "fragile" nature of the "cognitive" *Dasein* [46] – at least of contemporary beings-like-us. I also consider it fundamental in the analysis of fallacies from the point of view of military intelligence. A basic aspect of the human fallacious use of language regarding military effects is – so to say – the softness and gentleness which the constitutive capacity of fallacies to *conceal error* – especially when they involve hypothesis guessing – can grant. Being constitutively and easily unaware of our errors is very often intertwined with the self-conviction that we are not at all aggressive in our performed argumentations. Human beings use the so-called fallacies because they often work in a positive vital way: but when they are eco-cognitively fruitful, we cannot call them fallacies anymore. I contend that in this case we find ourselves inside a kind of *moral bubble*, that is very homomorphic with the epistemic bubble. Unawareness of our error is often accompanied by

[46] I am using here the word *Dasein* to refer to the actual and concrete existence of the cognitive endowments of a human agent. It is a German philosophical word famously used by Martin Heidegger in his magnum opus *Being and Time*. The word *Dasein* was used by several philosophers before Heidegger, with the meaning of "existence" or "presence". It is derived from da-sein, which literally means being-there/here, though Heidegger was adamant that this was an inappropriate translation of *Dasein*.

unawareness of the deceptive/aggressive character of our speeches (and behaviors). Woods continues: "*Proposition 11* (*Immunization*) Although a cognitive agent may well be aware of the Bubble Thesis and may accept it as true, the phenomenological structure of cognitive states precludes such awareness as a concomitant feature of our general cognitive awareness", and, consequently "Even when an agent X is in a cognitive state S in which he takes himself as knowing the Bubble Thesis to be true, S is immune to it as long as it is operative for X". In short, a skeptical conclusion derives that errors are unavoidable, their very nature lies embedded in their concealment; that is, in an epistemic bubble, "any act of error-detection and error-correction is subject in its own right to the concealedness of error" (*ibid.*)

In a sense, *there is nothing to correct*, even when we are aware of the error in reasoning we are performing. Analogously, there is nothing to complain about ourselves, even if in some sense we are aware of the possible deceptive character of the reasoning and hypothetical cognition we are performing. The kind of "awareness" that *has a priority* and that is *stable* is about the fact we are contending *our* opinions, which are endowed with an intellectual value simply because they are ours. Moreover, these opinions are endowed with an intrinsic moral value because they fit some moral perspective *we* agree with as individuals – a perspective that we usually share with various groups we belong to. The value human beings attribute to some of their opinion can also be merely subjective and isolated: obviously in this last case it can be easily seen as perverse from the perspective of others. Errors, and so deception and aggressiveness, tend to be a constitutive "occluding edge" of agent-based linguistic acts, and consequently they are not recognized or felt by the subjects that commit them, but only by others, in some cases because they are negatively affected by them or because they present ideal and sophisticated criteria of reasoning. It is from this perspective that we can also grasp the effective importance reserved by humans for so-called "intuition", where they simply reason in ways that are *typical* for them, and *typically* justified for them [Woods, 2010].

Finally, a quotation is noteworthy and self-evident

> *Proposition 12* (*Inapparent falsity*) The putative distinction between a merely apparent truth and a genuine truth collapses from the perspective of first-person awareness, i.e., it collapses within epistemic bubbles. *Corollary 12a.* As before, when in an epistemic bubble, cognitive agents always resolve the tension between only the apparently true and the genuinely true in favour of the genuinely true. *Corollary 12a* reminds us of the remarkable perceptiveness of Peirce [. . .] that the production of belief is the sole function of thought. What Peirce underlines is that enquiry stops when belief is attained and is wholly natural that this should be so. However, as Peirce was also aware, the propensity of belief to suspend thinking constitutes a significant economic advantage. It discourages the over-use of (often) scarce resources (*ibid.*).

Hence, truth is "fugitive" because one can never attain it without thinking that one has done so; but thinking that one has attained it is not attaining it and so cognitive agents lack the possibility to distinguish between merely apparent and genuine truth-apprehension: "One cannot have a reason for believing P without thinking one does. But thinking that one has a reason to believe P is not having a reason to believe P". It can be said that fallibilism in some sense acknowledges the perspective above and, because of its attention to the propensity to guess (and thus also to abduce) and to error-correction, it does not share the error-elimination paroxysm of the syllogistic tradition.

If humans are so inclined to disregard errors it is natural to especially devote to the so-called fallacious reasoning a kind of natural, light military role which becomes evident when more or less vital interests of various kinds are at stake. In this case arguments that embed fallacies can nevertheless aim at 1) defending and protecting ourselves and/or our group(s) – normally, human beings belong to various groups, as citizens, workers, members of the family, friends, etc. – groups which are always potential aggressive coalitions; 2) attacking, offending and harming other individuals and groups. Here, by way of example, it is well-known that gossip takes advantage of many fallacies, especially *ad hominem*, *ad baculum*, *ad misericordian*, *ad verecundiam*, *ad populum*, straw man, and begging the question.[47]

From this perspective gossip (full of guessed hypotheses about everyone and everything) which exploit the so-called fallacies contemplates

1. the telling of narratives that exemplify moral characters and situations and so inform and disseminate the moral dominant knowledge of a group (or subgroup) (a teaching and learning role which enforces the group as a coalition) *possibly* favoring its adaptivity,[48] while "at the same" time facilitating some disadvantage, persecution, and punishment
   a. of free riders inside the group (or inside the same subgroup as the arguer, or inside other subgroups of the same group as the arguer),
   and
   b. of alien individuals and groups presenting different moral and other conflicting perspectives.

The disseminating process of gossip shown above is *moral* and at the same time it secures the more or less *violent* persecution of free riders inside the group. Furthermore, the parallel process of protecting and defending ourselves and our groups, is *moral* and at the same time it secures aggression and *violence* against other, different, groups. In other words exploiting the so-called fallacies in gossip can be seen as *cooperative* because they carry moral knowledge, but it is at the same time *uncooperative* or even *conspiratory* because is triggers the violence embedded in

[47] Some considerations on the moral role of gossip are given in chapter three of my book [Magnani, 2007d] and in [Bardone and Magnani, 2009].

[48] Even if Wilson's evolutionary perspective in terms of *group selection* is questioned because of its strict Darwinist view of the development of human culture, a suggestion can be derived, especially if we reframe it in the theory of "cognitive niches": schemas of gossiping establish cognitive niches that can be adaptive but of course also *maladaptive* for the group [Wilson, 2002]. Human coalitions produce various standard gossip templates which exploit fallacies and that can be interpreted in terms of conflict-making cognitive niches. However it is unlikely these "military" schemas, which embed both moral and violent aspects, can directly establish appreciable selective pressures. I will describe this eco-cognitive perspective concerning the intertwining between morality and violence in the following chapter (section 8.6).

harming, punishing and mobbing. It has to be said that the type of violence perpetrated through the so-called fallacies in these cases is situated at an intermediate level – in between sanguinary violence and indifference. If the "supposed to be" – from our intellectual viewpoint – fallacies are eco-cognitively fruitful for the individual or the coalition, can we still call them fallacies, if we take into account their own perspective?

The hypotheses generated by the so-called fallacies in a dialectic interplay (but also when addressed to a non-interactive audience) are certainly conflict-makers[49] but they do not have to be conceived absolutely as a priori "deal-breakers" and "dialogue-breakers", like Woods very usefully notes.[50] I would contend that the potential deceptive and uncooperative aim of fallacies can be intertwined with pieces of both information and disinformation, logical valid and invalid inferences, other typical mistakes of reasoning like perceptual errors, faulty memories and misinterpreted or wrongly transmitted evidence, but fallacious argumentations still can be – at first sight paradoxically – "civil" ways/conventions for negotiating. That is, sending a so-called deceptive fallacy to a listener is much less violent than sending a bullet, even if it "can" enter – as violent linguistic behavior – a further causal chain of possible "more" violent results. Also in the case of potentially deceptive/uncooperative fallacious argumentation addressed to a non-interactive audience, the listener is "in principle" in the condition to disagree with and reject what is proposed (like in the case of deceptive and fallacious advertising or political and religious propaganda). The case of a mobbed person is more problematic, often it is impossible to prevent the violent effects of mobbing performed through gossip: any reaction of the mobbed ineluctably tends to confirm the reason adduced by the mobbing agencies as right. This is one of the reasons that explains how mobbing is considered a very violent behavior.

In sum, when an argument (related or not to the so-called fallacies) "perpetrated" by the proposer(s), is accepted (for example when gossip full of *ad hominem* and *ad populum* arguments is fortunate and assumed inside a group), it has been proved successful and so must have been – simply – a good argument. When the argument is rejected, often, but not necessarily, it happens because it has been recognized as a fallacy and an error of reasoning: Proposition 12 (Error relativity), in [Woods, 2010], clearly states: "Something may be an error in relation to the standards required for target attainment, in relation to the legitimacy of the target itself, or in relation to the agent's cognitive wherewithal for attaining it".

7.8.3 Abduction in Argument Evaluation and Assessment

I recently watched a talk show on television devoted to the case of a Catholic priest, Don Gelmini, accused of sexual abuse by nine guys hosted in a home for people with

[49] The relationship between arguer and audience is analyzed by [Tindale, 2005] in a contextual-based approach to fallacious reasoning.

[50] many chapters of [Woods, 2010] are devoted to an analysis of the argument *ad hominem* and contain a rich description of various cases, examples, and problems, which broaden my point.

addiction problems belonging to Comunità Incontro in Italy, a charitable organization now present worldwide, which he had founded many years before. Two pairs of journalists argued in favor and against Don Gelmini, by using many so-called fallacious arguments (mainly *ad hominem*) centered on the past of both the accused and the witnesses. I think the description of this television program is useful to illustrate the role of abduction in the filtration and evaluation of arguments, when seen as *distributed* in a real-life dialectical and rhetorical contexts. As an individual belonging to the audience, at the end of the program I concluded in favor of the *ad hominem* arguments (that I also "recognized" as fallacies) used by the journalists and so to the hypothesis which argued against Don Gelmini. Hence, I considered the data and gossip embedded in those fallacious reasoning describing the "immoral" and "judiciary" past of Don Gelmini more relevant than the ones which described the bad past of the witnesses. I was aware of being in the midst of a riddle of hypotheses generated by various arguments, and of course this was probably not the case of the average viewer, who may not have been trained in logic and critical thinking, but it is easy to see that this fact does not actually affect the rhetorical success or failure of arguments in real-time contexts, like it was occurring in my case: we are all in an "epistemic bubble" as listeners – compelled to think we know things even if we do not know them, a bubble that in this case forces you to quickly evaluate and pick up what you consider the best choice. I would like to put forward the idea that, at least in this case, the evaluation of the *ad hominem* arguments has to be seen as the fruit of an abductive appraisal, and that this abductive process is not rare in argumentative settings.

An analogy to the situation of trials in the common law tradition, as described by [Woods, 2009], can be of help. Like in the case of the judge in the trial, in our case the audience (and myself, as part of the audience)[51] was basically faced with the *circumstantial* evidence carried by the two clusters of *ad hominen* arguments, that is, faced with evidence from which a *fact* can be reasonably inferred but not directly proven.

In a situation of lack of information and knowledge, and so of constitutive "ignorance", abduction is usually the best cognitive tool human beings can adopt to relatively quickly reach explanatory, non-explanatory, and instrumental hypotheses/conjectures. Moreover, it is noteworthy that evidence – embedded in the *ad hominen* arguments – concerning the "past" (supposed) reprehensible behavior of the priest and of the witnesses were far from being reliable, probably chosen *ad hoc*, and deceptively supplied, that is, so to say, highly circumstantial for the judge/audience (and for me, in this case). I had to base my process of abductive evaluation regarding the fallacious dialectics between the two groups of journalists on a kind of sub-abductive process of *filtration* of the evidence provided, choosing what seemed to me the most reliable evidence in a more or less intuitive way, then I performed an abductive appraisal of all the data. The filtration strategy is of course

[51] Furthermore, like the jury in trials, an audience is on average composed of individuals who are not experts able to "overt calibration of performance to criteria", but instead ordinary – reasonable and untutored – people reasoning in the way ordinary people reason, a kind of "intuitive and unreflective reasoning" [Woods, 2009].

abductively performed guided by various "reasons", the conceptual ones, for example being based on conceptual judgments of credibility. However, these reasons were intertwined with more or less other reasons as more or less conscious emotional reactions, based on various feelings triggered by the entire distributed visual and auditory interplay between the audience and the scene, in itself full of body gestures, voices, and images (also variably and smartly mediated by the director of the program and the cameramen).[52]

In summary, I was able to abductively make a selection (selective abduction) between the two non-rival and incommensurable hypothetical narratives about the priest,[53] forming a kind of explanatory theory of them: the guessed – and quickly accepted without further testing[54]– theory of what was happening in that dialectics further implied the hypothesis for guilt with respect to the priest. That is, the *ad hominen* of the journalists that were speaking about the priest's past (he was for example convicted for four years because of bankruptcy fraud and some acceptable evidence – data, trials documents – was immediately after provided by the staff of the television program) appeared to me convincing, that is, no more negatively biased, but a plausible acceptable argument. Was it still a fallacy from my own actual eco-cognitive perspective? I do not think so: it still was and is a fallacy only from a special subset of my own eco-cognitive perspective, the intellectual/academic one! The "military" nature of the above interplay between contrasting *ad hominem* arguments is patent. Indeed, they were armed linguistic forces involving "military machines for guessing hypotheses" clearly aiming at forming an opinion in my mind (and in that of the audience) which I reached through an abductive appraisal, quickly able to explain one of the two narratives as more plausible. In the meantime I became part of the wide coalition of the individuals who strongly suspect Don Gelmini is guilty and that can potentially be engaged in further "armed" gossiping.[55]

Four additional remarks have to be added.

1. In special contexts where the so-called fallacies and various kinds of hypothetical reasoning are at play, both at the rhetorical and the dialectic level, the assessment of them can be established in a more general way, beyond specific cases.[56] An example is the case of a fallacy embedding patently false empirical data, which can easily be recognized as false at least by the standard intended audience; another example is when a fallacy is structured, from the argumentative point of view, in a way that renders it impossible to address the

[52] Along these lines, I might add that the also the journalists fallaciously discussing the case were concerned with the accounts they could trust and certainly emotions played a role in their inferences.

[53] Of course I could have avoided the choice, privileging indifference, thus stopping any abductive engagement.

[54] This means that the cycle of testing I have illustrated in chapter one, in the case of the ST-model of abduction (cf. section 1.3), is in this case absent.

[55] A vivid example of the aggressive military use of language is the so-called "poisoning the well", "[...] a tactic to silence an opponent, violating her right to put forward arguments on an issue both parties have agreed to discuss at the confrontation stage of a critical discussion" [Walton, 2006, p. 273]. "Poisoning the well" is often considered as a species of *ad hominem* fallacy.

[56] Examples are provided in [Tindale, 2005].

intended audience and in these cases the fallacy can be referred to as "always committed".

2. Not only abduction, but also other kinds of (supposed to be) fallacious argumentation can be further employed to evaluate arguments in dialectic situations like the ones I have quoted above, such as for instance *ad hominem, ad populum, ad ignorantiam*, etc., but also hasty generalization and deductive schemes.
3. The success of a so-called fallacy and of the so-called fallacious hypothetical arguments can also be seen from the perspective of the arguer in so far as she is able to guess an accurate abductive assessment of her actual or possible audience's character. From this perspective an argument is put forward and "shaped" according to an abductive hypothesis about the audience, which the arguer guesses on the basis of available data (internal, already stored in memory, external – useful cues derived from audience features and suitably picked up, and other intentionally sought information). Misjudging the audience would jeopardize the efficacy of the argument, which would consequently be a simple error of reasoning/argumentation. Obviously also in assessment made by the audience inferences which are less complicated than abduction can be exploited, like hasty generalization, etc.
4. As is clearly shown by the example of the priest, arguments are not only distributed, as I have contended in the subsection 7.8.1 above, but they are also embedded, nested, and intertwined in self-sustaining clusters, which individuate peculiar global "military" effects.

7.8.4 Narrative Abduction

Sentential abduction (cf. chapter two, section 1.4) is clearly active at the level of everyday natural language, when we generate more or less creative narratives. As in science, restoring coherence is sometimes important in *narratives* as well, many "stories" have to be believable and devoid of emotional and story world inconsistencies, and characters' unmotivated goals.

MINSTREL is a classical computer program that tells creative stories about King Arthur and his knights [Turner, 1994]. To be good, their stories must be consistent at many levels, and the characters and the world should act consistently and predictably. After having created a scene in which Galahad kills a dragon and observing that there is not explanation of why Galahad killed the dragon – this unexplained fact leads to a story inconsistency – MINSTREL adds new story scenes to correct the problem using knowledge about knights and transform and adapt strategies. The program has nine plans for maintaining story consistency and of course it is also endowed with routines which check inconsistencies. In this case inconsistencies constitute the driving force of the narrative creative abduction. Therefore storytelling can be considered as a kind of narrative abductive problem solving where the role of inconsistencies is central.

The heuristics that make MINSTREL creative are called Trasform-Recall-Adapt and perform creativity as an integrated process of search and adaptation; moreover, they are integrated with a kind of case-based (or "analogical") reasoning that 1) recalls a paste problem with the same features and its associated solution, 2) adapts and transform – generalization, specialization, mutation, recombination – the past narrative solution to the current problem, and then 3) assesses the result. One particularly widespread way of creating new story themes is to use generalizations and specializations on existing story themes.

Not only about King Arthur, there are many kinds of narratives, for instance scientists do not use only mathematics, diagrams, and experiments but also stories. As illustrated in chapter two, scientists use "experimental" narratives in a constructive and hypothesizing role. Moreover, they consent to construct and reconstruct experience, and to distribute it across a social communicative network of negotiations among the different scientists by means of the so-called construals.

In all the narratives, and especially in the narratives of detection, the problem of prompting abductive explanations of actions is widespread [Eco and Sebeok, 1983; Oatley, 1996]. I have quoted in chapter two (subsection 1.5.2) the passage from Peirce about the fact that all sensations participate in the nature of a unifying hypothesis, that is, in abduction, in the case of emotions too: in their cognitive theory of emotions largely based on Peirce's intuitions [Oatley and Johnson-Laird, 2002] are able to explain how the reader, in front of narratives, feels emotions as abductions.

[Gabbay and Woods, 2005, pp. 93-96] address the problem of narrative abduction by illustrating what they call "the rational model" of explanation, already stressed by the neo-Kantian, Dilthey, and, later on Collingwood, who argued against the scientific D-N model as appropriate in the case of *Geisteswissenschaften*: to explain an event we do not have to find a general law but rather specify conditions – reasons embedded in narratives and stories, not causes – that make us able to explain it.[57] They also think it is unlikely to find general laws in history and human sciences, a position further stressed in some recent hermeneutic developments [Ricoeur, 1977]. Of course in these cases of narrative abductive hypotheses the problem of their evaluation and comparison on the basis of observational data is open. The authors also doubt these kinds of explanations can always involve genuine conjectures and so they should not be considered abductive. A related problem concerns the role of abduction in interpretation of texts and in interpretation of interactive discourse for example in extracting/compreheding hidden meanings. Enthymeme resolution (also called the problem of missing-premises, as a possible kind of abductive inference, is treated in [Gabbay and Woods, 2005, pp. 289–312]), together with an illustration of the the role of abduction in translations, as related to the Quinean problem of the indeterminacy of translation.[58]

[57] Analogous considerations are provided in the case of the so-called functional (for example teleological and design) explanations, widespread in biology, explanations still supposed to be irreducible to the D-N model.

[58] On the mechanization of textual interpretation and the role of abduction cf. [Bruza *et al.*, 2006]. The auhtors introduce semantic space to the model-based reasoning and abduction community and illustrate its potential for principled, operational abduction by semi-automatically replicating the fish oil discovery in medical text.

Summary

In this chapter I have described the relation between induction and abduction in the light of an agent-based framework to the aim of clarifying their fallacious character and the role of their related ideal systems (logical and computational). From this point of view I have analyzed some inductive and abductive ways of reasoning that in the light of classical and informal logic are defined as fallacious, showing the fact they can nevertheless realize a kind of strategic "rationality". Agent-based fallacies and agent-based hypothetical fallacious reasoning can be redefined and considered as a good way of reasoning, at the same time involving social and moral/violent effects. After having illustrated the distinction between internal and external representations in the tradition of both logic programming and distributed reasoning, I have described some important aspects of manipulative abduction, which can be re-interpreted as a form of practical reasoning a better understanding of which can furnish a description of human beings as *hybrid reasoners* in so far they are users of ideal (logical) and computational agents, for example devoted to performing sophisticated inductions and abductions.

I argue that ideal logical agents have to be seen as "demonstrative environments", which in turn are suitable tools for the creation of logical/deductive models of abductive reasoning itself. They are the fruit of cognitive externalizations in objective logical systems, endowed with symbolic, abstract and rigorous cognitive features, which are also potentially exploitable at the computational level. Moreover, awareness of the importance of ideal and logical demonstrative environments allows better understanding of how heuristic procedures, which involve abductive steps, as Hintikka notes – are active in the deductive reasoning performed by real human agents. My philosophical and cognitive approach to abduction in the light of logical agents, depicted in this chapter, aims at revealing a new understanding of its fallacious role. My approach provides a very general model that conveys the multi-faceted nature of the so-called fallacies as eco-logical good or bad ways of reasoning, which are immediately intertwined with the basic human activity of coalition enforcement, exerting both moral and violent effects.

Abductive cognition is so pervasive, so much a part of human (and animal) life, that it is not easy to imagine how to live and develop without it. The activity of guessing hypotheses is a complicated cognitive mechanism, intertwined with the conjectural power assigned to fallacies, seen as being employed by people to perform various tasks, including those that are related to defensive, moral and violent commitments, as described at the end of this chapter. Drawing attention to this issue extends the analysis of abductive cognition. Here we return to the general aim of this book, to build an extended perspective on abductive cognition. A new understanding of it can be gained by looking at the traditions of physics and mathematics. Therefore, the next step in the process is to identify how a "physics of abduction" can be constructed: the morphodynamical results illustrated in the following chapter aim at reaching this target.

Chapter 8
Morphodynamical Abduction
Causation of Hypotheses by Attractors Dynamics

This last chapter is intended to clarify some central methodological aspects of morphodynamical abduction as regards *dynamical systems* and the *catastrophe theory*. Some problems arise in the classical computational approach to cognition in describing the most interesting abductive issues. A cognitive process (and thus abduction) is described by the manipulation of internal semiotic representations of external world. This view assumes a discrete set of representations fixed in discrete time jumps and, because of its *functionalist* character, cannot render the *embodied* dimension of cognition and the issue of anticipation and *causation* of a new hypothesis adequately. An integration of the traditional computational view with some ideas developed within the so-called dynamical approach and catastrophe theory can lead to important insights. What is the role of abduction in the dynamical system approach? What is the role of the "salient/pregnant" dichotomy with respect to abduction? What is embodied cognition from the point of view of its "physics"?

The first part of the chapter will reconsider and reevaluate some aspects of the tradition of dynamic systems and analyze the concept of abduction as a form of hypothetical reasoning that clarifies the process of anticipation and the concept of "attractor", also taking advantage of the results of the phenomenological tradition, which also introduces the important concept of "adumbration" (see also chapter four, subsection 4.7.2. This analysis permits a description of the abductive generation of new hypotheses in terms of a catastrophic rearrangement of the parameters responsible for the behavior of the system. As I have explained in the previous more methodological chapters of this book, I argue that abduction has to be observed in the whole distributed interplay between internal and external representations: the new approach laid out in this chapter is consistent with this perspective.

In the second part I will introduce issues derived from Thom's catastrophe theory. Some of the illustrated results can be seen in the light of chapter six, devoted to affordances and abduction, and chapter five, which stressed the relationships between instincts, abduction and hardwired aspects. Pregnance and salience are key concepts which provide further information about the instinctual/plastic abduction dichotomy. As they acquire their meaning in a very naturalistic intellectual framework, they are especially useful to propose a unified perspective on abductive

L. Magnani: Abductive Cognition, COSMOS 3, pp. 417–453.
springerlink.com

cognitive/psychic processes, seen as fundamental physico-biological events. The consequence for abduction is that we have to further stress the fact it is useful to see it as a semiophysical process, endowed with a profound eco-cognitive significance. For example, a pregnant stimulus can be termed "highly diagnostic", like in the case of the hardwired pregnancy occurring when our Peircean chicken (cf. chapter five) is looking for food and promptly reacts when perceiving it. The pregnance affects an organism, and the related abductive response is promptly triggered. Pregnances are genetically transmitted but can also be actively created for example through learning and high cognitive capacities, so that the door is opened to the formation of multiple forms of abductive intelligence. The fact that saliences and pregnances have to be understood by taking their semiotic character into account, which is also linked to various fundamental functions of natural human languages, is further illustrated.

The third part of the paper (section 8.6) is worth of further clarification. It plays the role of a sort of appendix which illustrates the so-called coalition enforcement hypothesis, which sees humans as self-domesticated animals engaged in the continuous activity of building morality, incorporating at the same time punishment policies. I think this hypothesis can be simply and well illustrated in the cognitive niche framework I have described in chapter six. Its main speculative value stresses the role in human and animal groups of more or less stable stages of cooperation through morality and related inexorable violence: morality and violence are seen as strictly intertwined with social and institutional aspects, implicit in the activity of cognitive niche construction. In chapter six I have tried to show that in the activity of niche construction *hypothetical thinking* (and so *abduction*) is fundamental; in the previous chapter I have tried to illustrate that hypothetical thinking is often embedded in various kinds of the so-called fallacious reasoning (which in turn constitutes a relevant part of the linguistic cognitive niches where human beings are embedded). To better grasp its philosophical status I think it is intriguing to stress that hypothetical cognition might be favored for reasons of – so to say – "social epistemology" and moral reasoning.[1] Indeed, in evolution, coalition enforcement works through the building of social cognitive niches seen as new ways of diverse human adaptation, where guessing hypotheses is central and where guessing hypotheses is occurring as it can, depending on the cognitive/moral options human beings adopt. Basically the coalition enforcement framework refers to cooperation between related and unrelated human animals to produce significant mutual benefit that exceeds costs and is thereby potentially adaptive for the cooperators.

The hypothesis allows us to see altruism in a new light as related both to Simon's idea of docility as socializability and to the violent behavior needed to defend and enforce group coalitions. To gain cooperation – to potentially control and manipulate free-riders inside the group and any threatening individuals from alien groups – human coalitions and most gregarious animal groups have first of all to take

[1] I plan to address the problem of the relationships between morality, violence, and knowledge in a forthcoming book. In this book I just consider some aspects of this important ethical and philosophical issue, which are particularly relevant to the idea of the cognitive niche (this chapter) and to certain roles of fallacies and abduction (chapter seven).

care – morally and more or less altruistically – of the individuals who cooperate. Abduction is still at stake, the direct consequence of coalition enforcement is the central role and development of cultural heritage (morality and sense of guilt included). The long-lived and abstract human sense of guilt represents a psychological adaptation to abductively anticipate an appraisal of a moral situation to avoid becoming a target of coalitional enforcement. I also stress the moral and violent effect played by human natural languages, a theme already explored in the previous chapter, devoted to the analysis of the relationships between language, logic, fallacies, and abduction. This "military" nature of linguistic communication is intrinsically "moral" (protecting the group by obeying shared norms), and at the same time "violent" (for example, harming or mobbing others – members of not of the group – still to protecting the group itself).

8.1 Abduction as Embodied Cognition

As we have seen in the previous chapters extra-theoretical aspects and manipulations of "external" objects in reasoning are very important. Paying attention to the *perceptual* and *manipulative* dimension of cognition, Peirce reminded us that real cognitive systems are not "isolated" and autonomous entities. Cognition is *embodied* and the interactions between mind and external environment are its central aspects. Knowledge is possible only by means of a constant exchange of information in a complex *distributed* system that crosses the boundary between the human and non-human agent and the surrounding environment.

As I have stressed in the previous chapters, traditional accounts in cognitive science in fact describe reasoning processes in terms of the so-called "computational" approach. The main idea is that cognition is the operation of a special mental "computer", located in the brain. Sensory organs deliver representations of the environment to that computer. Representations are *static* structures of *discrete* symbols; cognitive operations are transformations from one static symbol structure to the next; these transformations are discrete, effectively *instantaneous* and sequential (see [Port and van Gelder, 1995]). This approach, because of its *functionalist* character, cannot render the *embodied* dimension of cognition adequately.[2]

Another central problem is *time*. Cognitive processes, being complex structures in natural systems, unfold continuously and simultaneously in real time, while computational models specify a discrete sequence of states in arbitrary time periods. The notion of time is present in any intelligent activity (and so it is fundamental in abductive reasoning). Time is profoundly involved in human perception and understanding of the world. Things appear to remain in a particular state for a period of

[2] It is worth quoting Jerry Fodor's skeptical attitude with respect to abduction [Fodor, 2000]. Fodor's skepticism is basically related to the contention that Turing machines cannot duplicate humans' ability to perform abduction as inference to the best explanation, and it is also linked to his pessimism regarding the prospects of cognitive science. Criticisms about this thesis are illustrated in [Pinker, 2005]; [Rellihan, 2009] proposes a solution of this "riddle of abduction" by augmenting the computational theory of the mind to allow for non-computational processes, such as those indicated by classical associationists and contemporary connectionists.

time until a certain event happens, so we can say that time is central to reasoning about change and action (as we have seen in the case of abduction related to sense activity). The first task is to consider the different states or conditions of a thing and define how they are related; we can say that we know about a *causal* relationship when we can use our knowledge to define how the related system evolves. Temporal reference is highly integrated in scientific knowledge but this is also the case in human common sense reasoning, both verbal and model-based/manipulative. Humans are involved in managing time when coordinating with the environment. For example memories and many mental models seem to be organized around time – past events come to mind when reconstructed with the help of a temporal framework (chronologically). An alternative approach, then, should be based on the main argument that cognitive processes happen and unfold continuously over time and are constantly related to the external world.[3]

8.1.1 Discreteness and Cognition: Imitation vs. Intelligibility

Related to the problem of illustrating abduction taking advantage of the analysis of cognitive systems in dynamical terms (so abandoning the merely computational-representational framework) is the dichotomy between discreteness and continuity. It addresses the other related dichotomy between imitation (as an effect of the computational representation) and intelligibility (as the fruit of scientific knowledge). This subsection is devoted to better describe this important issue, at least from a general theoretical point of view.

As [Longo, 2009] illustrates, the digital machine (a discrete state machine) is first of all an alphabetic machine, made possible thanks to the human evolution to alphabetic natural language (of course we know it is also a logical and formal machine). This fact is at the root of the extremely powerful "discretization of knowledge" generated by the computational turn. At a linguistic level the alphabet already divides "continuous" language into insignificant atoms, forming letters. The letters

[3] From this point of view continuous quantities of physical processes could, at best, be approximated in digital terms since "[...] processes based on continuous quantities exceed the limits of the orthodox notion of computation" based on Church's thesis [Cordeschi and Frixione, 2007, p. 45]. These authors also provide an analysis of Marr's position [Marr, 1982], which tries to reconcile the information processing point of view and the ecological approach (on this approach cf. this book, chapter six). However, following [Petitot and Smith, 1990], we can say that Marr – coherently with an orthodox computational perspective – clarifies the situation in this respect: what Gibson considered as the extraction of invariants from the environment, can still be understood as a form of information processing based on calculation of symbolic mental representations explicable in computational terms. The algorithms employed are determined by objective properties of the environment so that their syntactic status is reconciled with ecological semantics. This perspective is coherent with the idea of logic programs seen as agents endowed with deductive databases considered as *theory presentations* from which logical consequences are derived, both in order to *internally* solve problems with the help of *theoretical sentences* and to assimilate new information from the *external* world of observations (*observational sentences*) (cf. chapter six, subsection 7.5.1). We have to remember that the part of the knowledge base, which includes observational sentences and theoretical sentences is called *world model*, but it is still considered a completely syntactic concept.

are in themselves without meaning but the meaning of letters and of their syntactical combinations immediately arises in competent humans who sense and manage them.

Let us illustrate the consequences of this discretization in the case of the mental representations of "concepts". As I have already illustrated in chapter three (subsection 3.4.2), we can hypothesize that an isomorphism exists between the mental mechanisms (where of course "phonemes" play a central role), which ensure the stability of a concept, and the physical and material mechanisms which ensure the stability of the actual object represented by the concept.[4] The *semantic depth* of a concept would be characterized by the time taken by the mental mechanisms of analysis to reduce this concept to the representative sign/word. The more complex the concept is, the more its stability needs regulatory mechanisms and the greater its semantic depth becomes, as obviously happens in the case of nouns, which refer to a substance or an actual object.

The philosophical and epistemological success of this tradition of discretization, already at work in western natural languages, has been huge in philosophical and logical thinking, from Democritus to Descartes and from the modern XIX century axiomatics to the computational turn. However, we are still faced with the fact that the elementary is actually very complex, as clearly testified by research into the dynamical systems theory, quantum physics, and biology. For example the cell, as the elementary component of living matter, has to be considered not only from the point of view of its components, but also as a whole. We are similarly confronted with the fact that digital machines are discrete machines (cf. chapter three, sections 3.1, 3.2, and 3.7), which unfortunately have serious expressive limits when considered as instruments of intelligibility regarding complex objects/systems like human cognitive systems.

The following few paragraphs aim to illustrate the fundamental epistemological gap between the idea of knowledge as simply mimetic *imitation*, embedded in digital machine "modeling capacity", and the idea of knowledge as production of *rational intelligibility*, this distinction being a core problem of the dynamical approach. As I have already illustrated in chapter three, Turing distinguished between

[4] The cognitive importance in humans of the internalization of phonemes has been recently stressed by some cognitive paleoanthropological research [Coolidge and Wynn, 2005]. The so called "enhanced working memory" (EWM) (and its executive functions) can be plausibly traced back about 30.000 years in hominids; the authors show that it seemed to coevolve with the birth of a *phonological storage capacity* along with consequent language and other modern reasoning abilities such as planning, problem solving/algorithm manipulation, analogy, modeling, holding inner representations, tool-use, and tool-making. In particular, an increase in phonological storage could also have aided cross-modal thinking (and so hypothetical cognition) and the social tasks caused by the need for coalition enforcement: "[...] enhanced phonological storage may have freed language from the laconic and its confinement to present tense and simple imperatives to rapidly-spoken speech and the use of future tense – the linking of past, present, and future, and the use of the subjunctive [...]. Although real enemy's actions might be anticipated, imaginary enemies could be envisioned and other intangible terrors could be given life. Great anxieties could arise with novel vistas (e.g. the meaning of life, thoughts of death, life after death, etc.)" (cit., p. 22) (on the concept of coalition enforcement and its moral and violent aspects cf. chapter eight, section 8.6).

the simple *imitation* power of machines and the *modeling* power of mathematics: "The double pendulum, a perfectly deterministic machine (it is only determined by two equations!) is sensitive to minor variations, below the threshold of observability: it is a typical chaotic deterministic system" [Longo, 2009]. Longo adds that this is a system sensitive to initial conditions. Its properties can be described with the mathematical rigor of deterministic chaos, where determination does not imply predictability, as in this system a process "[...] does not follow the same 'trajectory' despite being iterated with the same initial conditions, within the limit of physical measures" (*ibid.*). Longo stresses that, in the case of computational programs that simulate such a system, after having observed previous digital simulations, when we click on restart, re-launching the pendulum with the same initial data used before, it takes the same trajectory, already seen, once again. Unfortunately, this cannot be done with a "real" pendulum, as once re-initialized the pendulum never takes the same trajectory:

> And this because of "principles" which are inherent to Physics (modern Physics): physical measurement is always an interval and the (inevitable) variation, below the threshold of measurement, suffices to very quickly produce a different evolution. The analysis of the equations within the *continuum* leads to an understanding of this random aspect of chaos, whereas computational imitation makes it disappear completely, by the discrete nature of its data types. Only tricks and stratagems (pseudo-synchronization with distant watches, pseudo-random generators introduced ad hoc) can imitate, but not model, the physical phenomenon [Longo, 2009].

Only "mathematical" models can express the structure of physical causality regarding the system at play, digital simulation only *resembles* causality. A computer pseudo-generator of random numbers, is deterministic, but not chaotic, the 1s and 0s given by the machine are only seemingly distributed at *random*, unlike for example the real toss of a coin, which on the contrary is completely unpredictable. In the case of digital simulation, if you click restart while leaving all the machine parameters identical the random series generated will be identical to the previous one, thus realizing a process which is Laplacian and predictable. The problem is that the basis of digital data has a *discrete* topology. Physical measurement is always an interval that is well represented by continuous mathematics. In chaotic deterministic systems, a fluctuation below the interval of measurement generates greatly different evolutions of the system, so that we can think of unpredictability as resulting from the "friction between mathematics and the world" (cit.).

Only mathematical equations – which are certainly computable and thus endowed with predictive power – can appropriately model a deterministic chaotic system or a non-linear system while at the same time being unable to predict the evolution of the physical process at hand. Why? Because the measurement which enables us to pass from the physical process to the mathematical system is an interval that cannot provide an exact value. This does not prevent the fact that a mathematical physical theory preserves a *qualitative* epistemological force, in so far as it describes the structure of the physical causality of the system, even if it lacks the quantitative/predictive power. Of course, the gap between simulation and intelligibility is much larger in the case of virtual simulation of processes in living organisms, where variability is central. A simple example is given by the contrast between digital memory, which can reproduce experience exactly, pixel by pixel, and human memory, which is strongly abductive/interpretive.

In summary, digital simulation – even if extremely useful in biological science as well – is certainly epistemologically biased by its merely mimetic character. Furthermore, it is noteworthy that the use of mathematical invariance in biology is also very puzzling, notwithstanding for example the attempts by René Thom to grasp some morphogenesis phenomena using the singularity theory.[5] According to [Bailly and Longo, 2006; Bailly and Longo, 2009], to overcome these difficulties current mathematical and physical modeling should deal with the problem of making the phylogenetic and ontogenetic trajectories of living matter intelligible or "generic", that is, seeing them as evolution possibilities referring to temporal and interspecific "invariants", and not simply as specific geodesics (which concern individuals). The new notion of "extended critical situation" is proposed as the most appropriate candidate to fulfil the task. It is now necessary to illustrate some ideas about dynamical systems that are strictly related to what I call morphodynamical abduction.

8.1.2 Dynamical Systems

There is a theory able to describe the behavior of natural systems in time: the *dynamical systems theory* (DST), a widely used, powerful, and successful descriptive framework in natural science.[6] We can use the mathematical tools of dynamics to study cognition by thinking of a cognitive system not just as a computer but as a dynamical system, made of brain, body and external environment, mutually and simultaneously influencing and coevolving. To clarify some hidden processes in creative abductive reasoning we can try to integrate the dynamical and computational perspectives.[7]

---

[5] In section 8.3 below, I illustrate some aspects of Thom's catastrophe theory which are also very interesting from the perspective of abductive cognition.

[6] On the relationships, complementarity, and compatibility in the computationalism vs. dynamicism conflict, situated and embodied cognition, connectionism, and behavioral and evolutionary robotics cf. [Cordeschi and Frixione, 2007]. It is noteworthy that [Bechtel, 1998, p. 312] sees the role of dynamical explanations as fulfilling the ecological requirement imposed by Gibson and other eco-psychologists for describing cognitive systems (cf. this book, chapter six): "DST theorists [...] are also neo-Gibsonian".

[7] The recently edited book [Loula *et al.*, 2006] illustrates various ways of taking advantage of the dynamical approach (but also of research in embodied and distributed cognition, phenomenology, ethology, and semiotics) to build artificial cognition systems. Useful insights on agency (and its realization in AI) due to the dynamical system approach are illustrated in [Sørensen and Ziemke, 2007], who also refer to the other main theories belonging to this tradition of research: autonomous systems organizationally closed (autopoiesis) [Maturana and Varela, 1980; Varela *et al.*, 1991; Varela, 1979; Varela, 1997], theories of adaptive systems [Godfrey-Smith, 1998; Godfrey-Smith, 2002; Ziemke, 2007; Sterelny, 2004], and situated and embodied cognition [Clancey, 1997; Clark, 1997; Steels, 1994; Brooks, 1999].

The attribute "dynamical" refers to the way a system evolves (or behaves) in time. A "system" is a collection of correlated parts perceived as a single entity. Thus, a dynamical system is a system changing in time. It is the state of the system that changes, that is, the overall look of the system in a certain instant. A dynamical system is defined by the space of all possible states it can assume (state space), and by a rule (the dynamics), which determines the system evolution in a given instant with respect to any initial state [Hirsch, 1984]. It is possible to study the system behavior by analyzing change in its states.

In the history of sciences, one of the most important dynamical system is the Solar System. The Sun, the planets, and the satellites are the parts making up the system; the configurations and speeds they can assume make up the states. The main problem is to find the right dynamics with which it is possible to describe the evolution of the system and to make predictions (for example in the case of eclipses). Early descriptions (Ptolemy, Copernicus, Galileo, Brahe, Kepler) used mathematical models to study astronomical dynamics; then, beginning with Newton and Leibniz, differential equations became the main mathematical means used in describing systems under forces responsible for their evolution. These equations (in which both function of variables and their derivatives appear) are able to specify how a system evolves, in any moment, from a given state (with an accentuated sensibility to initial conditions). For instance, the differential equation

$$m\ddot{x} + kx = 0$$

describes the way an object with mass, hung from a spring, rebounds, defining the instantaneous acceleration $\ddot{x}$ in function of its position x; k and m are the constants relating to the tension of the spring and to the mass.

If a system can be described dynamically, it means it has $2n$ characteristics (position and velocity of each of the n particles composing the system) evolving simultaneously in time. These characteristics can be measured, at any given instant, and associated to a real number. The overall state of the system can then be thought of as an ordered set made of $2n$ real numbers, and the state space as isomorphic to a space of real numbers, the dimensions of which correspond to the different system characteristics (the *phase space*). The evolution of the system in time corresponds to a sequence of points, a *trajectory*, in the phase space. This sequence can usually be described mathematically as a function of time, considered an independent variable, giving a solution to the differential equations. A definition of "dynamical system" can then be a state-determined system with a numerical phase space and an evolution rule which can define trajectories in the space.

8.1.3 Attractors

Some dynamical systems are very complex and behave non-linearly and erratically, jumping from a point in the state space to another very different point in a brief time (like in the case of the states of the atmosphere). However, notwithstanding these sudden changes, a dynamical system has a series of states into which a system repeatedly falls, the so-called attractors, which are stationary (cf. Figure 8.1). A system can have a lot of attractors, contemplating more than a single stable state. The transition from an attractor to another can be viewed as analogous to a sort of *phase transition* (as in the case of water that that becomes ice: in dynamical systems small local changes could lead the system to a qualitatively different state).

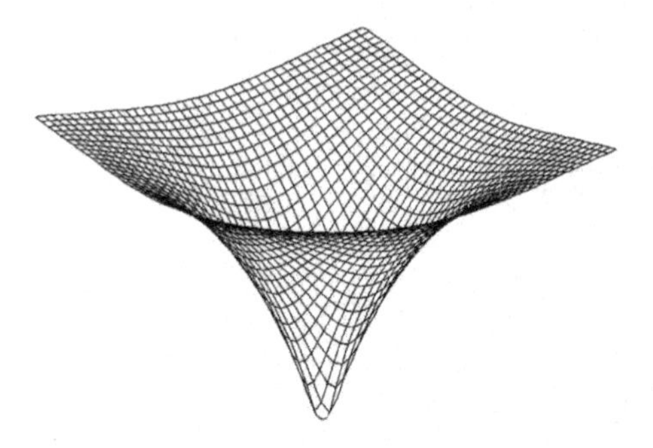

Fig. 8.1 An intuitive representation of the idea underlying the concept of attractor is useful in clarifying the idea. Think of a marble rolling on a plane as far as it falls in a hollow like that in the picture. The marble will rotate inside it; then it will reach the resting position, at the bottom. Attractor is the stationary point corresponding to that position.

In subsection 8.2.1 I will suggest a description of abduction in terms of attractors: creative abduction can be seen as the formation of a new attractor, so that in the case of cognitive systems attractors will represent the tendencies to produce hypothetical (for example interpretive/explanatory) models. As for now it is interesting to understand how attractors evolve and change as the system parameters gradually change. Dynamical systems can constitute the background in the dynamical approach to cognitive science: the difference between dynamical and computational systems is notable. Computational systems have states made by configurations of symbols and evolution rules, which specify the transformations between two different symbolic configurations (cf. below subsection 8.2). Numerical phase spaces possess a metric by which it is possible to determine the distance among points. If the phase space is dense enough, then it is possible to find a set of other points between two points, describing the state of the system at any given instant (the notion of time used to describe the system has the same properties of the *real* time).

This kind of description is not possible from a classical computational point of view: in this last case there is not a natural notion which can define the distance between two states of the system, and the concept of "time" is just a synonym of "order" $(t_1, t_2, \ldots)$. Then, it is not possible to specify the direction and the speed in the system evolution. But it is important to determine it, because real processes (including cognitive processes), in real world, occur in real time at a certain speed. Time is essential.

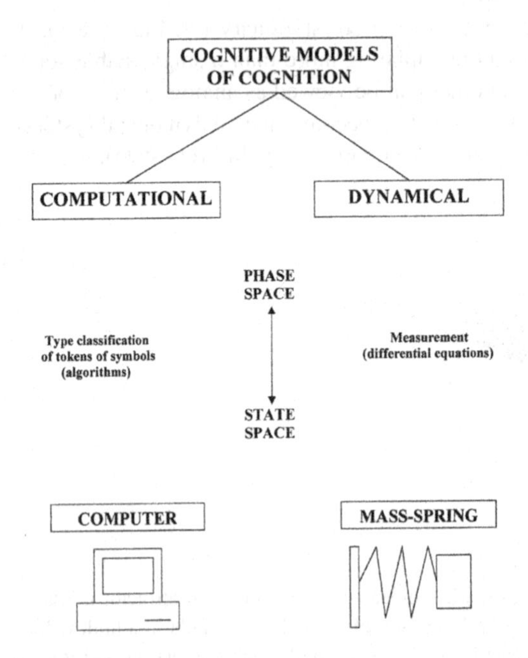

Fig. 8.2 Computational and dynamical modeling of cognition

8.1.4 Cognitive Processes as Super-Representational

Usually cognitive scientists distinguish cognitive processes from other natural processes by assuming they are dependent on *knowledge*, something that can be stored and used. The computational approach is centered on the idea that knowledge is in some way "represented", and that cognitive processes are nothing but manipulations of representations. As already illustrated above, mental representations are usually conceived as symbolic, in the sense that they have a combinatorial syntax and compositional semantics, that is, mental processes are "structure-sensitive" because they are defined by the implemented symbolic structure [Fodor and Pylyshyn, 1988]. Thus, it seems, there is a problem: how is it possible for a dynamical model, which does not require the notion of "representation", to describe cognitive processes, since they depend on knowledge? The fact that these models are not founded on manipulation of symbolic structures does not mean they do not admit them at all. Cognitive processes can be considered *super-representational*, they not only involve some kind of symbolic representation, but much more besides.

They are not so simple that we can describe them without the notion of representation, in fact, they are so complex that the simple notion of representation can only partly describe their functioning. In any case, a representational state is associable to many aspects of dynamical models: states, attractors, trajectories and bifurcations (sudden qualitative transformations). In a computational model the rules governing the system are defined on the basis of the entities with a representational status, whereas in a dynamical model rules are defined by numerical states. This means that a dynamical system can be "representational" even if its evolution rules are not specified by representations (for example, "representations" of objects stored in memory are nothing but configurations of attractors in the phase space of the system). As such, a dynamical system is also able to store knowledge which influences its behavior.[8]

8.1.5 Embodied Cognition and Qualitative Modeling

We have already seen that knowledge also involves pragmatic and "embodied" aspects. Advocates of computational approach claim that the cognitive system is made by the mind, considered a sort of control unit embedded in a body in turn embedded in an external environment. So the cognitive system interacts with the external world through the body: there are transducers which translate the physical interaction between body and environment in states defined by symbolic representations, medium of cognitive processes. What is important is that since the system works through representations, it is possible to consider it as autonomous, isolated with respect to the body and the external world. Its main function then is simply that of transforming representations (inputs) into other representations (outputs).[9]

On the other hand, dynamical systems are defined by multiple aspects, which simultaneously evolve in a temporal continuum and affect each other. Since nervous system, body and environment are continuously evolving, and affect each other, the cognitive system cannot be constituted simply by an isolated "brain": it is a single system in which the three subsystems are unified. The system does not interact cyclically with body and environment by means of symbolical inputs and outputs: internal and external processes are coupled in a way so that they interact constantly. Moreover, the process is not sequential, because all the aspects of the system constantly change simultaneously.

Surely there are many cognitive performances that exhibit a sequential behavior: for example the pronunciation of a sentence. But this "sequential aspect" is nothing but something that emerges in time, underlying the overall trajectory of the system, in which the rules of evolution do not specify a sequential change, but a mutual and simultaneous coevolution. Hence it is natural to think of the system evolution in terms of its movement inside the state space. Consequently the phase space is described numerically and it is possible to apply the notion of distance: an interesting point in the dynamical theory is just the fact that it permits to "geometrically" conceptualize cognitive processes. The distinctive character of a cognitive process developing in time depends on the disposition in space of the states through which it passes.

Notwithstanding the difficulties in fulfilling the main purpose of the dynamical approach (that is to build a quantitative model of cognitive processes), it is nevertheless interesting to develop mathematical models which express qualitatively

[8] On the controversial epistemological status of representations in the tradition of dynamicism cf. the observations given below in subsection 8.5.2.

[9] The idea is directly derived from the notion of "functionalism".

similar (and not quantitatively exact) behaviors of the studied phenomena. Qualitative modeling aims at building qualitative causal descriptions of the studied systems. A qualitative model represents the structure of a system, and gives a qualitative description of its overall behavior. Some dynamical properties, such as catastrophic jumps, oscillations, and chaotic behaviors due to variations in control parameters, can be examined without knowing the exact equation governing the system evolution, also thanks to the "geometrical" properties exhibited by this kind of modeling. The description of cognitive systems through attractors accords with the "compositional" aspects of qualitative modeling: the structure is represented as a set of devices interconnected by causal principles; the study of the relation between these devices permits the analysis of the state of the system.

8.2 Morphodynamical Abduction and Adumbrations

Morphodynamical abduction is abduction considered in the light of the mathematical framework described above. The main idea is that a complex system, as the cognitive one, and its transformations, can be described in terms of a configurational structure. That is, different mental states are defined by their geometrical relationships within a larger dynamical environment. This suggests that the system, at any given instant, possesses a general *morphology* we can study by observing how it changes and develops. The term *morphodynamics* refers to those theories aimed at explaining morphologies and iconic, schematic, *Gestalt*-like aspects of structures, whatever their underlying physical substrate may be, for example by using the mathematical theory of dynamical systems proposed by Thom [1980].[10]

As I have already said, my aim is to cognitively understand what happens during the process of "creation" of an hypothetical model, when from the disordered flow of "experience" an order emerges. I maintain that it is possible to obtain interesting indications by integrating the traditional computational models with some ideas coming from the dynamical approach entangled in the tradition of phenomenology (cf. also below subsection 8.2.1). As I have illustrated by introducing the notion of phase space it is possible to represent the evolution of a system "drawing" its trajectory in this $2n$-dimensional space (where n represents the number of components of the system).

The system remains in a state S_A, until the parameters by which it is influenced lead it to some unstable state. It is a catastrophic rearrangement inside the overall aspect of the system which changes the state S_A into the final one S_B. The system rests around an attractor, until it goes through unstable configurations that rapidly disappear and the system stabilizes around another attractor.

This model underlines the fact that the parameters involved (and their interactions) determine the behavior of a cognitive system. We can in fact represent the system as in Figure 8.3. By using a metaphor, we can consider the parameters (P_1,

[10] On the relevance of Thom's catastrophe theory to abduction cf. the following section 8.3. This theory graciously constructs a mathematics of various morphodynamical processes, included the cognitive ones.

P_2, ...) as "interacting" among themselves as "atoms under forces" (F_1, F_2, ...). Any parameter can act on the other ones, moving the overall cognitive "structure". Suddenly, in a certain instant, the system reaches stability and the activity stops. The values assumed by the parameters determine the overall configuration of the system in a given instant. In these terms it is easy to understand the evolution described by attractors. The parameters maintain the system in the same state S_A until some values disturb the overall equilibrium, and this leads the system to a new state S_B.

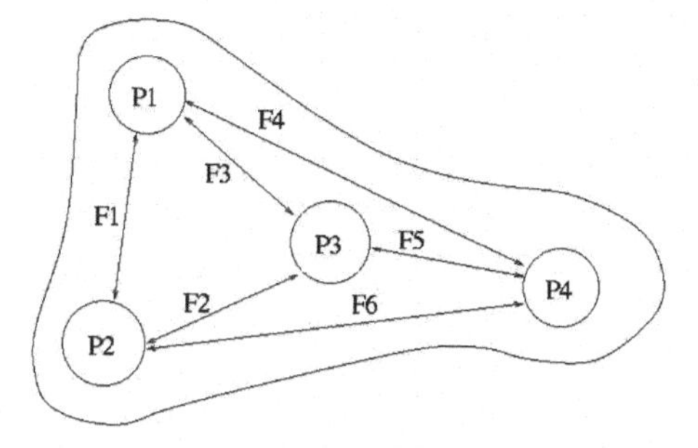

Fig. 8.3 Interactions between parameters

I have used an intuitive representation of the concept of attractor in the previous section. This was justified by identifying possible cognitive states with attractors in the state space (here depicted, to better visualize the situation, as a geometrical surface in which possible cognitive states interact) of the cognitive system. Like in the case of the intuitive representation of the relativistic conception of gravitation, we can see this surface as a flat horizontal rubber sheet (Figure 8.4A). If we imagine the cognitive system behaving as I have just illustrated, we can easily see it "determining" the shape of the rubber sheet (Figure 8.4B and 8.4C). If we see the cognitive system in the light of dynamics, these parameters are the variables which affect the behavior of the system: for example, in the epistemological case of scientific cognitive systems, background knowledge, results of observations and different perspectives on them, available resources, etc.[11] The parameters responsible for the behavior of the system determine the "weight" of the attractor, then the shape of the surface. I maintain that this process is assimilable to the notion of *anticipation* (see below) developed in Husserl's phenomenology.

It is possible to understand how the system behaves in purely transitory instants, or when it reaches a stable position because of an attractor. Figure 8.1 can be seen, temporarily, as describing the movements inside the states-surface system. If we represent the important parameters as in Figure 8.3, we can identify any point of the picture in Figure 8.1 with their different settings in time.

[11] Cf. below in subsection 8.2.1 the epistemological example – seen in a dynamical perspective – of the discovery of Neptune where the astronomers' use of the telescope can be seen as responsible for the modification of the aggregation of parameters.

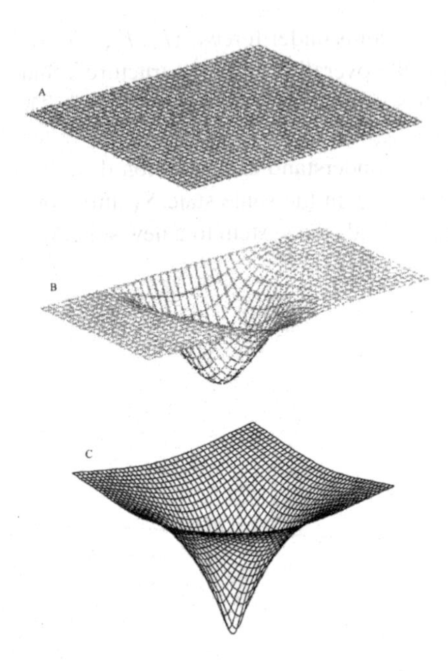

Fig. 8.4 Deformations in the cognitive space of the system

The system can assume any unstable configuration possible in that space, until it reaches the bottom (that is, configuration *C* in Figure 8.4). In order to understand the cognitive behavior of the system, it is possible to study the overall behavior of the system looking at any interesting space "shape" that leads it from point *A* to point *C*, passing through point *B* (see Figure 8.4).

From state *A* to *B* the space begins to be subject to a first deformation. If we imagine the diagrams as illustrating the behavior of a cognitive system, *A* represents a *tabula rasa* state, while the cognitive agent is still searching and no ideas influence its judgment. Suddenly something changes in the parameters, leading to the transitory state *B*. This state is just transitory and unstable: at this point a *bifurcation* happens. In fact, the system can "reject" state *B* by coming back to state *A* (the weight of the parameters is not yet irremediably heavy), or any other previous state. For example, a cognitive system faced with a perception judged as anomalous can choose not to activate an explanatory abduction. In subsection 2.1.1 of chapter two, this case was described by the situation in which, in front of ignorance problems, the agent does not abduce, but yields to her ignorance.

Otherwise, the system can persist in state *B* anticipating the abductive inference that leads to its final state *C* (here I am applying this dynamical description to the problem of abduction). This conformation can be changed only by drastic and catastrophic changes. To better understand creative abductive cognitive processes and processes of discovery in science, we can integrate this geometrical metaphor with the concepts of *anticipation* and *adumbration*.

8.2.1 Hypotheses Anticipation and Abduction

I have already illustrated in subsection 4.7.2 of chapter four how the philosophical tradition of phenomenology recognizes the important role of perceptual and kinesthetic abilities in the generation of "idealities" and "mental" constructs (of course this tradition employs a typical philosophical lexicon I am adopting in the following few passages). In this perspective perception is a "structured" *intentional* constitution of the external objects, established by the rule-governed activity of consciousness. Sensible schemata constituted in the temporal continual mutation of adumbrations mediate and structure the perception of material things immediately given. Appearances of pure lived experiences in "presentational perception" – obtained by direct acquaintance – only provide a constituting multiplicity of partial aspects (adumbrations), which need be further unified to establish they belong to a particular and single object (*noema*). When we see a – potential, we cannot foretell what it is – spherical form from one perspective, we are adumbrating it. To manage the adumbrative aspects of things as part of the visual field we have to glue different fillings-in of the visual field to construct the temporal continuum of perceptive adumbrations in a global space.

As I have already illustrated in chapter four the kinesthetic control of perception is also related to the problem of the generation of the objective notion of three-dimensional space, that in turn is at the core of the phenomenological constitution of a "thing" as a single body unified through the multiplicity of its appearances/adumbrations. The "meaning identity" of a thing is of course related to the continuous flow of adumbrations: given the fact that the incompleteness of adumbrations implies their synthetic consideration in a temporal way, a synthesis which is *kinesthetic* and involves eyes, body, and objects. Visual sensations are not sufficient to constitute objective spatiality and have to be intertwined with kinesthetic sensations relative to the movements of the perceiver's own body. The ego itself is only constituted thanks to the capabilities of movement and action. Kinesthetic controls play the role of *spatial* gluing operators. They are able to compose, in the case of visual field, different partial aspects – identifying them as belonging to the same object, that is constituting an ideal and transcendent "object". They are realized in the "pure consciousness" and are characterized by an intentionality which requires a temporal lapse of time.

Adumbrations are multiple and infinite, and there is a potential co-givenness of some of them (those potentially related to single objects). Adumbrations, as rough information that has to be further processed, influence the parameters governing the cognitive system, in the sense that they are responsible for its shifts in the state space. They are incomplete and partial so for the complete givenness of an object a temporal process is necessary. *Anticipations* are the explanatory operations necessary to manage adumbrations that have to be performed by objective transcendence. Just because defeasible, anticipations correspond to a kind of non-intuitive intentional expectation. When we see a spherical form from one perspective (as an adumbration), we will assume through explanatory abduction that it is effectively a sphere, but it could be also a hemisphere (an example already employed by Locke). Anticipations can be usefully seen as highly conjectural – and nonmonotonic – multimodal abductions, which can be withdrawn and replaced by other more plausible ones. Moreover, they constitute an activity of "generate and test" as a kind of "manipulative" cognition: indeed the finding of adumbrations involves kinesthetic controls, sometimes in turn involving manipulations of objects; but the activity of testing anticipations also implies kinesthetic controls and manipulations.

Finally, not all the anticipations are informationally equivalent and work like attractors for privileged individuations of objects: they foretell subsequent *new trends*. In this sense the whole activity is toward "the best anticipation", the one that can display the object in an optimal way. Prototypical adumbrations work like structural-stable systems, in the sense that they can "vary inside some limits", without altering the apprehension of the object. Like in the case of selective abduction, anticipations are able to select possible paths for constituting objects, actualizing them among the many that remain completely tacit. Like in the case of creative abduction, they can construct new explanatory ways of aggregating adumbrations, by delineating the constitution of new objects/things. In this case they originate interesting new "attractors" that give rise to new stable "conceptual" generalizations.

Let us give an astronomical example of anticipation stemming from analysis of the evolution of the cognitive "distributed" system expressing classical physics, a cognitive system that is intended to involve humans (scientists), non-humans (natural non biological entities), and artificial agents. New problems arose after Uranus was accepted as being a planet. The orbit of Uranus could not be accurately predicted from Newtonian theory. In fact, looking at the predicted orbit with an artifact in the form of a telescope with its own cognitive effect, it was not possible to observe any astronomical body. This was an interesting *anomaly* to be solved. To explain this inconsistency, Adams and Leverrier, in the first half of the nineteenth century, introduced the *ad hoc* hypothesis that this anomaly could be explained by postulating the existence of another still unobserved planet. This is a case of productive *ad hoc* hypothesis guessing, an abduction as an anticipation. In 1846 Galle decided to point his telescope in the direction indicated by the new hypothesis to effectively determine the existence of the planet. He actually "discovered" Neptune. It was the *adoption* of the *ad hoc* hypothesis and the *decision* to use an external artifact which were able to "prosthesize" scientists' cognitive skills and to lead the cognitive system to the new attractor, where the guessed hypothesis was tested and accepted, thanks to the subsequent empirical discovery.

Metaphorically we can say that the new hypothesis and the telescope (an external tool manipulated by the scientist), "bumped" against the existing attractor accounting for the beliefs regarding the orbit of Uranus as predicted by the Newtonian theory. This led to a catastrophic rearrangement of parameters; that is to the discovery of a new planet and to the development of a new conception of the solar system.

8.3 Abduction, Pregnances, Affordances

I have anticipated in subsection 8.1.1 that further insight regarding abduction in cognitive systems from a "dynamical" perspective can be found in Thom's theory of morphogenesis, based on the catastrophe theory. Some of the results illustrated in the following sections can be seen in the light of chapter six, devoted to affordances and abduction, and chapter five, which stressed the relationships between instincts, abduction and hardwired aspects.

As I have shown in subsection 5.2.1 of chapter five, an example of instinctual (and putatively "unconscious") abduction is given by the case of some animal embodied cognitive abilities. These abilities are in turn capable of leading to some appropriate behavior: as Peirce said, abduction even takes place when a new born chick picks up the right sort of corn. I have contended that this is an example of spontaneous abduction – analogous to the case of other hardwired unconscious/embodied abductive processes in humans: "The chicken you say pecks by instinct. But if you are going to think every poor chicken endowed with an innate tendency toward a positive truth, why should you think that to man alone this gift is denied? [Peirce, 1931-1958, 5.591]". I think the concept of pregnance, introduced by Thom [1975; 1980] on the basis of Wertheimer's Gestaltic concept of *Prägnanz*, can shed further light on a kind of morphodynamical "physics" of abduction, first of all in the case of its instinctual hardwired aspects.

In sum, pregnance and salience are key concepts which provide further information about the instinctual/plastic abduction dichotomy. As they acquire their meaning in a very naturalistic intellectual framework, they are useful to propose a unified perspective on abductive cognitive/psychic processes, seen as basic physico-biological events, endowed with a profound eco-cognitive significance. The pregnance affects an organism, and the related abductive/hypothetical response is promptly triggered. Pregnances are genetically transmitted but can also be actively created for example through learning and high cognitive capacities, through the formation of multiple forms of hypothetical intelligence.

8.3.1 Saliences

The complicated – and at first sight obscure – concept of *pregnance* is based on the concept of *salience*, which emerges in the dynamical framework of the "semiophysical" perspective. First of all we can say that in general, phenomenal discontinuities are perceived by organisms as *salient forms* (for example, in the auditive case, the eruption of a sound in the midst of silence),[12] that is, as contextual effects between forms. Discontinuities over there in the environment are basically *translated* into other more or less amplified discontinuities in the subjective sensorial state, as a kind of "echo" or "shock" within an organism of the physical environment. In the

[12] "The simplest feature is the punctual discontinuity geometrically represented by a point dividing the real straight line **R** into two half lines" [Thom, 1988, p. 3].

case of sensory systems, salience of course is at the basis of the first possibility of perceiving individuated forms.[13]

The term pregnance can be applied both to physical and biological phenomena. It can further clarify the distinction between the instinctual chicken abduction above and other plastically acquired abductive ways of cognition:

> So we will get this general pattern of a world made up of salient forms and pregnances – salient forms being objects, very often individuated, that are impenetrable to one another, and pregnances being occult qualities,[14] efficient virtues that emanate from source-forms and invest other salient forms in which they produce visible effects (that is the so-called "figurative" effects for the organisms invested) [Thom, 1988, p. 2].

Thom further says: "*Pregnances* are non-localized entities emitted and received by salient forms. When a salient form 'seizes' a pregnance, it is invaded by this pregnance and consequently undergoes transformations in its inner state which can in turn produce outward manifestations in its form: we call these *figurative effects*" [Thom, 1988, p. 16]. To clarify the two concepts of salience and pregnance the following two examples can be of some utility [the wide range of events covered by the two concepts is testified by the fact that the first example does not have any cognitive/psychological significance]: 1) an infection – pregnance – contaminates healthy subjects – that represent the "invested" form: salience. These subjects in turn re-emit the same infection – pregnance – into the environment. In this case pregnance has in itself a material/biological support (for example a virus) – as a mediator, which in turn is transmitted thanks to a suitable medium (for example air or blood); 2) worker honeybees communicate with each other by means of signs (through the iconic movements of a dance) – pregnance – that express the site where they have found food in order to inform the other cospecific individuals – the invested salience – about the location. In this second case the pregnance is transmitted – mediated – through ondulatory sounds and light signals and produces a neurobiological effect at the destination, that is, in other words – a "psychic" effect [of course we can use in this case the expression "psychic" only if we unorthodoxically and mentalistically admit honeybees are endowed with a kind of animal psyche – an example regarding a cat or a boy would have been more convincing for the reader. . .].[15] However, to better grasp the concept of pregnance further analysis is needed.

[13] In this case perception can be appropriately influenced by a certain form of "concept, that is to say a class of equivalence between forms referent to the same concept": the lack of the concept can annihilate the grasping of the individuated form, especially when analysis proceeds from the whole to the parts.

[14] This is just a metaphor, actually Thom thinks that pregnances are not occult and mysterious qualities, because they could be accounted for as fully explainable psychological phenomena in neurological and biological terms, and they can also be made intelligible through mathematical models. The description of the processes affected by pregnance activity aims at providing what Thom calls a "protophysics, source and reservoir of all permanent intuitions, of all those archetypal metaphors that have nourished man's imagination over the ages" [Thom, 1988, p. 3].

[15] Fields in physics are the true paradigm of *objective pregnances* in modern science, because in that case we are theoretically able to calculate their variation in space-time thanks to a mathematical description (based on an explicit geometrical definition of space-time) [Thom, 1988, p. 32].

8.4 Pregnances as Eco-Cognitive Forms

In general, in the case of salient forms, their impact on the organism's sensory apparatus "remains transient and short lived" [Thom, 1988, p. 2], so they do not have relevant long-term effects on the behavior of the organisms. In the case of cognitive events, if we adopt the perspective of the affordances described in chapter six, we can say that salient forms – contrary to pregnant forms, "afford" organisms without triggering *relevant* modifications either at the level of possible inner rumination or in terms of motor actions. Thom says that when salient forms carry "biological significance", like in the form of prey for the hungry predator, or the predator for its prey, or in the case of sex and fear, or when a salient form is invested by an infection, the reaction is much bigger and involves the freeing of hormones, emotive excitement and behavior (or an immune response in the case of the infection) devoted to attracting or repulsing the form: *salient forms of this type are called pregnant*.

In the *cognitive* case, *pregnances*, no matter whether due to innate releasing processes or to complicated, more or less stable internal learnt processes and representations, are triggered by a very small sensory stimulus (a stimulus "with a little figuration, an olfactory stimulus for instance" [Thom, 1988, p. 6]). Hence, they represent a relationship with certain *special* phenomenological aspects, that are of course only more or less stable and so can appear and disappear. At some times and in some cases the special sensitivity to pregnances is disregarded. Like in the case of affordances, this variability and transience can be seen at the level of the differences of pregnance sensitivity among organisms and also at the level of the same organism at successive stages of its cognitive and biological development. In terms of the hypothetical abductive framework described in this book we can say that a pregnant stimulus is – so to say – *highly diagnostic* and a trigger to initiate abductive cognition, like in the case of the hardwired pregnancy occurring regarding our Peircean chicken and its food, the chicken promptly reacts when perceiving it. When a pregnance affects an organism, the abductive reaction can be promptly triggered. Finally, we have to recall that the pregnant character of a form is always relative to a receiving subject (or group of subjects), just as we have seen in the eco-psychological case of affordances.

Pregnances can be abductively activated or created. When bell ringing is repeated often enough together with the exhibition of a piece of meat to a dog, thanks to Pavlovian conditioning the alimentary pregnance of meat spreads by contiguity to the salient auditive form, so that the salient form, in this case the sound of the bell, is invested by the alimentary pregnance of the meat; here the metaphor of the invasive fluid – even if exoteric – can be useful: "So we can look on a pregnance as an invasive fluid spreading through the field of perceived salient forms, the salient form acting as a 'fissure' in reality through which seeps the infiltrating fluid of pregnance" [Thom, 1988, p. 7].[16] The propagation can also occur through similarity, taking advantage of the mirroring force of some features. Once the reinforcement is

[16] To explain the formation of pregnances Thom exploits the classical Pavlovian perspective. More recent approaches take advantage of Hebbian [Hebb, 1949] and other more adequate learning principles and models, cf. for example [Loula *et al.*, 2009].

established, the bell, Thom says, refers *symbolically* in a more or less stable way, to the meat.[17] From this point of view the "symbolic activity" is seen as fundamentally linked to biological control systems in two ways: 1) it is an extension of their efficacy (new favorable cognitive abductive chances – new pregnances – are added); 2) an internal simulation concerning the relationship between the food and its index, the bell, is implemented, so that the door is opened to the formation of multiple forms of abductive semiotic cognition (and /or intelligence):

> The fact that initially, as in the Pavlovian schema, this stimulation is no more than a simple association, does not stop us from considering that we have the first tremors in the plastic and competent dynamic of the psychism of [the actant] of an external spatio-temporal liaison interpreted not without reason as causal. [...] Hence, from the beginning, the situation is not fundamentally different from that of language [...]. Only these fundamental "catastrophes" of biological finality have the power of generating the symbols in animals [Thom, 1980, pp. 268–269].

As I have already indicated Thom sees pregnances not only as innate endowments (like in the case of the basic ones seen in birds and mammals: hunger, fear, sexual desire), but also as related to higher-level cognitive capacities "When animal pregnance is generalized in the direction of human conceptualization 'conceptual' or individuating pregnances will be revealed, the nature of which is close to 'salience' " [Thom, 1988, p. 6]. At this point it would have to be clear that I maintain we can synthetically account for both these processes in terms of different kinds of abductive cognition. For example, Thom observes, reverberating the view of visual perception as semi-encapsulated I have illustrated in chapter five "[...] it is doubtful whether genetics alone would be able to code a *visual* form [...]. Whence the necessity of invoking cultural transmission, linked with the social or family organization of the community" [Thom, 1988, p. 10]. A cognitive situation also analogous to the one I have repeatedly stressed in chapter six in the case of "mediated" affordances (cf. subsection 6.4.1). In gregarious animals the signals (which also have to be seen as referring to the explanation of the origin of the "pregnance-mirroring" functions of human language) are a vector of pregnances in so far as they transfer a pregnance from one individual to another, or to several others. In such a way they favor teaching and learning, working to constitute the collective shared behavior needed for example to capture food and to ward off predators.

When an organism – through abductive cognition – traces back a symbolic reference to a "source" form [in the Thom's sense indicated above], often a motor reaction becomes necessary to bring satisfaction:

> In a social group, one individual's encounter with a source form *S* may give rise to a dilemma: whether to pursue the "individual interest" which consists in using the regulatory reflex that will result in selfish satisfaction, or to follow the altruistic community strategy by uttering the cry that will carry the pregnance *S* to the other members of the

[17] Of course extinction of pregnances through absence of reinforcement is possible, when an organism moves away for a long time from the source form or when the invested salient form is associated with another pregnant form still in absence of reinforcement.

community; such a cry is then the signal by which the signal P of S experienced by individual 1 can be transferred to another individual 2" [Thom, 1988, p. 12].[18]

An example is furnished by the case of a signal (or a proximal "clue", in Brunswik's terms), which transfers the pregnance of fear in birds, which further prompts the motion of taking flight but that also incurs the risk of attracting the predator's attention. Animals perceive the pregnant sign/clue (for example tracks or excreta of the predator), and then emit a further sign (cry) that mirrors that sign/clue and its pregnance.

8.4.1 Pregnances and Human Language

From the point of view of the functions of human language Thom sees the birth of the "genitive" as the syntactical form that denotes the proximity of a being whilst denying its immediate presence. This syntactical form permits us to emit and receive alarm calls[19] which provide individuals (and the group) with an adequate defense. From this perspective the presence of a pregnant sign associated with a form S can be considered as a fundamental kind of *concept* or class of equivalence between salient forms, which incorporates a primary, rudimentary and prompt abductive power.

As I have already stressed, the *cultural* acquisition of a sensitivity to source forms has to be hypothesized in both humans and various animals. In this case pregnance transmission occurs, beyond the hardwired cases, thanks to the presence of suitable artifactual cognitive niches (such as human natural languages), which function as *pregnance mediators*, where plastic teaching and learning is possible. These cognitive niches make plenty of cognitive tools available, that in turn make the organisms acquiring them able to pregnantly manage signs (which consequently gain a special "meaning"). This process is clearly illustrated by the description of various aspects of "plastic" – and not merely hardwired – cognitive skills in animal abduction (chapter five) and by the relevance of the "mediated" character of several affordances (chapter six). In these last two cases both cognitive skills and sensitivity to suitable affordances require cultural learning/training imbued in appropriate cognitive niches.

In section 5.4.2 of chapter five I have emphasized that fleeting and evanescent internal pseudorepresentations (beyond reflex-based innate releasing processes, trial and error or mere reinforcement learning) are needed to account for many animal "communication" performances even at the rudimentary level of chicken calls: Evans says that "[...] chicken calls produce effects by evoking representations of a class of eliciting events [food, predators, and presence of the appropriate receiver] [...]. The humble and much maligned chicken thus has a remarkably sophisticated

[18] Thom nicely adds that this kind of proto-moral conflict resonates with the more clearly "moral" conflict of civilized societies "This dilemma exists well and truly in our society. Witness the scruples most honest citizens have in making true declarations of their taxable revenues" [Thom, 1988, p. 12].

[19] Also, in many animals alarm calls/cries are the analogue of the second-person singular imperatives typical of human classical languages [Thom, 1980, p. 172].

system. Its calls denote at least three classes of external objects. They are not involuntary exclamations, but are produced under particular social circumstances" [Evans, 2002a, p. 321]: in Thom's words, these calls are of course pregnant signals which can be learnt. Chickens form separate representations when faced with different events and they are affected by prior experience (of food, for example). They are mainly due to internally developed plastic capacities to react to the environment, and can be thought of as the fruit of learning. In chapter five we have also seen that many animals go beyond the use of sound signals in their cognitive performances, they for example *reify* and delegate cognitive/semiotic roles to true pregnant external artificial "pseudorepresentations" (for example landmarks, urine-marks, etc.) which artificially modify the environment to later on afford themselves and other individuals of the group or of other species.

8.5 Semiotic Brains Make Up Signs: Mental and Mindless Semiosis through Abductive Anticipation

In the previous chapters of this book we have seen how various animals possess a subsystem consisting of neuron activities that mainly aims at furnishing models of the exterior space reflecting the interesting objects found there (for example prey, food, predators) and the position of the body – spatial representations – in relation to these objects, thus allowing appropriate and fruitful spatial competition. In section 3.3 in chapter three I have argued that in humans and many animals what I have called *semiotic brains* are brains that make up a series of signs and are engaged in making or manifesting or reacting to a series of signs. Through this semiotic activity they can be at the same time engaged in "being minds" and in thinking intelligently. Thom's example of the Pavlovian dog is striking, when the dog is conditioned to recognize the sound sign "the ringing of a bell" as a pregnant alimentary form "[...] this transformation only occurs in the psyche of the conditioned dog. Nothing in the phonic structure of the signal has changed intrinsically; the only objective aspect concerns the dog itself, the subject, whose psyche is going to run through a whole associative chain of the category Γ_p in order to reach the source form. (When this chain includes actions on the part of the subject, then we have the principle of training or dressage)" [Thom, 1988, p. 14]. Humans have conditioned the dog to acknowledge *material*, physical events (like the the ringing of a bell), which in the dog's "psyche", as Thom says, becomes a pregnant sign. This is a psyche that consequently manifests the existence of a kind of "mind" absorbed in reacting to the signs (and further able to make up and manifest other related signs),[20] that it is absorbed in – so to say – "being a mind".

It is important to stress again that Thom provides a mathematical model of many of these semiophysical processes. In the case of Pavlovian conditioning the ground state of the receiving subject can be seen as a non-absolute minimum, surrounded by a ring of basins the bottoms of which are lower than the ground state. The perception

[20] For example the dog can emit the cry for food, a salient form which is also a pregnant form for other dogs or for humans.

of a form as pregnant is such because it creates a "tunnel effect", well-known in quantum mechanics, which precipitates the representative point into a peripheral basin therefore liberating energy and constituting an excited state (cf. Figure 8.5). Attractors (see above subsection 8.1) are still at play, the indetermination of the pregnant form is due to the attraction held by it in a space of forms, where the excited states have to play a biological regulation and a regulative subsequent action looking for satisfaction (or pain and sex, in other cases).

Pregnance (which in the present Pavlovian case is just the memo of an earlier satisfaction) is the *abductive anticipation* of the satisfaction at play, an anticipation that implies a "virtual" affectivity (investment of the subject) that has an effect. The hungry animal is excited: once the food/prey appears, is caught and ingested, the "actual" affective satisfaction lifts the level of the excited state basin and possibly annihilates it so that the animal returns to the ground state. In the general case of the interplay predator/prey a distinction has to be made between potential investment by a pregnance (when the animal is looking for the prey) and actual investment "which comes in after the perception catastrophe" [Thom, 1988, pp. 15–16]. Following my semiotic interpretation it is clear that pregnant forms/signs are "dialogic" both at the intrasubjective (in some cases – humans for example – organisms are aware of them) and at the intersubjective level, while at the same time their conceptual significance can only be grasped in an eco-cognitive framework.

These perspective pregnances consist in intertwined internal and external semiotic processes. If a pregnant alimentary form "only occurs in the psyche of the conditioned dog", and in turn salient signals can be emitted and then seen as pregnant by other individuals, following Thom we can say that every element of the dog's psyche has something corresponding to it in the world. Peirce analogously said that "[...] man is an external sign. That is to say, the man and the *external sign* are identical, in the same *sense* in which the words *homo* and *man* are identical" [Peirce, 1931-1958, 5.314] (cf. chapter three, subsection 3.5.1): let us remember once more that for Peirce, and also from my perspective, signs include feeling, image, conception, and other representations, which means that model-based aspects of both animal and human cognition are central and constitute a significant

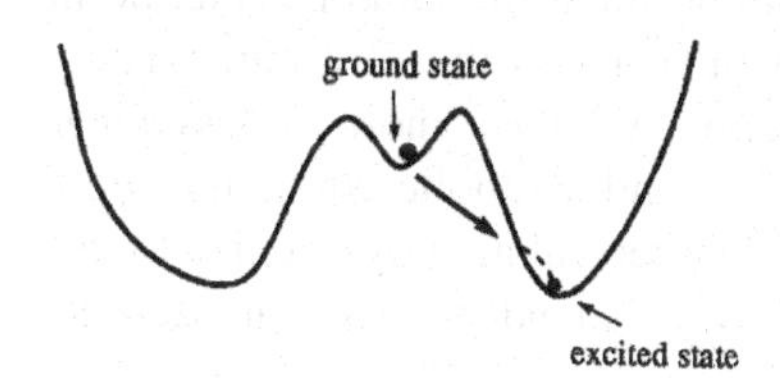

Fig. 8.5 Tunnel effect, due to the apparition of a pregnant form, on a sensitive subject. In [Thom, 1988, p. 15]. ©1988 Dunod, reprinted by permission.

phenomenal manifestation – pregnant – of organisms. In such a way material animal salient signals become available in a social process to other animals (and, in part, to humans) for pregnant interpretations and consequently guide our actions in a positive or negative way. In Thom's terms, salient forms become pregnant for the receiver because they mirror the object through resemblance/isomorphism [icons], contiguity [indexes], or conventionality [symbols], just as Peirce contended. The social interplay between saliences and pregnances established by Thom is a process of "sign action", or *semiosis*, and does not necessarily refer to an actual person or mind, but can also be attributed to an animal interpreter and, furthermore, to the obviously "mindless" cases of triadic semiosis (sign, object, interpretant), or even to plant behavior. Indeed we can recognize that a sign has been interpreted not because we have observed a mental action but by observing another related material sign. Finally, also unconscious or emotional interpretants are typically widespread in human beings (cf. chapter three, subsection 3.5.1).

As anything can be seen as a pregnant sign, the collection of potential signs may encompass virtually anything available within the agent, including all data gathered by its sensors. As I have already illustrated in the quoted subsection of chapter three, in the context of the science of complexity *pregnant semiosis* can be depicted as an emergent property of a system. Emergent properties constitute a certain class of higher-level properties, related in a certain way to the microstructure of a class of systems, that thus become able to produce, transmit, receive, compute and interpret signs of different kinds.

In summary, abduced pregnances mediate salient signs and work in a triple hierarchy: feelings, actions, and concepts. They are partially analogous to Peirce's "habits", that is, various generalities as pregnant responses to a sign. They are the effect of a sign process. In Peircean terms, the interpretant produced by the sign can lead to a feeling (*emotional* interpretant), or to a muscular or mental effort, that is an *energetic* interpretant of some kind of action – (not only outward, bodily action, but also purely inward exertions like those "mental soliloquies strutting and fretting on the stage of imagination" – [Colapietro, 2000, p. 142]). Of course sign(al)s are subjected to modifications, because they are always, so to say, incomplete, with the possibility of taking any particular feature previously unknown to their interpreters, so that their possible pregnant character can be further revised or withdrawn. In humans some signs (for example linguistic and/or symbolic in a Peircean sense) are pregnant for the receiver in so far as they are immediately endowed with abstract meaning and live in semiosis with a more or less long-term stability in the hierarchy of iconic, enactive, and symbolic communication. They give rise to "thoughts" related not only to intellectual activity but also to initiating ethical action, as a "modification of a person's tendencies toward action" [Peirce, 1931-1958, 5.476].[21]

[21] I have devoted part of chapter three to the clarification of the relationship between the semiotic approach and abduction, taking advantage of the Peirce's original results.

8.5.1 *Language Acquisition through Attunement and Parental Deixis*

In the subsection 5.7.3 in chapter five I have maintained that in ontogenetic affect attunement between human infants and their caregivers the interplay is mainly model-based and mostly iconic (taking advantage of the force of gestures and voice). Hence, meaningful words are present, but the semiotic propositional flow is fully "understood" only on the part of the adult, not on the part of the infant, where words and their meanings are simply being learnt. Let us focus on this problem related to language acquisition in the light of catastrophe theory. The child is, in a way, "invaded" by a great number of words, that can be characterized as *potentially pregnant* and their presence is commonly thought to be at the basis of the distinction between human and other animal communication. Prior to the model-based interplay related to face and gestures (that reaches the baby when it is just a few days old) there is the main (alimentary) pregnance, when the child is breast-fed. Following Thom, it can be hypothesized that in this process the pregnance of a baby's own body, step by step, becomes autonomous "if the organic coenaesthesis is linked up with the external image of the body (mirror stage)" [Thom, 1988, p. 23]. A similar process is occurring in the case of the formation of the assessment of the autonomy of external objects in so far as they are the object of action, so that we witness "the mother's original pregnance exfoliating into a great number of beings and objects" (*ibid.*).

The child receives some words uttered by the mother, who, at the same time, through *deixis*, emits a quantum of her pregnance to invest the object indicated (through contact or the pointing finger), thus coupling it with the sound. The heard word acts as a tunnel effect (cf. above subsection 8.5), and stabilizes, for individuating the pregnance of the object: "[...] the possibility the child has of causing the object to appear by saying the word reinforces the autonomy of this pregnance which eventually dissociates itself completely from that of the mother" (*ibid.*). Deixis, which externalizes a gesture related to a word, creates a transitory pregnance which invests the objects shown and, by conveying the designating word, invades the receiver's psychism through internalization and neural fixation (towards eight or nine months the child is able to reproduce the same behavior as the mother asking her the name of the object and pointing to it). It is in this way that, the "meaning" of the word is firmly grasped and internalized.

The syntactical aspects of natural languages,[22] such as for example the genitive – see subsection 8.4 above – or the management of the verb,[23] can be further explained in terms of archetypal morphological space-time processes susceptible of being described in mathematical topological terms. For instance "Upon hearing an order the cerebral dynamics suffers a specific stimulus s, which sends it into an unstable state of excitation. This state then evolves towards stability through its capture by the attractor A, whose excitation generates the motor execution of the order by coupling the

[22] Basic syntactical mechanisms are intended by Thom as simulated copies (defined on an abstract space) of the fundamental biological functions such as predation and sexuality.

[23] For example a verb transfers a pregnance from subject to object and so constitutes an attractor of the cerebral dynamics.

motor neurones" [Thom, 1980, p. 172]. For example, the verbs of feeling *to fear* and *to hope* express that an actant subject admits in internal co-ordinates "a morphology of the future, which is accepted with repulsion or attraction" (p. 211).[24]

The further ontogenetic syntactical development of language in humans is linked to the communication target – it is driven by social necessity – of extracting the "meaning" incorporated in pregnances received from humans and artifacts through various sensory systems based on ondulatory sounds, light, and direct contact: basically, auditory, visual [written texts], tactile [for example, Braille cells]. Obviously, the acquired linguistic syntactic competence allows humans to affect others with subjective pregnances at will. A linguistic sentence like "Peter sends John a letter by post" where the target is of course to *inform* the hearer, has the precise biological realization of precipitating "his logos into a new stationary state, a new *form*" [Thom, 1980, p. 183]. Indeed any message is in a state of morphological instability, which then stabilizes itself in the recipient in a locally stable network of behavior. It is interesting to stress that from Thom's semiophysical perspective the elementary mathematical structure (singularity elliptic umbilic) of this sentence also describes messages that are externalized without the help of uttered or written sentences. Semiotic information takes advantage of very different material realizations, for instance model-based and through artifacts. A special case of the same mathematical morphological structure is the extreme event of "communication" regarding a message that has the aim of capturing or destroying the receiver, here the message is a projectile, the messenger being a firearm used by an aggressive and violent human being.

In this sense we can say that vocal and written language is a tool exactly like a knife;[25] the material prolongation of an organ. It is a vector of pregnances of biological origin, the possible support of a subsequent action and the outward extension of an organic activity.[26]

Let us come back to the mother/child interplay related to language acquisition. Once the reciprocal use of a "word" abductively is stabilized, it constitutes the expected linguistic "attunement" to the mother, (the "communication" is established), and at the same time the appropriate (maternal) cognitive link to the external environment. The process is slow, the fruit of a progressive abductive activity of subsequent "linguistic (abductive) hypotheses" [words] externalized by the two agents, until the plausible and acceptable one is reached. In this process, the external manifestation [sound/word] of the baby is established as that commitment to the external world, which in turn testifies a possible shared coherent flux of communication with the mother. A side effect is also a sharing of affection that can be mediated by words

[24] An example of the morphological isomorphism between language and biological functions is the following: the verb as it appears in sentences like in "The cat eats the mouse" reproduces at the linguistic level the biological transition between the virtual investment of a subject by a pregnance and the satisfaction resulting from the act analogous to the one I have described above in the Pavlovian case. Of course when the "verb" is used to describe the process of inanimate nature, part of this process – intentionality and satisfaction – tends to disappear.

[25] Language as a tool is already a key point of the perspective established in the framework of distributed cognition, cf. chapter three, subsection 3.6.5.

[26] On the analysis of language in the systemic terms of catastrophe theory cf. [Thom, 1988, chapters two and eight] and [Petitot, 1985; Petitot, 1992; Petitot, 2003].

and which is also at the basis of further social expression of emotions in a language-based way. [Roberts, 2004] states that children form knowledge and expectations about the symbolic functioning of a particular word through routine events where model-based perceptual and manipulative aspects of cognition are predominant and furnish suitable constraints. Finally, in the general dialectic interplay of human adult communication the formation of new words can be traced back to the general interplay between source(s) – both biological and artificial – and receiver. Words are frequently used with meanings which are different to their original ones, so "tensions" are introduced, which can lead to potentially successful neologisms in further communication and to the extinction of old terms.

As a counterpart to the ontogenetic description above, the phylogenetic aspects of attunement have to be taken into account. From the phylogenetic perspective deixis can be traced back to the effect of light and heat given the fact that "Biological pregnances have turned little by little into objective pregnances, physical in modern terms" [Thom, 1988, p. 25]. It can be hypothesized that the abductive mechanisms of propagation and investments of subjective pregnances, which act on organisms like humans, have been mirrored and represented externally onto things, in the interplay between internal and external representations, with a considerable epistemic and practical success. I agree with Thom: he says it would have been easy for our ancestors to detect that light behaves like a pregnance which has its own source forms and which, when emitted, invests and transforms the objects on which it falls: objects which in turn emit a similar pregnance. We can say, adopting a different lexicon, following Jung, that first of all rough external representations (Jung says "observations"), merely reached through senses, were rapidly "internalized" and became "mythologized". "Summer and winter, the phases of the moon, the rainy season, and so forth, are in no sense allegories" of those primary observations, because they become "symbolic expressions of the inner" [i.e, pregnances] and "become accessible to man's consciousness by way of projection, that is mirrored in the events of nature" [Jung, 1968a, p. 6] (cf. also chapter three, subsection 3.8.1). Later on, a mechanism or subsequent projections of internal representations outside in the external environment and subsequent re-internationalizations are implemented.

8.5.2 "Discreteness" and Cognition

In the subsection 8.1.1 above I have stressed the role in cognitive science of the discrete/continuous dichotomy. Thom's analysis sheds further light on the cognitive role of discreteness seeing it as intrinsically intertwined with the emergence of saliences and pregnances. The discrete decomposition of naturally continuous phenomena is "not just an illusion of the mind" [Thom, 1988, p. 25], but can be grasped in terms of the distinction between salience and pregnance. This distinction is fundamental even if it sometimes tends to be blurred:

> [...] spreading and ramifying objects such as smoke from fire, or chemical diffusions (odors), are pregnances which sometimes have salient aspects; the phase transitions of matter, from salient solid to purely pregnant gas, provide an example of the modification of bodies which can be interpreted as due to conflicts between local pregnances

> [...]. (The form of a solid is a "figurative effect" of its "individuating pregnance" and at the same time its definition). So it is easy to understand why exceptional natural forms (*lusus naturae*), such as rocks resembling human faces, were endowed with a pregnance that is half religious, half magic, an undifferentiated pregnance in which Durkheim saw the origin of the modern notion of energy [Thom, 1988, p. 25].[27]

In the first case (magic), the imaginary world of pregnances is controlled by the will of man (or rather of certain men, magicians, expert in efficient practices); in the second (science), control is determined by the internal generativity of formal language describing external situations, a generativity over which man has no hold, once the initial conditions are laid down [Thom, 1988, p. 33].

Thom's observation about the fact that discrete decomposition of naturally continuous phenomena is "not just an illusion of the mind" [Thom, 1988, p. 25], allows me to add some final considerations about the controversial status of cognitive "representations" in the tradition of dynamic systems theories. If cognition is studied in the "continuous" interactions between brain, body, and environment, it is natural to be inclined to think that "representational states " cannot be subject to the so-called CCH hypothesis (cognitive continuity hypothesis), thus reaching an antirepresentationalist attitude. If discreteness and discontinuities at a higher level do not exclude continuous behavior, there would be no discontinuities and so representations should not exist. For example, [Garzón, 2008, pp. 288–290] argues that "In order to make conceptual progress we need a principled distinction which gets couched in terms of degree of continuity. The less discontinuous, the closer we are to nonrepresentational dynamics. Therefore, representations and computations deflate as we approach primagenesis, and they finally evaporate. [...] The lesson to be extracted from an antirepresentationalist reading of dynamic system theory is that coupled dynamic processes remain equally grounded at any age. There is no stage in development where cognition ceases to be embedded and embodied". Thus, antirepresentationalism is seen as a "probe bed for cognitive science research and philosophical inquiry". From this dynamic viewpoint – even if considered in the so-called *minimal* account framework – representations, and their computational descriptions can probably be dismissed when treating system states. Nevertheless, at the different epistemological and phenomenological level of other disciplines, such as philosophy, logic, psychology, etc. representations still display their gnoseological virtues. It can still be consistently and rationally contended that discrete decomposition of naturally continuous phenomena – like Thom says – is "not just an illusion of the mind".

[Clark, 2008, p. 165] contends that to see cognition as embodied, embedded, and even extended does not necessarily lead "[...] toward radical antirepresentationalism or the rejection of computational and functional approaches to understanding mind and intelligence", even if he acknowledges that these negative theses do

[27] In the subsection in 3.8.3 in chapter three I have described, from a psychoanalytic perspective, how not only natural pregnant objects like the above rocks resembling a human face, but also intentionally and creatively built artifacts such as cognitive/affective mediators, can be seen in the framework of the "psychic energy" flow. Those artifacts, in so far as they are aspects of human cultures and parts of the related cognitive niches, are clearly seen as "transformers of (psychic) energy".

appear in the literature on the embodied mind. On the contrary Clark contends, and of course I agree with him, that embodied agents are "[...] benefiting from multiple forms of internal and external representation, as deploying a variety of computational transformations defined over those representations, and as apt to participate in extended functional organizations that allow cognitive processes to spread productively across brain, body, and world. [...] Appeals to embodiment and cognitive extension go hand in hand with appeals to dynamics and to internal and external processes of representation and computation."

8.6 Hypothetical Cognition and Coalition Enforcement: Language, Morality, and Violence

The study on abductive cognition offered in this book demonstrates that the activity of guessing hypotheses touches on the important subject of morality and moral reasoning. In chapter six I have tried to show that in the activity of niche construction *hypothetical thinking* (and so *abduction*) is fundamental; in the previous chapter I have tried to illustrate that hypothetical thinking is often embedded in various kinds of the so-called fallacious reasoning (which in turn constitutes a relevant part of the linguistic cognitive niches where human beings are embedded). To better grasp its philosophical status I think it is intriguing to stress that hypothetical cognition might be favored for reasons of – so to say – "social epistemology" and moral reasoning.

In chapter four, in the framework of "distributed morality", a term coined in my recent book [Magnani, 2007d], I have illustrated the role of abduction in moral decision, both in deliberate and unconscious cases and its relationship with hardwired and trained emotions. The result emphasized in chapter five that abduction is partly explicable as a more or less "logical" operation related to the "plastic" cognitive endowments of all organisms and partly as a biological instinctual phenomenon, naturally led to the rediscovery of animals as "cognitive agents" but also as endowed with moral intrinsic value. I argued that people have to learn to be "respected as things" and sometimes, are. Various kinds of "things", and among them work of arts, institutions, symbols and of course animals, are now endowed with increasing intrinsic moral worth. Darwin noted that studying cognitive capacities in non-human animals possesses an "independent interest, as an attempt to see how far the study of the lower animals throws light on one of the highest psychical faculties of man" – the moral sense [Darwin, 1981]. Further, the problem of the abductive construction of extended cognitive niches, explored in chapter six, offers a chance to see the role in human cognition of management and correction of maladaptive artifactual niches, which immediately relates to the relationships between morality and knowledge in our technological world and to the role of the creative hypothetical reasoning employed in such a task.

In the previous chapter the analysis of the interplay between fallacies and abduction has acknowledged that: 1) abductive and other kinds of hypothetical reasoning are involved in dialectic processes, which are at play in both agent-based everyday and scientific settings; 2) they are strictly linked to so-called smart heuristics and to the fact that very often less information gives rise to better performance; 3) heuristics linked to hypothetical reasoning like "following the crowd", or social imitation, more or less linked to fallacious aspects which involve abductive steps, are often very effective. I stressed that these and other fallacies, are linked to what Réné Thom calls "military intelligence", which relates to the problem of the role of language in the so-called *coalition enforcement.* It is in this sense that I pointed out the importance of fallacies as "distributed military intelligence".

The aim of the following subsections is to clarify the idea of "coalition enforcement". This idea illustrates a whole theoretical background for interpreting the topics above concerning morality and hypothetical reasoning as well as my own position, which resorts to the hypothesis about the existence of a strict link between morality and violence. The theme is further linked to some of the issues dealt with in the previous sections of this chapter, where I value Thom's attention to the moral/violent role of what he calls "proto-moral conflicts": I contend that, for example, the fundamental function of language can only be completely seen in the light of its intrinsic *moral* (and at the same time *violent*) purpose, which is basically rooted in an activity of *military intelligence*.

8.6.1 Coalition Enforcement

The coalition enforcement hypothesis, put forward by [Bingham, 1999; Bingham, 2000], aims at providing an explanation of the "human uniqueness" and spectacular ecological dominance that is at the origin of human communication and language, and of the role of cultural heritage. From this perspective, and due to the related constant moral and punishing effect of coalition enforcement (which has an approximately two-million year evolutionary history), human beings can be fundamentally seen as *self-domesticated animals*. I think the main speculative value of this hypothesis consists in stressing the role of the more or less stable stages of cooperation through morality (and through the related inexorable violence). It implicitly shows how both *morality* and *violence* are strictly intertwined with the social and institutional aspects, implicit in the activity of cognitive niche construction. In evolution coalition enforcement works through the building of social *cognitive niches* as a new way of diverse human adaptation. Basically the hypothesis refers to cooperation between related and unrelated animals to produce significant mutual benefits that exceed costs and are potentially adaptive for the cooperators.[28]

In hominids cooperation in groups (which, contrary to the case of non-human animals, is largely independent from kinship) fundamentally derived from the need to

[28] The role of the so called "group mind" in cognition and group adaptation, which aims at demonstrating the possibility that groups engage in coordinated and cooperative cognitive processes, which far exceed the ability of individual thinking alone, is illustrated in [Wilson *et al.*, 2004]. The formation of appropriate groups which behave according to explicit and implicit more or less flexible rules of various types can be reinterpreted – beyond Wilson's strict and puzzling "direct" Darwinian version – as the "social" constitution of a cognitive niche, that is a cognitive modification of the environment which confronts the coevolutionary problem of selective pressure in an adaptive or maladaptive way.

detect, control, and punish social parasites, who for example did not share meat (also variously referred to as free riders, defectors, and cheaters). These social parasites were variously dealt with by killing or injuring them (and also by killing cooperators who refused to punish them) *from a distance* using projectile and clubbing weapons.[29] In such a way group cooperation (for example for efficient collective hunting and meat sharing through control of free riders) has been able to adaptively evolve and to render parasitic strategies no longer efficaciously adaptive. Through cooperation and remote killing, individual costs of punishing are greatly reduced and so is individual aggressiveness and violence, "perhaps" because violence is morally "distributed" in a more sustainable way: "Consistent with this view, contemporary humans are unique among top predators in being relatively placid in dealing with unrelated conspecific nonmates under a wide variety of circumstances" [Bingham, 1999, p. 140]. It has to be said that humans, contrary to non-human animals, exchange a fundamental and considerable amount of relatively reliable information with unrelated conspecifics [Bingham, 1999, p. 144].

Here again, the role of docility is worth citing. In Simon's work, docility is related to the idea of socializability, and to altruism in the sense that one cannot be altruistic if he or she is not docile. However, the most important concept is docility and not altruism, because docility is the condition of the possibility of the emergence of altruism. I strongly believe, in the light of the coalition enforcement hypothesis, that moral altruism can be correctly seen as a subproduct of (or at least intertwined with) the violent behavior needed to "morally" defend and enforce coalitions. I have said that groups need to detect, control and punish social parasites, that for example do not share meat, by killing or injuring them (and any cooperators who refuse to carry out punishment) and to this aim they have to gain the cooperation of other potential punishers. This explains altruistic behavior (and the related cognitive endowments which make it possible, such as affectivity, empathy and other non violent aspects of *moral* inclinations) which can be used in order to reach cooperation. To control free-riders inside the group and guard against threat from other alien groups, human coalitions and most gregarious animal groups, have to take care of the individuals who cooperate. It is from this perspective that we can explain, as I have said above, quoting Bingham, why contemporary humans are not only violent but *also* very docile and "[...] unique among top predators in being relatively placid in dealing with unrelated conspecific nonmates under a wide variety of circumstances".[30]

---

[29] In this case injuring and killing are cooperative and remote (and at the same time they are "cognitive" activities). According to the coalition enforcement hypothesis, the avoidance of proximal conflict reduces risks for the individuals. Of course cooperative morality that generates "violence" against unusually "violent" and aggressive free riders and parasites can be performed in other weaker ways, such as through denial of future access to a resource, injuring a juvenile relative, gossiping to persecute dishonest communication and manipulative behaviors within groups or war waged by some groups against less cooperative ones, etc. On the moral/violent nature of gossip and fallacies cf. [Bardone and Magnani, 2009].

[30] On the intrinsic moral character of human communities – with behavioral prescriptions, social monitoring, and punishment of deviance – for much of their evolutionary history cf. [Wilson, 2002, p. 62] and [Boehm, 1999].

The problem here is twofold. First, people delegate data acquisition to their experience, to external cultural resources and to other individuals, as I have illustrated in the sections 3.1-3.6 in chapter three. Second, people generally put their trust in others in order to learn. I have already illustrated how a big cortex, speech, rudimentary social settings and primitive material culture furnished the conditions for the birth of the mind as a universal machine (cf. section 3.7, chapter three). I contended that a big cortex can provide an evolutionary advantage only in the presence of a massive store of meaningful information and knowledge on external supports that only a developed (even if small) community of human beings can possess. If we consider high-level consciousness as related to a high-level organization of the human cortex, its origins can be related to the active role of environmental, social, linguistic, and cultural aspects. It is in this sense that docile interaction lies at the root of our social (and neurological) development. It is obvious that docility is related to the development of cultures, morality, cultural availability and to the quality of cross-cultural relationships. Of course, the type of cultural dissemination and possible cultural enhancements affect the chances that human collectives have to take advantage of docility and thus to increase or decrease their fitness.

As I have already noted, the direct consequence of coalition enforcement is the development and the central role of cultural heritage (morality and sense of guilt included), that is of those *cognitive niches* as new ways of arriving at diverse human adaptations, I have described in chapter six of this book. In this perspective the long-lived and abstract human sense of guilt represents a psychological adaptation, "abductively" anticipating an appraisal of a moral situation to avoid becoming a target of coalitional enforcement. Again, we have to recall that Darwinian processes are involved not only in the genetic domain but also in the additional cultural domain, through the selective pressure activated by modifications in the environment brought about by cognitive niche construction. According to the theory of cognitive niches, coercive human coalition as a fundamental cognitive niche constructed by humans becomes itself a major element of the selective environment and thus imposes constraints (designed by *extragenetic information*, cf. section 6.1.4, chapter six of this book) on its members.[31]

Usually it is said that Darwinian processes operating on genetic information produce human minds whose properties somehow include generation of the novel, complex adaptive design reflected in human material artifacts sui generis. However, these explanations

[31] Some empirical evidence seems to support the coalition enforcement hypothesis: fossils of *Homo* (but not in australopithecines) show, on observation of skeletal adaptations, how selection developed an astonishing competence in humans relating to the controlled and violent use of projectile and clubbing weapons (bipedal locomotion, the development of gluteus maximus muscle and its capacity to produce rotational acceleration, etc.). The observed parallel increase in cranial volume relates to the increased social cooperation based on the receipt, use and transmission of "extragenetic information". Moreover, physiological, evolutionary, and obstetric constraints on brain size and structure indicate that humans can individually acquire a limited amount of extragenetical information, that consequently has to be massively stored and made available in the external environment.

> [...] fail to explain human uniqueness. If building such minds by the action of Darwinian selection on genetic information were somehow possible, this adaptation would presumably be recurrent. Instead, it is unique to humans. Before turning to a possible resolution of this confusion, two additional properties of human technological innovation must be recalled. First, its scale has recently become massive with the emergence of behaviorally modern humans about 40,000 years ago. Second, the speed of modern human innovation is unprecedented and sometimes appears to exceed rates achievable by the action of Darwinian selection on genetic information [Bingham, 1999, p. 145].

Hence, a fundamental role in the evolution of "non" genetic information has to be hypothesized. Appropriately, coalition enforcement implies the emergence of novel extragenetic information, such as large scale mutualistic information exchange – including both linguistically and model-based supported communication – between unrelated kin. As mentioned above, it is noteworthy that extragenetic information plays a fundamental role in terms of ideas transmitted (cultural/moral aspects), behavior, and resources embedded in artifacts (ecological inheritance). It is easy to acknowledge that this information can be stored in human memory – in various ways, both at the level of long-lived neural structures that influence behavior, and at the level of external devices (cognitive niches), which are transmitted indefinitely and are thus potentially immortal – but also independently of small kinship groups. Moreover, transmission and selection of extragenetic information is at least partially independent of an organism's biological reproduction.

8.6.2 The Role of Abduction in the Moral/Violent Nature of Language

In a study concerning language as adaptation the cognitive scientist and evolutionary psychologist [Pinker, 2003] says: "[...] a species that has evolved to rely on information should thus also evolve a means to exchange that information. Language multiplies the benefit of knowledge, because a bit of know-how is useful not only for its practical benefits to oneself but as a trade good with others". The expression "trade good" seems related to a moral/economical function of language: let us explore this issue in the light of the coalition enforcement hypothesis.

Taking advantage of some ideas brought up by Thom's catastrophe theory (cf. the previous sections of this chapter) on how natural syntactical language is seen (cf. subsections 8.5.1 8.4.1) as the fruit of social necessity,[32] its fundamental function can only be seen clearly if also linked to an intrinsic *moral* (and at the same time *violent*) aim which is basically rooted in a kind of *military intelligence*. Thom says language can simply and efficiently transmit *vital* pieces of information about the fundamental biological oppositions (life, death – good, bad). It is from this perspective that we can clearly see how human language – even at the level of more complicated syntactical expressions – always carries information (pregnancies) about

[32] A view that is shared by other approaches in cognitive science, illustrated in chapter three, cf. the case of Mithen (subsection 3.3), the so-called Machiavellian hypothesis (Whiten, Byrne, and Dunbar, subsection 3.3.2, Clark (section 3.4.2), Donald, Logan, and Deacon (subsection 3.6.5).

moral qualities of persons, things, and events. Qualities which are always directly or indirectly related to the needs of the individual and/or of the group/coalition for survival.

I have illustrated in the previous subsection that in human or pre-human groups the appearance of coalitions dominated by a central leader quickly leads to the need for surveillance of surrounding territory to monitor prey and watch for enemies with the potential to jeopardize the survival of the coalition.

This is an idea shared by Thom who believes that language becomes a fundamental tool for granting stability and favoring the indispensable manipulation of the world "thus the localization of external facts appeared as an essential part of social communication" [Thom, 1988, p. 26], a performance that is already realized by naming[33] (the containing relationship) in divalent structures: "X is in Y is a basic form of investment (the localizing pregnance of Y invests X). When X is invested with a ubiquitous biological quality (favorable or hostile), then so is Y" (*ibid.*). A divalent syntactical structure of language becomes fundamental if a *conflict* between two outside agents has to be reported. The trivalent syntactical structure subject/verb/object forges a salient "messenger" form that conveys the pregnance between subject and recipient. In sum, the usual abstract functions of syntactic languages, such as conceptualization, appear strictly intertwined with the basic *military* nature of communication.

I contend that this military nature of linguistic communication is intrinsically "moral" (protecting the group by obeying shared norms), and at the same time "violent" (for example, killing or mobbing to protect the group). This basic moral/violent effect can be traced back to past ages; currently, when we see the "prehuman" use of everyday natural language in current mobbers to express linguistic strategic communications "against" the mobbed target (inside the group or in alien threatening groups). These linguistic strategic communications are often performed thanks to hypothetical reasoning, abductive or not. In this case the use of natural language can take advantage of efficient hypothetical cognition through gossip, fallacies etc., but also of the moral/violent exploitation of apparently more respectable and sound truth-preserving and "rational" inferences. The narratives used in a dialectic and rhetorical setting qualify the mobbed individual and its behavior in a way that is usually thought of by the mobbers themselves (and by the individuals of their coalition/group) as moral, neutral, objective and justified while at the same time hurting the mobbed individual in various ways. Violence is very often subjectively dissimulated and paradoxically thought of as the act of performing just, objective moral judgments and of persecuting moral targets. *De facto* the mobbers' coordinated narratives harm the target, very often without an appreciable awareness of the violence performed.

This human linguistic behavior is clearly made intelligible when we analogously see it as echoing the antipredatory behavior which "weaker" groups of animals (birds, for example) perform, for example through the use of suitable alarm calls and aggressive threats. Of course such behavior is mediated in humans through

[33] It is important to stress that pregnant forms receiving names, loose their alienating character.

socially available ideologies (more or less endowed with moral ideas) and systems of culture. Ideologies can be seen as fuzzy and ill-defined cultural mediators that spread pregnances which invest all those who put their faith in them and stabilize and reinforce the coalitions/groups: "[...] the follower who invokes them at every turn (and even out of turn) is demonstrating his allegiance to an ideology. After successful uses the ideological concepts are extended, stretched, even abused", so that their meaning slowly changes in imprecise (and "ambiguous", Thom says)[34] ways, as we have seen happens in the application of the archetypical principles of mobbing behavior.

I have illustrated the speculative hypothesis of collective unconscious in chapter three (subsection 3.5.2). It is that part of the individual unconscious we share with others humans shaped by evolution which wired archetypes into it, like the "scapegoat" (mobbing) mechanism, I have just mentioned. Here a paroxysm of violence focuses on an arbitrary sacrificial victim and a unanimous antipathy would, mimetically, grow against him. The process leading to ultimate bloody violence which was, for example, widespread in ancient and barbarian societies, is mainly carried out in current social groups through linguistic communication. Following [Girard, 1977; Girard, 1986] we can say that in the case of ancient social groups the extreme brutal elimination of the victim (still visible in some subgroups of current rich societies) would reduce the appetite for violence that had possessed everyone just a moment before, leaving the group suddenly appeased and calm thus achieving equilibrium in the related social organization (a sacrifice-oriented social organization which maybe repugnant to us but is no less a society for this reason). This kind of archaic brutal behavior is still present in civilized human conduct in rich countries, almost always implicit and unconscious, for example in racist and mobbing behavior. Given the fact that this kind of behavior is widespread and partially unconsciously performed, it is easy to understand how it can be implicitly "learned" in infancy and still implicitly embedded in that individual's *cultured unconscious* of ideological character we humans collectively share with others. I strongly believe that the analysis of this archaic mechanism (and of other similar, "moral/ideological/violent" mechanisms) might shed new light on what I call the basic equivalence between engagement in morality and engagement in violence, amazingly almost always hidden from the awareness of the human individual agents.

It is worth mentioning in conclusion the way Thom accounts for this social phenomena in terms of pregnancies. "Mimetic desire", in which Girard roots the violent and aggressive behavior (and the scapegoat mechanism) of human beings [Girard, 1986] can be seen as the act of appropriating a desired object which imbues that object with a pregnance "the same pregnance as that which is associated with the act by which 'satisfaction' is obtained" [Thom, 1988, p. 38]. Of course this pregnance can be propagated by imitation through the mere sensory sight of "superior" individuals in whom it is manifest (or through exposure to linguistic descriptions and narratives about her/him): "In a sense, the pleasure derived from looking forward to

[34] From this perspective the huge moral/violent exploitation of equivocal fallacies in ideological discussions, oratories, and speeches is obvious and clearly explainable, as I have illustrated above (chapter seven, section 7.8).

a satisfaction can surpass that obtained from the satisfaction itself. This would have been able to seduce societies century after century (their pragmatic failure in real terms having allowed them to escape the indifference that goes with satiety as well as the ordeal of actual existence)" (*ibid.*). "Mimetic desire"[35] is indeed a template of behavior which can be picked up from appropriate cultural systems, available over there, as part of the external cognitive niches built by many human collectives and gradually externalized (and always transmitted through teaching and learning) during the centuries, as fruitful ways of favoring social control over coalitions. Indeed mimetic desire triggers envy and violence but at the same time the perpetrated violence causes a reduction in appetite for violence, leaving the group suddenly appeased and calm, thus achieving equilibrium in the related social organization through a *moral effect*, that is at the same time a *carrier of violence*.

Summary

In the conclusion to chapter five I have argued that the cognitive importance of abduction means that we must take great care over its intellectual treatment. Analysis in the first two Peircean sections of the chapter shows that in classical philosophical studies on the subject great intellectual sophistication was already at play. Peirce teaches us something interesting about well-known conflicting views on abduction: 1) there is no sharp contrast between the idea of abduction (both creative and selective) as perception and the idea of abduction as something that pertains to logic, both aspects are inferential in themselves, and fruit of sign activity, indeed, 2) a suitable meaning of the concept of inference does not have to be exhausted by its logical aspects, but instead referred to the effect of various sensorial activities, thus 3) the dichotomy instinct vs. inference represents a conflict we can overcome simply by observing that the work of abduction is partly explicable as a biological phenomenon and partly as a "logical" operation related to the "plastic" cognitive endowments of all organisms.

The result that abduction is partly explicable as a biological instictual phenomenon and partly as a more or less "logical" operation related to the"plastic" cognitive endowments of all organisms naturally leads to the analysis of what I have called *morphodynamical abduction*, from the dynamical systems and catastrophe theory perspective. This perspective is contrasted with the classical computational approach to cognition that describes a cognitive process (and so abduction) by the manipulation of internal semiotic representations of the external world. To grasp the role of abduction in the dynamical system I have provided an analysis of the concepts of anticipation, adumbration, and attractor and also of the salient/pregnant dichotomy. This chapter's analysis has permitted a description of the abductive

[35] Mimetic desire is related to envy: when we are attracted to something the others have but that we cannot acquire because others already possess it, we experience an offense which generates envy. In this Girard's perspective envy is a mismanagement of desire and it is of capital importance for the moral life of both communities and individuals. As a reaction to offense, envy easily causes violent behavior".

generation of new hypotheses in terms of a catastrophic rearrangement of the parameters responsible for the behavior of the system. This approach fits well with the distributed cognition perspective, explained in the previous more methodological chapters of this book, where abduction is in turn encountered in the sophisticated distributed interplay between internal and external representations.

The main aim of the sections devoted to catastrophe theory is to demonstrate that pregnances and saliences provide further help in increasing knowledge about abductive "hypothesis generation" at the level of both instinctual behavior and representation-oriented behavior, where nonlinguistic features drive a "plastic" model-based cognitive role. Furthermore, in terms of dynamic systems and of Thom's mathematical modeling we reach an initial outline of a "physics of abduction", where its cognitive essence is seen in a whole unified naturalistic framework where all phenomena and thus also cognition, gain a primal eco-physical significance.

Finally, I would like to highlight the so-called coalition enforcement hypothesis illustrated in this chapter, which describes humans as self-domesticated animals engaged in a continuous activity of building morality, an activity that at the same time incorporates punishment policies, where hypothetical (and abductive) cognition is centra. I have anticipated that natural language is a fundamental mediator of this enforcement activity in cognitive niches and in the previous chapter, devoted to the analysis of abduction (and induction) from the perspective of logic, this "military" nature of linguistic communication has been further explored, also taking advantage of the analysis of the special cognitive role performed by fallacies.

In summary, in light of the considerations I outline in this chapter it can be said that an important facet of abductive cognition, which needs be emphasized and studied, resorts to its semiotic and eco-physical status. In this case we give up the cognitive centrality of discreteness, due to its computational privileging, and we see it as intertwined with the proper continuity of physical systems. However, Thom says that the discrete decomposition of naturally continuous phenomena is "not just an illusion of the mind" [Thom, 1988, p. 25], but can be grasped in terms of the distinction between salience and pregnance: this distinction is fundamental even if it sometimes tends to be blurred.

Conclusion

In place of a formal conclusion, I offer here a sort of summary and a few final comments about what I see as the most important elements of *Abductive Cognition.* My hope is to have established a compelling rationale for making a serious commitment to further increasing knowledge, both logico-epistemological and cognitive, on abduction. It is certainly through new knowledge on this important way of thinking that we can increase our chances of success in maintaining the epistemological commitment to studying human "rational" attitudes, keeping research lively and fruitful. At the same time, given the importance of hypothetical reasons in human cognition, study on abduction helps to maintain the importance of scientific mentality, also beyond the strictly intellectual scientific community. The knowledge about abductive cognition I passionately endorsed in this book also aims at supplying researchers with the rational poise required to handle controversial issues regarding creative and hypothetical reasoning, now and in the future.

As I said in the Preface, in 1998 Jaakko Hintikka had already stressed the philosophical importance of abduction in "What is abduction? The fundamental problem of contemporary epistemology". At that time many articles about abductive reasoning were already available, especially related to the history of philosophy, artificial intelligence (for example, in the case of knowledge-based medical systems and in planning), and logic. Those articles already characterized the field as constitutively interdisciplinary: 1) articles written by the historians of philosophy and researchers studying the semiotics related to the work of Charles Sanders Peirce; 2) articles concerning research on abductive reasoning stimulated by the development of artificial intelligence, which immediately recognized its importance; 3) articles written by logicians engaged in building logical non-standard models of abduction (such as nonmonotonic systems), in turn promoted by the new AI ideas.

In that intellectual atmosphere I published the pioneering book *Abduction, Reason, and Science* (Kluwer Academic/Plenum Publishers, New York, 2001), in which I presented a new interdisciplinary perspective on abduction. The book illustrated a novel philosophical and cognitive perspective able to introduce an unconventional approach taking advantage of the research available in epistemology, cognitive science and artificial intelligence and proposing an interdisciplinary treatment. The book also introduced the following main concepts: creative and selective abduction, model-based and manipulative abduction, and the notion of the epistemic mediator, all concepts which were then usefully exploited and discussed within the international scientific community. It seems the book also opened up new research opportunities, still to be further exploited in international scientific debate, and which I am honored to offer again through this new book from a fresher and wider perspective. Over the last decades I have benefited from several international relationships with other researchers, been involved in and taken advantage of the direct organization of international events, of the publication of edited books, conference proceedings, and articles in international journals and I expect this book to enhance this type of social activity which I consider so precious to academic work.

Guessing hypotheses is one of the most important cognitive activities of human beings and of various animal species. We have seen in this book that this cognitive activity occurs in a multimodal way, which involves all sensormotor modalities and happens with the help of visualizations, simulations, analogies, emotions, etc. Consequently, multimodal abductive cognition in humans is fundamentally hybrid, and does not occur only with the help of natural or artificial languages, as supposed from the traditional epistemological perspectives. Its hybridity is also at the root of several high-level creative skills. All sensorimotor modalities are involved in this performance. Culture and science represent the main fruit of that important human endowment – i.e. the capacity to guess hypotheses – which has also presented an astonishing evolutionary success. I always thought the concept of abduction, already studied by Charles Sanders Peirce in XIX Century, is the best philosophical and intellectual tool to afford the study of that human capacity which is often very creative. In the second half of the last century this concept suddenly acquired a great importance in logic, cognitive science, artificial intelligence, neuroscience and biology demonstrating both a ground-breaking nature and a strong interdisciplinary character.

Studying multimodality embedded in model-based reasoning further offers the chance to stress the role of models in human reasoning and thus in the case of hypothesis formation capabilities. Furthermore, joining abduction and model-based reasoning in this book allowed me to clearly acknowledge the fact that hypothesis guessing is not only multimodal but also distributed. Cognitive agents distribute in – and delegate to – natural entities and artifacts various cognitive roles. In this way the activity of building what current biological research calls "cognitive niches" plays an important role also, and especially, in abductive cognition. Thus the study of abductive cognition opens up new horizons and opportunities for research, mainly because of its interdisciplinary nature, which crosses the boundaries between different fields of research, but also because the consideration of model-based and abductive aspects of reasoning permits an analysis of cognition capable to understand it in a broader way: 1) cognition is shown as an activity which involves models and so we can acknowledge that it is present in various organisms, not only in humans; 2) cognition is distributed, so the door is opened to the fundamental analysis of epistemic and cognitive roles of artifacts; 3) "creative" – abductive – aspects of cognition have to be studied and possibly enhanced, given the ethical importance of knowledge in our technological world.[1]

I tried to shed new light on explanatory, non-explanatory, instrumental, selective and creative aspects of abduction with the aim of unifying philosophical, logical, computational, neurological, cognitive (and eco-cognitive), and epistemological issues, not excluding their role in common sense. I strongly believe that the study of abduction can clarify and unify issues and problems belonging to different disciplines. I often mention in this book that an interesting and still neglected point of contention about human reasoning is whether or not concrete manipulations of external objects and embodied forms of reasoning influence the generation of hypotheses. The book tries to fill this gap stressing the role of manipulative abduction, showing how we can find methods of constructivity based on external models, cognitive epistemic mediators, and hybrid representations in scientific and everyday reasoning (but also in technological innovation).

While many results were already available, especially in the field of logical and computational models, other research topics needed to be further explored for this book. First of all at the cognitive level: further research was needed on the problem of abductive cognition in animals, already envisaged by Peirce, other investigation ranged from the role of "affordances" in clarifying some aspects of abduction, to the recent concept of "cognitive niche", as a way of reinterpreting innovative abductive thinking from a wide evolutionary perspective. I also tried to focus attention on other central issues: philosophical (further insight on Peircean and other classical texts), epistemological (especially related to the evolutionist background), logical (the problem of the agent-based approach and the status of fallacious reasoning), neurological (how to see abductive cognition as a neuro-multimodal process), and physico/mathematical (studying morphodynamical abduction).

Those who are suspicious of interdisciplinarity – this book is mainly interdisciplinary – invoke the conviction that we must "focus" and restrict our research in order to honor the target of producing scientific results, which can only be reached through specialist commitment. Even if it is undoubtedly true that specialist research is an optimal way to reach scientific answers, that conviction becomes dogmatic and intellectually dangerous when used to proscribe different approaches. Here in the increasingly complex twenty-first century, that conviction alone can sometimes seem too simplistic and general to do us much good in the face of the excessive specialization of research and of the lack of unifying and broader perspectives. But we can build on that dogmatic conviction: by moderating its mandate – and by encouraging researchers of all non scientific disciplines to seriously take into consideration the results of scientific research – we can begin to make peace with that exclusive mentality that inevitably transforms other approaches into "deviant" approaches. Indeed, defending the interdisciplinary approach we are simply re-engaged in one of the basic tenets of the philosophical mentality, now enriched by a naturalistic commitment, which acknowledges the relevance of scientific results. By endorsing this view I am

[1] I was recently engaged in stressing the importance of studying the role of artifacts from both a cognitive and an ethical point of view when I wrote my recent book *Morality in a technological World. Knowledge as Duty* (Cambridge University Press, Cambridge, 2007).

just following the lesson of the neopositivist philosophers who already taught, at the beginning of the last century, the importance of scientific research and results but "also" the importance of a basic profound philosophical interdisciplinary openness. This commitment always characterized the life of that part of philosophical reflection which is interested in the structure of knowledge and of cognition. How could we deal with a concept like abductive cognition abandoning the philosophical spirit of openness to every kind of knowledge, and without taking into account the results of many modern disciplines?

Thinking of abductive cognition as a valuable and central topic of contemporary epistemology and cognitive science is also a modest but useful way to navigate the murky waters of that "virtuosity" (the term in taken from John Woods, an expert in abductive reasoning and fallacies) of many modern logical technicalities aiming at modeling abduction, which are looking for a more fruitful cultural life. Faced with these virtuosities, even as a trained logician, and not only a philosopher, I felt myself compelled to approach abduction in a broader way. I think my extended method is appropriate if we wish to live as philosophers of cognition, now and, especially, in the future, still committed to scientific mentality and to rational methods, in that future when mythical and mystical knowledge – i.e. that knowledge which is for example always hidden in religion, everyday thinking, and mass-media – could be more common than today and intertwined with an unwelcome triumph of conflictual and violent contrasting and irreducible beliefs. Conflicts and violence that tomorrow could otherwise increase, in the absence of a more widespread intellectual dedication to scientific mentality and "notwithstanding" the further welcome growth of specialization in science and technology.

Furthermore, the increasing hybridization between the human and the artificial, a process fuelled, of course, by scientific knowledge, makes an interdisciplinary philosophical "rational" attitude to human and animal hypothetical cognition increasingly relevant to our everyday lives; similarly, it demands greater understanding of this new hybrid human condition. I would also hope that readers come away from *Abductive Cognition* with a renewed appreciation for the astonishing human capacity of guessing more or less creative hypotheses, which I believe addresses important issues about the relevance of cognition and knowledge in our technological era. Overall, I would say that the study on how we as individuals deal (or fail to deal) with the attainment of appropriate hypothetical knowledge needs much greater attention in scientific research and in the interdisciplinary arena of philosophical discussion. Such work will, in turn, contribute to psychological research, which is often a resource overlooked in discussion of cognitive issues.

Abductive cognition is so pervasive, so much a part of human (and animal) life, that it is not easy to imagine how to live and flourish without it. Guessing hypotheses is a complicated cognitive mechanism, intertwined with the role of what have traditionally been called fallacies. These being available to people to perform various tasks, including those related to their defensive, moral, and violent commitments, as described in the closing chapters of this book. Stressing this issue has been a further extension of the analysis of abductive cognition. Acknowledging our "condition" as "hypothetical beings" ("Popperian creatures", as Daniel Dennett says), and increasing knowledge about this astounding capacity to guess, is a form of accepting responsibility - it will hopefully weaken errors and could be of help in improving our freedom and the ownership of our destinies.

Let us begin.

Lexicon of Abductive Cognition

Abductive logic programming. Abductive logic programming (ALP) is a high level knowledge-representation framework that can be used to solve problems declaratively based on abductive reasoning. It extends normal logic programming by allowing some predicates to be incompletely defined, declared as abducible predicates. Problem solving is accomplished by deriving hypotheses on these abducible predicates (abductive hypotheses) as solutions of problems to be solved. These problems can be either observations that need to be explained (as in classical abduction) or goals to be achieved (as in normal logic programming). It can be used to resolve problems in diagnosis, planning, natural language and machine learning. It has also been used to handle negation as failure as a way of expressing some forms of abductive reasoning.

Affordances and abduction. An affordance is a resource or chance that the environment presents to the specific organism, such as the availability of water or of finding recovery and concealment. From an abductive perspective, affordances as environmental chances can be related to the variable (degree of) abducibility of a semiotic configuration of "signs": a chair affords sitting in the sense that the action of sitting is a result of a sign activity in which we perceive some physical properties (flatness, rigidity, etc.), and therefore we can ordinarily infer (in Peircean sense) – abduce – that a possible way to cope with a chair is sitting on it.

AKM-schema. The AKM-schema illustrates the explanatory dimension of abduction. That is, it describes abduction as the kind of reasoning which aims at generating hypotheses, which are related to considerations of plausibility, relevance and characteristicness. In this perspective abduction is immediately considered a generation/selection of "plausible" hypotheses. However, the AKM model does not exhaust the concept of abduction, which also presents a non-explanatory and instrumental dimension, possibly considered as not intrinsically consequentialist.

Animal abduction. Many animals – traditionally considered "mindless" organisms – make up a series of signs and are engaged in making, manifesting or reacting to a series of signs. Through this semiotic activity they are engaged in "being cognitive agents" and therefore in thinking "intelligently". An important effect of this semiotic activity is a continuous process of "hypothesis generation" that can be seen at the level of both instinctual behavior, as a kind of "hardwired" cognition, and representation-oriented behavior, where nonlinguistic pseudothoughts drive a plastic model-based cognitive role. This activity is at the root of a variety of abductive performances.

Anomalies and abduction. Anomaly refers to unexpected and uncharacteristic events that create problems which demand a solution. Anomalous events are marked by both an epistemic disadvantage and an emotional rating. Philosophers of science in the last century have illustrated that inconsistencies and anomalies often play an important role in the growth of scientific knowledge. Hence, contradictions and inconsistencies are fundamental in abductive reasoning, and abductive reasoning is appropriate for "governing" inconsistencies.

Anticipation as abduction. In the tradition of phenomenology anticipations correspond to a kind of an intentional expectation. When we see a spherical form from one perspective, we will assume that it is effectively a sphere, but it could be also a hemisphere. Anticipations share with other kinds of abductions various features: they are highly conjectural and nonmonotonic, so wrong anticipations have to be replaced by other plausible ones. Moreover, they constitute an activity of "generating and testing" as a kind of action-based cognition. Not all the anticipations are informationally equivalent and work like attractors for privileged individuations of objects. In this sense the whole activity is toward "the best anticipation", the one that can display the object in an optimal way.

Brain plasticity and abduction. Cognitive performances are to a large extent neurally prespecified. Research in genetics shows substantial evidence for genetic pre-wiring of a great deal of brain structure in both human and non-human animals. However, it is worth to note that many neuroanatomical modules need considerable input from the external world (already at the level of utero biochemical factors), fine tuning, and experience-driven maintenance to get their "restructured" functional adult abilities, and also to preserve the structure itself. In this sense pre-wiring does not imply hardwiring, but the so-called brain plasticity" endowments have limits and genome-directed constraints. Instinct vs. inference represents a conflict we can overcome simply by observing that the work of abduction is partly explicable as a biological phenomenon and partly as a more or less "logical" operation related to "plastic" cognitive endowments of all organisms.

Chance and abduction. Humans can be considered chance seekers as far as they are continuously engaged in a process – mainly abductive – of building up and then extracting latent possibilities to uncover new valuable information and knowledge. That is, humans like other organisms do not simply live their environment, but they actively shape and change it looking for suitable chances. Hence, in this perspective chances constitute that "information" which is not stored internally in memory or already available in an external reserve, but that has to be "extracted" and then picked up upon occasion.

Creative abduction. Creative abduction is the process in which a completely new hypothesis is created. An example of creative abduction is scientific discovery: the discovery of a new disease and the manifestations it causes in the field of medical knowledge or the Kepler's discovery of the elliptic orbit of the planets.

Deduction and abduction. Deduction is that kind of truth-preserving reasoning which starts from reasons and looks for consequences. In doing so deduction does not produce any new idea. New ideas are due to abduction that involves the generation and evaluation of hypotheses. In the process of hypothesis generation and evaluation, deduction plays a fundamental role in deriving the hypotheses consequences, which, in turn, are compared with the available data by induction.

Diagnosis. Conjectures can be either the fruit of an abductive selection in a set of pre-stored hypotheses or the creation of new ones. "Creative" abduction deals for example with the whole field of the growth of scientific knowledge. This is irrelevant in medical diagnosis where instead the task is to "select" from an encyclopedia of pre-stored diagnostic entities. In this case abduction selects hypotheses; from these hypotheses consequences are derived by deduction that are compared with the available data by induction. Often the information available does not allow a physician to make a precise diagnosis. Therefore, she has to get further data, or even try some different manipulations to uncover symptoms otherwise hidden.

Distributed military intelligence and abduction. According to the so-called coalition enforcement hypothesis, language is basically rooted in a kind of military intelligence. This perspective especially stresses its intrinsic moral (and at the same time violent) nature of language (and of abductive and other hypothetical forms of reasoning and arguments intertwined with the propositional/linguistic level). Language efficiently transmits vital pieces of information about the fundamental biological opponents (life, death – good, bad) and so it is intrinsically involved in moral/violent activities. In this process abductive ways of making hypotheses of various kind plays a key role: its cognitive essence can be seen in a whole unified naturalistic framework where abductive cognition gains a fundamental eco-physical significance, which also nicely includes some aspects related to a kind of "social epistemology". This conceptual framework also sheds further light on some fundamental dialectical and rhetorical roles played by the so-called fallacies, which are of great importance to stress some basic aspects of human abductive cognition.

Embodied and distributed cognition. Extra-theoretical aspects and manipulations of "external" objects in reasoning are very important in hypothetical reasoning. Paying attention to the perceptual and manipulative dimension of cognition, Peirce reminded us that human cognitive systems are not "isolated" and autonomous entities. Cognition is embodied and the interactions between brains, bodies, and external environment are its central aspects. Cognition is occurring taking advantage of a constant exchange of information in a complex distributed system that crosses the boundary between humans, artifacts, and the surrounding environment. This interplay is especially manifest and clear in various aspects of abductive cognition.

Emotion as abduction. Emotion furnishes immediate abductive appraisals of the bodily states, and provides a kind of explanation of them. Happiness, sadness, fear, anger, disgust, and surprise all can be viewed as judgments about a person's general state; a man who unexpectedly comes across a tiger on the loose, for example, would be understandably afraid because the large carnivore threatens his instinct to stay alive. In this sense, all emotions are connected to goal accomplishment: people become angry when they are thwarted, for instance, and feel pleased when they are successful.

Fallacies and abduction. In the perspective of classical logic abduction reasoning can be defined as fallacious. More precisely, abduction is classified as the fallacy of affirming the consequent. In the light of an agent-based approach to logic the fallacious character of abduction an be further clarified and weakened: abduction is recognized as a very precious method of explanation and discovery in science and in everyday reasoning. Hence, in an agent-based perspective this kind of fallacious reasoning can be redefined and considered as a fundamental and good way of reasoning.

Fallacy of affirming the consequent. Abduction appears to be a formal fallacy that can be recognized in the perspective of classical logic: it is the fallacy of affirming the consequent. Basically it is an invalid inference, which assumes the form: "if A, then B, B; then A. From the classical logic point of view this inference is fallacious and not truth preserving.

GW-schema. The GW-schema provides the opportunity to illustrate non-explanatory and instrumentalist aspects of abductive cognition. Examples of the non-explanatory features of abduction are for example present in logic and mathematical reasoning. Furthermore, physics often aims at discovering physical dependencies which can be considered explanatorily undetermined. In this case abduction also exhibits an instrumental aspect.

Habit as abduction. Habits are related to abductions, which are automatically triggered apparently without any conscious involvement. They can derive from evolutionary established instinctual hard-wired behaviors, but they can also be learned and subsequently stabilized.

Heuristic strategies. We can usefully see selective and creative abduction as performed by the application of "heuristic procedures" (often called "heuristic strategies") that involve all kinds of good and bad inferential moves, beyond the simple mechanical application of rules. It is only by means of these heuristic procedures that the acquisition of new truths is guaranteed. Also Peirce's mature view on creative abduction as a kind of inference seems to stress the strategic component of reasoning. Strategic rules are smart rules, even if they fail in individual cases, and show a propensity for cognitive success. When there is a problem to solve, we usually face several possibilities (hypotheses) and we have to create or select the suitable one. In this situation we can rely on heuristics, that are for example rules of thumb expressed in sentential terms and that help us reach satisfactory choices without considering all the possibilities.

Hybrid abducers. The various abductive activities human agents are engaged in drastically rely on external supports. For instance, human agents create mimetic external representations which mirror concepts and problems that are already represented in the brain and need to be enhanced, solved, further complicated, etc. so they can sometimes creatively give rise to new concepts and meanings. In the process of distributing the mind into the material world around them, human agents develop new abductive skills, which would not be exhibited without the continuous interplay with the various external supports. In doing so, human agents can be described as hybrid abducers. Organic agents are spontaneous inducers and abducers but human beings also construct logical and computational (ideal) systems both able to mimic human inductions and abductions and to create new and more "rational" ways of inducing and abducing. These systems are in turn used by organic (human) agents: they consequently have to be seen as hybrid reasoners.

Ignorance-preserving abductive cognition. Abduction is a procedure in which something that lacks classical epistemic virtue is accepted because it has virtue of another kind. For example: let S be the standard that you are not able to meet (e.g., that of mathematical proof). It is possible that there is a lesser epistemic standard S' (e.g., having reason to believe) that you do meet. Focusing attention on this cognitive aspect of abduction we can contend that abduction (basically seen as a scant-resource strategy, which proceeds in absence of knowledge) presents an ignorance preserving (but also an ignorance mitigating) character. Of course it is not at all necessary, or frequent, that the abducer be wholly in the dark, that his ignorance be total. It needs not be the case, and typically isn't, that the abducer's choice of a hypothesis is a blind guess, or that nothing positive can be said of it beyond the role it plays in the subjunctive attainment of the abducer's original target (although sometimes this is precisely so). Abductive reasoning is a response to an ignorance-problem: one has an ignorance-problem when one has a cognitive target that cannot be attained on the basis of what one currently knows. Ignorance problems trigger one or other of three responses. In the first case, one overcomes one's ignorance by attaining some additional knowledge (subduance). In the second instance, one yields to one's ignorance (at least for the time being) (surrender). In the third instance, one abduces, and so has a positive and reasoned basis for new action even if in the presence of the constitutive ignorance.

Induction and abduction. Since the time of John Stuart Mill, the name given to the most important kinds of conjectural non deductive reasoning has been induction, considered as an aggregate of many methods for discovering generalizations and casual relationships. Consequently induction in its widest sense is an ampliative process of the generalization of knowledge. By using induction it is possible to synthesize individual statements into general laws – inductive generalizations – in a defeasible way, but it is also possible to confirm or discount hypotheses. In the case of the relationship between abduction, deduction, and induction, illustrated by Peirce and exploited in the ST-model, induction does not deal with the problem of individuating the ways of "generating" inductive hypotheses but refers to a logic of hypothesis "evaluation". Abduction creates or selects hypotheses; from these hypotheses consequences are derived by deduction that are compared with the available data by induction. In the ST-model induction is used as the process of reducing the uncertainty of established hypotheses by comparing their consequences with observed facts.

Inference to the best explanation. There are two main epistemologico/cognitive meanings of the word abduction: (1) abduction that only generates plausible hypotheses (selective or creative) – this is the meaning of abduction accepted in my epistemological ST-model – and (2) abduction considered as inference to the best explanation, that also evaluates hypotheses by induction. In the latter sense the classical meaning of abduction as inference to the best explanation (for instance in medicine, to the best diagnosis – selective abduction) is described by the complete abduction-deduction-induction cycle.

Instinct as abduction. An example of instinctual (and putatively "unconscious") abduction is given by the case of animal embodied kinesthetic/motor abilities, capable of leading to some appropriate cognitive behavior; Peirce says abduction even takes place when a new born chick picks up the right sort of corn. This is another example, so to say, of spontaneous abduction – analogous to the case of some unconscious/embodied abductive processes in humans. Therefore, instinct vs. inference represents a conflict we can overcome simply by observing that the work of abduction is partly explicable as a biological phenomenon and partly as a more or less "logical" operation related to "plastic" cognitive endowments of all organisms.

Instrumental abduction. Abduction exhibits an instrumental and strategic aspect, for instance, when intertwined with the exquisite epistemological problem of the role of unfalsifiable hypotheses in scientific reasoning. In this case, an abductive hypothesis can be highly implausible from the "propositional" point of view and nevertheless it can be adopted for its instrumental virtues, such as in the Newtonian case of action-at-a-distance. Highly implausible hypotheses from the "propositional" point of view can be conjectured because of their high "instrumental" plausibility, where a different role of characteristicness is at stake. We have to note that in some sense all abductions embed instrumental factors. In the general case, one accepts because doing so enables ones target to be attained, notwithstanding that lacks the relevant epistemic virtue. However, in cases such as Newtons, is selected notwithstanding that it is considered to be epistemically hopeless.

Logical abductive agents. Logical agents have to be seen as ideal "demonstrative environments", which in turn are suitable tools for the creation of logical/deductive models of abductive reasoning itself. They are the fruit of cognitive externalizations in objective logical systems, endowed with symbolic, abstract and rigorous cognitive features, which are also potentially exploitable at the computational level.

Manipulative abduction. Manipulative abduction is a process in which a hypothesis is formed and evaluated resorting to a basically extra-theoretical and extra-sentential behavior that aims at creating communicable accounts of new experiences to integrate them into previously existing systems of experimental and linguistic (theoretical) practices. Manipulative abduction represents a kind of redistribution of the epistemic and cognitive effort to manage objects and information that cannot be immediately represented or found internally. An example of manipulative abduction is the case of the human use of the construction of external diagrams in geometrical reasoning, useful to make observations and "experiments" to transform one cognitive state into another for example to discover new properties and theorems.

Model-based abduction. Inaugurating the new semiotic perspective, Peirce stated that all thinking is in signs, and signs can be icons, indices, or symbols and that all inference is a form of sign activity, where the word sign includes "feeling, image, conception, and other representation". In this light it can be maintained that a considerable part of the abductive processes is model-based. That is, a considerable part of hypothesis creation and selection is model-based: it is occurring in the middle of a relationship between brains and model-based aspects of external objects and tools that have received cognitive and/or epistemological delegations.

Moral deliberation and abduction. We usually base our guessing of hypotheses on incomplete information, and so we are facing nonmonotonic inferences: we reach defeasible conclusions from limited information, and these conclusions are always withdrawable. It is in this sense that both explanatory and instrumental abductive reasoning constitute a possible useful model of "practical reasoning": ethical deliberations are always adopted on the basis of incomplete information and on the basis of the creation or selection of particular abduced hypotheses which play the role of reasons. Hence, ethical deliberation shares some aspects with hypothetical cognition.

Morphodynamical abduction. Morphodynamical abduction is abduction considered in the light of a physico-mathematical framework. We can see a cognitive system as a complex system, where its transformations can be described in terms of a configurational structure. That is, different mental states are defined by their geometrical relationships within a larger dynamical environment. This suggests that the system, at any given instant, possesses a general morphology we can study by observing how it changes and develops. In this perspective the reconsideration and reevaluation of some aspects of the tradition of dynamic systems can provide a new analysis of the concept of abduction as a form of hypothetical reasoning that clarifies the process of anticipation and the concept of "attractor", also taking advantage of the results of the phenomenological tradition. This analysis permits a description of the abductive generation of new hypotheses in terms of a catastrophic rearrangement of the parameters responsible for the behavior of the system.

Multimodal abduction. Multimodal abduction depicts hybrid aspects of abductive reasoning. Abductive inference can be visual as well as verbal, and consequently we have to acknowledge the sentential, model-based, and manipulative nature of abduction. Both evidence and hypotheses can be represented using various sensory modalities. Some basic aspects of this constitutive hybrid nature of multimodal abduction involve words, sights, images, smells, etc. but also kinesthetic experiences and other feelings such as pain, and thus all sensory modalities.

Narrative abduction. Narrative abduction is the process underlying the generation of more or less creative and new narratives. For example narratives are important in science when scientists do not use only mathematics, diagrams, and experiments but also stories. Scientists use "experimental" narratives in a constructive and hypothesizing way. Narrative abductions consent to construct and reconstruct experience, and to distribute it across a social communicative network of negotiations among the different scientists by means of the so-called construals. Moreover, in all narratives, and especially in the narratives of detection, the problem of prompting abductive explanations of actions is widespread. Peirce contended that all sensations participate in the nature of a unifying hypothesis, that is, in abduction, in the case of emotions too: recent cognitive theories of emotions, largely based on Peirce's intuitions, are able to explain how the reader, in front of narratives, feels emotions as abductions.

Neuroabduction. Considering a neural structure as a set of neurons, connections, and spiking behaviors, and their interplay, and the behavior of neurons as patterns of activation, neuroabduction is a process in which one neural structure representing the explanatory (or non-explanatory or instrumental) target generates another neural structure that constitutes a hypothesis.

Niche construction and abduction. The theory of cognitive niches describes the environment as a sort of "global market" that provides living creatures with unlimited possibilities. Indeed, not all the possibilities that the environment offers can be exploited by the human and non-human animals that act on it. For instance, the environment provides organisms with water to swim in, air to fly in, flat surfaces to walk on, and so on. However, no creatures are fully able to take advantage of all of them. Through the activity of niche construction all organisms try to modify their surroundings in order to better exploit those elements that suit them and eliminate or mitigate the effect of the negative ones. This activity is highly related to the hypothetical virtues provided by abductive cognition. That is, the relevant aspects of the environment are appropriately abductively selected and/or reconstructed so as to turn the local environment – inert from a cognitive point of view – into a cultural mediating structure able to deliver suitable chances for behavior control. Through cognitive niche construction organisms not only influence the nature of their world, but also in part determine the Darwinian selection pressure to which they and their descendants are exposed (and of course the selection pressures to which other species are subjected). It is in this perspective that any abduced hypothesis potentially introduces a change (and an opportunity) in the semiotic processes to advance new perspectives in the coevolution of the organism and the environment.

Non-explanatory abduction. Abduction is not intrinsically explanationist, like for example its description in terms of inference to the best explanation would suggest. If we maintain that E explains E' *only if* the first implies the second, certainly the reverse does not hold. This means that various cases of abduction are consequentialist but not explanationist. Other cases are neither consequentialist nor explanationist. Non-explanatory modes of abduction are clearly exploited in the "reverse mathematics" where propositions can be taken as axioms because they support the axiomatic proofs of target theorems. Furthermore, often in physics the target is the discovery of physical dependencies which we consider explanatorily undetermined. In this case abduction can exhibit an instrumental aspect. Non-explanatory abductions deal with those situations in which the plausibility of certain hypotheses involves less "propositional" and more "strategical" and instrumental aspects, so that propositional plausibility is lower and strategic plausibility higher. These cases are far from the clear ones of explanatory abduction that are for example occurring in science and in various kinds of diagnosis.

Nonmonotonic reasoning. A logical system is monotonic if the function *Theo* that relates every set of wffs to the set of their theorems holds the following property: for every set of premises S and for every set of premises S', $S \subseteq S'$ implies $Theo(S) \subseteq Theo(S')$. Traditional deductive logics are monotonic: intuitively, adding new premises (axioms) will never invalidate old conclusions. In a nonmonotonic system, when axioms, or premises, increase, their theorems do not. For example, following this deductive nonmonotonic view of abduction, we can stress the fact that in actual abductive medical reasoning, when we increase symptoms and patients' data [premises], we are compelled to abandon previously derived plausible diagnostic hypotheses [theorems], as already – epistemologically – illustrated by the ST-model. If new information emerges, hypotheses not previously considered can be suggested and a new cycle takes place. In this case the nonmonotonic character of abductive reasoning is clear and arises from the "classical" logical unsoundness of the inference rule: it draws defeasible conclusions from incomplete information.

Perception as abduction. Perception is viewed by Peirce as a fast and uncontrolled knowledge-production procedure. Perception, in this philosophical perspective, is seen as a vehicle for the instantaneous retrieval of knowledge that possibly was previously structured in our mind through more structured inferential processes. By perception, knowledge constructions are so instantly reorganized that they become habitual and diffuse and do not need any further testing. By perception, knowledge constructions are so instantly reorganized that they become habitual and diffuse and do not need any further testing: "[...] a fully accepted, simple, and interesting inference tends to obliterate all recognition of the uninteresting and complex premises from which it was derived" [Peirce, 1931-1958, 7.37]. Many visual stimuli – that can be considered the "premises" of the involved abduction – are ambiguous, yet people are adept at imposing order on them. For example we readily form such hypotheses as that an obscurely seen face belongs to a friend of ours, because we can thereby explain what has been observed. This kind of image-based hypothesis formation can be considered as a form of what I have called visual (or iconic) abduction.

Plastic cognition and abduction. See **Brain plasticity and abduction.**

Pregnancy as abduction. In terms of the hypothetical abductive framework we can say that a pregnant stimulus is highly diagnostic and a trigger to initiate abductive cognition, like in the case of the hardwired pregnancy occurring regarding a chicken and its food: the chicken promptly reacts when perceiving it. When a pregnance affects an organism, the abductive reaction can be promptly triggered.

Selective abduction. Selective abduction is the process in which a hypothesis is abductively selected from a pre-stored encyclopedia of "abducibles". An example of selective abduction is diagnostic reasoning in medicine: it starts from patient data that is abstracted into clinical features to be explained. Then, selective abduction generates plausible explanatory diagnostic hypotheses.

Sentential abduction. Sentential abduction – related to logic and to verbal/symbolic inferences – refers to those situations in which a hypothesis is formed and evaluated relying to the sentential aspects of natural or artificial languages, like in the case of logic. Sentential abduction can be rendered in different ways. For example, in the Peircean syllogistic framework abduction is considered like something propositional and as a type of fallacious reasoning. The sentential models of abduction are powerful but limited, because they do not capture various reasoning tasks which involve creative, model-based, and manipulative aspects.

ST-model of abduction. The so-called Select and Test Model (ST-model) is an epistemological model of medical reasoning, which can be described in terms of the classical notions of abduction, deduction and induction; it describes the different roles played by such basic inference types in developing various kinds of medical reasoning (diagnosis, therapy planning, monitoring). The model is consistent with the Peircean view about the various stages of scientific inquiry in terms of "hypothesis" generation (abduction), deduction (prediction), and induction. The model has been used to implement medical knowledge-based systems of medical reasoning.

Visual Abduction. See **Perception as abduction**.

References

[Abe *et al.*, 2006] Abe, A., Ozaku, H.I., Kuwahara, N., Kogure, K.: Cooperation between abductive and inductive nursing risk management. In: Tsumoto, S., Clifton, C.W., Zhong, N., Wu, X., Liu, J., Wah, B.W., Cheung, Y.-M. (eds.) Sixth IEEE International on Data Mining Workshops, pp. 705–708. IEEE Computer Society Press, Los Alamitos (2006)

[Abe, 2009] Abe, A.: Cognitive chance discovery. In: Stephanidis, C. (ed.) Universal Access in Human-Computer Interaction - Addressing Diversity, V International Conference, UAHCI 2009, pp. 315–323. Springer, Berlin (2009); Held as Part of HCI International 2009, San Diego, CA, USA, July 19-24, Proceedings, Part I (2009)

[Ackermann, 1989] Ackermann, R.: The new experimentalism. British Journal for the Philosophy of Science 40, 185–190 (1989)

[Addis and Gooding, 2008] Addis, T.R., Gooding, D.: Simulation methods for an abductive system in science. Foundations of Science 13(1), 37–52 (2008); Special issue edited by Magnani, L.: Tracking irrational sets. Science, technology, ethics. In: Proceedings of the International Conference Model-Based Reasoning in Science and Engineering - MBR 2004

[Adelman *et al.*, 2006] Adelman, L., Yeo, C., Miller, S.L.: Understanding the effects of computer displays and time pressure on the performance of distributed teams. In: Kirlik, A. (ed.) Human-Technology Interaction. Methods and Models for Cognitive Engineering and Human-Computer Interaction, pp. 43–54. Oxford University Press, Oxford (2006)

[Adolphs, 2006] Adolphs, R.: How do we know the minds of others? Domain specificity, simulation, and enactive social cognition. Brain Research 1079, 25–35 (2006)

[Agree and Chapman, 1990] Agree, P., Chapman, D.: What are plans for? In: Maes, P. (ed.) Designing Autonomous Agents, pp. 17–34. The MIT Press, Cambridge (1990)

[Alchourrón *et al.*, 1985] Alchourrón, C., Gärdenfors, P., Makison, P.: On the theory of logic change: partial meet functions for contractions and revision. Journal of Symbolic Logic 50, 510–530 (1985)

[Aliseda, 1997] Aliseda, A.: Seeking Explanations: Abduction in Logic, Philosophy of Science and Artificial Intelligence. PhD thesis, Institute for Logic, Language and Computation, Amsterdam (1997)

[Aliseda, 2000] Aliseda, A.: Abduction as epistemic change: a Peircian model in artificial intelligence. In: Flach, P., Kakas, A. (eds.) Abductive and Inductive Reasoning: Essays on Their Relations and Integration, pp. 45–58. Kluwer Academic Publishers, The Netherlands (2000)

[Aliseda, 2006] Aliseda, A.: Abductive Reasoning. Logical Investigations into Discovery and Explanation. Springer, Berlin (2006)

[Allen *et al.*, 2005] Allen, J.S., Bruss, J., Damasio, H.: The aging brain: the cognitive reserve hypothesis and hominid evolution. Americam Journal of Human Biology 17, 673–689 (2005)

[Aloimonos *et al.*, 1988] Aloimonos, J., Weiss, I., Bandyopadhyay, A.: Active vision. International Journal of Computer Vision 1, 333–356 (1988)

[Alpert and Simms, 2002] Alpert, P., Simms, E.L.: The relative advantages of plasticity and fixity in different environments: when is it good for a plant to adjust? Evolutionary Ecology 16, 285–297 (2002)

[Ambrosio, 2007] Ambrosio, C.: Iconicity and Homomorphism in Picasso's Guernica: a Study of Creativity Across the Boundaries. PhD thesis, University of London (2007)

[Anderson and Belnap, 1975] Anderson, A., Belnap, N.: Entailment. Princeton University Press, Princeton (1975)

[Anderson, 1986] Anderson, D.R.: The evolution of Peirce's concept of abduction. Transactions of the Charles S. Peirce Society 22(2), 45–58 (1986)

[Anderson, 1987] Anderson, D.R.: Creativity and the Philosophy of Charles S. Peirce. Claredon Press, Oxford (1987)

[Anderson, 2003] Anderson, M.L.: Embodied cognition: a field guide. Artificial Intelligence 149(1), 91–130 (2003)

[Andler *et al.*, 2006] Andler, D., Ogawa, Y., Okada, M., Watanabe, S. (eds.): Reasoning and Cognition, vol. 153. Keio University Press, Tokyo (2006)

[Angelis, 1964] De Angelis, E.: Il metodo geometrico nella filosofia del Seicento. Istituto di Filosofia, Università degli Studi di Pisa, Pisa (1964)

[Anquetil and Jeannerod, 2007] Anquetil, T., Jeannerod, M.: Simulated actions in the first and in the third person perspective. Brain Research 1130, 125–129 (2007)

[Aravindan and Dung, 1995] Aravindan, C., Dung, P.M.: Knowledge base dynamics, abduction, and databases updates. Journal of Applied Non-Classical Logics 5, 51–76 (1995)

[Arrighi and Ferrario, 2008] Arrighi, C., Ferrario, R.: Abductive reasoning, interpretation and collaborative processes. Foundations of Science 13(1), 75–87 (2008); Special issue edited by Magnani, L.: Tracking irrational sets. Science, technology, ethics. In: Proceedings of the International Conference on Model-Based Reasoning in Science and Engineering - MBR 2004 (2004)

[Aunger, 2002] Aunger, R.: The Electric Meme. A New Theory of How We Think. The Free Press, New York (2002)

[Aust *et al.*, 2005] Aust, U., Apfalter, W., Huber, L.: Pigeon categorization: classification strategies in a non-linguistic species. In: Grialou, P., Longo, G., Okada, M. (eds.) Images and Reasoning, pp. 183–204. Keio University, Tokyo (2005)

[Ayim, 1974] Ayim, M.: Retroduction: the rational instinct. Transaction of the Charles S. Peirce Society 10(1), 34–43 (1974)

[Backhtin, 1982] Backhtin, M.: The Dialogic Imagination: Four Essays by M. M. Bakhtin. The University of Texas Press, Austin, TX (1982)

[Bacon, 2000] Bacon, F.: The New Organon (1620). Cambridge University Press, Cambridge (2000)

[Bailer-Jones, 1999] Bailer-Jones, D.M.: Tracing the development of models in the philosophy of science. In: Magnani, L., Nersessian, N.J., Thagard, P. (eds.) Model-Based Reasoning in Scientific Discovery, pp. 23–40. Kluwer Academic/Plenum Publishers, New York (1999)

[Bailly and Longo, 2006] Bailly, F., Longo, G.: Mathématiques et science de la nature. La singularité physique du vivant. Hermann, Paris (2006)

[Bailly and Longo, 2009] Bailly, F., Longo, G.: Causes and symmetries. The continuum and the discrete in mathematical modelling. In: Giorello, G. (ed.) More Geometrico. Elsevier, New York (2009)

[Balcetis and Dunning, 2006] Balcetis, E., Dunning, D.: See what you want to see: motivational influences on visual perception. Journal of Personality and Social Psychology 91(4), 612–625 (2006)

[Balda and Kamil, 2002] Balda, R.P., Kamil, A.C.: Spatial and social cognition in corvids: an evolutionary approach. In: Bekoff, M., Allen, C., Burghardt, M. (eds.) The Cognitive Animal. Empirical and Theoretical Perspectives on Animal Cognition, pp. 129–134. The MIT Press, Cambridge (2002)

[Ballard, 1991] Ballard, D.H.: Animate vision. Artificial Intelligence 48, 57–86 (1991)

[Banerjee, 2006] Banerjee, B.: A layered abductive inference framework for diagramming group motions. Logic Journal of the IGPL 14(2), 363–378 (2006)

[Barbaras, 1999] Barbaras, R.: The movement of the living as the originary foundation of perceptual intentionality. In: Petitot, J., Varela, F.J., Pachoud, B., Roy, J.-M. (eds.) Naturalizing Phenomenology, pp. 525–538. Stanford University Press, Stanford (1999)

[Barbieri, 2008] Barbieri, M. (ed.): Introduction to Biosemiotics. The New Biological Synthesis. Springer, Berlin (2008)

[Bardone and Magnani, 2009] Bardone, E., Magnani, L.: The appeal of gossiping fallacies. Reasoning and its social roots (2009) (Forthcoming)

[Bardone and Secchi, 2006] Bardone, E., Secchi, D.: The distributed cognition approach to rationality: getting the framework. Presented at the 2006 Academy of Management Annual Meeting, Atlanta (2006)

[Barrett and Kurzban, 2006] Barrett, H.C., Kurzban, R.: Modularity in cognition: framing the debate. Psychological Review 113(3), 628–647 (2006)

[Barsalou *et al.*, 2003] Barsalou, L.W., Niedenthal, P.M., Barbey, A.K., Ruppert, J.M.: Social embodiment. In: Ross, B.H. (ed.) The Psychology of Learning and Motivation, vol. 43, pp. 43–92. Academic Press, San Diego (2003)

[Barwise and Etchemendy, 1990] Barwise, J., Etchemendy, J.: Visual information and valid reasoning visualization in mathematics. In: Zinnermann, W. (ed.) Mathematical Association of America, Washington, DC (1990)

[Batens, 2006] Batens, D.: A diagrammatic proof search procedure as part of a formal approach to problem solving. In: Magnani, L. (ed.) Model-Based Reasoning in Science and Engineering, pp. 265–284. College Publications, London (2006)

[Baum, 2006] Baum, E.B.: What Is Thought? The MIT Press, Cambridge (2006)

[Bauters, 2007] Bauters, M.: Mediation seen through the sensory eye: an alternative to the "old" and "new" media paradigm. International Journal of Applied Semiotics 6(1), 79–101 (2007)

[Bavelier and Neville, 2002] Bavelier, D., Neville, H.J.: Cross-modal plasticity: where and how? Nature Review 3, 443–452 (2002)

[Bechtel, 1998] Bechtel, W.: Representations and cognitive explanations: assessing the dynamicist challenge in cognitive science. Cognitive Science 22, 295–318 (1998)

[Beers, 1997] Beers, R.: Expressions of mind in animal behavior. In: Mitchell, W., Thomson, N.S., Miles, H.L. (eds.) Anthropomorphism, Anecdotes, and Animals, pp. 198–209. State University of New York Press, Albany (1997)

[Ben Jacob *et al.*, 2006] Ben Jacob, E., Shapira, Y., Tauber, A.I.: Seeking the foundation of cognition in bacteria. From Schrödinger's negative entropy to latent information. Physica A 359, 495–524 (2006)

[Ben-Ze'ev, 2000] Ben-Ze'ev, A.: The Subtlety of Emotions. The MIT Press, Cambridge (2000)

[Benacerraf and Putnam, 1964] Benacerraf, P., Putnam, H. (eds.): Philosophy of Mathematics. Selected Readings. Prentice Hall, Englewood Cliffs (1964)

[Bermúdez, 2003] Bermúdez, J.L.: Thinking without Words. Oxford University Press, Oxford (2003)

[Bessant, 2000] Bessant, B.: On the relationships between induction and abduction: a logical point of view. In: Flach, P., Kakas, A. (eds.) Abductive and Inductive Reasoning: Essays on Their Relation and Integration, pp. 77–88. Kluwer Academic, Dordrecht (2000)

[Beth, 1956–1957] Beth, E.: Über lockes "allgemeines dreieck". Kant-Studien 48, 361–380 (1957)

[Beth, 1965] Beth, E.W.: The Foundations of Mathematics. North Holland Publishing Company, Amsterdam (1965)

[Bharathan and Josephson, 2006] Bharathan, V., Josephson, J.R.: Belief revision controlled by meta-abduction. Logic Journal of the IGPL 14(2), 271–286 (2006)

[Bhatta and Goel, 1997] Bhatta, S.R., Goel, A.K.: A functional theory of design patterns. In: Proceedings of IJCAI 1997, pp. 294–300 (1997)

[Bickerton, 1990] Bickerton, D.: Language and Species. University of Chicago Press, Chicago (1990)

[Bingham, 1999] Bingham, P.M.: Human uniqueness: a general theory. The Quarterly Review of Biology 74(2), 133–169 (1999)

[Bingham, 2000] Bingham, P.M.: Human evolution and human history: a complete theory. Evolutionary Anthropology 9(6), 248–257 (2000)

[Biot, 1821] Biot, J.-B.: On the magnetism impressed on metals by electricity in motion. Quarterly Journal of Science 11, 281–290 (1821); Read at the public setting of the Academy of Sciences (April 2, 1821)

[Bjiorklunf, 2006] Bjiorklunf, D.F.: Mother knows best: epigenetic inheritance, maternal effects, and the evolution of human intelligence. Developmental Review 26, 213–242 (2006)

[Blades, 1997] Blades, M.: Research paradigms and methodologies for investigating children's wayfinding. In: Foreman, N., Gillett, R. (eds.) A Handbook of Spatial Research Paradigms and Methodologies, vol. I, pp. 103–129. Psychology Press, Taylor and Francis, Hove, East Sussex (1997)

[Bloch and Morange, 1997] Bloch, H., Morange, F.: Research paradigms and methodologies for investigating children's wayfinding. In: Foreman, N., Gillett, R. (eds.) A Handbook of Spatial Research Paradigms and Methodologies, vol. I, pp. 15–40. Psychology Press, Taylor and Francis, Hove, East Sussex (1997)

[Boden, 1992] Boden, M.A.: The Creative Mind: Myths and Mechanisms. Basic Books, New York (1992)

[Boehm, 1999] Boehm, C.: Hierarchy in the Forest. Harvard University Press, Cambridge (1999)

[Bogdan, 2003] Bogdan, R.J.: Minding Minds. The MIT Press, Cambridge (2003)

[Boneh *et al.*, 1996] Boneh, D., Dunworth, C., Lipton, R.J., Sgall, J.: On the computational power of DNA, discrete applied mathematics. Computational Molecular Biology 71, 79–94 (1996)

[Boutilier and Becher, 1995] Boutilier, C., Becher, V.: Abduction as belief revision. Artificial Intelligence 77, 43–94 (1995)

[Bovet, 1997] Bovet, J.: Long-distance travels and homing: dispersal, migrations, excursions. In: Foreman, N., Gillett, R. (eds.) A Handbook of Spatial Research Paradigms and Methodologies, vol. II, pp. 239–269. Psychology Press, Taylor and Francis, Hove, East Sussex (1997)

[Bowden and Jung-Beeman, 2003] Bowden, E.M., Jung-Beeman, M.: Aha! Insight experience correlates with solution activation in the right hemisphere. Psychonomic Bulletin & Review 10(3), 730–737 (2003)

[Brent, 2000] Brent, J.: A brief introduction to the life and thought of Charles Sanders Peirce. In: Muller, J., Brent, J. (eds.) Peirce, Semiosis, and Psychoanalysis, pp. 1–14. John Hopkins, Baltimore (2000)

[Brewer and Lambert, 2001] Brewer, W.F., Lambert, B.L.: The theory-ladenness of observation and the theory-ladenness of the rest of the scientific process. Philosophy of Science 68, S176–S186 (2001); Proceedings of the PSA 2000 Biennal Meeting

[Brewka, 1989] Brewka, G.: Preferred subtheories: an extended logical framework for default reasoning. In: Proceedings IJCAI 1989, Detroit, MI, pp. 1043–1048. Kluwer Academic Publishers, Dordrecht (1989)

[Bringsjiord, 2000] Bringsjiord, S.: Is (Gödelian) model-based deductive reasoning computational? Philosophica 61, 51–76 (2000); Special Issue on "Abduction and Scientific Discovery"

[Brittan, 1978] Brittan, G.G.: Kant's Theory of Science. Princeton University Press, Princeton (1978)

[Brogaard, 1999] Brogaard, B.: A Peircean theory of decision. Synthese 118, 383–401 (1999)

[Brooks and Stein, 1994] Brooks, R.A., Stein, L.: Building brains for bodies. Autonomous Robots 1, 7–25 (1994)

[Brooks, 1991] Brooks, R.A.: Intelligence without representation. Artificial Intelligence 47, 139–159 (1991)

[Brooks, 1999] Brooks, R.A.: Cambrian Intelligence: the Early History of the New AI. The MIT Press, CAmbridge (1999)

[Brunswik, 1943] Brunswik, E.: Organismic achievement and envornmental probability. Psychological Review 50, 255–272 (1943)

[Brunswik, 1952] Brunswik, E.: The Conceptual Framework of Psychology. University of Chicago Press, Chicago (1952)

[Brunswik, 1955] Brunswik, E.: Representative design and probabilistic theory in a functional psychology. Psychological Review 62, 193–217 (1955)

[Bruza *et al.*, 2006] Bruza, P.D., Cole, R.J., Song, D., Bari, Z.: Towards operational abduction from a cognitive perspective. Logic Journal of the IGPL 14(2), 161–179 (2006)

[Buchanan, 1985] Buchanan, B.G.: Steps toward mechanizing discovery. In: Schaffner, K.F. (ed.) Discovery and Diagnosis in Medicine, pp. 94–114. University of California Press, Berkeley (1985)

[Burgeois-Gironde, 2006] Burgeois-Gironde, S.: Dual processes theories of the mind and rationality assumptions. In: Andler, D., Ogawa, Y., Okada, M., Watanabe, S. (eds.) Reasoning and Cognition, pp. 37–52. Keio University Press, Tokyo (2006)

[Butterworth, 1999] Butterworth, B.: The Mathematical Brain. MacMillan, New York (1999)

[Bylander *et al.*, 1991] Bylander, T., Allemang, D., Tanner, M.C., Josephson, J.R.: The computational complexity of abduction. Artificial Intelligence 49, 25–60 (1991)

[Byrne and Whiten, 1988] Byrne, R., Whiten, A.: Machiavellian Intelligence. Oxford University Press, Oxford (1988)

[Byrne *et al.*, 2006] Byrne, D., Kirlik, A., Fick, C.S.: Kilograms matter: rational analysis, ecological rationality, and closed-loop modeling of interactive cognition and behavior. In: Kirlik, A. (ed.) Human-Technology Interaction. Methods and Models for Cognitive Engineering and Human-Computer Interaction, pp. 267–286. Oxford University Press, Oxford (2006)

[Byrne, 1847] Byrne, O.: The First Six Books of the Elements of Euclid in which Coloured Diagrams and Symbols are Used Instead of Letters for the Greater Ease of Learners. William Pickering, London (1847)

[Callagher, 2005] Callagher, S.: How the Body Shapes the Mind. Oxford University Press, Oxford (2005)

[Callon and Latour, 1992] Callon, M., Latour, B.: Don't throw the baby out with the bath school! A reply to Collins and Yearley. In: Pickering, A. (ed.) Science as Practice and Culture, pp. 343–368. The University of Chicago Press, Chicago (1992)

[Callon, 1994] Callon, M.: Four models for the dynamics of science. In: Jasanoff, S., Markle, G.E., Petersen, J.C., Pinch, T.J. (eds.) Handbook of Science and Technology Studies, pp. 29–63. Sage, Los Angeles (1994)

[Callon, 1997] Callon, M.: Society in the making: the study of technology as a tool for sociological analysis. In: Bjiker, W.E., Hughes, T.P., Pinch, T. (eds.) The Social Construction of Technological Systems, pp. 83–106. The MIT Press, Cambridge (1997)

[Cangelosi, 2007] Cangelosi, A.: Adaptive agent modeling of distributed language: investigations on the effects of cultural variation and internal action representations. Language Sciences 29, 633–649 (2007)

[Carey *et al.*, 1996] Carey, D.P., Harvey, M., Milner, A.D.: Visuomotor sensitivity for shape and orientation in a patient with visual form agnosia. Neuropsychologia 34(5), 329–337 (1996)

[Carey, 1985] Carey, S.: Conceptual Change in Childhood. The MIT Press, Cambridge (1985)

[Carnap, 1950] Carnap, R.: Logical Foundations of Probability. Routledge and Kegan Paul, London (1950)

[Carnielli, 2006] Carnielli, W.: Surviving abduction. Logic Journal of the IGPL 14(2), 237–256 (2006)

[Carr, 1981] Carr, D.: Introduction to the The Origin of Geometry. In: McCormick, P., Elliston, F.A. (eds.) Husserl. Shorter Works, pp. 25–153. University of Notre Dame Press, Notre Dame (1981)

[Carruthers, 2002a] Carruthers, P.: The cognitive function of language. Behavioral and Brain Sciences 25(6), 657–674 (2002)

[Carruthers, 2002b] Carruthers, P.: The roots of scientific reasoning: infancy, modularity and the art of tracking. In: Carruthers, P., Stich, S., Siegal, M. (eds.) The Cognitive Basis of Science, pp. 73–95. Cambridge University Press, Cambridge (2002)

[Cellucci, 2005] Cellucci, C.: Mathematical discourse vs. mathematical intuition. In: Cellucci, C., Gillies, D. (eds.) Mathematical Reasoning and Heuristics, pp. 138–166. King's College Publications, London (2005)

[Chalmers, 1999] Chalmers, A.F.: What is this Thing Called Science. Hackett, Indianapolis (1999)

[Changeux and Dehaene, 1989] Changeux, J.-P., Dehaene, S.: Neuronal models of cognitive function. Cognition 33, 63–109 (1989)

[Châtelet, 1993] Châtelet, G.: Les enjeux du mobile. Seuil, Paris (1993); English translation by Shore, R., Zagha, M.: Figuring Space: Philosophy, Mathematics, and Physics. Kluwer Academic Publishers, Dordrecht (2000)

[Chemero and Silberstein, 2008] Chemero, A., Silberstein, M.: Defending extended cognition. In: CogSci 2008, XXX Annual Conference of the Cognitive Science Society, Chicago, IL (2008) CD-Rom

[Chemero, 2003] Chemero, A.: An outline of a theory of affordances. Ecological Psychology 15(2), 181–195 (2003)

[Chi *et al.*, 1981] Chi, M.T.H., Feltovich, P.J., Glaser, R.: Categorization and representation of physics problems by experts and novices. Cognitive Science 5, 121–152 (1981)

[Chiesa *et al.*, 2006] Della Chiesa, A., Pecchia, T., Tommasi, L.: Multiple landmarks, the encoding of enviromental geometry and the spatial logics of a dual brainlogic and human reasoning: an assessment of deduction. Animal Cognition 9, 281–293 (2006)

[Chomsky, 1986] Chomsky, N.: Knowledge of Language. Its Nature, Origins, and Use. Praeger, New York (1986)

[Chrisley, 1995] Chrisley, R.L.: Taking embodiment seriously: non conceptual content and robotics. In: Ford, K.M., Glymour, C., Hayes, P.J. (eds.) Android Epistemology, pp. 141–166. The MIT Press, Cambridge (1995)

[Churchland, 1988] Churchland, P.M.: Perceptual plasticity and theoretical neutrality: a reply to Jerry Fodor. Philosophy of Science 55, 167–187 (1988)

[Clancey, 1997] Clancey, W.J.: Situated Cognition: on Human Knowledge and Computer Representations. Cambridge University Press, Cambridge (1997)

[Clancey, 2002] Clancey, W.J.: Simulating activities: relating motives, deliberation, and attentive coordination. Cognitive Systems Research 3(1-4), 471–500 (2002)

[Clark and Chalmers, 1998] Clark, A., Chalmers, D.J.: The extended mind. Analysis 58, 10–23 (1998)

[Clark, 1978] Clark, K.L.: Negation as failure. In: Gallaire, H., Minker, J. (eds.) Logic and Data Bases, pp. 94–114. Plenum, New York (1978)

[Clark, 1997] Clark, A.: Being There: Putting Brain, Body, and World Together Again. The MIT Press, Cambridge (1997)

[Clark, 2003] Clark, A.: Natural-Born Cyborgs. Minds, Technologies, and the Future of Human Intelligence. Oxford University Press, Oxford (2003)

[Clark, 2005] Clark, A.: Material symbols: from translation to co-ordination in the constitution of thought and reason. In: Bara, B., Barsalou, L., Bucciarelli, M. (eds.) CogSci 2005, XXVII Annual Conference of the Cognitive Science Society, Stresa (2005) CD-Rom

[Clark, 2006] Clark, A.: Language, embodiment, and the cognitive niche. Trends in Cognitive Science 10(8), 370–374 (2006)

[Clark, 2008] Clark, A.: Supersizing the Mind. Embodiment, Action, and Cognitive Extension. Oxford University Press, Oxford (2008)

[Clowes and Morse, 2005] Clowes, R.W., Morse, A.: Scaffolding cognition with words. In: Berthouze, L., Kaplan, F., Kozima, H., Yano, H., Konczak, J., Metta, G., Nadel, J., Sandini, G., Stojianov, G., Balkenius, C. (eds.) Proceedings of the Fifth International Workshop on Epigenetic Robotics: Modeling Cognitive Development in Robotic Systems, Nara, pp. 101–105 (2005)

[Cohn *et al.*, 2002] Cohn, A.G., Magee, D.R., Galata, G., Hogg, D.C., Hazarika, S.M.: Towards an architecture for cognitive vision using qualitative spatio-temporal representations and abduction. In: Freksa, C., Brauer, W., Habel, C., Wender, K.F. (eds.) Spatial Cognition III. LNCS (LNAI), vol. 2685, pp. 232–248. Springer, Heidelberg (2003)

[Colapietro, 2000] Colapietro, V.: Futher consequences of a singular capacity. In: Muller, J., Brent, J. (eds.) Peirce, Semiosis, and Psychoanalysis, pp. 136–158. John Hopkins, Baltimore (2000)

[Colapietro, 2005] Colapietro, V.: Conjectures concerning an uncertain faculty. Semiotica 153(1/4), 413–430 (2005)

[Cole, 1996] Cole, M.: Cultural Psychology. Harvard University Press, Cambridge (1996)

[Colton, 1999] Colton, S. (ed.): AI and Scientific Creativity. Proceedings of the AISB99 Symposium on Scientific Creativity, Society for the Study of Artificial Intelligence and Simulation of Behaviour. Edinburgh College of Art and Division of Informatics, University of Edinburgh, Edinburgh (1999)

[Console and Saitta, 2000] Console, L., Saitta, L.: On the relations between abductive and inductive explanations. In: Flach, P., Kakas, A. (eds.) Abductive and Inductive Reasoning: Essays on Their Relation and Integration, pp. 133–151. Kluwer Academic, Dordrecht (2000)

[Console and Torasso, 1991] Console, L., Torasso, P.: A spectrum of logical definitions of model-based diagnosis. Computational Intelligence 7(3), 133–141 (1991)

[Console *et al.*, 1991] Console, L., Theseider Duprè, D., Torasso, P.: On the relationships between abduction and deduction. Journal of Logic and Computation 1(5), 661–690 (1991)

[Cook, 2002] Cook, R.G.: Same-different concept formation in pigeons. In: Bekoff, M., Allen, C., Burghardt, M. (eds.) The Cognitive Animal. Empirical and Theoretical Perspectives on Animal Cognition, pp. 229–238. The MIT Press, Cambridge (2002)

[Coolidge and Wynn, 2005] Coolidge, F.L., Wynn, T.: Working memory, its executive functions, and the emergence of modern thinking. Cambridge Archealogical Journal 5(1), 5–26 (2005)

[Cordeschi and Frixione, 2007] Cordeschi, R., Frixione, M.: Computationalism under attack. In: Marraffa, M., De Caro, M., Ferretti, M. (eds.) Cartographies of the Mind, pp. 37–49. Springer, Berlin (2007)

[Cornuéjols *et al.*, 2000] Cornuéjols, A., Tiberghien, A., Collet, G.: A new mechanism for transfer between conceptual domains in scientific discovery and education. Foundations of Science 5(2), 129–155 (2000); Special Issue, Magnani, L., Nersessian, N.J., Thagard, P.: Model-based Reasoning in Science: Learning and Discovery

[Corruble and Ganascia, 1997] Corruble, V., Ganascia, J.-G.: Induction and the discovery of the causes of scurvy: a computational reconstruction. Artificial Intelligence 91, 205–223 (1997)

[Coz and Pietrsykowski, 1986] Coz, P.T., Pietrsykowski, T.: Causes for events: their computation and application. In: Siekmann, J.H. (ed.) CADE 1986. LNCS, vol. 230, pp. 608–621. Springer, Heidelberg (1986)

[Crist, 2002] Crist, E.: The inner life of eartworms: Darwin's argument and its implications. In: Bekoff, M., Allen, C., Burghardt, M. (eds.) The Cognitive Animal. Empirical and Theoretical Perspectives on Animal Cognition, pp. 3–8. The MIT Press, Cambridge (2002)

[Cross and Thomason, 1992] Cross, C., Thomason, R.H.: Conditionals and knowledge-base update. In: Gärdenfors, P. (ed.) Belief Revision, pp. 247–275. Cambridge University Press, Cambridge (1992)

[Cunningham, 1988] Cunningham, D.J.: Affordance and abduction: a semiotic view of cognition. Paper presented at the Annual Meeting of the American Educational Research Association, New Orleans, LA (1988)

[Curley and Keverne, 2005] Curley, J.P., Keverne, E.B.: Genes, brains and mammalian social bonds. Trends in Cognitive Science 20(10), 561–566 (2005)

[Dagher *et al.*, 1999] Dagher, A., Owen, A.M., Boecker, H., Brooks, D.J.: Mapping the network for planning. Brain 122, 1973–1987 (1999)

[Damasio, 1994] Damasio, A.R.: Descartes' Error. Putnam, New York (1994)

[Damasio, 1999] Damasio, A.R.: The Feeling of What Happens. Harcourt Brace, New York (1999)

[Darden, 1991] Darden, L.: Theory Change in Science: Strategies from Mendelian Genetics. Oxford University Press, Oxford (1991)

[Dartnall, 2004] Dartnall, T.: Epistemology, emulators, and extended mind. Behavioral and Brain Sciences 27, 401–402 (2004); Open Peer Commentary to R. Grush, The emulation theory of representation: motor control, imagery, and perception

[Dartnall, 2005] Dartnall, T.: Does the world leak into the mind? Active externalism, "internalism" and epistemology. Cognitive Science 29, 135–143 (2005)

[Darwin, 1981] Darwin, C.: The Descent of Man and Selection in Relation to Sex (1871). Princeton University Press, Princeton (1981)

[Darwin, 1985] Darwin, C.: The Formation of Vegetable Mould, through the Action of Worms with Observations on Their Habits (1881). University of Chicago Press, Chicago (1985)

[Davidson, 2001] Davidson, D.: Subjective, Intersubjective, Objective. Clarendon Press, Oxford (2001)

[Davies and Goel, 2000] Davies, J., Goel, A.K.: A computational theory of visual analogical transfer, Technical Report, Georgia Institute of Technology, Atlanta, GA (2000)

[Davies *et al.*, 2009] Davies, J., Goel, A.K., Yaner, P.W.: Proteus: visual analogy in problem solving. Journal of Artificial Intelligence Research (2009) (Forthcoming)

[Davis, 1972] Davis, W.H.: Peirce's Epistemology. Nijhoff, The Hague (1972)

[Davy, 821] Davy, H.: On the magnetic phenomena produced by electricity. Philosophical Transactions 111, 7–19, 821

[Dawkins, 1977] Dawkins, R.: The Selfish Gene. Oxford University Press, Oxford (1977)

[Dawkins, 1982] Dawkins, R.: The Extended Phenotype. Oxford University Press, Oxford (1982)

[Dawkins, 2004] Dawkins, R.: Extended phenotype - but not extended. A reply to Laland, Turner and Jablonka. Biology and Philosophy 19, 377–397 (2004)

[Day *et al.*, 2003] Day, R.L., Laland, K., Odling-Smee, F.J.: Rethinking adaptation. The niche-construction perspective. Perspectives in Biology and Medicine 46(1), 80–95 (2003)

[de Cheveigné, 2006] de Cheveigné, A.: Hearing, action, and space. In: Andler, D., Ogawa, Y., Okada, M., Watanabe, S. (eds.) Reasoning and Cognition, pp. 253–264. Keio University Press, Tokyo (2006)

[de Kleer *et al.*, 1990] de Kleer, J., Mackworth, A.K., Reiter, R.: Characterizing diagnoses. In: Proceedings AAAI 1990, pp. 324–330. Kluwer Academic Publishers, Boston (1990)

[Deacon, 1995] Deacon, T.: The Symbolic Species. W. W. Norton, New York (1995)

[Debrok, 1997] Debrok, G.: The artful riddle of abduction. In: Rayo, M., Gimate-Welsh, A., Pellegrino, P. (eds.) Fifth International Congress of the International Association for Semiotic Studies: Semiotics Bridging Nature and Culture, Editorial Solidaridad, México, p. 230 (1997)

[Decety and Grèzes, 2006] Decety, J., Grèzes, J.: The power of simulation: imagining one's own and other behavior. Brain in Research 1079, 4–14 (2006)

[Decety, 1996] Decety, J.: Do imagined and executed actions share the same neural substrate? Cognitive Brain Research 3, 87–93 (1996)

[Degani *et al.*, 2006] Degani, A., Shafto, M., Kirlik, A.: What makes vicarious functioning work? Exploring the geometry of human-technology interaction. In: Kirlik, A. (ed.) Human-Technology Interaction. Methods and Models for Cognitive Engineering and Human-Computer Interaction, pp. 179–196. Oxford University Press, Oxford (2006)

[Dehaene *et al.*, 1999] Dehaene, S., Spelke, E., Pinel, P., Stanescu, R., Tsivkin, S.: Sources of mathematical thinking: behavioral and brain imaging evidence. Science 284(5416), 970–974 (1999)

[Dehaene, 1997] Dehaene, S.: The Number Sense. Oxford University Press, Oxford (1997)

[Dennett, 1984] Dennett, D.: Elbow Room. The Variety of Free Will Worth Wanting. The MIT Press, Cambridge (1984)

[Dennett, 1987] Dennett, D.: The Intentional Stance. The MIT Press, Cambridge (1987)

[Dennett, 1991] Dennett, D.: Consciousness Explained. Little, Brown, and Company, New York (1991)

[Dennett, 1995] Dennett, D.: Darwin's Dangerous Idea. Simon and Schuster, New York (1995)

[Dennett, 2003] Dennett, D.: Freedom Evolves. Viking, New York (2003)

[Derbyshire *et al.*, 2006] Derbyshire, N., Ellis, R., Tucker, M.: The potentiation of two components of the reach-to-grasp action during object categorisation in visual memory. Acta Psychologica 122, 74–98 (2006)

[Derrida, 1978] Derrida, J.: Introduction to the "The Origin of Geometry". In: Derrida, J. (ed.) Edmund Husserl's "The Origin of Geometry", pp. 25–153. Nicolas Hays, Stony Brook, NY (1978); Translation by J. P. Leavy

[Desmurget and Grafton, 2002] Desmurget, M., Grafton, S.: Forward modeling allows feedback control for fast reaching movements. Trends in Cognitive Sciences 4, 423–431 (2002)

[Dimopoulos and Kakas, 1996] Dimopoulos, Y., Kakas, A.C.: Abduction and inductive learning. In: De Raedt, L. (ed.) Advances in Inductive Logic Programming, pp. 144–171. IOS Press, Amsterdam (1996)

[Dobbs, 1983] Dobbs, B.J.T.: The Foundations of Newton's Alchemy: or, "The Hunting of the Greene Lyon". Cambridge University Press, Cambridge (1983)

[Domenella and Plebe, 2005] Domenella, R.G., Plebe, A.: Can vision be computational? In: Magnani, L., Dossena, R. (eds.) Computing, Philosophy and Cognition, pp. 227–242. College Publications, London (2005)

[Donald, 1991] Donald, M.: The Origin of the Modern Mind. Harvard University Press, Cambridge (1991)

[Donald, 1998] Donald, M.: Hominid enculturation and cognitive evolution. In: Renfrew, C., Mellars, P., Scarre, C. (eds.) Cognition and Material Culture: the Archaeology of External Symbolic Storage, pp. 7–17. The McDonald Institute for Archaeological Research, Cambridge (1998)

[Donald, 2001] Donald, M.: A Mind So Rare. The Evolution of Human Consciousness. Norton, London (2001)

[Donohue, 2005] Donohue, K.: Niche construction through phenological plasticity: life history dynamics and ecological consequences. New Phytologist 166, 83–92 (2005)

[Dorigo and Colombetti, 1998] Dorigo, M., Colombetti, M.: Robot Shaping. An Experiment in Behavior Engineering. The MIT Press, Cambridge (1998)

[Dossena and Magnani, 2007] Dossena, R., Magnani, L.: Mathematics through diagrams: microscopes in non-standard and smooth analysis. In: Magnani, L., Li, P. (eds.) Model-Based Reasoning in Science, Technology, and Medicine, pp. 193–213. Springer, Berlin (2007)

[Dourish, 2001] Dourish, P.: Where the Action Is. In: The Foundations of Embodied Interaction. The MIT Press, Cambridge (2001)

[Doyle, 1979] Doyle, J.: A truth maintenance system. Artificial Intelligence 12, 231–272 (1979)

[Doyle, 1988] Doyle, J.: Constructive belief and rational representation. Computational Intelligence 5, 1–11 (1988)

[Doyle, 1992] Doyle, J.: Reason maintenance and belief revision: foundations versus coherence theories. In: Gärdenfors, P. (ed.) Knowledge in Flux, pp. 2–51. The MIT Press, Cambridge (1992)

[Dretske, 1988] Dretske, F.: Knowledge and the Flow of Information. The MIT Press, Cambridge (1988)

[Dummett, 1993] Dummett, M.: The Origins of Analytical Philosophy. Duckworth, London (1993)

[Dunbar, 1995] Dunbar, K.: How scientists think: online creativity and conceptual change in science. In: Sternberg, R.J., Davidson, J.E. (eds.) The Nature of Insight, pp. 365–395. The MIT Press, Cambridge (1995)

[Dunbar, 1998] Dunbar, R.: The social brain hypothesis. Evolutionary Anthropology 6, 178–190 (1998)

[Dunbar, 1999] Dunbar, K.: How scientists build models in vivo science as a window on the scientific mind. In: Magnani, L., Nersessian, N.J., Thagard, P. (eds.) Model-Based Reasoning in Scientific Discovery, pp. 85–99. Kluwer Academic/Plenum Publishers, New York (1999)

[Dunbar, 2003] Dunbar, R.: The social brain: mind, language, and society in evolutionary perspective. Annual Review of Anthropology 32, 163–181 (2003)

[Eco and Sebeok, 1983] Eco, U., Sebeok, T.A.: The Sign of Three. Holmes, Dupin, Peirce. Indiana University Press, Bloomington (1983)

[Edelman and Tononi, 2000] Edelman, G.M., Tononi, G.: A Universe of Consciousness: How Matter Becomes Imagination. Basic Books, New York (2000)

[Edelman, 1987] Edelman, G.M.: Neural Darwinism. Basic Books (HarperCollins), New York (1987)

[Edelman, 1989] Edelman, G.M.: The Remembered Present. Basic Books, New York (1989)

[Edelman, 1993] Edelman, G.M.: Wider than the Sky: the Phenomenal Gift of Consciousness. Yale University Press, New Haven (1993)

[Edelman, 2006] Edelman, G.M.: Second Nature. Brain Science and Human Knowledge. and London. Yale University Press, New Haven (2006)

[Einstein, 1961] Einstein, A.: Relativity: the Special and the General Theory. Crown, New York (1961)

[Ellis, 1995] Ellis, R.D.: Questioning Consciousness: the Interplay of Imagery, Cognition, and Emotion in the Human Brain. John Benjamins, Amsterdam (1995)

[Elveton, 2005] Elveton, R.: What is embodiment? In: Magnani, L., Dossena, R. (eds.) Computing, Philosophy and Cognition, pp. 243–258. College Publications, London (2005)

[Enfield, 2005] Enfield, N.: The body as a cognitive artifact in kinship representations: hand gestures diagrams by speakers of Lao. Current Anthropology 46, 51–81 (2005)

[Evans *et al.*, 2005] Evans, P.D., Gilbert, S.L., Mekel-Bobrov, N., Vallender, E.J., Anderson, J.R., Vaez-Azizi, L.M., Tishkoff, S.A., Hudson, R.R., Lahan, B.T.: Microcephalin, a gene regulating brain size, continues to evolve adaptively in humans. Science 309(9), 1717–1720 (2005)

[Evans, 2002a] Evans, C.S.: Cracking the code. Communication and cognition in birds. In: Bekoff, M., Allen, C., Burghardt, M. (eds.) The Cognitive Animal. Empirical and Theoretical Perspectives on Animal Cognition, pp. 315–322. The MIT Press, Cambridge (2002)

[Evans, 2002b] Evans, J.S.B.T.: Logic and human reasoning: an assessment of deduction paradigm. Psychological Bulletin 128(8), 978–996 (2002)

[Fajtlowicz, 1988] Fajtlowicz, S.: On conjectures of Graffiti. Discrete Mathematics 72, 113–118 (1988)

[Falkenhainer, 1990] Falkenhainer, B.C.: A unified approach to explanation and theory formation. In: Shrager, J., Langley, P. (eds.) Computational Models of Scientific Discovery and Theory Formation, pp. 157–196. Morgan Kaufmann, San Mateo (1990)

[Fann, 1970] Fann, K.T.: Peirce's Theory of Abduction. Nijhoff, The Hague (1970)

[Faraday, 1821-1822] Faraday, M.: Historical sketch on electromagnetism, 1821-1822. Annals of Philosophy 18, 195–200, 274–290; 19, 107–121

[Fauconnier and Turner, 2003] Fauconnier, G., Turner, M.: The Way We Think. Basic Books, New York (2003)

[Fauconnier, 2005] Fauconnier, G.: Compression and emergent structure. Language and Linguistics 6(4), 523–538 (2005)

[Fetzer, 1990] Fetzer, J.K.: Artificial Intelligence: Its Scope and Limits. Kluwer Academic Publisher, Dordrecht (1990)

[Feyerabend, 1975] Feyerabend, P.: Against Method. Verso, London (1975)

[Finke *et al.*, 1992] Finke, R.A., Ward, T.B., Smith, S.M.: Creative Cognition. Theory, Research, and Applications. The MIT Press, Cambridge (1992)

[Finke, 1990] Finke, R.A.: Creative Imagery: Discoveries and Invention in Visualization. Erlbaum, Hillsdale (1990)

[Fischer, 2001] Fischer, H.R.: Abductive reasoning as a way of worldmaking. Foundations of Science 6(4), 361–383 (2001)

[Flach and Kakas, 2000a] Flach, P., Kakas, A.: Abduction and induction: background and issues. In: Flach, P., Kakas, A. (eds.) Abductive and Inductive Reasoning: Essays on Their Relation and Integration, pp. 1–29. Kluwer Academic, Dordrecht (2000)

[Flach and Kakas, 2000b] Flach, P., Kakas, A. (eds.): Abductive and Inductive Reasoning: Essays on Their Relation and Integration. Kluwer Academic Publishers, Dordrecht (2000)

[Flach and Warren, 1995] Flach, J.M., Warren, R.: Active psychophysics: the relation between mind and what matters. In: Flach, J.M., Hancock, J., Caird, P., Vincente, K. (eds.) Global Perspectives on the Ecology of Human-Machine Systems, pp. 189–209. Erlbaum, Hillsdale (1995)

[Fleck, 1996] Fleck, J.: Informal information flow and the nature of expertise in financial service. International Journal of Technology Management 11(1/2), 104–128 (1996)

[Fodor and Pylyshyn, 1988] Fodor, J., Pylyshyn, Z.: Connectionism and cognitive architecture: a critical analysis. Cognition 28, 3–71 (1988)

[Fodor, 1983] Fodor, J.: The Modularity of the Mind. The MIT Press, Cambridge (1983)

[Fodor, 1984] Fodor, J.: Observation reconsidered. Philosophy of Science 51, 23–43 (1984); Reprinted in [Goldman,1993, pp. 119–139]

[Fodor, 1987] Fodor, J.: Psychosemantics. The MIT Press, Cambridge (1987)

[Fodor, 2000] Fodor, J.: The Mind Doesn't Work That Way. The MIT Press, Cambridge (2000)

[Føllesdal, 1999] Føllesdal, D.: Gödel and Husserl. In: Petitot, J., Varela, F.J., Pachoud, B., Roy, J.-M. (eds.) Naturalizing Phenomenology, pp. 385–400. Stanford University Press, Stanford (1999)

[Forbus, 1984] Forbus, K.D.: Qualitative process theory. Artificial Intelligence 24, 85–168 (1984)

[Forbus, 1986] Forbus, K.D.: Interpreting measurements of physical systems. In: Proceedings of the Fifth National Conference on Artificial Intelligence, pp. 113–117. Morgan Kaufmann, San Francisco (1986)

[Foreman and Gillett, 1997] Foreman, N., Gillett, R. (eds.): A Handbook of Spatial Research Paradigms and Methodologies, vol. 2. Psychology Press, Taylor and Francis, Hove, East Sussex (1997)

[Francis and Barlow, 1993] Francis, R., Barlow, G.W.: Social control of primary sex determination in the Midas cichlid. Proceedings of the National Academy of Sciences, USA 90, 10673–10675 (1993)

[Frankfurt, 1958] Frankfurt, H.: Peirce's notion of abduction. Journal of Philosophy 55, 593–597 (1958)

[Franklin, 2005] Franklin, I.R.: Exploratory experiments. Philosophy of Science 72, 888–899 (2005)

[Franklin, 2007] Franklin, L.R.: Bacteria, sex, and systematics. Philosophy of Science 74, 69–95 (2007)

[Freska, 2000] Freska, C.: Spatial cognition. In: López De Mántaras, R., Saitta, L. (eds.) ECAI 2004 Proceedings of the 16th European Conference on Artificial Intelligence, pp. 1122–1128. IOS Press, Amsterdam (2004)

[Freud, 1916] Freud, S.: Leonardo da Vinci: a Study in Sexuality. Brill, New York (1916)

[Freud, 1953-1974] Freud, S.: The Standard Edition of the Complete Psychological Works of Sigmund Freud. Hogarth Press, London (1953-1974); Translated by J. Strachey in collaboration with Freud, A., et al

[Friedman and Simpson, 2000] Friedman, H., Simpson, S.: Issues and problems in reverse mathematics. Computability Theory and its Applications: Contemporary Mathematics 257, 127–144 (2000)

[Friedman, 1992] Friedman, M.: Kant and the Exact Sciences. Harvard University Press, Cambridge (1992)

[Frith and Dolan, 1996] Frith, C., Dolan, R.: The role of the prefrontal cortex in higher cognitive functions. Cognitive Brain Research 5, 175–181 (1996)

[Fugelsang *et al.*, 2005] Fugelsang, J.A., Roser, M.E., Corballis, P.M., Gazzaniga, M.S., Dunbar, K.N.: Brain mechanisms underlying perceptual causality. Animal Learning and Behavior 24(1), 41–47 (2005)

[Fusaroli and Vandi, 2009] Fusaroli, R., Vandi, C.: Language as an object of perception. Gestaltic rationalities for semantic description. Gestalt Theory Journal (2009) (Forthcoming)

[Fusaroli, 2007] Fusaroli, R.: A Peircean contribution to the contemporary debate on perception: the sensorimotor theory and diagrams. Presented at the 9th World Congress of IASS/AIS, Communication: Understanding/Misunderstanding, Helsinki (June 2007)

[Fyhn *et al.*, 2004] Fyhn, M., Molden, S., Witter, M.P., Moser, E.I., Moser, M.-B.: Spatial representation in the entorhinal cortex. Science 305(5688), 1258–1264 (2004)

[Gabbay and Kruse, 2000] Gabbay, D.M., Kruse, R.: Abductive Reasoning and Learning. In: Ketner, K.L. (ed.) Handbook of Defeasible Reasoning and Uncertainty Management Systems, vol. 4. Kluwer Academic Publishers, Dordrecht (2000)

[Gabbay and Woods, 2001] Gabbay, D.M., Woods, J.: The new logic. Logic Journal of the IGPL 9(2), 141–174 (2001)

[Gabbay and Woods, 2005] Gabbay, D.M., Woods, J.: The Reach of Abduction. A Practical Logic of Cognitive Systems, vol. 2. North-Holland, Amsterdam (2005)

[Gabbay and Woods, 2006] Gabbay, D.M., Woods, J.: Advice on abductive logic. Logic Journal of the IGPL 14(1), 189–220 (2006)

[Gabbay and Woods, 2008] Gabbay, D., Woods, J.: Resource-origins of nonmonotonicity. Studia Logica 88, 85–112 (2008); Special Issue, Edited by Leitgeb, H.: Psychologism in Logic?

[Gabbay, 2002] Gabbay, D.M.: Abduction in labelled deductive systems. In: Gabbay, D.M., Kruse, R. (eds.) Handook of Defeasible Reasoning and Uncertainty Management Systems, pp. 99–153. Kluwer Academic Publishers, Dordrecht (2002)

[Gabbay, 2007] Gabbay, D.M.: The Leverhulme lectures on logics, Draft (2007)

[Galilei, 1914] Galilei, G.: Dialogues Concerning Two New sciences [1638]. In: Discoveries and Opinions of Galileo. Macmillan, New York (1914); Translated by Crew, H., De Salvo, A.: Reprinted by Prometheus Book, Buffalo, NY (1991)

[Galilei, 1957] Galilei, G.: The Starry Messenger [1610]. In: Discoveries and Opinions of Galileo, pp. 23–58. Doubleday, New York (1957); Translated and edited by S. Drake

[Galilei, 1989] Galilei, G.: Dialogues Concerning the Two Chief World Systems (1632). In: Matthews, M.R. (ed.) The Scientific Background to Modern Philosophy. Hackett, Indianapolis (1989); Translated by S. Drake

[Gallese, 2005] Gallese, V.: Embodied simulation: from neurons to phenomenal experience. Phenomenology and Cognitive Science 4, 23–48 (2005)

[Gallese, 2006] Gallese, V.: Intentional attunement: a neurophysiological perspective on social cognition and its disruption in autism. Brain Research 1079, 15–24 (2006)

[Gallistel, 1990] Gallistel, C.R.: The Organization of Learning. The MIT Press, Cambridge (1990)

[Ganascia, 2008] Ganascia, J.-G.: Reconstructing true wrong inductions. AI Magazine 29(2), 57–65 (2008)

[Garbarini and Adenzato, 2004] Garbarini, F., Adenzato, M.: At the root of embodied cognition: cognitive science meets neurophysiology. Brain and Cognition 56, 100–106 (2004)

[Gärdenfors, 1988] Gärdenfors, P.: Knowledge in Flux. The MIT Press, Cambridge (1988)

[Gärdenfors, 1992] Gärdenfors, P. (ed.): Belief Revision. Cambridge University Press, Cambridge (1992)

[Gärling *et al.*, 1997] Gärling, T., Selart, M., Böök, A.: Investigating spatial choice and navigation in large-scale environments. In: Foreman, N., Gillett, R. (eds.) A Handbook of Spatial Research Paradigms and Methodologies, vol. I, pp. 153–175. Psychology Press, Taylor and Francis, Hove, East Sussex (1997)

[Garzón, 2008] Garzón, F.C.: Towards a general theory of antirepresentationalism. The British Journal for the Philosophy of Science 59, 259–292 (2008)

[Gasser, 2004] Gasser, M.: The origins of arbitrariness in language. In: Forbus, K.D., Gentner, D., Regier, T. (eds.) CogSci 2004, XXVI Annual Conference of the Cognitive Science Society, Chicago, IL (2004) CD-Rom

[Gatti and Magnani, 2005] Gatti, A., Magnani, L.: On the representational role of the environment and on the cognitive nature of manipulations. In: Magnani, L., Dossena, R. (eds.) Computing, Philosophy and Cognition, pp. 227–242. College Publications, London (2005)

[Gaver, 1991] Gaver, W.W.: Technology affordances. In: CHI 1991 Conference Proceedings, pp. 79–84 (1991)

[Gazzaniga, 1995] Gazzaniga, M.: The Cognitive Neuroscience, 2nd edn. The MIT Press, Cambridge (1995)

[Gazzaniga, 2005] Gazzaniga, M.: The Ethical Brain. Dana Press, New York (2005)

[Gebhard *et al.*, 2003] Gebhard, U., Nevers, P., Bilmann-Mahecha, E.: Moralizing trees: anthropomorphism and identity in children's relationships to nature. In: Clayton, S., Opotow, S. (eds.) Identity and the Natural Environment: the Psychological Significance of Nature, pp. 91–111. The MIT Press, Cambridge (2003)

[Gelertner, 1959] Gelertner, H.: Realization of a geometry theorem proving machine. In: International Conference on Information Processing, Paris, Unesco House, pp. 273–282 (1959)

[Gentner *et al.*, 1997] Gentner, D., Brem, S., Ferguson, R., Wolff, P., Markman, A.B., Forbus, K.: Analogy and creativity in the work of Johannes Kepler. In: Ward, T.B., Ward, S.M., Vaid, J. (eds.) Creative Thought: an Investigation of Conceptual Structures and Processes, pp. 403–459. American Psychological Association, Washington (1997)

[Gentner, 1982] Gentner, D.: Are scientific analogies metaphors? In: Miall, D.S. (ed.) Metaphor: Problems and Perspectives, pp. 107–132. Harvester, Brighton (1982)

[Gentner, 1983] Gentner, D.: Structure-mapping: a theoretical framework for analogy. Cognitive Science 7, 155–170 (1983)

[Gibson, 1951] Gibson, J.J.: What is a form? Psychological Review 58, 403–413 (1951)

[Gibson, 1979] Gibson, J.J.: The Ecological Approach to Visual Perception. Houghton Mifflin, Boston (1979)

[Giedymin, 1982] Giedymin, J.: Science and Convention. In: Essays on Henri Poincaré's Philosophy of Science and the Conventionalist Tradition. Pergamon Press, Oxford (1982)

[Giere, 1988] Giere, R.N.: Explaining Science: a Cognitive Approach. University of Chicago Press, Chicago (1988)

[Giere, 1999] Giere, R.N.: Using models to represent reality. In: Magnani, L., Nersessian, N.J., Thagard, P. (eds.) Model-Based Reasoning in Scientific Discovery, pp. 41–57. Kluwer Academic Publishers, New York (1999)

[Giere, 2006] Giere, R.N.: The role of agency in distributed cognitive systems. Philosophy of Science 73, 710–719 (2006)

[Gigerenzer and Brighton, 2009] Gigerenzer, G., Brighton, H.: Homo heuristicus: why biased minds make better inferences. Topics in Cognitive Science 1, 107–143 (2009)

[Gigerenzer and Selten, 2002] Gigerenzer, G., Selten, R.: Bounded Rationality. The Adaptive Toolbox. The MIT Press, Cambridge (2002)

[Gigerenzer and Todd, 1999] Gigerenzer, G., Todd, P.: Simple Heuristics that Make Us Smart. Oxford University Press, Oxford (1999)

[Ginsberg, 1987] Ginsberg, M.L. (ed.): Readings in Nonmonotonic Reasoning. Morgan Kaufmann, San Francisco (1987)

[Girard, 1977] Girard, R.: Violence and the Sacred [1972]. Johns Hopkins University Press, Baltimore (1977)

[Girard, 1986] Girard, R.: The Scapegoat [1982]. Johns Hopkins University Press, Baltimore (1986)

[Girard, 1987] Girard, R.: Linear logic. Theoretical Computer Science 50(1), 1–102 (1987)

[Girard, 2006] Girard, J.-L.: Lectures on essentialism and incompleness. In: Okada, M. (ed.) Towards New Logic and Semantics. Franco-Japanaise Collaborative Lectures on Philosophy of Logic, pp. 71–123. Keyo University, Tokyo (2006)

[Girard, 2007] Girard, J.-L.: La logique, d'Aristote aux algèbres d'opérateurs. In: Okada, M. (ed.) Essays in the Foudantions of Logical and Phenomenological Studies, pp. 45–66. Keyo University, Tokyo (2007)

[Glas, 2009] Glas, E.: Abduction and model-based mathematical discovery (2009) (Forthcoming)

[Glasgow and Papadias, 1992] Glasgow, J.I., Papadias, D.: Computational imagery. Cognitive Science 16, 255–394 (1992)

[Glenberg and Kaschak, 2003] Glenberg, A.M., Kaschak, M.P.: The body's contribution to language. In: Ross, B.H. (ed.) The Psychology of Learning and Motivation, vol. 43, pp. 93–126. Academic Press, San Diego (2003)

[Glymour *et al.*, 1987] Glymour, C., Scheines, R., Spirtes, P., Kelly, K.: Discovering Causal Structure. Academic Press, San Diego (1987)

[Gödel, 1944] Gödel, K.: Russell's mathematical logic. In: Schilpp, P. (ed.) The Philosophy of Bertrand Russell, pp. 123–153. The Tudor Publishing Company, New York (1944); Reprinted in [Benacerraf and Putnam, 1964], pp. 258–273

[Gödel, 1990a] Gödel, K.: Remarks before the Princeton bicentennial conference on problems in mathematics. In: Feferman, S., Dawson Jr., J.W., Kleene, S.C., Moore, G.H., Solovay, R.M., van Heijenoort, J. (eds.) Kurt Gödel: Collected Works, pp.150–153. Oxford University Press, Oxford (1990); Publications 1938-1974

[Gödel, 1990b] Gödel, K.: What is Cantor's continuum problem? [1947]. In: Feferman, S., Dawson Jr., J.W., Kleene, S.C., Moore, G.H., Solovay, R.M., van Heijenoort, J. (eds.) Kurt Gödel Collected Works, pp. 176–187. Oxford University Press, Oxford (1990); Publications 1938-1974. Originally published in The American Mathematical Monthly 54, 515–525. Revised Edition in [benacerraf-putnam, 1964], pp. 258–273

[Godfrey-Smith, 1998] Godfrey-Smith, P.: Complexity and the Function of Mind in Nature. Cambridge University Press, Cambridge (1998)

[Godfrey-Smith, 2002] Godfrey-Smith, P.: Environmental complexity and the evolution of cognition. In: Sternberg, R., Kaufman, K. (eds.) The Evolution of Intelligence, pp. 233–249. Lawrence Erlbaum Associates, Mawhah (2002)

[Godges and Lindheim, 2006] Godges, B.H., Lindheim, O.: Carrying babies and groceries: the effect of moral and social weight on caring. Ecological Psychology 18(2), 93–111 (2006)

[Goldman, 1993] Goldman, A.I. (ed.): Readings in Philosophy and Cognitive Science. Cambridge University Press, Cambridge (1993)

[Goldstein, 2006] Goldstein, W.M.: Introduction to Brunswikian theory and mehod. In: Kirlik, A. (ed.) Human-Technology Interaction. Methods and Models for Cognitive Engineering and Human-Computer Interaction, pp. 10–26. Oxford University Press, Oxford (2006)

[Gomes *et al.*, 200] Gomes, A., El-Hani, C.N., Gudwin, R., Queiroz, J.: Toward emergence of meaning processes in computers from Peircian semiotics. Mind and Society 6, 173–187 (2000)

[Gonzalez and Haselager, 2005] Gonzalez, M.E.Q., Haselager, W.F.G.: Creativity: surprise and abductive reasoning. The British Journal for the Philosophy of Science 153(1), 325–341 (2005)

[Gooding and Addis, 1999] Gooding, D., Addis, T.R.: A simulation of model-based reasoning about disparate phenomena. In: Magnani, L., Nersessian, N.J., Thagard, P. (eds.) Model-Based Reasoning in Scientific Discovery, pp. 103–123. Kluwer Academic/Plenum Publishers, New York (1999)

[Gooding and Addis, 2008] Gooding, D., Addis, T.R.: Modelling experiments as mediating models. Foundations of Science 13(1), 17–35 (2008); Special issue edited by Magnani, L.: Tracking irrational sets. Science, technology, ethics. Proceedings of the International Conference Model-Based Reasoning in Science and Engineering - MBR 2004

[Gooding, 1990] Gooding, D.: Experiment and the Making of Meaning. Kluwer, Dordrecht (1990)

[Gooding, 1996] Gooding, D.: Creative rationality: towards an abductive model of scientific change. Philosophica 58(2), 73–102 (1996)

[Gooding, 2006] Gooding, D.: Visual cognition: where cognition and culture meet. Philosophy of Science 73, 688–698 (2006)

[Gopnik and Meltzoff, 1997] Gopnik, A., Meltzoff, A.: Words, Thoughts and Theories (Learning, Development, and Conceptual Change). The MIT Press, Cambridge (1997)

[Gorman, 1998] Gorman, M.E.: Transforming Nature. Ethics, Invention and Discovery. Kluwer, Dordrecht (1998)

[Gould and Vrba, 1982] Gould, S.J., Vrba, E.S.: Exaptation - a missing term in the science of form. Paleobiology 8(1), 4–15 (1982)

[Gould, 2002] Gould, J.L.: Can honey bees create cognitive maps. In: Bekoff, M., Allen, C., Burghardt, M. (eds.) The Cognitive Animal. Empirical and Theoretical Perspectives on Animal Cognition, pp. 41–46. The MIT Press, Cambridge (2002)

[Gould, 2007] Gould, J.L.: Animal artifacts. In: Margolis, E., Laurence, S. (eds.) Creations of the Mind. Theories of Artifacts and Their Representation, pp. 249–266. Oxford University Press, Oxford (2007)

[Granger, 2006] Granger, R.: Engines of the brain. The computational instruction set of human cognition. AI Magazine 27(2), 15–31 (2006)

[Grasshoff and May, 1995] Grasshoff, G., May, M.: From historical case studies to systematic methods of discovery. In: AAAI Symposium on Systematic Methods of Scientific Discovery, Technical Report SS-95-03. AAAI Press, Menlo Park, CA (1995)

[Grau, 2002] Grau, J.W.: Learning and memory without a brain. In: Bekoff, M., Allen, C., Burghardt, M. (eds.) The Cognitive Animal. Empirical and Theoretical Perspectives on Animal Cognition, pp. 77–88. The MIT Press, Cambridge (2002)

[Greenberg, 1974] Greenberg, M.J.: Euclidean and Non-Euclidean Geometries. Freeman and Company, New York (1974)

[Greene and Haidt, 2002] Greene, J., Haidt, J.: How (and where) does moral judgment work? Trends in Cognitive Science 6(12), 517–523 (2002)

[Greeno, 1994] Greeno, J.G.: Gibson's affordances. Psychological Review 101(2), 336–342 (1994)

[Gregory, 1987] Gregory, R.L.: Perception as hypothesis. In: Gregory, R.L. (ed.) The Oxford Companion to the Mind, pp. 608–611. Oxford University Press, New York (1987)

[Grialou and Okada, 2005] Grialou, P., Okada, M.: Questions on two cognitive models of deductive reasoning. In: Grialou, P., Longo, G., Okada, M. (eds.) Images and Reasoning, pp. 31–67. Keio University, Tokyo (2005)

[Grime and Mackey, 2002] Grime, J.P., Mackey, J.M.L.: The role of plasticity in resource capture by plants. Evolutionary Ecology 16, 299–307 (2002)

[Gruen, 2002] Gruen, L.: The morals of animal minds. In: Bekoff, M., Allen, C., Burghardt, M. (eds.) The Cognitive Animal. Empirical and Theoretical Perspectives on Animal Cognition, pp. 437–442. The MIT Press, Cambridge (2002)

[Grünbaum, 1984] Grünbaum, A.: The Foundations of Psychoanalysis. A Philosophical Critique. University of California Press, Berkeley (1984)

[Grush, 2004a] Grush, R.: The emulation theory of representation: motor control, imagery, and perception. Behavioral and Brain Sciences 27, 377–442 (2004)

[Grush, 2004b] Grush, R.: Further explorations of the empirical and theoretical aspects of the emulation theory. Behavioral and Brain Sciences 27, 425–435 (2004); Author's Response to Open Peer Commentary to R. Grush, The emulation theory of representation: motor control, imagery, and perception

[Grush, 2007] Grush, R.: Agency, emulation and other minds. Cognitive Semiotics, 49–67 (2007)

[Habermas, 1968] Habermas, J.: Erkenntnis und Interesse. Suhrkamp, Frankfurt am Main (1968); Knowledge and Human Interests (1971); translated by J. J. Shapiro. Beacon Press, Boston

[Hacking, 1983] Hacking, I.: Representing and Intervening. In: Introductory Topics in the Philosophy of Natural Science. Cambridge University Press, Cambridge (1983)

[Hage, 2000] Hage, J.: Dialectical models in artificial intelligence and law. Artificial Intelligence and Law 8, 137–172 (2000)

[Hammond and Steward, 2001] Hammond, K.R., Steward, T.R. (eds.): The Essential Brunswik. Beginnings, Explications, Applications. Oxford University Press, Oxford (2001)

[Hammond *et al.*, 1987] Hammond, K.R., Hamm, R.M., Grassia, J., Pearson, T.: Direct comparison of intuitive and analytical cognition in expert judgment. IEEE Transactions on Systems, Man, and Cybernetics SMC-17, 753–770 (1987)

[Hansen, 2002] Hansen, H.H.: The straw thing of fallacy theory: the standard definition of "fallacy". Argumentation 16(2), 133–155 (2002)

[Hanson, 1958] Hanson, N.R.: Patterns of Discovery. An Inquiry into the Conceptual Foundations of Science. Cambridge University Press, London (1958)

[Harman, 1965] Harman, G.: The inference to the best explanation. Philosophical Review 74, 88–95 (1965)

[Harman, 1968] Harman, G.: Enumerative induction as inference to the best explanation. Journal of Philosophy, 65(18):529–533 (1968)

[Harman, 1973] Harman, G.: Thought. Princeton University Press, Princeton (1973)

[Harris, 1989] Harris, R.: How does writing restructure thought? Language and Communication 9(2/3), 99–106 (1989)

[Harris, 1999] Harris, T.: A hierarchy of models and electron microscopy. In: Magnani, L., Nersessian, N.J., Thagard, P. (eds.) Model-Based Reasoning in Scientific Discovery, pp. 139–148. Kluwer Academic/Plenum Publishers, New York (1999)

[Hasselmo, 2005] Hasselmo, M.E.: The role of hippocampal regions CA3 and CA1 in matching entorhinal input with retrieval of associations between objects and context: theoretical comment on Lee et al. Behavioral Neuroscience 119(1), 342–345 (2005)

[Heath, 1992] Heath, T.L.: The Thirteen Books of Euclid's Elements, vol. 3. Cambridge University Press, Cambridge (1992)

[Hebb, 1949] Hebb, D.O.: The Organization of Behavior. John Wiley, New York (1949)

[Heeffer, 2007] Heeffer, A.: Abduction as a strategy for concept formation in mathematics: Cardano postulating a negative. In: Pombo, O. (ed.) International Meeting Abduction and the Process of Scientific Discovery, pp. 179–194. Centro de Filosofia das Ciências da Universidade de Lisboa, Lisbon (2007)

[Heeffer, 2009] Heeffer, A.: The emergence of symbolic algebra as a shift in pre-dominant models. Foundations of science 13(2), 149–161 (2009); Special issue Model-based reasoning in science and engineering, edited by Magnani, L.

[Hegarty and Kozhevnikov, 2000] Hegarty, M., Kozhevnikov, M.: Spatial abilities, working memories, and mechanical resoning. Technical report, Department of Psychology, University of California, Santa Barbara, CA (2000)

[Heidegger, 1926] Heidegger, M.: Being and Time. Blackwell, Oxford (1926); translated by Macquarrie, J., Robinson, E.

[Heinrich, 2002] Heinrich, B.: Raven consciousness. In: Bekoff, M., Allen, C., Burghardt, M. (eds.) The Cognitive Animal. Empirical and Theoretical Perspectives on Animal Cognition, pp. 47–52. The MIT Press, Cambridge (2002)

[Hempel, 1966] Hempel, C.G.: Philosophy of Natural Science. Prentice-Hall, Englewood Cliffs (1966)

[Hendricks and Faye, 1999] Hendricks, F.V., Faye, J.: Abducting explanation. In: Magnani, L., Nersessian, N.J., Thagard, P. (eds.) Model-Based Reasoning in Scientific Discovery, pp. 271–294. Kluwer Academic/Plenum Publishers, New York (1999)

[Henst *et al.*, 2006] Van Der Henst, J.-B., Bujakowska, K., Ciceron, C., Noveck, I.A.: How to make a participant logical: the role of premise presentation in a conditional reasoning task. In: Andler, D., Ogawa, Y., Okada, M., Watanabe, S. (eds.) Reasoning and Cognition, pp. 7–18. Keio University Press, Tokyo (2006)

[Herman, 2002] Herman, L.M.: Exploring the cognitive world of the bottlenosed dolphin. In: Bekoff, M., Allen, C., Burghardt, M. (eds.) The Cognitive Animal. Empirical and Theoretical Perspectives on Animal Cognition, pp. 275–284. The MIT Press, Cambridge (2002)

[Hershberger, 1986] Hershberger, W.A.: An approach through the looking glass. Animal Learning and Behavior 14, 443–451 (1986)

[Hesslow, 2002] Hesslow, G.: Conscious thought as simulation of behaviour and perception. Trends in Cognitive Sciences 6, 242–247 (2002)

[Himmelbach and Karnath, 2005] Himmelbach, M., Karnath, H.-O.: Dorsal and ventral stream interaction: contributions from optic ataxia. Journal of Cognitive Neuroscience 17(4), 632–640 (2005)

[Himmelbach *et al.*, 2006] Himmelbach, M., Karnath, H.-O., Perenin, M.T., Franz, V.H., Stockmeier, K.: A general deficit of the "automatic pilot" with posterior parietal cortex lesions? Neuropsychologia 44(13), 2749–2756 (2006)

[Hintikka and Remes, 1974] Hintikka, J., Remes, U.: The Method of Analysis. Its Geometrical Origin and Its General Significance. Reidel, Dordrecht (1974)

[Hintikka, 1973] Hintikka, J.: Logic, Language-Games and Information. Clarendon Press, London (1973)

[Hintikka, 1997] Hintikka, J.: The place of C. S. Peirce in the history of logical theory. In: Brunning, J., Forster, P. (eds.) The Rule of Reason. The Philosophy of Charles Sanders Peirce, University of Toronto Press, Toronto (1997)

[Hintikka, 1998] Hintikka, J.: What is abduction? The fundamental problem of contemporary epistemology. Transactions of the Charles S. Peirce Society 34, 503–533 (1998)

[Hirsch, 1984] Hirsch, M.W.: The dynamical systems approach to differential equations. Bulletin of the American Mathematical Society 11, 1–64 (1984)

[Hoffman, 1998] Hoffman, D.D.: Visual Intelligence: How We Create What We See. Norton, New York (1998)

[Hoffmann, 1999] Hoffmann, M.H.G.: Problems with Peirce's concept of abduction. Foundations of Science 4(3), 271–305 (1999)

[Hoffmann, 2003] Hoffmann, M.H.G.: Peirce's diagrammatic reasoning as a solution of the learning paradox. In: Debrock, G. (ed.) Process Pragmatism: Essays on a Quiet Philosophical Revolution, pp. 121–143. Rodopi Press, Amsterdam (2003)

[Hoffmann, 2004] Hoffmann, M.H.G.: How to get it. Diagrammatic reasoning as a tool for knowledge development and its pragmatic dimension. Foundations of Science 9, 285–305 (2004)

[Holyoak and Thagard, 1995] Holyoak, K.J., Thagard, P.: Mental Leaps. Analogy in Creative Thought. The MIT Press, Cambridge (1995)

[Holyoak and Thagard, 1997] Holyoak, K.J., Thagard, P.: The analogical mind. American Psychologist 52(1), 35–44 (1997)

[Hookway, 1992] Hookway, C.: Peirce. Routledge and Kegan Paul, London (1992)

[Howells, 1996] Howells, J.: Tacit knowledge, innovation and technology transfer. Technology Analysis & Strategic Management 8(2), 91–105 (1996)

[Humphrey, 2002] Humphrey, N.: The Mind Made Flesh. Oxford University Press, Oxford (2002)

[Husserl, 1931] Husserl, E.: Ideas. General Introduction to Pure Phenomenology [First book, 1913]. Northwestern University Press, London (1931); Translated by W.R. Boyce Gibson

[Husserl, 1970] Husserl, E.: The Crisis of European Sciences and Transcendental Phenomenology [1954]. George Allen & Unwin and Humanities Press, London (1970); Translated by Carr, D.

[Husserl, 1973] Husserl, E.: Ding und Raum: Vorlesungen. Nijhoff, The Hague (1907); Husserliana 16, edited by Claesges, U.

[Husserl, 1978] Husserl, E.: The Origin of Geometry. In: Derrida, J. (ed.) Edmund Husserl's "The Origin of Geometry", pp. 157–180. Nicolas Hays, Stony Brooks, NY (1939); Translated by D. Carr. Originally published in [Husserl, 1970], pp. 353–378

[Hutchins, 1995] Hutchins, E.: Cognition in the Wild. The MIT Press, Cambridge (1995)

[Hutchins, 2005] Hutchins, E.: Material anchors for conceptual blends. Journal of Pragmatics 37, 1555–1577 (2005)

[Iacoboni, 2003] Iacoboni, M.: Understanding intentions through imitation. In: Johnson Frey, S.H. (ed.) Taking Action. Cognitive Neuroscience Perspectives on Intentional Acts, pp. 107–138. The MIT Press, Cambridge (2003)

[Inoue and Sakama, 2006] Inoue, K., Sakama, C.: Abductive equivalence in first-order logic. Logic Journal of the IGPL 14(2), 391–406 (2006)

[Ippolito and Tweney, 1995] Ippolito, M.F., Tweney, R.D.: The inception of insight. In: Sternberg, R.J., Davidson, J.E. (eds.) The Nature of Insight. The MIT Press, Cambridge (1995)

[Irvine, 1989] Irvine, A.: Epistemic logicism and Russell's regressive method. Philosophical Studies 55, 303–327 (1989)

[Jablonka and Lamb, 2005] Jablonka, E., Lamb, M.J.: Evolution in Four Dimensions. Genetic, Epigenetic, Behavioral, and Symbolic Variation in the History of Life. The MIT Press, Cambridge (2005)

[Jackson, 1989] Jackson, P.: Propositional abductive logic. In: Proceedings of the 7th AISB, pp. 89–94 (1989)

[Jamesion, 2002] Jamesion, D.: Cognitive ethology and the end of neuroscience. In: Bekoff, M., Allen, C., Burghardt, M. (eds.) The Cognitive Animal. Empirical and Theoretical Perspectives on Animal Cognition, pp. 69–76. The MIT Press, Cambridge (2002)

[Jeannerod and Pacherie, 2004] Jeannerod, M., Pacherie, E.: Agent, simulation, and self-identification. Mind and Language 19(2), 113–146 (2004)

[Jeannerod, 2001] Jeannerod, M.: Neural simulation of action. NeuroImage 14, S103–S109 (2001)

[Jerne, 1967] Jerne, N.K.: Antibodies and learning: selection versus instruction. In: Quarton, G.C., et al. (eds.) Neurosciences, pp. 200–205. Rockfeller University Press, New York (1967)

[Johnson-Laird, 1983] Johnson-Laird, P.N.: Mental Models. Harvard University Press, Cambridge (1983)

[Johnson-Laird, 1988] Johnson-Laird, P.N.: The Computer and the Mind: an Introduction to Cognitive Science. Harvard University Press, Cambridge (1988)

[Johnson-Laird, 1993] Johnson-Laird, P.N.: Human and Machine Thinking. Erlbaum, Hillsdale (1993)

[Johnson, 1956] Johnson, M.: Moral Imagination. Implications of Cognitive Science in Ethics. The Chicago University Press, Chicago, IL (1956)

[Jones and Scaife, 2000] Jones, S., Scaife, M.: Animated diagrams: an investigation into the cognitive effects of using animation to illustrate dynamic processes. In: Anderson, M., Cheng, P., Haarslev, V. (eds.) Diagrams 2000. LNCS (LNAI), vol. 1889, pp. 231–244. Springer, Heidelberg (2000)

[Josephson and Josephson, 1994] Josephson, J.R., Josephson, S.G. (eds.): Abductive Inference. Computation, Philosophy, Technology. Cambridge University Press, Cambridge (1994)

[Josephson, 1998] Josephson, J.R.: Abduction-prediction model of scientific inference reflected in a prototype system for model-based diagnosis. Philosophica 61(1), 9–17 (1998); Special issue on Abduction and Scientific Discovery, edited by Magnani, L., Nersessian, N.J., Thagard, P.

[Josephson, 2000] Josephson, J.R.: Smart inductive generalizations are abductions. In: Flach, P., Kakas, A. (eds.) Abduction and Induction, pp. 31–44. Kluwer Academic, Dordrecht (2000)

[Jung-Beeman *et al.*, 2004] Jung-Beeman, M., Bowden, E.M., Haberman, J., Frymiare, J.L., Arambel-Liu, S., Greenblatt, R., Reber, P., Kounios, J.: Neural activity when people solve verbal problems with insight. PLoS Biology 2(4), 500–510 (2004)

[Jung, 1967] Jung, C.G.: Two kinds of thinking. In: The Collected Works of C. G. Jung, pp. 7–33. Princeton University Press, Princeton (1967); Translated by Hull, R.F.C.: 2nd edn., vol. 5

[Jung, 1968a] Jung, C.G.: Archetypes and the collective unconscious. In: The Collected Works of C. G. Jung, pp. 3–41. Princeton University Press, Princeton (1968); Translated by Hull, R.F.C.: 2nd edn., vol. 9

[Jung, 1968b] Jung, C.G.: Concerning Mandala symbolism [1950]. In: The Collected Works of C. G. Jung, pp. 355–384. Princeton University Press, Princeton (1968); Translated by Hull, R.F.C.: 2nd edn., vol. 9

[Jung, 1968c] Jung, C.G.: A study in the process of individuation. In: The Collected Works of C. G. Jung, pp. 290–354. Princeton University Press, Princeton (1968); Translated by Hull, R.F.C.: 2nd edn., vol. 9

[Jung, 1972a] Jung, C.G.: On psychic energy. In: The Collected Works of C. G. Jung, pp. 3–66. Princeton University Press, Princeton (1972); Translated by Hull, R.F.C.: 2nd edn., vol. 8

[Jung, 1972b] Jung, C.G.: On the nature of the psyche. In: The Collected Works of C. G. Jung, pp. 159–236. Princeton University Press, Princeton (1972); Translated by Hull, R.F.C.: 2nd edn., vol. 8

[Jung, 1972c] Jung, C.G.: The transcendent function. In: The Collected Works of C. G. Jung, pp. 67–91. Princeton University Press, Princeton (1972); Translated by Hull, R.F.C.: 2nd edn., vol. 8

[Kahn, 2003] Kahn Jr., P.H.: The development of environmental moral identity. In: Clayton, S., Opotow, S. (eds.) Identity and the Natural Environment: the Psychological Significance of Nature, pp. 113–134. The MIT Press, Cambridge (2003)

[Kakas and Riguzzi, 1997] Kakas, A., Riguzzi, F.: Learning with abduction. In: Džeroski, S., Lavrač, N. (eds.) ILP 1997. LNCS (LNAI), vol. 1297, pp. 181–188. Springer, Heidelberg (1997)

[Kakas *et al.*, 1993] Kakas, A., Kowalski, R.A., Toni, F.: Abductive logic programming. Journal of Logic and Computation 2(6), 719–770 (1993)

[Kant, 1902–1983] Kant, I.: Kant's gesammelte Schriften (Herausgegeben von der Akademie der Wissenschaften), vol. 33. Reimer - De Gruyter, Berlin (1902–1983)

[Kant, 1929] Kant, I.: Critique of Pure Reason. MacMillan, London (1929); Translated by N. Kemp Smith, originally published 1787, reprint 1998

[Kant, 1966] Kant, I.: Prolegomena to Any Future Metaphysics (1783). Manchester University Press, Manchester (1966); Translation by P. G. Lucas, third impression

[Kant, 1968] Kant, I.: Inaugural dissertation on the forms and principles of the sensible and intelligible world. In: Kerferd, G.B., Walford, D.E. (eds.) Kant. Selected Pre-Critical Writings, pp. 45–92. Manchester University Press, Manchester (1770); Translated by G.B. Kerferd and D.E. Walford (1968), Also translated by J. Handyside. I. Kant, Kant's Inaugural Dissertation and Early Writings on Space. Open Court, Chicago, IL, pp. 35–85 (1929)

[Kapitan, 1990] Kapitan, T.: In what way is abductive inference creative? Transactions of the Charles S. Peirce Society 26(4), 449–512 (1990)

[Kapitan, 1997] Kapitan, T.: Peirce and the structure of adductive inference. In: Houser, N., Roberts, D.D., van Evra, J. (eds.) Studies in the Logic of Charles Sanders Peirce, pp. 477–496. Indiana University Press, Bloomington (1997)

[Kaplan *et al.*, 2000] Kaplan, H.K., Hill, K., Lancaster, J., Hurtado, M.: A theory of human life history evolution: diet, intelligence, and longevity. Evolutionary Anthropology 9(4), 155–185 (2000)

[Karmiloff-Smith, 1992] Karmiloff-Smith, A.: Beyond Modularity: a Developmental Perspective on Cognitive Science. The MIT Press, Cambridge (1992)

[Katsuno and Mendelzon, 1992] Katsuno, H., Mendelzon, A.: On the difference between updating a knowledge base and revising it. In: Gärdenfors, P. (ed.) Belief Revision, pp. 183–203. Cambridge University Press, Cambridge (1992)

[Kirlik, 1998] Kirlik, A.: The Ecological Expert: acting to create information to guide action. In: Human Interaction with Complex Systems (HICS 1998) Proceedings. Fourth Annual Symposium, Dayton, OH, Symposium, Dayton, OH, pp. 22–25. IEEE Computer Society Press, Los Alamitos (1998)

[Kirlik, 2006a] Kirlik, A.: Abstracting situated action: implications for cognitive modeling and interface design. In: Kirlik, A. (ed.) Human-Technology Interaction. Methods and Models for Cognitive Engineering and Human-Computer Interaction, pp. 212–226. Oxford University Press, Oxford (2006)

[Kirlik, 2006b] Kirlik, A. (ed.): Human-Technology Interaction. Methods and Models for Cognitive Engineering and Human-Computer Interaction. Oxford University Press, Oxford (2006)

[Kirlik, 2007] Kirlik, A.: Reiventing intelligence for an invented world. In: Sternberg, R.J., Preiss, D.D. (eds.) Intelligence and Technology: the Impact of Tools on the Nature and Development of Human Abilities, pp. 105–134. Lawrence Erlbaum Associates, Mawhah (2007)

[Kirsh and Maglio, 1994] Kirsh, D., Maglio, P.: On distinguishing epistemic from pragmatic action. Cognitive Science 18, 513–549 (1994)

[Kirsh, 1995] Kirsh, A.: The intelligent use of space. Artificial Intelligence 73, 31–68 (1995)

[Klahr and Dunbar, 1988] Klahr, D., Dunbar, K.: Dual space search during scientific reasoning. Cognitive Science 12, 1–48 (1988)

[Knierim, 2007] Knierim, J.J.: The matrix in your head. Scientific American Mind, 46–48 (June/July 2007)

[Knoblich and Flach, 2001] Knoblich, G., Flach, R.: Predicting the effects of actions: interactions of perception and action. Psychological Science 12(6), 467–472 (2001)

[Knuuttila and Honkela, 2005] Knuuttila, T., Honkela, T.: Questioning external and internal representations: the case of scientific models. In: Magnani, L., Dossena, R. (eds.) Computing, Philosophy and Cognition, pp. 209–226. College Publications, London (2005)

[Koestler, 1964] Koestler, A.: *The Act of Creation*. MacMillan, New York (1964)

[Kolodner, 1993] Kolodner, J.L.: Case-Based Reasoning. Morgan Kaufmann, San Mateo (1993)

[Konolige, 1992] Konolige, K.: Abduction versus closure in causal theories. Artificial Intelligence 53, 255–272 (1992)

[Kosfeld *et al.*, 2005] Kosfeld, M., Heinrichs, M., Zak, P.J., Fischbacher, U., Fehr, E.: Oxytocin increases trust in humans. Nature 435, 673–676 (2005)

[Koslowski, 1996] Koslowski, B.: Theory and Evidence. The Development of Scientific Reasoning. The MIT Press, Cambridge (1996)

[Kosslyn and Koenig, 1992] Kosslyn, S.M., Koenig, O.: Wet Mind, the New Cognitive Neuroscience. Free Press, New York (1992)

[Kowalski, 1979] Kowalski, R.A.: Logic for Problem Solving. Elsevier, New York (1979)

[Kowalski, 1994] Kowalski, R.A.: Logic without model theory. In: Gabbay, D.M. (ed.) What is a Logical System?, pp. 35–71. Oxford University Press, Oxford (1994)

[Krauss *et al.*, 1999] Krauss, S., Martignon, L., Hoffrage, U.: Simplifying Bayesian inference: the general case. In: Magnani, L., Nersessian, N.J., Thagard, P. (eds.) Model-Based Reasoning in Scientific Discovery, pp. 165–179. Kluwer Academic/Plenun Publishers, New York (1999)

[Kruijff, 2005] Kruijff, G.-J.-M.: Peirce's late theory of abduction: a comprehensive account. Semiotica 153(1/4), 431–454 (2005)

[Kuhn, 1962] Kuhn, T.S.: The Structure of Scientific Revolutions. University of Chicago Press, Chicago (1962); 2nd expanded edn. (1970)

[Kuhn, 1991] Kuhn, D.: The Skills of Argument. Cambridge University Press, Cambridge (1991)

[Kuhn, 1996] Kuhn, D.: Is good thinking scientific thinking? In: Olson, D.R., Torrance, N. (eds.) Modes of Thought. Explorations in Culture and Cognition, pp. 261–281. Cambridge University Press, Cambridge (1996)

[Kuipers, 1999] Kuipers, T.A.F.: Abduction aiming at empirical progress of even truth approximation leading to a challenge for computational modelling. Foundations of Science 4, 307–323 (1999)

[Kuipers, 2000] Kuipers, T.: From Instrumentalism to Constructive Realism. On Some Relations between Confirmation, Empirical Progress and Truth Approximation. Kluwer Academic Publisher, Dordrecht (2000)

[Kulkarni and Simon, 1988] Kulkarni, D., Simon, H.A.: The process of scientific discovery: the strategy of experimentation. Cognitive Science 12, 139–176 (1988)

[Lachapelle *et al.*, 2006] Lachapelle, J., Faucher, L., Poirier, P.: Cultural evolution, the Baldwin effect, and social norms. In: Gontier, N., Van Bendegem, J.P., Aerts, D. (eds.) Evolutionary Epistemology, Language and Culture, pp. 313–334. Springer, Berlin (2006)

[Ladrière, 1981] Ladrière, J.: La philosophie des mathématiques de Kurt Gödel. Epistemologia 4(1), 287–311 (1981)

[Lakatos, 1970] Lakatos, I.: Falsification and the methodology of scientific research programs. In: Lakatos, I., Musgrave, A. (eds.) Criticism and the Growth of Knowledge, pp. 365–395. The MIT Press, Cambridge (1970)

[Lakatos, 1976] Lakatos, I.: Proofs and Refutations. The Logic of Mathematical Discovery. Cambridge University Press, Cambridge (1976)

[Laland and Brown, 2006] Laland, K.N., Brown, G.R.: Niche construction, human behavior, and the adaptive-lag hypothesis. Evolutionary Anthropology 15, 95–104 (2006)

[Laland *et al.*, 2000] Laland, K.N., Odling-Smee, F.J., Feldman, M.W.: Niche construction, biological evolution and cultural change. Behavioral and Brain Sciences 23(1), 131–175 (2000)

[Laland *et al.*, 2001] Laland, K.N., Odling-Smee, F.J., Feldman, M.W.: Cultural niche construction and human evolution. Journal of Evolutionary Biology 14, 22–33 (2001)

[Laland *et al.*, 2005] Laland, K.N., Odling-Smee, F.J., Feldman, M.W.: On the breath and significance of niche construction: a reply to Griffiths, Okasha and Sterelny. Biology and Philosophy 20, 37–55 (2005)

[Lambert, 1786] Lambert, J.H.: Theorie der Parallellinien. Magazin für die reine und angewandte Mathematik 2-3, 137–164, 325–358 (1786); Written about 1766; posthumously published by J. Bernoulli

[Landy and Goldstone, 2007a] Landy, D., Goldstone, R.L.: Formal notations are diagrams: evidence from a production task. Memory and Cognition 35(8), 2033–2040 (2007)

[Landy and Goldstone, 2007b] Landy, D., Goldstone, R.L.: How abstract is symbolic thought? Journal of Experimental Psychology. Learning, Memory, and Cognition 33(4), 720–733 (2007)

[Langley *et al.*, 1987] Langley, P., Simon, H.A., Bradshaw, G.L., Zytkow, J.M.: Scientific Discovery. Computational Explorations of the Creative Processes. The MIT Press, Cambridge (1987)

[Lanzola *et al.*, 1990] Lanzola, G., Stefanelli, M., Barosi, G., Magnani, L.: Neoanemia: a knowledge-based system emulating diagnostic reasoning. Computer and Biomedical Research 23, 560–582 (1990)

[Lassègue, 1998] Lassègue, J.: Turing. Les Belles Lettres, Paris (1998)

[Lassègue, 1999] Lassègue, J.: Turing entre formel et forme; remarque sur la convergence des perspectives morphologiques. Intellectica 35(2), 185–198 (1999)

[Latour, 1987] Latour, J.: Science in Action: How to Follow Scientists and Engineers through Society. Harvard University Press, Cambridge (1987)

[Latour, 1988] Latour, J.: The Pasteurization of France. Harvard University Press, Cambridge (1988)

[Law, 1993] Law, J.: Modernity, Myth, and Materialism. Blackwell, Oxford (1993)

[Leake, 1992] Leake, D.B.: Evaluating Explanations. A Content Theory. Erlbaum, Hillsdale (1992)

[Lederman and Klatzky, 1990] Lederman, S.J., Klatzky, R.: Haptic exploration and object representation. In: Goodale, M.A. (ed.) Vision and Action: the Control of Grasping, pp. 98–109. Ablex, Norwood (1990)

[Lee *et al.*, 2005] Lee, I., Hunsaker, M.R., Kesner, R.P.: The role of hyppocampal subregions in detecting spatial novelty. Behavioral Neuroscience 119(1), 145–153 (2005)

[Leibniz, 1949] Leibniz, G.W.: New Essays Concerning Human Understanding (1690). Cambridge University Press, Cambridge (1949); Translated and edited by P. Remnant and J. Bennett

[Lenat, 1982] Lenat, D.: Discovery in mathematics as heuristic search. In: Davis, R., Lenat, D.B. (eds.) Knowledge-Based Systems in Artificial Intelligence. McGraw Hill, New York (1982)

[Lestel and Herzfeld, 2005] Lestel, D., Herzfeld, C.: Topological ape: knots tying and untying and the origins of mathematics. In: Grialou, P., Longo, G., Okada, M. (eds.) Images and Reasoning, pp. 147–162. Keio University, Tokyo (2005)

[Levesque, 1989] Levesque, H.J.: A knowledge-level account of abduction. In: Proceedings of the Eleventh IJCAI, Los Altos, CA, pp. 1061–1067 (1989)

[Lévi-Bruhl, 1923] Lévi-Bruhl, L.: Primitive Mentality. Beacon Press, Boston (1923)

[Levi, 1996] Levi, I.: For the Sake of the Argument. Ramsey Test Conditionals, Inductive Inference, and Nonmonotonic Reasoning. Cambridge University Press, Cambridge (1996)

[Levy, 1997] Levy, S.H.: Peirce's theoremic/corollarial distinction and the interconnections between mathematics and logic. In: Houser, N., Roberts, D.D., van Evra, J. (eds.) Studies in the Logic of Charles Sanders Peirce, pp. 85–110. Indiana University Press, Bloomington (1997)

[Lewis-Williams, 2002] Lewis-Williams, D.: The Mind in the Cave. Thames and Hudson, London (2002)

[Leyton, 1999] Leyton, M.: Simmetry, Causality, Mind. The MIT Press, Cambridge (1999)

[Leyton, 2001] Leyton, M.: A Generative Theory of Shape. Springer, Berlin (2001)

[Leyton, 2006] Leyton, M.: The Structure of Paintings. Springer, Berlin/New York (2006)

[Liben, 2001] Liben, L.S.: Thinking through maps. In: Gattis, M. (ed.) Spatial Schemas and Abstract Thought, pp. 45–78. The MIT Press, Cambridge (2001)

[Lindsay *et al.*, 1980] Lindsay, R.K., Buchanan, B., Feingenbaum, E., Lederberg, J.: Applications of Artificial Intelligence for Organic Chemistry: the Dendral Project. McGraw Hill, New York (1980)

[Lindsay, 1994] Lindsay, R.K.: Understanding diagrammatic demonstrations. In: Ram, A., Eiselt, K. (eds.) Proceedings of the 16th Annual Conference of the Cognitive Science Society, Paris, pp. 572–576. Erlbaum, Hillsdale (1994)

[Lindsay, 1998] Lindsay, R.K.: Using diagrams to understand geometry. Computational Intelligence 9(4), 343–345 (1998)

[Lindsay, 2000a] Lindsay, R.K.: Playing with diagrams. In: Anderson, M., Cheng, P., Haarslev, V. (eds.) Diagrams 2000. LNCS (LNAI), vol. 1889, pp. 300–313. Springer, Heidelberg (2000)

[Lindsay, 2000b] Lindsay, R.K.: Using spatial semantics to discover and verify diagrammatic demonstrations of geometric propositions. In: O'Nuallian, S. (ed.) Spatial Cognition. Proceedings of the 16th Annual Conference of the Cognitive Science Society, pp. 199–212. John Benjamins, Amsterdam (2000)

[Lipton, 2004] Lipton, P.: Inference to the Best Explanation. New Revised edn. Routledge, London (2004); Originally published in 1991

[Lloyd, 1987] Lloyd, J.W.: Foundations of Logic Programming. Springer, Berlin (1987)

[Lobachevsky, 1829-1830, 1835-1838] Lobachevsky, N.I.: Zwei geometrische Abhandlungen, aus dem Russischen bersetzt, mit Anmerkungen und mit einer Biographie des Verfassers von Friedrich Engel. B. G. Teubner, Leipzig (1829-1830) (1835-1838)

[Lobachevsky, 1891] Lobachevsky, N.I.: Geometrical Researches on the Theory of Parallels [1840]. University of Texas, Austin (1891)

[Lobachevsky, 1897] Lobachevsky, N.I.: The "Introduction" to Lobachevsky's New Elements of Geometry. Transactions of Texas Academy 2, 1–17 (1897); Translated by G.B. Halsted. Originally published in: Lobachevsky, N.I.: Novye nachala geometrii. Uchonia sapiski Kasanskava Universiteta 3, 3–48 (1835)

[Lobachevsky, 1929] Lobachevsky, N.I.: Pangeometry or a Summary of Geometry Founded upon a General and Rigorous Theory of Parallels. In: Smith, D.E. (ed.) A Source Book in Mathematics, pp. 360–374. McGraw Hill, New York (1929)

[Logan, 2000] Logan, R.K.: The Sixth Language. Learning a Living in the Computer Age. Stoddart Publishing, Toronto (2000)

[Logan, 2006] Logan, R.K.: The extended mind model of the origin of language and culture. In: Gontier, N., Van Bendegem, J.P., Aerts, D. (eds.) Evolutionary epistemology, Language and Culture, pp. 149–167. Springer, Berlin (2006)

[Longo, 2002] Longo, G.: Laplace, Turing, et la géométrie impossible du "jeu de l'imitation": aléas, determinisme et programmes dans le test de Turing. Intellectica 35(2), 131–161 (2002)

[Longo, 2005] Longo, G.: The cognitive foundations of mathematics: human gestures in proofs and mathematical incompleteness of formalisms. In: Grialou, P., Longo, G., Okada, M. (eds.) Images and Reasoning, pp. 105–134. Keio University, Tokyo (2005)

[Longo, 2009] Longo, G.: Critique of computational reason in the natural sciences. In: Gelenbe, E., Kahane, J.-P. (eds.) Fundamental Concepts in Computer Science. Imperial College Press/World Scientific, London (2009)

[Loula *et al.*, 2006] Loula, A., Gudwin, R., Queiroz, J. (eds.): Artificial Cognition Systems. Idea Group Publishers, Hershey (2006)

[Loula *et al.*, 2009] Loula, A., Gudwin, R., El-HaniD., C.N., Queiroz, J.: Emergence of self-organized symbol-based communication in artificial creatures. Cognitive Systems Research (2009) (Forthcoming)

[Love, 2004] Love, N.: Cognition and the language myth. Language Sciences 26, 525–544 (2004)

[Lozano *et al.*, 2006] Lozano, S., Martin Hard, B., Tversky, B.: Perspective taking promotes action understanding and learning. Journal of Experimental Psychology 32(6), 1405–1421 (2006)

[Luan *et al.*, 2006] Luan, S., Magnani, L., Dai, G.: Algorithms for computing minimal conflicts. Logic Journal of the IGPL 14(2), 391–406 (2006)

[Lukasiewicz, 1970] Lukasiewicz, J.: Creative elements in science [1912], pp. 12–44. North Holland, Amsterdam (1970)

[Lukaszewicz, 1970] Lukaszewicz, W.: Non-Monotonic Reasoning. Formalization of Commonsense Reasoning. Horwood, Chichester (1970)

[Lungarella and Sporns, 2005] Lungarella, M., Sporns, O.: Information self-structuring: key principles for learning and development. In: Proceedings 2005 IEEE International Conference on Development and Learning, pp. 25–30 (2005)

[Määttänen, 1997] Määttänen, P.: Intelligence, agency, and interaction. In: Grahne, G. (ed.) Sixth Scandinavian Conference on Artificial Intelligence, pp. 52–58. IOS Press, Amsterdam (1997)

[Määttänen, 2009] Määttänen, P.: Habits as vehicles of cognition. In: Proceedings of the International Conference "Applying Peirce", Helsinki, Finland (2009) (Forthcoming)

[Mach, 1905] Mach, E.: Erkenntnis und Irrtum. Skizzen zur Psychologie der Forschung. Iohann Ambrosius Barth, Leipzig (1905); Translation by Foulkes, P.: Knowledge and Error: Sketches on the Psychology of Enquiry. Reidel, Dordrecht (1975)

[Maddalena, 2005] Maddalena, G.: Abduction and metaphysical realism. Semiotica 153(1/4), 243–259 (2005)

[Magnani and Bardone, 2005] Magnani, L., Bardone, E.: Designing human interfaces. The role of abduction. In: Magnani, L., Dossena, R. (eds.) Computing, Philosophy and Cognition, pp. 131–146. College Publications, London (2005)

[Magnani and Bardone, 2008] Magnani, L., Bardone, E.: Sharing representations and creating chances through cognitive niche construction. The role of affordances and abduction. In: Iwata, S., Oshawa, Y., Tsumoto, S., Zhong, N., Shi, Y., Magnani, L. (eds.) Communications and Discoveries from Multidisciplinary Data, pp. 3–40. Springer, Berlin (2008)

[Magnani and Belli, 2006] Magnani, L., Belli, E.: Agent-based abduction: being rational through fallacies. In: Magnani, L. (ed.) Model-Based Reasoning in Science and Engineering. Cognitive Science, Epistemology, Logic, pp. 415–439. College Publications, London (2006)

[Magnani and Dossena, 2005] Magnani, L., Dossena, R.: Perceiving the infinite and the infinitesimal world: unveiling and optical diagrams and the construction of mathematical concepts. Foundations of Science 10, 7–23 (2005)

[Magnani and Gennari, 1997] Magnani, L., Gennari, R.: Manuale di logica. Guerini, Milan (1997)

[Magnani and Li, 2007] Magnani, L., Li, P. (eds.): Model-Based Reasoning in Science, Technology, and Medicine. Springer, Berlin (2007)

[Magnani and Piazza, 2005] Magnani, L., Piazza, M.: Morphodynamical abduction: causation by attractors dynamics of explanatory hypotheses in science. Foundations of Science 10, 107–132 (2005)

[Magnani *et al.*, 1994] Magnani, L., Civita, S., Previde Massara, G.: Visual cognition and cognitive modeling. In: Cantoni, V. (ed.) Human and Machine Vision: Analogies and Divergences, pp. 229–243. Plenum Publishers, New York (1994)

[Magnani *et al.*, 2002a] Magnani, L., Nersessian, N.J., Pizzi, C. (eds.): Logical and Computational Aspects of Model-Based Reasoning. Kluwer Academic Publishers, Dordrecht (2002)

[Magnani *et al.*, 2002b] Magnani, L., Piazza, M., Dossena, R.: Epistemic mediators and chance morphodynamics. In: Abe, A. (ed.) Proceedings of PRICAI 2002 Conference, Working Notes of the 2nd International Workshop on Chance Discovery, Tokyo, pp. 38–46 (2002)

[Magnani, 1988] Magnani, L.: Epistémologie de l'invention scientifique. Communication and Cognition 21, 273–291 (1988)

[Magnani, 1991] Magnani, L.: Epistemologia applicata. Marcos y Marcos, Milan (1991)

[Magnani, 1992] Magnani, L.: Abductive reasoning: philosophical and educational perspectives in medicine. In: Evans, D.A., Patel, V.L. (eds.) Advanced Models of Cognition for Medical Training and Practice, pp. 21–41. Springer, Berlin (1992)

[Magnani, 1996] Magnani, L.: Visual abduction: philosophical problems and perspectives. In: AAAI Spring Symposium, pp. 21–24. American Association for Artificial Intelligence, Stanford (1996); Comment to R. Lindsay, Generalizing from diagrams

[Magnani, 1997] Magnani, L.: Ingegnerie della conoscenza. Introduzione alla filosofia computazionale. Marcos y Marcos, Milan (1997)

[Magnani, 1999] Magnani, L.: Inconsistencies and creative abduction in science. In: AI and Scientific Creativity. Proceedings of the AISB 1999 Symposium on Scientific Creativity, Edinburgh, pp. 1–8. Society for the Study of Artificial Intelligence and Simulation of Behaviour, University of Edinburgh (1999)

[Magnani, 2001a] Magnani, L.: Limitations of recent formal models of abductive reasoning. In: Kakas, A., Toni, F. (eds.) Abductive Reasoning. KKR-2, 17th International Joint Conference on Artificial Intelligence (IJCAI 2001), pp. 34–40. Springer, Berlin (2001)

[Magnani, 2001b] Magnani, L.: Abduction, Reason, and Science. Processes of Discovery and Explanation. Kluwer Academic/Plenum Publishers, New York (2001)
[Magnani, 2001c] Magnani, L.: Philosophy and Geometry. Theoretical and Historical Issues. Kluwer Academic Publisher, Dordrecht (2001)
[Magnani, 2002a] Magnani, L.: Epistemic mediators and model-based discovery in science. In: Magnani, L., Nersessian, N.J. (eds.) Model-Based Reasoning: Science, Technology, Values, pp. 305–329. Kluwer Academic/Plenum Publishers, New York (2002)
[Magnani, 2002b] Magnani, L.: Thinking through doing, external representations in abductive reasoning. In: AISB 2002 Symposium on AI and Creativity in Arts and Science. Imperial College, London (2002)
[Magnani, 2005] Magnani, L.: Creativity and the disembodiment of mind. In: Gervás, P., Pease, A., Veale, T. (eds.) Proceedings of CC 2005, Computational Creativity Workshop, IJCAIO 2005, Edinburgh, pp. 60–67 (2005)
[Magnani, 2006a] Magnani, L. (ed.): Abduction, Practical Reasoning, and Creative Inferences in Science. Oxford University Press, Oxford (2006); Special Issue of the Logic Journal of the IGPL
[Magnani, 2006b] Magnani, L.: Disembodying minds, externalising minds: how brains make up creative scientific reasoning. In: Magnani, L. (ed.) Model-Based Reasoning in Science and Engineering, Cognitive Science, Epistemology, Logic, pp. 185–202. College Publications, London (2006)
[Magnani, 2006c] Magnani, L.: Mimetic minds. Meaning formation through epistemic mediators and external representations. In: Loula, A., Gudwin, R., Queiroz, J. (eds.) Artificial Cognition Systems, pp. 327–357. Idea Group Publishers, Hershey (2006)
[Magnani, 2006d] Magnani, L.: Multimodal abduction. External semiotic anchors and hybrid representations. Logic Journal of the IGPL 14(2), 107–136 (2006)
[Magnani, 2006e] Magnani, L.: Prefiguring ethical chances: the role of moral mediators. In: Oshawa, Y., Tsumoto, S. (eds.) Chance Discoveries in Real World Decision Making: Data-based Interaction of Human and Artificial Intelligence, pp. 205–229. Springer, Berlin (2006)
[Magnani, 2006f] Magnani, L. (ed.): Model-Based Reasoning in Science and Engineering. College Publications, London (2006)
[Magnani, 2007a] Magnani, L.: Abduction and cognition in human and logical agents. In: Artemov, S., Barringer, H., Garcez, A., Lamb, L., Woods, J. (eds.) We Will Show Them: Essays in Honour of Dov Gabbay, vol. II, pp. 225–258. College Publications, London (2007)
[Magnani, 2007b] Magnani, L.: Animal abduction. From mindless organisms to artifactual mediators. In: Magnani, L., Li, P. (eds.) Model-Based Reasoning in Science, Technology, and Medicine, pp. 3–37. Springer, Berlin (2007)
[Magnani, 2007c] Magnani, L.: Semiotic brains and artificial minds. How brains make up material cognitive systems. In: Gudwin, R., Queiroz, J. (eds.) Semiotics and Intelligent Systems Development, pp. 1–41. Idea Group Inc., Hershey (2007)
[Magnani, 2007d] Magnani, L.: Morality in a Technological World. Knowledge as Duty. Cambridge University Press, Cambridge (2007)
[Magnani, 2009a] Magnani, L.: Mindless abduction. From animal guesses to artifactual mediators. In: Proceedings of the International Conference "Applying Peirce", Helsinki, Finland (2009) (Forthcoming)
[Magnani, 2009b] Magnani, L.: Multimodal abduction in knowledge development. In: International Workshop on Abductive and Inductive Knowledge Development, pp. 21–26 (2009); IJCAI 2009, Pasadena, USA, Preworkshop Proceedings

[Malle *et al.*, 2001] Malle, B.F., Moses, L.J., Baldwin, D.A.: Introduction: the significance of intentionality. In: Malle, B.F., Moses, L.J., Baldwin, D.A. (eds.) Intentions and Intentionality. Foundations of Social Cognition, pp. 1–24. The MIT Press, Cambridge (2001)

[Mancosu, 1996] Mancosu, P.: Philosophy of Mathematics and Mathematical Practice in the Seventeenth Century. Oxford University Press, Oxford (1996)

[Mandler, 2007] Mandler, J.M.: The conceptual foundations of animals and artifacts. In: Margolis, E., Laurence, S. (eds.) Creations of the Mind. Theories of Artifacts and Their Representation, pp. 191–210. Oxford University Press, Oxford (2007)

[Marcus, 2004] Marcus, G.F.: The Birth of the Mind. How a Tiny Number of Genes Creates the Complexity of Human Thought. Basic Books, New York (2004)

[Marcus, 2006] Marcus, G.F.: Cognitive architecture and descent with modification. Cognition 101, 443–465 (2006)

[Margolis and Laurence, 2007] Margolis, E., Laurence, S. (eds.): Creations of the Mind. Theories of Artifacts and Their Representation. Oxford University Press, Oxford (2007)

[Marr, 1982] Marr, D.: Vision. Freeman, New York (1982)

[Matthews *et al.*, 2002] Matthews, G., Zeidner, M., Roberts, R.D.: Emotional Intelligence. Science and Myth. The MIT Press, Cambridge (2002)

[Maturana and Varela, 1980] Maturana, H., Varela, F.J.: Autopoiesis and Cognition. Reidel, Dordrecht (1980)

[Mayo, 1996] Mayo, D.: Error and the Growth of Experimental Knowledge. The University of Chicago Press, Chicago (1996)

[McGrenere and Ho, 2000] McGrenere, J., Ho, W.: Affordances: clarifying and evolving a concept. In: Proceedings of Graphics Interface, Montreal, Quebec, Canada, May 15-17, 2000, pp. 179–186 (2000)

[Meheus and Batens, 2006] Meheus, J., Batens, D.: A formal logic for abductive reasoning. Logic Journal of the IGPL 14(1), 221–236 (2006)

[Meheus *et al.*, 2002] Meheus, J., Verhoeven, L., Van Dyck, M., Provijn, D.: Ampliative adaptive logics and the foundation of logic-based approaches to abduction. In: Magnani, L., Nersessian, N.J., Pizzi, C. (eds.) Logical and Computational Aspects of Model-Based Reasoning, pp. 39–71. Kluwer Academic Publishers, Dordrecht (2002)

[Meheus, 1999] Meheus, J.: Model-based reasoning in creative processes. In: Magnani, L., Nersessian, N.J., Thagard, P. (eds.) Model Based Reasoning in Scientific Discovery, pp. 199–217. Plenum Publishers/Kluwer Academic, New York (1999)

[Meltzoff and Brooks, 2001] Meltzoff, A.N., Brooks, R.: "Like me" as a building block for understanding other minds: bodily acts, attention, and intention. In: Malle, B., Moses, L., Baldwin, D. (eds.) Intentions and Intentionality. Foundations of Social Cognition, pp. 171–191. The MIT Press, Cambridge (2001)

[Menary, 2007] Menary, R.: Writing as thinking. Language Sciences 29, 621–632 (2007)

[Michalski, 1993] Michalski, R.S.: Inferential theory of learning as a conceptual basis for multistrategy learning. Machine Learning 11, 111–151 (1993)

[Milner and Goodale, 1995] Milner, A.D., Goodale, M.A.: The Visual Brain in Action. Oxford University Press, Oxford (1995)

[Milner *et al.*, 2001] Milner, A.D., Dijkerman, H.C., Pisella, L., McIntosh, R.D., Tilikete, C., Vighetto, A., Rossetti, Y.: Grasping the past: delay can improve visuomotor performance. Current Biology 11, 1896–1901 (2001)

[Milton, 2006] Milton, K.: Diet and primate evolution. Scientific American 16(2), 22–29 (2006)

[Minnameier, 2004] Minnameier, G.: Peirce-suit of truth. Why inference to the best explanation and abduction ought not to be confused. Erkenntnis 60, 75–105 (2004)

[Minski, 1985] Minski, M.: The Society of Mind. Simon and Schuster, New York (1985)

[Mitchell, 2002] Mitchell, R.W.: Kinesthetic-visual matching, imitation, and self-recognition. In: Bekoff, M., Allen, C., Burghardt, M. (eds.) The Cognitive Animal. Empirical and Theoretical Perspectives on Animal Cognition, pp. 345–352. The MIT Press, Cambridge (2002)

[Mithen, 1996] Mithen, S.: The Prehistory of the Mind. A Search for the Origins of Art, Religion, and Science. Thames and Hudson, London (1996)

[Mithen, 1999] Mithen, S.: Handaxes and ice age carvings: hard evidence for the evolution of consciousness. In: Hameroff, A.R., Kaszniak, A.W., Chalmers, D.J. (eds.) Toward a Science of Consciousness III. The Third Tucson Discussions and Debates, pp. 281–296. MIT Press, Cambridge (1999)

[Mithen, 2007] Mithen, S.: Creation of pre-modern human minds: stone tool manufacture. In: Margolis, E., Laurence, S. (eds.) Creations of the Mind. Theories of Artifacts and Their Representation, pp. 289–311. Oxford University Press, Oxford (2007)

[Modell, 2003] Modell, A.H.: Imagination and the Meaningful Brain. The MIT Press, Cambridge (2003)

[Moll *et al.*, 2002] Moll, J., de Oliveira-Souza, R., Eslinger, P.J., Bramati, I.E., Mourão-Miranda, J., Andreiuolo, P.D., Pessoa, L.: The neural correlates of moral sensitivity: a functional magnetic resonance imaging investigation of basic and moral emotions. The Journal of Neuroscience 22(7), 2730–2736 (2002)

[Moll *et al.*, 2005] Moll, J., Zahn, R., de Oliveira-Souza, R., Krueger, F., Grafman, J.: The neural basis of human moral cognition. Nature Reviews Neuroscience 6, 799–809 (2005)

[Monekosso *et al.*, 2004] Monekosso, N., Remagnino, P., Ferri, F.J.: Learning machines for chance discovery. In: Abe, A., Oehlmann, R. (eds.) The First European Workshop on Chance Discovery, Valencia, Spain, pp. 84–93 (2004)

[Moore and Oaksford, 2002] Moore, S.C., Oaksford, M. (eds.): Emotional Cognition. From Brain to Behaviour. John Benjamins, Amsterdam (2002)

[Moorjani, 2000] Moorjani, A.: Peirce and psychopragmatics. In: Muller, J., Brent, J. (eds.) Peirce, Semiosis, and Psychoanalysis, pp. 102–121. John Hopkins, Baltimore (2000)

[Morgan and Morrison, 1999] Morgan, M.S., Morrison, M. (eds.): Models as Mediators. Perspectives on Natural and Social Science. Cambridge University Press, Cambridge (1999)

[Moxon and Wills, 1999] Moxon, E.R., Wills, C.: DNA microsatellites: agents of evolution. Scientific American 280(1), 94–99 (1999)

[Moxon *et al.*, 1994] Moxon, E.R., Rainey, P.B., Nowak, M.A., Lenski, R.E.: Adaptive evolution of highly mutable loci in pathogenic bacteria. Current Biology 4, 24–33 (1994)

[Nara and Ohsawa, 2004] Nara, Y., Ohsawa, Y.: Knowing melting pot. Emerging topics based on scenario communication on technical foresights. In: Abe, A., Oehlmann, R. (eds.) The First European Workshop on Chance Discovery, Valencia, pp. 114–121 (2004)

[Natsoulas, 2004] Natsoulas, T.: To see things is to perceive what they afford: James J. Gibson's concept of affordance. Mind and Behaviour 2(4), 323–348 (2004)

[Nersessian and Chandrasekharan, 2009] Nersessian, N.J., Chandrasekharan, S.: Hybrid analogies in conceptual innovation in science. Cognitive Systems Research Journal 435, 673–676 (2009); Special Issue on Integrative Analogy (Forthcoming)

[Nersessian and Patton, 2009] Nersessian, N.J., Patton, C.: Model-based reasoning in interdisciplinary engineering: cases from biomedical engineering research laboratories. In: Meijers, A.W.M. (ed.) The Handbook of the Philosophy of Technology and Engineering Sciences. Springer, Berlin (2009) (forthcoming)

[Nersessian *et al.*, 1997] Nersessian, N.J., Griffith, T.W., Goel, A.: Constructive modeling in scientific discovery. Technical report, Georgia Institute of Technology, Atlanta, GA (1997)

[Nersessian, 1992] Nersessian, N.J.: How do scientists think? Capturing the dynamics of conceptual change in science. In: Giere, R.N. (ed.) Cognitive Models of Science, Minnesota Studies in the Philosophy of Science, pp. 3–44. University of Minnesota Press, Minneapolis (1992)

[Nersessian, 1995] Nersessian, N.J.: Should physicists preach what they practice? Constructive modeling in doing and learning physics. Science and Education 4, 203–226 (1995)

[Nersessian, 1998] Nersessian, N.J.: Kuhn and the cognitive revolution. Configurations 6, 87–120 (1998)

[Nersessian, 1999a] Nersessian, N.J.: Inconsistency, generic modeling, and conceptual change in science. In: Meheus, J. (ed.) Inconsistency in Science, pp. 197–211. Kluwer Academic Publishers, Dordrecht (1999)

[Nersessian, 1999b] Nersessian, N.J.: Model-based reasoning in conceptual change. In: Magnani, L., Nersessian, N.J., Thagard, P. (eds.) Model-based Reasoning in Scientific Discovery, pp. 5–22. Kluwer Academic Publishers, Dordrecht (1999)

[Netz, 1999] Netz, R.: The Shaping of Deduction in Greek Mathematics. Cambridge University Press, Cambridge (1999)

[Newell *et al.*, 1957] Newell, A., Shaw, J.C., Simon, H.A.: Empirical explorations of the logic theory machine: a case study in heuristic. In: Proceedings of the Western Joint Computer Conference (JCC 11), Los Angeles, February 1957, pp. 218–239 (1957)

[Newton, 1721] Newton, I. (ed.): Opticks: or a Treatise of the Reflections, Refractions, Inflections, and Colours of Light. G. Bell & Sons, London (1721); Reprinted from the fourth edition with a foreword by A. Einstein and an introduction by E.T. Whittaker, G. Bell, London (1931)

[Niiniluoto, 1999] Niiniluoto, I.: Abduction and geometrical analysis. Notes on Charles S. Peirce and Edgar Allan Poe. In: Magnani, L., Nersessian, N.J., Thagard, P. (eds.) Model Based Reasoning in Scientific Discovery, pp. 239–254. Plenum Publishers/Kluwer Academic, New York (1999)

[Ninio, 1989] Ninio, J.: L'empreinte des senses. Odile Jacob, Paris (1989)

[Noë, 2005] Noë, A.: Action in Preception. The MIT Press, Cambridge (2005)

[Noë, 2006] Noë, A.: Experience of the world in time. Analysis 66(1), 26–32 (2006)

[Norman, 1988] Norman, D.A.: The Psychology of Everyday Things. Basic Books, New York (1988)

[Norman, 1993] Norman, D.A.: The Invisible Computer. The MIT Press, Cambridge (1993)

[Norman, 2002] Norman, J.: Two visual systems and two theories of perception: an attempt to reconcile the constructivist and ecological approaches. Behavioral and Brain Sciences 25, 73–144 (2002)

[Novoplansky, 2002] Novoplansky, A.: Developmental plasticity in plants: implications of non-cognitive behavior. Evolutionary Ecology 16, 177–188 (2002)

[Nussbaum, 2001] Nussbaum, M.C.: Upheavals of Thought. The Intelligence of Emotion. Cambridge University Press, Cambridge (2001)

[Oatley and Johnson-Laird, 2002] Oatley, K., Johnson-Laird, P.N.: Emotion and reasoning to consistency. In: Moore, S.C., Oaksford, M. (eds.) Emotional Cognition, pp. 157–181. Johns Benjamins, Amsterdam (2002)

[Oatley, 1992] Oatley, K.: Best Laid Schemes: the Psychology of Emotions. Cambridge University Press, Cambridge (1992)

[Oatley, 1996] Oatley, K.: Inference in narrative and science. In: Olson, D., Torrance, N. (eds.) Modes of Thought. Explorations in Culture and Cognition, pp. 123–140. Cambridge University Press, Cambridge (1996)

[Oberman *et al.*, 2005] Oberman, L.M., Hubbard, E.H., McCleery, J.P., Altschuler, E., Ramachandran, V.S., Pineda, J.A.: EEG evidence for mirror neurons dysfunction in autism spectrum disorders. Cognitive Brain Research 24, 190–198 (2005)

[Odling-Smee *et al.*, 2003] Odling-Smee, F.J., Laland, K.N., Feldman, M.W.: Niche Construction. The Neglected Process in Evolution. Princeton University Press, Princeton (2003)

[Odling-Smee, 1988] Odling-Smee, F.J.: The Role of Behavior in Evolution. Cambridge University Press, Cambridge (1988)

[Okada and Simon, 1997] Okada, T., Simon, H.A.: Collaborative discovery in a scientific domain. Cognitive Science 21, 109–146 (1997)

[O'Keefe and Nadel, 1978] O'Keefe, J., Nadel, S.: The Hippocampus as a Cognitive Map. Oxford University Press, Oxford (1978)

[O'Keefe, 1999] O'Keefe, J.: Kant and sea-horse: an essay in the neurophilosophy of space. In: Elian, N., McCarthy, R., Brewer, B. (eds.) Spatial Representation. Problems in Philosophy and Psychology, pp. 43–64. Oxford University Press, Oxford (1999)

[O'Regan and Noë, 2001] O'Regan, J.K., Noë, A.: A sensorimotor approach to vision and visual perception. Behavioral and Brain Sciences 24, 939–973 (2001)

[O'Rorke and Ortony, 1992] O'Rorke, P., Ortony, A.: Abductive explanations of emotions. In: Proceedings of the 14th Annual Conference of the Cognitive Science Society. Erlbaum, Hillsdale (1992)

[O'Rorke *et al.*, 1990] O'Rorke, P., Morris, S., Schulemburg, D.: Theory formation by abduction: a case study based on the chemical revolution. In: Shrager, J., Langley, P. (eds.) Computational Models of Scientific Discovery and Theory Formation, pp. 197–224. Morgan Kaufmann, San Mateo (1990)

[O'Rorke, 1994] O'Rorke, P.: Abduction and explanation-based learning: case studies in diverse domains. Computational Intelligence 66, 311–344 (1994)

[Otte and Panza, 1999] Otte, M., Panza, M.: Analysis and Synthesis in Mathematics. Kluwer Academic Publisher, Dordrecht (1999)

[Ourston and Mooney, 1994] Ourston, D., Mooney, R.J.: Theory refining combining analytical and empirical methods. Artificial Intelligence 66, 311–344 (1994)

[Overgaard and Grünbaum, 2007] Overgaard, S., Grünbaum, T.: What do weather watchers see? Perceptual intentionality and agency. Cognitive Semiotics, 8–31 (2007)

[Paavola *et al.*, 2006] Paavola, S., Hakkarainen, K., Sintonen, M.: Abduction with dialogical and trialogical means. Logic Journal of the IGPL 14(1), 137–150 (2006)

[Paavola, 2004] Paavola, S.: Abduction through grammar, critic and methodeutic. Transactions of the Charles S. Peirce Society 40(2), 245–270 (2004)

[Paavola, 2005] Paavola, S.: Peircean abduction: instinct or inference? Semiotica 153(1/4), 131–154 (2005)

[Paccalin and Jeannerod, 2000] Paccalin, C., Jeannerod, M.: Changes in breathing during observation of effortful actions. Brain in Research 862, 194–200 (2000)

[Pachoud, 1999] Pachoud, B.: The teleological dimension of perceptual and motor intentionality. In: Petitot, J., Varela, F.J., Pachoud, B., Roy, J.-M. (eds.) Naturalizing Phenomenology, pp. 196–219. Stanford University Press, Stanford (1999)

[Papi, 1992] Papi, F. (ed.): Animal Homing. Chapman and Hall, London (1992)

[Patokorpi, 2007] Patokorpi, E.: Logic of Sherlock Holmes in technology enhanced learning. Educational Technology and Society 10(1), 171–185 (2007)

[Pearl, 1988] Pearl, J.: Probabilistic Reasoning in Intelligent Systems. Morgan Kaufmann, San Mateo (1988)

[Pease *et al.*, 2005] Pease, A., Colton, S., Smaill, A., Lee, J.: A model of Lakatos's philosophy of mathematics. In: Magnani, L., Dossena, R. (eds.) Computing, Philosophy and Cognition, pp. 57–85. College Publications, London (2005)

[Peirce, 1931-1958] Peirce, C.S.: Collected Papers of Charles Sanders Peirce. Harvard University Press, Cambridge (1931-1958); vol. 1-6, Hartshorne, C., Weiss, P. (eds.); vols. 7-8, Burks, A.W. (ed)

[Peirce, 1955a] Peirce, C.S.: Abduction and induction. In: Buchler, J. (ed.) Philosophical Writings of Peirce, pp. 150–156. Dover, New York (1955)

[Peirce, 1955b] Peirce, C.S.: The fixation of belief. In: Buchler, J. (ed.) Philosophical Writings of Peirce, pp. 5–22. Dover, New York (1955)

[Peirce, 1955c] Peirce, C.S.: Visual cognition and cognitive modeling. In: Buchler, J. (ed.) Philosophical Writings of Peirce, pp. 302–305. Dover, New York (1955)

[Peirce, 1966] Peirce, C.S.: The Charles S. Peirce Papers: Manuscript Collection in the Houghton Library. The University of Massachusetts Press, Worcester, MA (1966), Annotated Catalogue of the Papers of Charles S. Peirce. Numbered according to Richard S. Robin. Available in the Peirce Microfilm edition. Pagination: CSP = Peirce / ISP = Institute for Studies in Pragmaticism

[Peirce, 1976] Peirce, C.S.: The New Elements of Mathematics by Charles Sanders Peirce. Mouton/Humanities Press, The Hague-Paris/Atlantic Higlands (1976); vo. I-IV, Eisele, C. (ed.)

[Peirce, 1986] Peirce, C.S.: Pragmatism as a Principle and Method of Right Thinking. The 1903 Harvard Lectures on Pragmatism. State University of New York Press, Albany (1986); edited by Turrisi, P.A., Peirce, C.S.: Lectures on Pragmatism, Cambridge, MA, March 26 - May 17 (1903); Reprinted in [Peirce, 1992-1998, II, pp. 133–241]

[Peirce, 1987] Peirce, C.S.: Historical Perspectives on Peirce's Logic of Science: a History of Science. Mouton, Berlin (1987); vols. I-II, edited by C. Eisele

[Peirce, 1992-1998] Peirce, C.S.: The Essential Peirce. Selected Philosophical Writings. Indiana University Press, Bloomington (1992-1998); vol. 1 (1867-1893), ed. by N. Houser & C. Kloesel; vol. 2 (1893-1913) ed. by the Peirce Edition Project

[Peirce, 2005] Peirce, C.S.: Reasoning and the Logic of Things: the 1898 Cambridge Conferences Lectures by Charles Sanders Peirce. Harvard University Press, Amsterdam (2005); Edited by K. L. Ketner

[Pennock, 1999] Pennock, R.T.: Tower of Babel. The Evidence Against the New Creationism. The MIT Press, Cambridge (1999)

[Pennock, 2000] Pennock, R.T.: Can Darwinian mechanisms make novel discoveries? Learning from discoveries made by evolving neural network. Foundations of Science 5(2), 225–238 (2000)

[Pericliev and Valdés-Pérez, 1998] Pericliev, V., Valdés-Pérez, R.E.: Automatic componential analysis of kinship semantics with a proposed structural solution to the problem of multiple models. Anthropological Linguistics 40(2), 272–317 (1998)

[Petit, 1999] Petit, J.-L.: Constitution by movement: Husserl in the light of recent neurobiological findings. In: Petitot, J., Varela, F.J., Pachoud, B., Roy, J.-M. (eds.) Naturalizing Phenomenology, pp. 220–244. Stanford University Press, Stanford (1999)

[Petitot and Smith, 1990] Petitot, J., Smith, B.: New foundations of qualitative physics. In: Tiles, J.E., McKee, G.T., Dean, C.G. (eds.) Evolving Knowledge in Natural Science and Artificial Intelligence, pp. 231–249. Pitman Publishing, London (1990)

[Petitot *et al.*, 1999] Petitot, J., Varela, F.J., Pachoud, B., Roy, J.-M. (eds.): Naturalizing Phenomenology. Stanford University Press, Stanford (1999)

[Petitot, 1985] Petitot, J.: Les catastrophes de la parole: de Roman Jakobson à René Thom. Maloine, Paris (1985)

[Petitot, 1992] Petitot, J.: Physique du sens. Éditions du CNRS, Paris (1992)

[Petitot, 1999] Petitot, J.: Morphological eidetics for a phenomenology of perception. In: Petitot, J., Varela, F.J., Pachoud, B., Roy, J.-M. (eds.) Naturalizing Phenomenology, pp. 330–371. Stanford University Press, Stanford (1999)

[Petitot, 2003] Petitot, J.: Morphogenesis of Meaning [1985]. Peter Lang, Bern (2003); Translated by Franson Manjali

[Piaget and Inhelder, 1956] Piaget, J., Inhelder, B.: The Child's Conception of Space. Routledge and Kegan Paul, London (1956)
[Piaget, 1952] Piaget, J.: The Origins of Intelligence in Children. International University Press, New York (1952)
[Piaget, 1974] Piaget, J.: Adaption and Intelligence. University of Chicago Press, Chicago (1974)
[Picard, 1994] Picard, R.W.: Affective Computing. The MIT Press, Cambridge (1994)
[Pickering and Garrod, 2006] Pickering, M.J., Garrod, S.: Do people use language production to make predictions during comprehension? Trends in Cognitive Science 11(3), 105–110 (2006)
[Pickering, 1995] Pickering, A.: The Mangle of Practice. Time, Agency, and Science. The University of Chicago Press, Chicago (1995)
[Pigliucci *et al.*, 2006] Pigliucci, M., Murren, C.J., Schlichting, C.D.: Phenotypic plasticity and evolution by genetic assimilation. The Journal of Experimental Biology 209, 2362–2367 (2006)
[Piller, 2000] Piller, C.: Doing what is best. The Philosophical Quarterly 50(199), 208–226 (2000)
[Pinker, 1994] Pinker, S.: The Language Instinct. William Morrow, New York (1994)
[Pinker, 1997] Pinker, S.: How the Mind Works. W. W. Norton, New York (1997)
[Pinker, 2003] Pinker, S.: Language as an adaptation to the cognitive niche. In: Christiansen, M.H., Kirby, S. (eds.) Language Evolution, pp. 16–37. Oxford University Press, Oxford (2003)
[Pinker, 2005] Pinker, S.: So how does the mind work? Mind and Language 20(1), 1–24 (2005)
[Pipers *et al.*, 2006] Pipers, J.R., Oudejans, R.R.D., Bakker, F.C., Beek, P.J.: The role of anxiety in perceiving and realizing affordances. Ecological Psychology 18(3), 131–161 (2006)
[Pisella *et al.*, 2006] Pisella, L., Binkofski, F., Lasek, K., Toni, I., Rossetti, Y.: No double-dissociation between optic ataxia and visual agnosia: multiple sub-streams for multiple visuo-manual integration. Neuropsychologia 44(13), 2734–2748 (2006)
[Pizzi, 2006] Pizzi, C.: Gestalt effects in counterfactual and abductive inference. Logic Journal of the IGPL 14(2), 257–270 (2006)
[Pizzi, 2007] Pizzi, C.: Abductive inference and iterated conditionals. In: Magnani, L., Li, P. (eds.) Model-Based Reasoning in Science, Technology, and Medicine, pp. 365–382. Springer, Berlin (2007)
[Plato, 1977] Plato: Plato in Twelve Volumes. Harvard University Press, Cambridge (1977); vol. II, Laches, Protagoras, Meno, Euthydemus, with an English translation by W. R. M. Lamb
[Plotkin, 1993] Plotkin, H.: Darwin, Machines and the Nature of Knowledge. Harvard University Press, Cambridge (1993)
[Poincaré, 1902] Poincaré, H.: La science et l'hypothèse. Flammarion, Paris (1902); English translation by W. J. G. [only initials indicated] (1958) Science and Hypothesis, with a Preface by Larmor, J. The Walter Scott Publishing Co., New York (1905); Also reprinted in Essential Writings of Henri Poincaré. Random House, New York (2001)
[Poincaré, 1905] Poincaré, H.: La valeur de la science. Flammarion, Paris (1905); English translation by G.B. Halsted, The Value of Science. Dover Publications, New York (1958). Also reprinted in Essential Writings of Henri Poincaré. Random House, New York (2001)
[Polanyi, 1966] Polanyi, M.: The Tacit Dimension. Routledge & Kegan Paul, London (1966)
[Polger, 2006] Polger, T.W.: Natural Minds. The MIT Press, Cambridge (2006)

[Polger, 2008] Polger, T.W.: Two confusions concerning the multiple realization. Philosophy of Science 75, 537–547 (2008)

[Poole and Rowen, 1990] Poole, D., Rowen, G.M.: What is an optimal diagnosis. In: Proceedings of the 6th Conference on Uncertainty in AI, pp. 46–53 (1990)

[Poole, 1988] Poole, D.: A logical framework for default reasoning. Artificial Intelligence 36, 27–47 (1988)

[Poole, 1991] Poole, D.: Representing diagnostic knowledge for probabilistic Horn abduction. In: Proceedings IJCAI 1991, NSW, Sydney, pp. 1129–1135 (1991)

[Popper, 1959] Popper, K.R.: The Logic of Scientific Discovery. Hutchinson, London (1959)

[Popper, 1963] Popper, K.R.: Conjectures and Refutations. The Growth of Scientific Knowledge. Routledge and Kegan Paul, London (1963)

[Port and van Gelder, 1995] Port, R.F., van Gelder, T. (eds.): Mind as Motion. Explorations in the Dynamics of Cognition. The MIT Press, Cambridge (1995)

[Powers, 1973] Powers, W.T.: Behavior: the Control of Perception. Aladine, Chicago (1973)

[Prigogine and Stengers, 1984] Prigogine, I., Stengers, I.: Order out of Chaos. Bantam, London (1984)

[Queiroz and Merrell, 2005] Queiroz, J., Merrell, F. (eds.): Abduction: Between Subjectivity and Objectivity, vol. 153. Walter de Gruyter, Berlin (2005); Special Issue of the Journal Semiotica

[Quine, 1960] Quine, W.V.O.: Word and Object. Cambridge University Press, Cambridge (1960)

[Quine, 1969] Quine, W.V.O.: Natural kinds. In: Ontological Relativity and other Essays. Columbia University Press, New York (1969)

[Quine, 1979] Quine, W.V.O.: Philosophy of Logic. Prentice-Hall, Englewood Cliffs (1979)

[Raab and Gigerenzer, 2005] Raab, M., Gigerenzer, G.: Intelligence as smart heuristics. In: Sternberg, R.J., Prets, J.E. (eds.) Cognition and Intelligence. Identifying the Mechanisms of the Mind, pp. 188–207. Cambridge University Press, Cambridge (2005)

[Rachels, 1999] Rachels, J.: The Elements of Moral Philosophy. McGraw Hill College, Boston Burr Ridge (1999)

[Rader and Vaughn, 2000] Rader, N., Vaughn, L.: Infant reaching to a hidden affordance: evidence for intentionality. Infant Behavior and Development 23, 531–541 (2000)

[Raedt and Bruynooghe, 1991] De Raedt, L., Bruynooghe, M.: A multistrategy interactive concept learner and theory revision system. In: Proceedings of the 1st International Workshop on Multistrategy Learning, Harpers Ferry, pp. 175–190 (1991)

[Raftopoulos, 2001a] Raftopoulos, A.: Is perception informationally encapsulated? The issue of theory-ladenness of perception. Cognitive Science 25, 423–451 (2001)

[Raftopoulos, 2001b] Raftopoulos, A.: Reentrant pathways and the theory-ladenness of perception. Philosophy of Science 68, S187–S189 (2001); Proceedings of PSA 2000 Biennal Meeting

[Rajamoney, 1993] Rajamoney, S.A.: The design of discrimination experiments. Machine Learning 12, 185–203 (1993)

[Ram *et al.*, 1995] Ram, A., Wills, L., Domeshek, E., Nersessian, N.J., Kolodner, J.L.: Understanding the creative mind: a review of Margaret Boden's Creative Mind. Artificial Intelligence 79, 111–128 (1995)

[Ramachandran and Hirstein, 1997] Ramachandran, V.S., Hirstein, W.: Three laws of qualia: what neurology tells us about the biological functions of consciousness. Journal of Consciousness Studies 4, 429–457 (1997)

[Ramoni *et al.*, 1992] Ramoni, M., Stefanelli, M., Magnani, L., Barosi, G.: An epistemological framework for medical knowledge-based systems. IEEE Transactions on Systems, Man, and Cybernetics 22(6), 1361–1375 (1992)

[Ramsey, 1927] Ramsey, F.P.: Facts and propositions. Aristotelian Society, Supplementary Volume 7, 152–170 (1927)

[Ramus, 2006] Ramus, F.: Genes, brain, and cognition: a roadmap for the cognitive scientist. Cognition 101, 247–269 (2006)

[Ray, 2007] Ray, O.: Automated abduction in scientific discovery. In: Magnani, L., Li, P. (eds.) Model-Based Reasoning in Science, Technology, and Medicine, pp. 103–116. Springer, Berlin (2007)

[Reed, 1988] Reed, E.S.: James J. Gibson and the Psychology of Perception. Yale University Press, New Haven (1988)

[Reichenbach, 1938] Reichenbach, H.: Experience and Prediction. University of Chicago Press, Chicago (1938)

[Reilly, 1970] Reilly, F.E.: Charles Peirce's Theory of Scientific Method. Fordham University Press, New York (1970)

[Reiter and de Kleer, 1991] Reiter, R., de Kleer, J.: Foundations of assumption-based truth maintenance systems: preliminary report. In: Proceedings AAAI 1987, Seattle, pp. 183–188 (1991)

[Reiter, 1987] Reiter, R.: A theory of diagnosis from first principles. Artificial Intelligence 32, 57–95 (1987)

[Rellihan, 2009] Rellihan, M.J.: Fodor's riddle of abduction. Philosophical Studies 144, 313–338 (2009)

[Rescher, 1995] Rescher, N.: Peirce on abduction, plausibility, and efficiency of scientific inquiry. In: Rescher, N. (ed.) Essays in the History of Philosophy, pp. 309–326. Avebury, Aldershot (1995)

[Rest *et al.*, 1999] Rest, J., Narvaez, N., Bebeau, M.J., Thoma, S.J.: Postconventional Moral Thinking. A Neo-Kohlbergian Approach. Lawrence Erlbaum, Mahwah (1999)

[Reyes *et al.*, 2006] Reyes, A.L., Aliseda, A., Nepomuceno, A.: Towards abductive reasoning in first order logic. Logic Journal of the IGPL 14(2), 287–304 (2006)

[Reznikoff, 1987] Reznikoff, I.: Sur la dimension sonore des grottes à peintures du paléolithique. C. R. Académie des Sciences 304(2), 353–356 (1987)

[Richerson and Boyd, 2005] Richerson, P.J., Boyd, R.: Not by Genes Alone. How Culture Trasformed Human Evolution. The University of Chicago Press, Chicago (2005)

[Ricoeur, 1977] Ricoeur, P.: The Rule of Metaphor: Multi-Disciplinary Studies in the Creation of Meaning in Language. University of Toronto Press, Toronto (1977); Translated by R. Czerny with K. McLaughlin and J. Costello. Originally published in 1975

[Rivas and Burghardt, 2002] Rivas, J., Burghardt, G.M.: Crotalomorphysm: a metaphor for understanding anthropomorphism by omission. In: Bekoff, M., Allen, C., Burghardt, M. (eds.) The Cognitive Animal. Empirical and Theoretical Perspectives on Animal Cognition, pp. 9–18. The MIT Press, Cambridge (2002)

[Rivera and Rossi Becker, 2007] Rivera, F.D., Rossi Becker, J.: Abduction-induction (generalization) processes of elementary majors on figural patterns in algebra. Journal of Mathematical Behavior 26, 140–155 (2007)

[Rizzolatti *et al.*, 1988] Rizzolatti, G., Carmada, R., Gentilucci, M., Luppino, G., Matelli, M.: Functional organization of area 6 in the macaque monkey. Experimental Brain Research 71, 491–507 (1988)

[Roberts, 2001] Roberts, W.A.: Spatial representation and the use of spatial code in animals. In: Gattis, M. (ed.) Spatial Schemas and Abstract Thought, pp. 15–44. The MIT Press, Cambridge (2001)

[Roberts, 2004] Roberts, L.D.: The relation of children's early word acquisition to abduction. Foundations of Science 9(3), 307–320 (2004)

[Robinson, 1966] Robinson, A.: Non-Standard Analysis. North Holland, Amsterdam (1966)

[Rock, 1982] Rock, I.: Inference in perception. PSA. Proceedings of the Biennial Meeting of the Philosophy of Science Association 2, 525–540 (1982)

[Roesler, 1997] Roesler, A.: Perception and abduction. In: Rayo, M., Gimate-Welsh, A., Pellegrino, P. (eds.) Fifth International Congress of the International Association for Semiotic Studies: Semiotics Bridging Nature and Culture, México, p. 226, Editorial Solidaridad (1997) Abstract

[Roitblat, 2002] Roitblat, H.L.: The cognitive dolphin. In: Bekoff, M., Allen, C., Burghardt, M. (eds.) The Cognitive Animal. Empirical and Theoretical Perspectives on Animal Cognition, pp. 183–188. The MIT Press, Cambridge (2002)

[Romdhane and Ayeb, 2009] Romdhane, L.B., Ayeb, R.: An evolutionary-based algorithm for abductive reasoning with application to medical diagnosis. Artificial Intelligence in Medicine (2009) (Forthcoming)

[Rose, 1993] Rose, S.: The Future of the Brain. The Promise and Perils of Tomorrow's Neuroscience. Oxford University Press, Oxford (1993)

[Rosenfeld, 1988] Rosenfeld, B.A.: A History of Non-Euclidean Geometry. Evolution of the Concept of Geometric Space. Springer, Berlin (1988)

[Rothrock and Kirlik, 2006] Rothrock, L., Kirlik, A.: Inferring fast and frugal heuristics from human judgment data. In: Kirlik, A. (ed.) Human-Technology Interaction. Methods and Models for Cognitive Engineering and Human-Computer Interaction, pp. 131–148. Oxford University Press, Oxford (2006)

[Roussel, 1987] Roussel, P.: PROLOG: Manual de référence et d'utilisation. Group d'Intelligence Artificielle, Université d'Aix-Marseille, Luminy (1987)

[Russell, 1919] Russell, B.: Introduction to Mathematical Philosophy. Allen and Unwin Ltd., London (1919)

[Russell, 1973] Russell, B.: The regressive method of discovering the premises of mathematics [1907]. In: Lackey, D. (ed.) Essays in Analysis, pp. 45–66. George Allen and Unwin, London (1973)

[Saccheri, 1920] Saccheri, G.: Euclides Vindicatus. Euclid Freed of Every Fleck. Open Court, Chicago (1920); Translated by G.B. Halsted. Originally published as: Euclides ab omni naevo vindicatus, Ex Typographia Pauli Antonii Montani, Mediolani (Milan) (1733)

[Sachs, 2002] Sachs, T.: Consequences of the inherent developmental plasticity of organ and tissue relations. Evolutionary Ecology 16, 243–265 (2002)

[Saidel, 2002] Saidel, E.: Animal minds, human minds. In: Bekoff, M., Allen, C., Burghardt, M. (eds.) The Cognitive Animal. Empirical and Theoretical Perspectives on Animal Cognition, pp. 53–58. The MIT Press, Cambridge (2002)

[Salmon, 1990] Salmon, W.C.: Four Decades of Scientific Explanation. University of Minnesota Press, Minneapolis (1990)

[Saltzman, 1995] Saltzman, E.L.: Dynamics and coordinate systems in skilled sensorimotor activity. In: Port, R.F., van Gelder, T. (eds.) Mind as Motion. Explorations in the Dynamics of Cognition, vol. 2, pp. 69–100. The MIT Press, Cambridge (1995)

[Santaella, 2005] Santaella, L.: Abduction: the logic of guessing. Semiotica 153(1/4), 175–198 (2005)

[Sathian, 2004] Sathian, K.: Modality, quo vadis? Behavioral and Brain Sciences 27, 413–414 (2004); Open Peer Commentary to R. Grush, The emulation theory of representation: motor control, imagery, and perception

[Scarantino, 2003] Scarantino, A.: Affordances explained. Philosophy of Science 70, 949–961 (2003)

[Schank and Abelson, 1987] Schank, R., Abelson, R.: Scripts, Plans, Goals and Understanding. Erlbaum, Hillsdale (1987)

[Schank, 1982] Schank, R.: Dynamic Memory: a Theory of Learning in Computers and People. Cambridge University Press, Cambridge (1982)
[Schantz, 2004] Schantz, R. (ed.): The Externalist Challenge. De Gruyter, Berlin (2004)
[Schrödinger, 1992] Schrödinger, E.: What is life? With "Mind and Matter" and "Autobiographical Sketches". Cambridge University Press, Cambridge (1992); Originally published in (1944)
[Schurz, 2008] Schurz, G.: Patterns of abduction. Synthese 164(2), 201–234 (2008)
[Schusterman *et al.*, 2002] Schusterman, R.J., Reichmuth Kastak, C., Kastak, D.: The cognitive sea lion: meaning and memory in laboratory and nature. In: Bekoff, M., Allen, C., Burghardt, M. (eds.) The Cognitive Animal. Empirical and Theoretical Perspectives on Animal Cognition, pp. 217–228. The MIT Press, Cambridge (2002)
[Scott and Markovitch, 1993] Scott, P.D., Markovitch, S.: Experience selection and problem choice in an exploratory learning system. Machine Learning 12, 49–67 (1993)
[Searle, 2001] Searle, J.: Rationality in Action. The MIT Press, Cambridge (2001)
[Secchi, 2007] Secchi, D.: A theory of docile society: the role of altruism in human behavior. Journal of Academy of Business and Economics 7(2), 446–461 (2007)
[Seth and Baars, 2005] Seth, A.K., Baars, B.J.: Neural Darwinism and consciousness. Consciousness and Cognition 14, 140–168 (2005)
[Shanahan, 1989] Shanahan, M.: Prediction is deduction but explanation is abduction. In: Proceedings IJCAI 1989, Detroit, MI, pp. 1140–1145 (1989)
[Shanahan, 2005] Shanahan, M.: Perception as abduction: turning sensory data into meaningful representation. Cognitive Science 29, 103–134 (2005)
[Shangmin *et al.*, 2007] Shangmin, L., Magnani, L., Dai, G.: An algebraic approach to model-based diagnosis. In: Magnani, L., Li, P. (eds.) Model-Based Reasoning in Science, Technology, and Medicine, pp. 103–116. Springer, Berlin (2007)
[Shapiro, 2004] Shapiro, L.A.: The Mind Incarnate. The MIT Press, Cambridge (2004)
[Shelley, 1996] Shelley, C.P.: Visual abductive reasoning in archaeology. Philosophy of Science 63(2), 278–301 (1996)
[Shelley, 1999] Shelley, C.P.: Multiple analogies in archaeology. Philosophy of Science 66, 579–605 (1999)
[Shen, 1993] Shen, W.M.: Discovery as autonomous learning from the environment. Machine Learning 12, 143–165 (1993)
[Sherry, 1988] Sherry, D.S.: Food storage, memory, and marsh tits. Animal Behavior 30, 631–633 (1988)
[Shettleworth, 2002] Shettleworth, S.J.: Spatial behavior, food storing, and the modular mind. In: Bekoff, M., Allen, C., Burghardt, M. (eds.) The Cognitive Animal. Empirical and Theoretical Perspectives on Animal Cognition, pp. 123–128. The MIT Press, Cambridge (2002)
[Shimojima, 2002] Shimojima, A.: A logical analysis of graphical consistency proof. In: Magnani, L., Nersessian, N.J., Pizzi, C. (eds.) Logical and Computational Aspects of Model-Based Reasoning, pp. 93–116. Kluwer Academic Publishers, Dordrecht (2002)
[Shin, 2002] Shin, S.-J.: The Iconic Logic of Peirce's Graphs. The MIT Press, Cambridge (2002)
[Shrager and Langley, 1990] Shrager, J., Langley, P. (eds.): Computational Models of Scientific Discovery and Theory Formation. Morgan Kaufmann, San Mateo (1990)
[Shunn and Klahr, 1995] Shunn, C., Klahr, D.: A 4-space model of scientific discovery. In: AAAI Symposium Systematic Methods of Scientific Discovery, Technical Report SS-95-03. AAAI Press, Menlo Park (1995)
[Simina and Kolodner, 1995] Simina, M.D., Kolodner, J.L.: Opportunistic reasoning: a design perspective. In: Proceedings of the 17th Annual Cognitive Science Conference (1995)

[Simon, 1955] Simon, H.A.: A behavioral model of rational choice. The Quarterly Journal of Economics 69, 99–118 (1955)
[Simon, 1965] Simon, H.A.: The logic of rational decision. British Journal for the Philosophy of Science 16, 169–186 (1965)
[Simon, 1969] Simon, H.A.: The Sciences of Artificial. The MIT Press, Cambridge (1969)
[Simon, 1977] Simon, H.A.: Models of Discovery and Other Topics in the Methods of Science. Reidel, Dordrecht (1977)
[Simon, 1993] Simon, H.A.: Altruism and economics. American Economic Review 83(2), 157–161 (1993)
[Simpson, 1999] Simpson, S.G.: Subsystems of Second Order Arithmetic. Springer, Berlin (1999); 2nd edn. Association for Symbolic Logic (forthcoming)
[Sinha, 2006] Sinha, C.: Epigenetics, semiotics, and the mysteries of the organism. Biological Theory 1(2), 112–115 (2006)
[Sinnott-Armstrong, 1996] Sinnott-Armstrong, W.: Moral skepticism and justification. In: Sinnott-Armstrong, W., Timmons, M. (eds.) Moral Knowledge? New Readings in Moral Epistemology, pp. 3–48. Oxford University Press, Oxford (1996)
[Skagestad, 1993] Skagestad, P.: Thinking with machines: intelligence augmentation, evolutionary epistemology, and semiotic. The Journal of Social and Evolutionary Systems 16(2), 157–180 (1993)
[Skyrms, 2008] Skyrms, B.: Presidential address: Signals. Philosophy of Science 75, 489–500 (2008)
[Sommer, 2003] Sommer, R.: Trees and human identity. In: Clayton, S., Opotow, S. (eds.) Identity and the Natural Environment: the Psychological Significance of Nature, pp. 79–204. The MIT Press, Cambridge (2003)
[Sørensen and Ziemke, 2007] Sørensen, M.H., Ziemke, T.: Agency without agency? Cognitive Semiotics, 102–124 (2007)
[Sovrano *et al.*, 2005] Sovrano, V.A., Bisazza, A., Vallortigara, G.: How fish do geometry in large and in small spaces. Animal Cognition 10(1), 47–54 (2005)
[Spelke, 1990] Spelke, E.S.: Principles of object segregation. Cognitive Science 14, 29–56 (1990)
[Steels, 1994] Steels, L.: The artificial life roots of artificial intelligence. Artificial Life 1, 75–100 (1994)
[Stefanelli and Ramoni, 1992] Stefanelli, M., Ramoni, M.: Epistemological constraints on medical knowledge–based systems. In: Evans, D.A., Patel, V.L. (eds.) Advanced Models of Cognition for Medical Training and Practice, pp. 3–20. Springer, Berlin (1992)
[Stenning, 2000] Stenning, K.: Distinctions with differences: comparing criteria for distinguishing diagrammatic from sentential systems. In: Anderson, M., Cheng, P., Haarslev, V. (eds.) Theory and Application of Diagrams, pp. 132–148. Springer, Berlin (2000)
[Stephanou and Sage, 1987] Stephanou, H., Sage, A.: Perspectives in imperfect information processing. IEEE Transactions on Systems, Man, and Cybernetics 17, 780–798 (1987)
[Sterelny, 2003] Sterelny, K.: Thought in a Hostile World. The Evolution of Human Cognition. Blackwell, Oxford (2003)
[Sterelny, 2004] Sterelny, K.: Externalism, epistemic artefacts and the extended mind. In: Schantz, R. (ed.) The Externalist Challenge, pp. 239–254. De Gruyter, Berlin (2004)
[Sterelny, 2005] Sterelny, K.: Made by each other: organisms and their environment. Biology and Philosophy 20, 21–36 (2005)
[Stern, 1985] Stern, D.N.: The Interpretation World of Infants. Academic Press, New York (1985)
[Stjernfelt, 2007] Stjernfelt, F.: Diagrammatology. An Investigation on the Borderlines of Phenomenology, Ontology, and Semiotics. Springer, Berlin (2007)

[Stoffregen, 2003] Stoffregen, T.A.: Affordances as properties of the animal-environment system. Ecological Psychology 15(3), 115–134 (2003)

[Stroyan, 2005] Stroyan, K.D.: Uniform continuity and rates of growth of meromorphic functions. In: Luxemburg, W.J., Robinson, A. (eds.) Contributions to Non-Standard Analysis, pp. 47–64. North-Holland, Amsterdam (2005)

[Suarez, 1999] Suarez, M.: Theories, models, and representations. In: Magnani, L., Nersessian, N.J., Thagard, P. (eds.) Model-Based Reasoning in Scientific Discovery, pp. 75–83. Kluwer Academic/Plenum Publishers, New York (1999)

[Sutton, 2006] Sutton, J.: Distributed cognition: domains and dimensions. Pragmatics & Cognition 14(2), 235–247 (2006); Special Issue on "Distributed Cognition" edited by S. Harnad and I. E. Dror

[Svensson and Ziemke, 2004] Svensson, H., Ziemke, T.: Making sense of embodiment: simulation theories and the sharing of neural circuitry between sensorimotor and cognitive processes. In: Forbus, K.D., Gentner, D., Regier, T. (eds.) CogSci 2004, XXVI Annual Conference of the Cognitive Science Society, Chicago, IL (2004) CD-Rom

[Swanson and Smalheiser, 1997] Swanson, D.R., Smalheiser, N.R.: An interactive system for finding complementary literatures: a stimulus to scientific discovery. Artificial Intelligence 91(2), 183–203 (1997)

[Swoboda and Allwein, 2002] Swoboda, N., Allwein, G.: A case study of the design and implementation of heterogeneous reasoning systems. In: Magnani, L., Nersessian, N.J., Pizzi, C. (eds.) Logical and Computational Aspects of Model-Based Reasoning, pp. 3–20. Kluwer Academic Publishers, Dordrecht (2002)

[Számadó and Szathmáry, 1997] Számadó, S., Szathmáry, E.: Patterns of life: intertwining identity and cognition. Brain and Cognition 34, 72–87 (1997)

[Szentagothai and Arbib, 1975] Szentagothai, J., Arbib, M.A.: Conceptual Models of Neural Organization. The MIT Press, Cambridge (1975)

[Tall, 2001] Tall, D.: Natural and formal infinities. Educational Studies in Mathematics 48, 199–238 (2001)

[Thagard and Litt, 2008] Thagard, P., Litt, A.: Models of scientific explanation. In: Sun, R. (ed.) Cambridge Handbook of Computational Psychology, pp. 549–564. Cambridge University Press, Cambridge (2008)

[Thagard and Millgram, 2001] Thagard, P., Millgram, E.: Inference to the best plan: a coherence theory of decision. In: Ram, A., Leake, D.B. (eds.) Goal-Driven Learning, pp. 439–454. The MIT Press, Cambridge (2001)

[Thagard and Shelley, 1997] Thagard, P., Shelley, C.P.: Abductive reasoning: logic, visual thinking, and coherence. In: Dalla Chiara, M.L., Doets, K., Mundici, D., van Benthem, J. (eds.) Logic and Scientific Methods, pp. 413–427. Kluwer, Dordrecht (1997)

[Thagard and Verbeurgt, 1998] Thagard, P., Verbeurgt, K.: Coherence as constraint satisfaction. Cognitive Science 22, 1–24 (1998)

[Thagard, 1987] Thagard, P.: The best explanation: criteria for theory choice. Journal of Philosophy 75, 76–92 (1987)

[Thagard, 1988] Thagard, P.: Computational Philosophy of Science. The MIT Press, Cambridge (1988)

[Thagard, 1989] Thagard, P.: Explanatory coherence. Behavioral and Brain Sciences 12(3), 435–467 (1989)

[Thagard, 1992] Thagard, P.: Conceptual Revolutions. Princeton University Press, Princeton (1992)

[Thagard, 1996] Thagard, P.: Mind. Introduction to Cognitive Science. The MIT Press, Cambridge (1996)

[Thagard, 1997a] Thagard, P.: Coherent and creative conceptual combinations, Technical Report, Department of Philosophy, Waterloo University, Waterloo, Ontario, Canada (1997)

[Thagard, 1997b] Thagard, P.: Collaborative knowledge. Noûs 31, 242–261 (1997)

[Thagard, 2000] Thagard, P.: Coherence in Thought and in Action. The MIT Press, Cambridge (2000)

[Thagard, 2001] Thagard, P.: How to make decisions: coherence, emotion, and practical inference. In: Millgram, E. (ed.) Varieties of Practical Reasoning, pp. 355–371. The MIT Press, Cambridge (2001)

[Thagard, 2002a] Thagard, P.: How molecules matter in mental computation. Philosophy of Science 69, 429–446 (2002)

[Thagard, 2002b] Thagard, P.: The passionate scientist: emotion in scientific cognition. In: Carruthers, P., Stich, S., Siegal, M. (eds.) The Cognitive Basis of Science, pp. 235–250. Cambridge University Press, Cambridge (2002)

[Thagard, 2005] Thagard, P.: How does the brain form hypotheses? Towards a neurologically realistic computational model of explanation. In: Thagard, P., Langley, P., Magnani, L., Shunn, C. (eds.) Symposium Generating explanatory hypotheses: mind, computer, brain, and world, Stresa, Italy (2005); Cognitive Science Society, CD-Rom. Proceedings of the 27th International Cognitive Science Conference

[Thagard, 2006] Thagard, P.: Coherence, truth, and the development of scientific knowledge. Philosophy of Science 73, 28–47 (2006)

[Thagard, 2007] Thagard, P.: Abductive inference: from philosophical analysis to neural mechanisms. In: Feeney, A., Heit, E. (eds.) Inductive Reasoning: Experimental, Developmental, and Computational Approaches, pp. 226–247. Cambridge University Press, Cambridge (2007)

[Theissier, 2005] Theissier, B.: Protomathematics, perception and the meaning of mathematical objects. In: Grialou, P., Longo, G., Okada, M. (eds.) Images and Reasoning, pp. 135–145. Keio University, Tokyo (2005)

[Thelen, 1995] Thelen, E.: Time-scale dynamics and the development of an embodied cognition. In: Port, R.F., van Gelder, T. (eds.) Mind as Motion. Explorations in the Dynamics of Cognition, vol. 2, pp. 69–100. The MIT Press, Cambridge (1995)

[Thinus-Blanc *et al.*, 1997] Thinus-Blanc, C., Save, E., Poucet, B.: Animal spatial cognition and exploration. In: Foreman, N., Gillett, R. (eds.) A Handbook of Spatial Research Paradigms and Methodologies, vol. II, pp. 59–86. Psychology Press, Taylor and Francis, Hove, East Sussex (1997)

[Thom, 1975] Thom, R.: Stabilité structurelle et morphogénèse. Essai d'une théorie générale des modèles. Reading, Benjamin, MA (1975); Translated by Fowler, D.H.: Structural Stability and Morphogenesis; an Outline of a General Theory of Models. W. A. Benjamin, Reading

[Thom, 1980] Thom, R.: Modèles mathématiques de la morphogenèse. Christian Bourgois, Paris (1980); Translated by Brookes, W.M., Rand, D.: Mathematical Models of Morphogenesis. Ellis Horwood, Chichester (1983)

[Thom, 1988] Thom, R.: Esquisse d'une sémiophysique. InterEditions, Paris (1988); Translated by Meyer, V.: Semio Physics: a Sketch. Addison Wesley, Redwood City (1990)

[Thomas, 1999] Thomas, H.J.: Are theories of imagery theories of imagination? An active perception approach to conscious mental content. Cognitive Science 23(2), 207–245 (1999)

[Thompson and Mooney, 1994] Thompson, C.A., Mooney, R.J.: Inductive learning for abductive diagnosis. In: Proceedings of the Twelfth National Conference on Artificial Intelligence, Seattle, WA, pp. 664–669 (1994)

[Thomson, 1999] Thomson, A.: Critical Reasoning in Ethics. Routledge, London (1999)

[Tiercelin, 2005] Tiercelin, C.: Abduction and the semiotic of perception. Semiotica 153(1/4), 389–412 (2005)

[Tindale, 2005] Tindale, C.W.: Hearing is believing: a perspective-dependent view of the fallacies. In: van Eemeren, F., Houtlosser, P. (eds.) Argumentative Practice, pp. 29–42. John Benjamins, Amsterdam (2005)

[Tindale, 2007] Tindale, C.W.: Fallacies and Argument Appraisal. Cambridge University Press, Cambridge (2007)

[Tirassa *et al.*, 1998] Tirassa, M., Carassa, A., Geminiani, G.: Describers and explorers: a method for investigating cognitive maps. In: Nualláin, S.Ó. (ed.) Spatial Cognition. Foundations and Applications, pp. 19–31. John Benjamins, Amsterdam (1998)

[Tolman *et al.*, 1946] Tolman, E.C., Ritchie, B.F., Kalish, D.: Studies in spatial learning II. Place learning versus response learning. Journal of Experimental Psychology 37, 385–392 (1946)

[Tomasello and Call, 1997] Tomasello, M., Call, J.: Primate Cognition. Oxford University Press, New York (1997)

[Tooby and DeVore, 1987] Tooby, J., DeVore, I.: The reconstruction of hominid behavioral evolution through strategic modeling. In: Kinzey, W.G. (ed.) Primate Models of Hominid Behavior, pp. 183–237. Suny Press, Albany (1987)

[Torretti, 1978] Torretti, R.: Philosophy of Geometry from Riemann to Poincaré. Reidel, Dordrecht (1978)

[Torretti, 2003] Torretti, R.: Review of L. Magnani. Philosophy and Geometry: Theoretical and Historical Issues. Kluwer Academic Publishers, Dordrecht (2001); Studies in History and Philosophy of Modern Physics 34b(1), 158–160 (2003)

[Toth, 1991] Toth, I.: Essere e non essere: il teorema induttivo di Saccheri e la sua rilevanza ontologica. In: Magnani, L. (ed.) Conoscenza e Matematica, pp. 87–156. Marcos y Marcos, Milan (1991); Translated from German by A. Marini

[Trafton *et al.*, 2005] Trafton, J.G., Trickett, S.B., Mintz, F.E.: Connecting internal and external representations: spatial transformations of scientific visualizations. Foundations of Science 10, 89–106 (2005)

[Trickett and Trafton, 2007] Trickett, S.B., Trafton, J.G.: "What if[. . .]": the use of operational simulations in scientific reasoning. Cognitive Science 31, 843–875 (2007)

[Truesdell, 1984] Truesdell, C.: An Idiot Fugitive Essay on Science: Criticism, Training, Circumstances. Springer, Berlin (1984)

[Tucker and Ellis, 2006] Tucker, M., Ellis, R.: On the relations between seen objects and components of potential actions. Journal of Experimental Psychology 24(3), 830–846 (2006)

[Tummolini *et al.*, 2006] Tummolini, L., Castelfranchi, C., Pacherie, E., Dokic, J.: From mirror neurons to joint actions. Cognitive Systems Research 7, 101–112 (2006)

[Turing, 1950] Turing, A.M.: Computing machinery and intelligence. Mind 49, 433–460 (1950)

[Turing, 1969] Turing, A.M.: Intelligent machinery [1948]. In: Meltzer, B., Michie, D. (eds.) Machine Intelligence, vol. 5, pp. 3–23. Edinburgh University Press, Edinburgh (1969)

[Turing, 1992] Turing, A.M.: Collected Works of Alan Turing. In: Ince, D.C. (ed.) Mechanical Intelligence. Elsevier, Amsterdam (1992)

[Turner, 1994] Turner, S.R.: The Creative Process: a Computer Model of Storytelling and Creativity. Erlbaum, Hillsdale (1994)

[Turner, 2004] Turner, J.S.: Extended phenotypes and extended organisms. Biology and Philosophy 19, 327–352 (2004)

[Turner, 2005] Turner, P.: Affordance as context. Interacting with Computers 17, 787–800 (2005)

[Turrisi, 1990] Turrisi, P.A.: Peirce's logic of discovery: abduction and the universal categories. Transactions of the Charles S. Peirce Society 26, 465–497 (1990)

[Turvey and Shaw, 2001] Turvey, M.T., Shaw, R.E.: Toward an ecological physics and a physical psychology. In: Solso, R.L., Massaro, D.W. (eds.) The Science of the Mind: 2001 and Beyond, pp. 144–169. Oxford University Press, Oxford (2001)

[Tversky *et al.*, 2006] Tversky, B., Agrawala, M., Heiser, J., Lee, P., Hanrahan, P., Phan, D., Stolte, C., Daniel, M.-P.: Cognitive design principles: from cognitive models to computer models. In: Magnani, L. (ed.) Model-Based Reasoning in Science and Engineering. College Publications, London (2006)

[Tversky, 2001] Tversky, B.: Spatial schemas in depiction. In: Gattis, M. (ed.) Spatial Schemas and Abstract Thought, pp. 79–112. The MIT Press, Cambridge (2001)

[Tweney, 1990] Tweney, R.D.: Five questions for computationalists. In: Shrager, J., Langley, P. (eds.) Computational Models of Scientific Discovery and Theory Formation, pp. 471–484. Morgan Kaufmann, San Mateo (1990)

[Tweney, 2006] Tweney, R.D.: Abductive seeing (2006) (unpublished manuscript)

[Vakalopoulos, 2005] Vakalopoulos, C.: A scientific paradigm for consciousness: a theory of premotor relations. Medical Hypotheses 65(4), 766–784 (2005)

[Valdés-Pérez, 1999] Valdés-Pérez, R.E.: Principles of human computer collaboration for knowledge discovery in science. Artificial Intelligence 107(2), 335–346 (1999)

[Vallortigara *et al.*, 2005] Vallortigara, G., Feruglio, M., Sovrano, V.A.: Reorientation by geometric and landmark information in environments of different sizes. Developmental Science 8(5), 393–401 (2005)

[van Benthem, 1996] van Benthem, J.: Exploring Logical Dynamics. CSLI Publications, Stanford (1996)

[van Gelder and Port, 1995] van Gelder, T., Port, R.F.: It's about time: an overview of the dynamical approach to cognition. In: Port, R.F., Gelder, T. (eds.) Mind as Motion. Explorations in the Dynamics of Cognition, pp. 1–44. The MIT Press, Cambridge (1995)

[van Gelder, 1999] van Gelder, T.: Wooden iron? Husserlian phenomenology meets cognitive science. In: Petitot, J., Varela, F.J., Pachoud, B., Roy, J.-M. (eds.) Naturalizing Phenomenology, pp. 245–265. Stanford University Press, Stanford (1999)

[van Helden, 1989] van Helden, A.: Galileo, telescopic astronomy, and the Copernican system. In: Taton, R., Wilson, C. (eds.) Planetary Astronomy from Renaissance to the Rise of Astrophysics. Part A: Tycho Brahe to Newton, pp. 81–105. Cambridge University Press, Cambridge (1989)

[Vandi, 2007] Vandi, C.: Beyond metaphors and icons: towards a perception-action model for graphic user interfaces. Presented at the Ninth World Congress of IASS/AIS, Communication: Understanding/Misunderstanding, Helsinki (June 2007)

[Varela *et al.*, 1991] Varela, F.J., Thomson, E., Rosch, E.: The Embodied Mind. In: Cognitive Science and Human Perspective. The MIT Press, Cambridge (1991)

[Varela, 1979] Varela, F.J.: Principles of Biological Autonomy. Elsevier, New York (1979)

[Varela, 1997] Varela, F.J.: Patterns of life: intertwining identity and cognition. Brain and Cognition 34, 72–87 (1997)

[Vicente, 2003] Vicente, K.J.: Beyond the lens model and direct perception: toward a broader ecological psychology. Ecological Psychology 15(3), 241–267 (2003)

[Vico, 1968] Vico, G.: The New Science of Giambattista Vico. Cornell University Press, Ithaca (1968); Revised translation of the third edition (1744) by T. G. Bergin and M. H. Fisch

[Vygotsky, 1978] Vygotsky, L.S.: Mind in Society. The Development of Higher Psychological Processes. Harvard University Press, Cambridge (1978)

[Vygotsky, 1986] Vygotsky, L.S.: Thought and Language. The MIT Press, Cambridge (1986)

[Wagar and Thagard, 2004] Wagar, B.M., Thagard, P.: Spiking Phineas Gage: a neurocomputational theory of cognitive-affective integration in decision making. Psychological Review 111(1), 67–79 (2004)

[Walton, 2004] Walton, D.: Abductive Reasoning. The University of Alabama Press, Tuscaloosa (2004)

[Walton, 2006] Walton, D.: Poisoning the well. Argumentation 20, 273–307 (2006)

[Walton, 2008] Walton, D.: Witness Testimony Evidence. Cambridge University Press, Cambridge (2008)

[Wang, 1987] Wang, H.: Reflections on Kurt Gödel. The MIT Press, Cambridge (1987)

[Warren, 1995] Warren, W.H.: Constructing an econiche. In: Flach, J., Hancock, P., Caird, J., Vicente, K.J. (eds.) Global Perspective on the Ecology of Human-Machine Systems, pp. 210–237. Lawrence Erlbaum Associates, Hillsdale (1995)

[Wason, 1960] Wason, P.C.: On the failure to eliminate hypotheses in a conceptual task. Quarterly Journal of Experimental Psychology 23, 63–71 (1960)

[Watanabe, 2006] Watanabe, S.: Do animals have "theory"? - Naïve biology in pigeons. In: Andler, D., Ogawa, Y., Okada, M., Watanabe, S. (eds.) Reasoning and Cognition, pp. 205–212. Keio University Press, Tokyo (2006)

[Weiskopf, 2008] Weiskopf, D.A.: Patrolling the mind's boundaries. Erkenntnis 68, 265–276 (2008)

[Wells, 2002] Wells, A.J.: Gibson's affordances and Turing's theory of computation. Ecological Psychology 14(3), 141–180 (2002)

[West-Eberhard, 2003] West-Eberhard, M.J.: Developmental Plasticity and Evolution. Oxford University Press, Oxford/New York (2003)

[Wheeler, 2004] Wheeler, M.: Is language an ultimate artifact? Language Sciences 26, 693–715 (2004)

[Whiten and Byrne, 1988] Whiten, A., Byrne, R.: Tactical deception in primates. Behavioral and Brain Sciences 12, 233–273 (1988)

[Whiten and Byrne, 1997] Whiten, A., Byrne, R.: Machiavellian Intelligence II: Evaluations and Extensions. Cambridge University Press, Cambridge (1997)

[Wilcox and Jackson, 2002] Wilcox, S., Jackson, R.: Jumping spider tricksters: deceit, predation, and cognition. In: Bekoff, M., Allen, C., Burghardt, M. (eds.) The Cognitive Animal. Empirical and Theoretical Perspectives on Animal Cognition, pp. 27–34. The MIT Press, Cambridge (2002)

[Wilson *et al.*, 2004] Wilson, D.S., Timmel, J.J., Miller, R.R.: Cognitive cooperation. When the going gets tough, think as a group. Human Nature 15(3), 1–15 (2004)

[Wilson, 1993] Wilson, J.Q.: The Moral Sense. Free Press, New York (1993)

[Wilson, 2002] Wilson, D.S.: Evolution, morality and human potential. In: Scher, S.J., Rauscher, F. (eds.) Evolutionary Psychology. Alternative Approaches, pp. 55–70. Kluwer Academic Publishers, Dordrecht (2002)

[Wilson, 2004] Wilson, R.A.: Boundaries of the Mind. Cambridge University Press, Cambridge (2004)

[Wilson, 2005] Wilson, R.A.: Collective memory, group minds, and the extended mind thesis. Cognitive Processing 6, 227–236 (2005)

[Windsor, 2004] Windsor, W.L.: An ecological approach to semiotics. Journal for the Theory of Social Behavior 34(2), 179–198 (2004)

[Winsberg, 1999] Winsberg, E.: The hierarchy of models in simulation. In: Magnani, L., Nersessian, N.J., Thagard, P. (eds.) Model-Based Reasoning in Scientific Discovery, pp. 255–269. Kluwer Academic/Plenum Publishers, New York (1999)

[Winston, 1980] Winston, P.H.: Learning and reasoning by analogy. Communications of ACM 23(12) (1980)

[Wirth, 1997] Wirth, U.: Abductive inference in semiotics and philosophy of language: Peirce's and Davidson's account of interpretation. In: Rayo, M., Gimate-Welsh, A., Pellegrino, P. (eds.) 5th International Congress of the International Association for Semiotic Studies: Semiotics Bridging Nature and Culture, p. 232. Editorial Solidaridad, México (1997); Abstract

[Wirth, 1999] Wirth, U.: Abductive reasoning in Peirce's and Davidson's account of interpretation. Transactions of the Charles S. Peirce Society 35(1), 115–127 (1999)

[Wirth, 2005] Wirth, U.: Abductive reasoning in Peirce and Davidson. Semiotica 153(1/4), 1999–2008 (2005)

[Witt *et al.*, 2006] Witt, J.K., Proffitt, D.R., Epstein, W.: Tool use affects perceived distance, but only when you intend to us. Journal of Experimental Psychology 32(6), 1405–1421 (2006)

[Woods, 2004] Woods, J.: The Death of Argument. Kluwer Academic Publishers, Dordrecht (2004)

[Woods, 2005] Woods, J.: Epistemic bubbles. In: Artemov, S., Barringer, H., Garcez, A., Lamb, L., Woods, J. (eds.) We Will Show Them: Essays in Honour of Dov Gabbay, vol. II, pp. 731–774. College Publications, London (2005)

[Woods, 2007] Woods, J.: The concept of fallacy is empty. In: Magnani, L., Li, P. (eds.) Model-Based Reasoning in Science, Technology, and Medicine, pp. 69–90. Springer, Berlin (2007)

[Woods, 2009] Woods, J.: Ignorance, inference and proof: abductive logic meets the criminal law. In: Proceedings of the International Conference Applying Peirce, Helsinki, Finland (2009) (forthcoming)

[Woods, 2010] Woods, J.: Seductions and Shortcuts: Error in the Cognitive Economy (2010) (forthcoming)

[Yamazaki *et al.*, 2006] Yamazaki, Y., Okanoya, K., Iriki, A.: Development of logical and illogical inference. In: Andler, D., Ogawa, Y., Okada, M., Watanabe, S. (eds.) Reasoning and Cognition, pp. 63–74. Keio University Press, Tokyo (2006)

[Young, 2006] Young, G.: Are different affordances subserved by different neural pathways? Brain and Cognition 62, 134–142 (2006)

[Zeigarnik *et al.*, 1997] Zeigarnik, A.V., Valdés-Pérez, R.E., Temkin, O.N., Bruk, L.G., Shalgunov, S.I.: Computer-aided mechanism elucidation of acetylene hydrocarboxylation to acrylic acid based on a novel union of empirical and formal methods. Organometallics 16(14), 3114–3127 (1997)

[Zhang and Patel, 2006] Zhang, J., Patel, V.L.: Distributed cognition, representation, and affordance. Cognition & Pragmatics 14(2), 333–341 (2006)

[Zhang, 1997] Zhang, J.: The nature of external representations in problem solving. Cognitive Science 21(2), 179–217 (1997)

[Ziemke, 2007] Ziemke, T.: What's life got to do with it? In: Chella, A., Manzotti, R. (eds.) Artificial Consciousness, pp. 48–66. Imprint Academic, Exeter (2007)

[Zlatev, 2007] Zlatev, J.: Embodiment, language, and mimesis. In: Ziemke, T., Zlatev, J., Frank, R. (eds.) Body, Language and Mind I: Embodiment, pp. 297–338. De Gruyter, Berlin (2007)

[Zwaan, 2004] Zwaan, R.A.: The immersed experiencer: toward an embodied theory of language comprehension. In: Ross, B.H. (ed.) The Psychology of Learning and Motivation. 44, pp. 35–62. Academic Press, New York (2004)

[Zytkow and Fischer, 1996] Zytkow, J., Fischer, P.: Incremental discovery of hidden structure: applications in theory of elementary particles. In: Proceedings of AAAI 1996, Stanford, pp. 150–156. The AAAI Press, Menlo Park (1996)

[Zytkow, 1992] Zytkow, J. (ed.): Proceedings of the ML 1992 Workshop on Machine Discovery (MD 1992). National Institute for Aviation Research, The Wichita State University (1992)

[Zytkow, 1997] Zytkow, J.: Creating a discoverer: autonomous knowledge seeking agent. In: Zytkow, J. (ed.) Machine Discovery, pp. 253–283. Kluwer, Dordrecht (1997); reprinted from Foundations of Science 2, 253–283 (1995/1996)

[Zytkow, 1999] Zytkow, J.: Scientific modeling: a multilevel feedback process. In: Magnani, L., Nersessian, N.J., Thagard, P. (eds.) Model-Based Reasoning in Scientific Discovery, pp. 311–235. Kluwer Academic/Plenum Publishers (1999)

Index

Cognitive Systems Monographs

Edited by R. Dillmann, Y. Nakamura, S. Schaal and D. Vernon

Vol. 1: Arena, P.; Patanè, L. (Eds.)
Spatial Temporal Patterns for
Action-Oriented Perception
in Roving Robots
425 p. 2009 [978-3-540-88463-7]

Vol. 2: Ivancevic, T.T.; Jovanovic, B.;
Djukic, S.; Djukic, M.; Markovic, S.
Complex Sports Biodynamics
326 p. 2009 [978-3-540-89970-9]

Vol. 3: Magnani, L.
Abductive Cognition
534 p. 2009 [978-3-642-03630-9]